ECHO SIGNAL PROCESSING

THE KLUWER INTERNATIONAL SERIES IN ENGINEERING AND COMPUTER SCIENCE

ECHO SIGNAL PROCESSING

Dennis W. Ricker
The Pennsylvania State University

KLUWER ACADEMIC PUBLISHERS
Boston / Dordrecht / London

Distributors for North, Central and South America:
Kluwer Academic Publishers
101 Philip Drive
Assinippi Park
Norwell, Massachusetts 02061 USA
Telephone (781) 871-6600
Fax (781) 681-9045
E-Mail: kluwer@wkap.com

Distributors for all other countries:
Kluwer Academic Publishers Group
Post Office Box 322
3300 AH Dordrecht, THE NETHERLANDS
Telephone 31 786 576 000
Fax 31 786 576 474
E-Mail: services@wkap.nl

Electronic Services <http://www.wkap.nl>

Library of Congress Cataloging-in-Publication Data

Ricker, Dennis W.
Echo Signal Processing / Dennis W. Ricker.
p.cm. —(The Kluwer International Series in Engineering and Computer Science; SECS 725)
Includes bibliographical references and index.
ISBN 1-4020-7395-X (alk. paper)
1. Signal Processing. 2. Antennas and Propagation. I. Title.
Series.

Printed on acid-free paper.

Printed in the United States of America.

Contents

List of Figures

List of Tables

Foreword

This book presents basic and advanced topics in the areas of signal theory and processing as applied to acoustic echo-location (sonar). It is written at the advanced undergraduate or graduate level, and assumes that the reader is conversant with the concepts and mathematics associated with introductory graduate courses in signal processing such as linear and complex algebra, Fourier analysis, probability, advanced calculus, and linear system theory. The material is presented in a tutorial fashion as a logical development starting with basic principles and leading to the development of topics in detection and estimation theory, waveform design, echo modeling, scattering theory, and spatial processing. Examples are provided throughout the book to illustrate important concepts and especially important relationships are boxed. The book addresses the practical aspects of receiver and waveform design, and therefore should be of interest to the practicing engineer as well as the student. Although much of the book is applicable to the general echo-location problem that includes radar, its emphasis is on acoustic echo location especially in regard to time mapping and the wideband or wavelet description of Doppler. Introductory signal theory material is included in the first chapter to provide a foundation for the material covered in the later chapters. A consistent notational convention is observed throughout the book so that the various mathematical entities are readily identified. This is described in the glossary and symbol list.

Sonar signals in comparison to radar signals are much more susceptible to motion induced distortion because the ratio of attainable to propagation speed is much higher. The kinematic echo formation process is covered in the second chapter where the concepts of time mapping, space-time ray cones, and generalized trajectory diagrams are introduced to describe the distortion of the signal time base. Linear time mapping or Doppler is the dominant effect but higher order time base distortion effects occur especially when propagation is not collinear. It is then necessary to employ the iterative algorithms described in the chapter to duplicate the distortion effects for receiver synthesis. Doppler frequency shift is a narrowband phenomenon arising from linear time mapping and only approximates the Doppler effect. Receivers for large time-bandwidth waveforms must also account for envelope dilation as well as Doppler shift in order to avoid performance degradation due to receiver mismatch. Design criteria indicating the tradeoffs between performance and receiver complexity are discussed. A comparison between time mapping in the acoustic and electromagnetic realms illustrates the difference between signal propagation in free space and in an acoustic medium.

Detection and parameter estimation are treated as a maximum likelihood problem. This is not a new concept and is well covered in standard signal processing texts but the approach to the derivation of the likelihood ratio and its interpretation is specifically tailored to the acoustic echo-location problem. The classic derivation of the likelihood ratio is presented in terms of a binary hypothesis decision policy that minimizes a cost objective such as the Baye's risk. The Karhunen-Loeve expansion for stochastic signals is utilized to form the likelihood ratio from the continuous data time series (echo return). This provides a discrete countable representation of a continuous waveform, and allows for the definition, in a limiting sense, of the required probability density functions. Furthermore, it is shown that the ambiguity function associated with echo location arises naturally from the maximum likelihood criterion, and that its form depends upon the fundamental assumptions regarding the echo formation process and the interfering noise.

A discussion and comparison of the properties of wideband (wavelet domain) and narrowband ambiguity functions is a principal contribution of the book. These include the familiar auto ambiguity functions that arise from the maximum likelihood criterion detection of deterministic and fluctuating point scatterers in white Gaussian noise. They are treated as a special case of the more general cross ambiguity functions that arise both from the maximum likelihood criterion for detection in correlated noise and the maximization of the detection index when interrogating spread scatterers.

The specification and design of echo location waveforms is introduced by first describing some of the traditional waveforms such as the CW and LFM and the concept of pulse compression. While waveform design has been extensively treated in a radar context there is a need to discuss more recent developments motivated by the requirements of acoustic echo location. These include the design of waveforms from a wideband or time dilation (wavelet) point of view and the associated issues of Doppler (scale) tolerance, dilation sensitivity, and simultaneous delay-Doppler (time scale) resolution. Original material treating the Wigner-Ville distribution as a tool for waveform design is introduced.

The book includes an extensive treatment of what are called noise-like (imaging) waveforms. These are PRN sequences and the classes of frequency shift keyed (FSK), Costas, and congruential codes that have high delay-time scale resolution capabilities. The algebraic derivation of the various classes of hop codes, the properties of their nearly ideal 'thumbtack' ambiguity functions, and a comparison with PRN sequences are provided.

The detection of doubly spread scattering processes is addressed upon establishing the background for point scatterer detection and estimation, and having introduced and discussed auto and cross ambiguity functions. Spread scattering theory is presented from a signal processing point of view by modeling the random scattering process as a time varying linear filter with an associated scattering function that describes its delay-Doppler/time dilation energy dispersion properties. It is shown that the expected response of a narrow or wideband matched filter receiver to a spread echo or interference process such as reverberation is the convolution of the signal ambiguity and scattering functions. This relationship is used to guide the choice of signal waveforms and as a mechanism for the derivation of useful expressions for the processing gain of matched filters operating in a clutter or reverberation limited environment. The discussion of scattering functions also includes their use as a means for the introduction of prior information for the formulation of optimal detection strategies such as the optimal mismatched receiver and the estimator correlator. The relationship of the scattering function to the time-frequency correlation function serves as an introduction to the problem of time and frequency selective fading and the use of diversity processing to improve detection performance under fading conditions.

Beam formation, array processing, bearing estimation, and the generalized space-time maximum likelihood receiver are discussed in the last chapter. Spatially spread scattering is described by a generalized scattering function that is a function of delay, Doppler and bearing. The maximum likelihood approach as well as monopulse processing are discussed as alternatives for bearing estimation.

The material in this book some of which is original is drawn from a wide selection of open literature sources. The book progresses in an orderly fashion from a discussion of basic signal processing principles, to the kinematic formation of echos and the fundamentals of detection and estimation for simple point scattering processes. Ambiguity functions that arise from the maximum likelihood criterion for Gaussian processes are given a chapter of their own followed by a chapter on the various echo location waveforms and their design. The groundwork established in the earlier chapters serves as the basis for the discussion of the detection of doubly spread processes and the extension to spatial processing, beam forming, and bearing estimation in the last two chapters.

Acknowledgments

I would like to thank all who contributed to this endeavor including: Dave Drumheller of ONR who contributed ideas, suggestions, and material for inclusion in the book, my colleagues at the Applied Research Laboratory (ARL) of the Pennsylvania State University for providing a stimulating environment and uncountable discussions regarding sonar signal processing, Professor Edward Titlebaum at the University of Rochester for his insight regarding waveform and processor design, Eric Schott and Steve Harp at ARL for their help in fathoming the intricacies of LaTeX, Anthony Cutezo at ARL for help with editing and figures, Jack Sharer at ARL for his aid in implementing many of the computations, and Andy Ward for generating a large number of the illustrations. I would like to thank my wife Beth for her patience and encouragement and the Applied Research Laboratory and the Office of Naval Research for their support throughout this project.

D.W. Ricker
Applied Research Laboratory
Pennsylvania State University

Chapter 1

BASIC SIGNAL THEORY

1.1 INTRODUCTION

The purpose of this chapter is to establish the basic definitions and terminology to be used throughout this book and to briefly review material required for the understanding of echo-location from an engineering and mathematical perspective. Accordingly, this chapter contains standard signal processing material that would be covered at the senior undergraduate or early graduate level in electrical engineering. Most of it may be found in standard texts some of which are listed in the references. It is assumed that the reader has some background in linear systems, probability and stochastic processes, vector matrix notation, and is familiar with the conventions used to denote random variables and vectors.

1.2 SIGNALS, POWER, AND ENERGY

For echo-location, a signal is a function of time and space representing a real physical property such as electro magnetic field strength in radar, acoustic sound pressure level in sonar or, when electronically converted, a voltage or current. Echo signals arise from *interrogation* by a transmitted signal composed of a unit energy *waveform* that has been amplified with a power amplifier or transmitter and imbued with the required energy. Once transmitted, the signal is propagated, scattered or reflected, and received as an echo in an echo-location or interrogation cycle. The geometry is called *monostatic* if the transmitter and receiver spatially coincide and *bistatic* if they do not.

"Signal" is a generic term used throughout this book to represent the time history of a physical quantity in the immediate context of interest. That is, it can refer to the signal that is transmitted, that which is scattered or observed at the point of reflection, or that which is received. The waveform is the unit energy time function that is the basis for the transmitted signal. Signals can be known (deterministic) quantities such as the time history of the voltage driving a transmitter, or they may be random such as the sound pressure received from a moving complex scatterer in a noisy environment. The goal of an echo-location system be it radar or sonar is to *detect* the presence of particular reflectors (targets) and to *estimate* parameters associated with target geometry and kinematics i.e., range, bearing, and velocity. Signals or echos received from these "desired" objects are usually accompanied by *interference* in the form of echos received from other interfering sources called clutter in radar and reverberation in sonar. Moreover, random noise independent of the interrogation process always accompanies the received echos contributing to the overall interference.

In this book, signals are denoted by the symbol s with different fonts and sizes denoting the various signal forms under discussion. Signals are real quantities but are conveniently treated mathematically as complex quantities from which the real quantities are derived. Accordingly continuous time signals $s(t)$ are represented as complex functions $s(t) = s_r(t) + js_i(t)$ where $j = \sqrt{-1}$ which is the engineering rather than mathematical notation.

A signal denoted as $s(t)$ has *instantaneous power*

$$P_s(t) = |s(t)|^2 = s(t)s^*(t) \geq 0$$

where $*$ represents conjugation, i.e., $(a + jb)^* = a - jb$. This follows from the definition used in electrical systems where power is proportional to the square of the current or voltage. The *average signal power* is formally given by

$$\bar{P}_s = \lim_{T\to\infty} \frac{1}{T} \int_{-T/2}^{T/2} |s(t)|^2 dt. \tag{1.1}$$

Signals for which the average power is finite are called *power signals* . The energy E_s expended by the physical process represented by the signal is the power integrated over time $E_s = T\bar{P}_s$ and if $E_s < \infty$, the signal is called an *energy signal.* The use of the limits in (1.1) accounts for cases where the integrals are improper and must be defined in a limiting sense. For example, one could deal with a signal that has an integrable singularity at some specific time or is nonzero for all time (infinite time duration). Generally speaking, such signals serve only as abstractions and are defined for mathematical convenience. In practice, signals and interference processes associated with echo-location have finite time duration, power, and energy. Consequently, the explicit use of the limit symbol when defining energy or power is dropped. Unless otherwise noted, the integrals will be written with infinite limits and their definition as improper integrals is implicit.

Echo-location *waveforms* are piecewise continuous functions $f(t)$ of finite duration T that represent the functional form of the signal generated in the transmitter. In order to separate issues concerning waveform properties and signal energy, they are defined as unit energy functions such that

$$\int_{-T/2}^{T/2} |f(t)|^2 dt = 1.$$

A signal comprised of the waveform with non-unit energy E_s is written as

$$s(t) = \sqrt{E_s} f(t)$$

so that

$$\int_{-T/2}^{T/2} |s(t)|^2 dt = E_s.$$

A signal is said to be periodic if for some time interval T_p,

$$s(t) = s(t + T_p) \quad (T_p \leq T). \tag{1.2}$$

T_p is called the *period* of the signal and is interpreted as the smallest interval for which (1.2) is true. Signals that are not periodic are called *aperiodic*. Echo-location signals of finite duration can only accommodate a period shorter than or equal to their duration however they can be composed of a segment of a function with a longer period.

1.3 FOURIER REPRESENTATION OF SIGNALS

1.3.1 Orthonormal Expansions and Fourier Transforms

Recall from physics, linear algebra, and analytic geometry that is possible to resolve a vector in N-dimensional Euclidean space into a weighted sum of elementary unit vectors directed along each axis. These unit vectors are perpendicular or *orthogonal* to each other and their inner product is zero. Furthermore, this set of unit vectors is called *complete* because all vectors in the space can be written as the weighted sum of them. They constitute a *basis* set for the space although it is not unique [1,2]. These concepts apply as well to echo-location signals since integrable piecewise continuous functions can be expanded in terms of a sum of elementary functions. This is known as *generalized Fourier analysis* and power and energy signals defined on an interval $t_0 \leq t \leq T$ can be expressed as the expansion

$$s(t) = \mathbf{s}^{\mathrm{T}}\boldsymbol{\xi}(t) = \sum_{i=1}^{\infty} s_i \xi_i(t), \quad (t_0 \leq t \leq T) \tag{1.3}$$

where $\xi_i(t)$, $1 \leq i \leq \infty$ are a *complete* set of orthonormal functions (CON set) and the components of $\boldsymbol{\xi}(t)$. The components of the signal vector are the complex coefficients s_i called the *generalized Fourier coefficients* [1] and are computed from

$$s_i = \int_{t_0}^{t_0+T} s(t)\xi_i^*(t)dt.$$

Orthonormality of the constituent functions is defined as the property

$$\int_{t_0}^{t_0+T} \xi_i(t)\xi_j^*(t)dt = \begin{cases} 1 & \text{if } i = j, \\ 0 & \text{if } i \neq j. \end{cases}$$

This represents the inner product of orthogonal functions each of which has unit energy on the finite interval T.

There are many sets of functions that may be used for signal expansion, where the choice depends upon the signal form and the application. Generally the best set results in an expansion that approximates the signal time series as closely as desired with the fewest terms. This is the issue of completeness that is considered in greater detail in Chapter 3.

Perhaps the most common orthonormal functions are the Fourier kernels for periodic signals

$$\xi_i(t) = \exp jn\omega t \quad n = 0, \pm 1, \pm 2, \cdots$$

[1]The word "generalized" is used to distinguish expansion coefficients in general from the specific Fourier (spectral) coefficients generated by the kernels exp $j\omega_i t$

where $\omega = 2\pi/T$. For these, the expansion (1.3) and coefficients are

$$s(t) = \sum_{n=-\infty}^{\infty} s_n \exp jn\omega t, \quad (t_0 \leq t \leq T)$$
$$s_n = \frac{1}{T}\int_{t_0}^{t_0+T} s(t)\exp(-jn\omega t)dt \tag{1.4}$$

and is called the *Fourier series* (FS) representation. This is valid for all values of t for periodic signals but applies to aperiodic signals only for $t_0 \leq t \leq t_0 + T$. The Fourier series is extended for aperiodic signals and all t by defining the continuous *Fourier transform pair* and the operator notation $\mathcal{F}$ and $\mathcal{F}^{-1}$

$$s(t) = \frac{1}{2\pi}\int_{-\infty}^{\infty} S(\omega)\exp(j\omega t)d\omega = \mathcal{F}^{-1}\, S(\omega)$$
$$S(\omega) = \int_{-\infty}^{\infty} s(t)\exp(-j\omega t)dt = \mathcal{F}\, s(t). \tag{1.5}$$

The functions $s(t)$ and $S(\omega)$ are respectively the time and frequency representations of the signal and $S(\omega)$ is called the signal *spectrum*. The first relationship of the pair transforms from the frequency to the time domain and is called the *inverse Fourier transform*. The second does the reverse and is called the *forward Fourier transform* or usually just the *Fourier transform* (FT). It is used extensively for the analysis and synthesis of echo-location processors as well as in many other signal processing applications. Throughout this book, the convention of denoting signals in the time and frequency domains in lower and upper case fonts respectively as for $s(t)$ and $S(\omega)$ is adopted. Discrete Fourier series coefficients arc denoted by upper case subscripted symbols such as S_k or S_k for deterministic and random quantities respectively.

1.3.2 Fourier Transform Properties

Fourier transforms and their manipulation is central to the development of much of the theory associated with the processing of echo-location signals and it is assumed that the transform and its inverse exist for the bounded piecewise continuous power and energy signals of interest. These may represent any physical quantity associated with echo-location, but for convenience they may be considered to be voltages. Then, $|s(t)|^2$ has the units of volts^2 or power (Watts), and integrating in time results in the energy in Watt-sec. or Joules. The FT (1.5) has the units of volts-sec. Therefore $|S(\omega)|^2$ = watts-sec.^{2} = Joules/Hz and is called the *energy spectral density*. Integrating over frequency yields

$$\frac{1}{2\pi}\int_{-\infty}^{\infty} |S(\omega)|^2 d\omega \quad \text{Joules-sec.} \times \text{Hz=Joules}$$

and thus integration in both the time and frequency domains yields an energy measure. The relationship between them is derived by writing the time domain energy as

$$E_s = \int_{-\infty}^{\infty} |s(t)|^2 dt$$

and substituting the inverse transform (1.5) for $s(t)$. This yields

$$\begin{aligned} E_s &= \int_{-\infty}^{\infty} \left(\frac{1}{2\pi}\int_{-\infty}^{\infty} S(\omega)\exp(j\omega t)d\omega\right) s^*(t)dt \\ &= \frac{1}{2\pi}\int_{-\infty}^{\infty}\left(\int_{-\infty}^{\infty} s^*(t)\exp(j\omega t)dt\right) S(\omega)s\omega \\ &= \frac{1}{2\pi}\int_{-\infty}^{\infty} |S(\omega)|^2 d\omega \end{aligned}$$

which is the frequency domain energy.

Integration of $|s(t)|^2$ and $S(\omega)|^2$ in their respective domains produce equivalent measures of signal energy. This important property of Fourier transforms and series is *Parseval's Theorem* and is a statement of the completeness of the Fourier expansion. More generally, it is the property of complete signal representations that all of the energy in the original time series is recoverable by integrating over the domain of the representation.

The *Schwarz inequality*

$$\left|\int_{-\infty}^{\infty} x(t)y^*(t)dt\right|^2 \leq \int_{-\infty}^{\infty} |x(t)|^2 dt \int_{-\infty}^{\infty} |y(t)|^2 dt.$$

is an important relationship that holds for bounded piecewise continuous functions $x(t)$ and $y(t)$ that possess a Fourier transform. If the functions are periodic, the relationship holds on the interval $t \in [t_0,\, t_0 + T]$. The left hand integral is known as an *inner product* and the inequality extends to vector quantities as well. The inner product of two complex vectors $\mathbf{x}$, $\mathbf{y}$ is $\mathbf{x}^H\mathbf{y}$ and the Schwartz inequality becomes

$$|\mathbf{x}^H\mathbf{y}|^2 \leq |\mathbf{x}|^2|\mathbf{y}|^2$$

where the notation $(\bullet)^{\mathrm{H}}$ indicates the *Hermitian* or conjugate transpose of the vector or matrix. The inner product for either continuous or discrete quantities f, g is often written simply as $< f, g >$.

Further details relating to the properties of Fourier transforms are found in several texts some examples of which are listed in the references. A brief summary of FT properties is provided in Table (1.1).

1.4 THE COMPLEX SIGNAL REPRESENTATION

Echo-location signals that are transmitted, propagated, scattered, and received represent physical quantities and are real functions of time. However, most signals are sinusoidal in nature and it is mathematically convenient for the purpose of analysis to treat the them as complex functions from which the real signals are derived. This is analogous to the way real voltages, currents, and impedances are treated in classical AC network theory, a subject familiar to most electrical engineers. The situation is more involved than that of simple AC circuits because echo-location signals are of finite duration and may have considerable complexity that is reflected in the signal bandwidth. It is therefore necessary to examine the role of signal spectrum and bandwidth upon the relationship between real signals and their complex representations.

The complex signal form typically encountered in sonar and radar is the amplitude shaded complex exponential

$$s(t) = a(t) \exp\left(j\Psi(t)\right) \tag{1.6}$$

where $a(t)$ is a real amplitude function often called a window or shading function. The phase function $\Psi(t) = \psi(t) + \omega_0 t$ is comprised of the time dependent phase modulation $\psi(t)$ and a linear term corresponding to a constant radian frequency ω_0.

$$s(t) = a \ \exp\ j(\omega_0 t + \psi)$$

is the special case of a complex sinusoid with constant phase ψ and period $2\pi/\omega_0$. Monochromatic signals with constant parameters or with at most slowly varying amplitudes $a(t)$ are the simplest echo-location signals and are often called *continuous wave* (CW) or *pure tone* (PT) signals. If

$$\omega_0 >> \max_{t \in T} \left| \frac{d\psi(t)}{dt} \right|,$$

Time domain $s(t)$		Frequency domain $S(\omega)$	
if $s(t)$ is:		then $S(\omega)$ is:	
complex,	$s(t) = s_r(t) + js_i(t)$	complex,	$S(\omega) = S_r(\omega) + jS_i(\omega)$
even	$s(t) = s(-t)$	even,	$S(\omega) = S(-\omega)$
odd,	$s(t) = -s(-t)$	odd,	$S(\omega) = -S(-\omega)$
conj-even	$s^*(-t) = s(t)$	real,	$S^*(\omega) = S(\omega),\ S_i(\omega) = 0$
conj-odd	$s^*(-t) = -s(t)$	imag	$S^*(\omega) = -S(\omega),\ S_r(\omega) = 0$
real,	$s_i(t) = 0$	conj-even,	$S(\omega) = S^*(-\omega)$
even,	$s_r(t) = s_r(-t)$	real-even,	$S_r(\omega) = S_r(-\omega),\ S_i(\omega) = 0$
odd,	$s_r(t) = -s_r(-t)$	imag-odd,	$S_i(\omega) = -S_i(-\omega),\ S_r(\omega) = 0$

a. Qualitative Properties

Operation	Time domain	Frequency domain
$\mathcal{F}$	$s(t)$	$S(\omega)$
time scale	$s(\beta t)$	$\beta^{-1}S(\omega)$
time delay	$s(t-\tau)$	$S(\omega)\exp(-j\omega\tau)$
differentiation	$d^n s(t)/dt^n$	$(j\omega)^n S(\omega)$
integration	$\int_{-\infty}^{t} s(z)dz$	$(1/j\omega)S(\omega) + \pi S(0)\delta(\omega)$
multiplication	$r(t)s(t)$	$(1/2\pi)R(\omega) * S(\omega)$
convolution	$r(t) * s(t)$	$R(\omega)S(\omega)$

b. Operational Properties

Table 1.1: Fourier transform properties

the amplitude and phase functions are said to *modulate* the CW component and $a(t)$ and $\psi(t)$ are called the *amplitude modulation* (AM) and *phase modulation* (PM) respectively [3–5]. The term *frequency modulation* (FM) is often used to represent the role of $\psi(t)$ but strictly speaking, frequency modulation is the time derivative of phase.

Complex exponential notation lends itself to the convenient graphical interpretation known as *phasor representation* derived from the application of Euler's Theorem, specifically

$$a(t)\exp j\Psi(t) = a(t)(\cos\Psi(t) + j\sin\Psi(t)).$$

This resolves the exponential into the 2-d Cartesian form displaying the signal as a vector in the complex plane. The vector rotates counterclockwise about the origin at the instantaneous frequency

$$\frac{d\Psi(t)}{dt} = \omega(t) \quad \text{(radians per second)}.$$

The amplitude function $a(t)$ represents the length of the vector (phasor) at time t, and $\Psi(t)$ is the total radian angle swept by the vector on the interval [0,t]. Signals that have significant energy only within a finite band $\omega \in [\pi W_1, \pi W_2]$ and $\omega \neq 0$, are called *modulated bandpass* signals and are written as

$$s(t) = \tilde{s}\ \exp\ j\omega_0 t.$$

where, $\tilde{s} = a(t)\exp j\psi(t)$ is the *complex envelope*, $\exp\omega_0 t$ is the monochromatic *carrier* term, and ω_0 is the carrier frequency [6–8]. Complex envelopes are often called *baseband* signals. The information inherent in a signal is coded in the complex envelope that modulates the carrier (a term that dates from the early days of radio) reflecting its role is to "carry" the information in a frequency band suitable for transmission and reception. For example the information (such as it is), transmitted by AM and FM radio is coded by the continuous amplitude and phase (frequency) functions $a(t)$ and $\psi(t)$ respectively.

If a real signal $s_r(t)$ is equivalent to the real part of a complex signal $s(t)$ denoted by $\Re\, s(t)$, the spectra must be related in a specific manner. To show this, let $S_r(\omega)$ represent the Fourier transform of $s_r(t)$. Then since $s_r^*(t) = s_r(t)$,

$$S_r(\omega) = \mathcal{F}s_r(t) = \mathcal{F}\,s_r^*(t) = \int_{-\infty}^{\infty} s_r^*(t)\exp(-j\omega t)\,dt = S_r^*(-\omega)$$

and the magnitude of the real signal spectrum is symmetric about ω_0. $s_r(t)$ is recovered from its spectrum by the Fourier inversion formula

$$\begin{aligned}
s_r(t) &= \frac{1}{2\pi}\int_{-\infty}^{\infty} S_r(\omega)\exp(j\omega t)\,d\omega \\
&= \frac{1}{2\pi}\int_{-\infty}^{0} S_r(\omega)\exp(j\omega t)\,d\omega + \frac{1}{2\pi}\int_{0}^{\infty} S_r(\omega)\exp(j\omega t)\,d\omega \\
&= \frac{1}{2\pi}\int_{0}^{\infty} S_r^*(\omega)\exp(-j\omega t)\,d\omega + \frac{1}{2\pi}\int_{0}^{\infty} S_r(\omega)\exp(j\omega t)\,d\omega \\
&= \frac{1}{2\pi}\Re\int_{0}^{\infty} 2S_r(\omega)\exp(j\omega t)\,d\omega.
\end{aligned}$$

This indicates that all of the information about the real signal is carried by one side of its spectrum. If $\mathcal{F}s(t) = S(\omega)$, the assertion that $s_r(t) = \Re\, s(t)$ is the same as asserting that

$$\begin{aligned}
s_r(t) &= \frac{1}{2\pi}\Re\int_{-\infty}^{\infty} S(\omega)\exp(j\omega t)\,d\omega \\
&= \frac{1}{2\pi}\Re\left(\int_{-\infty}^{0} S(\omega)\exp(j\omega t)\,d\omega + \int_{0}^{\infty} S(\omega)\exp(j\omega t)\,d\omega\right)
\end{aligned}$$

and by changing the variable of integration in the first integral, the relationship is expressed as

$$\begin{aligned}
\frac{1}{2\pi}\Re\int_{0}^{\infty} 2S_r(\omega)\exp(j\omega t)\,d\omega = \frac{1}{2\pi}\Re\int_{0}^{\infty} S(-\omega)\exp(-j\omega t)\,d\omega \\
+ \frac{1}{2\pi}\Re\int_{0}^{\infty} S(\omega)\exp(j\omega t)\,d\omega.
\end{aligned}$$

A sufficient (but not necessary) condition for equality is that the complex signal be *analytic*, that is, it has a one-sided spectrum such that

$$S(\omega) = \begin{cases} 2S_r(\omega) & \text{if } \omega > 0, \\ 0 & \text{if } \omega < 0. \end{cases} \tag{1.7}$$

Both the real and corresponding complex signals are then defined by the same frequency distribution.

The expression of real signals in terms of analytic complex signals considerably simplifies both computational and practical aspects of the processing of echo-location signals [9]. For example, operations involving convolution and correlation are central to much of the signal processing involved in echo-location. These are vastly simplified by employing the analytic equivalent of a real signal since the convolution of an analytic signal with the complex conjugate of another and the correlation between two different analytic signals (without conjugation) are zero [10]

The complex analytic signal is denoted by $s_a(t)$ and is expressed as

$$s_a(t) = s_r(t) + j\, s_h(t)$$

where $s_h(t)$ is a real function called the *Hilbert Transform* of $s_r(t)$ [11]. It is defined by the integral relationship

$$s_h(t) = \int_{-\infty}^{\infty} \frac{s_r(\tau)}{(t-\tau)} d\tau$$

and has the property that its spectrum is

$$S_h(\omega) = \begin{cases} jS_r(\omega) & \text{if } \omega < 0, \\ -jS_r(\omega) & \text{if } \omega > 0 \end{cases} \tag{1.8}$$

insuring that the analytic signal spectrum has the one-sided property (1.7). It is possible to define a complex signal

$$s(t) = s_r(t) + js_i(t)$$

where $s_i(t)$ is not the Hilbert transform. If $s(t)$ is given by (1.6) the amplitude function must be

$$a(t) = \sqrt{s_r(t)^2 + s_i(t)^2}$$

and

$$\Psi(t) = \tan^{-1}\left(\frac{s_i(t)}{s_r(t)}\right).$$

Assume that $s_r(t) = b(t) \cos \Theta(t)$ then, if

$$s_r(t) = \Re\left[a(t) \exp j\Psi(t) = a(t) \cos \Psi(t)\right],$$

which is equivalent to

$$s_r(t) = \left(\sqrt{b^2(t) \cos^2 \Theta(t) + (s_i(t)^2}\right) \cos \Psi(t),$$

the amplitude and phase functions are

$$\begin{aligned} a(t) &= \sqrt{b^2(t) \cos^2 \Theta(t) + (s_i(t)^2} \\ \Theta(t) &= \Psi(t). \end{aligned} \tag{1.9}$$

The equivalence of $b(t)\cos\Theta(t)$ and $a(t)\cos\Psi(t)$ requires that $b(t) = a(t)$ and $\Theta(t) = \Psi(t)$, but $a(t)$ is given by (1.9) and

$$a(t) = b(t) \Rightarrow s_i(t) = b(t)\sin\Theta(t).$$

This is the Hilbert transform of $b(t)\cos\Theta(t) = s_r(t)$ under the assumption that the real amplitude $b(t) = a(t)$ is slowly varying and has a spectrum distinct from that of tho phaoc function [10].

The conclusion to be drawn from this is that in order to express a real signal $s_r(t)$ in the form $a(t)\cos\Psi(t)$ which is the real part of the complex signal $s(t) = a(t)\exp j\Psi(t)$, $s(t)$ must be analytic. Moreover, the spectral shape is maintained. This is illustrated by determining the real signal spectrum corresponding to the real part of an arbitrary baseband complex envelope $\tilde{s}(t)$. The real signal is

$$\begin{aligned} s_r(t) &= \frac{1}{2}(\tilde{s}(t) + \tilde{s}^*(t)) \\ &= \frac{1}{2}\left(\frac{1}{2\pi}\right)\int_{-\infty}^{\infty}\left(\tilde{S}(\omega)\exp(j\omega t)\,d\omega + \tilde{S}^*(\omega)\exp(-j\omega t)\right)\,d\omega. \end{aligned}$$

Changing the sign of ω in the second argument results in

$$s_r(t) = \frac{1}{2}\left(\frac{1}{2\pi}\right)\int_{-\infty}^{\infty}\left(\tilde{S}(\omega) + \tilde{S}^*(-\omega)\right)\exp(j\omega t)\,d\omega$$

and

$$S_r(\omega) = \frac{1}{2}(\tilde{S}(\omega) + \tilde{S}^*(-\omega)).$$

This is equivalent to (1.7) for analytic complex signals but not otherwise as the spectrum becomes folded as illustrated in Figures 1.1a,b. The real spectrum corresponding to the non-analytic envelope of Figure 1.1a is the sum of the positive and negative frequency spectra resulting in the new spectral distribution as indicated in Figure 1.1b. (The sum is a complex sum, thus the addition as indicated by the shaded region is just an illustration). The analytic signal corresponding to the new frequency distribution is shown in Figure 1.1c. When a carrier term is added to the envelope, the analytic spectrum is shifted as indicated in Figure 1.1d. The original non-analytic baseband envelope of Figure 1.1a is converted to a bandpass analytic form by the simple expedient of multiplying by a carrier term $\exp j\omega_0 t$ with ω_0 sufficiently large to shift all of the negative frequencies to the positive side as shown in Figure 1.1e. Now, the real signal retains the spectral form of the envelope as shown in Figure 1.1f.

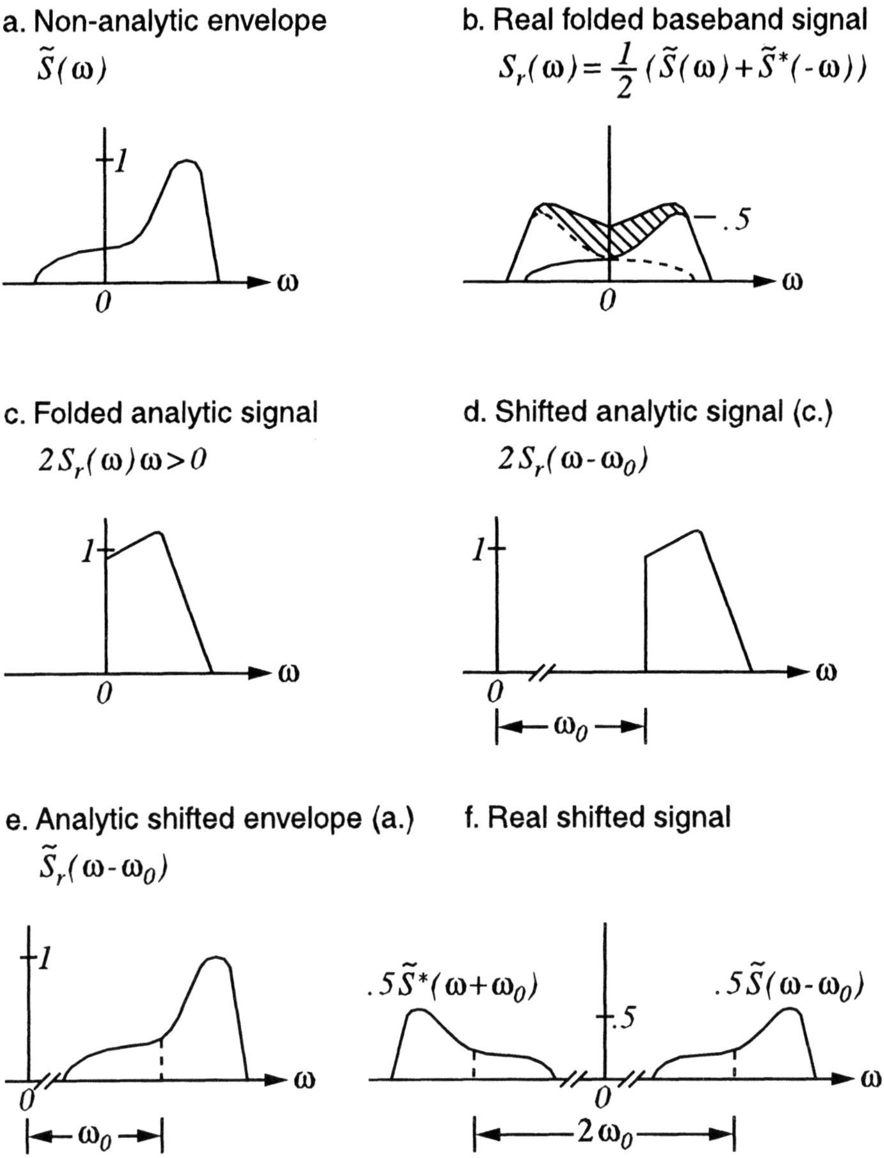

Figure 1.1: Complex envelope spectra

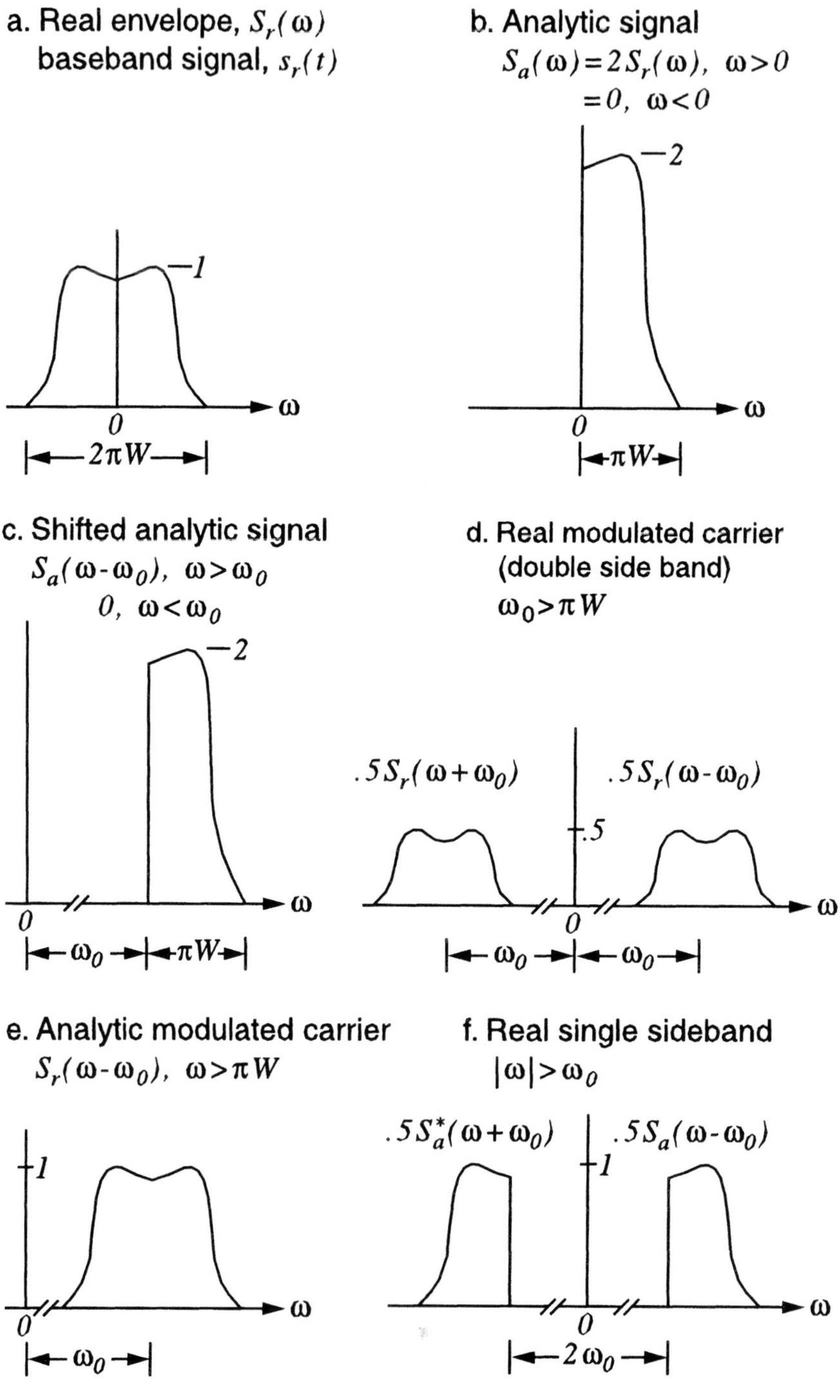

Figure 1.2: Spectral Representations

When the baseband envelope is real to begin with as in Figure 1.2a, the corresponding analytic signal spectrum consists of the positive half of the real spectrum as shown in Figure 1.2b. Forming the shifted analytic signal $s_a(t) \exp j\omega_0 t$ produces what is called an analytic *single sideband* (SSB) signal with a spectrum shown in Figure 1.2c [12]. These are often used to conserve bandwidth in radio communications because all of the information is contained in one half of the spectrum. The real bandpass signal corresponding to $s_r(t)$ is formed by multiplying it by $\cos \omega_0 t$ producing the real *double sideband* (DSB) spectrum shown in Figure 1.2d. This is the spectrum associated with ordinary radio broadcasting. It is also representative of real echo-location signal spectra.

The analytic modulated carrier spectrum of Figure 1.2e corresponds to the real envelope time function multiplied by the complex carrier to form $s_r(t) \exp j\omega_0 t$ where $\omega_0 \geq \pi W$. The real counterpart to the analytic SSB signal with spectrum as shown in Figure 1.2f is obtained by adding the conjugate and halving the sum forming

$$(s_a(t) \exp j\omega_0 t + s_a^*(t) \exp -j\omega_0 t) /2$$

that is equivalent to

$$s_r(t) \cos \omega_0 t - s_r(t) \sin \omega_0 t.$$

The figures indicate spectra with finite band limits but echo-location signals are of finite duration and consequently their spectra extend without limit. This is the essence of the *Paley-Weiner Theorem* [3] that states that no signal can be simultaneously time and band limited. Strictly speaking, an analytic signal is therefore an impossibility. Practically, echo-location waveforms usually have a low bandwidth to carrier ratio and waveform functions are used that contribute negligible spectral energy outside of a specified bandwidth. This is often accomplished by tailoring the amplitude modulation envelopes $a(t)$ through the use of what are know as *window functions*.

Modulated signals for which $\pi W << \omega_0$ are called *narrowband* signals. Their spectra are concentrated in the vicinity of the carrier and they tend to be predominantly sinusoidal in nature. Another way of looking at this is that narrowband signals have a baseband envelope spectrum that is distinct from that of the carrier or in other words, the carrier is sufficiently large so as to preclude any spectral fold-over as illustrated in Figure 1.1a,b. Most practical echo-location waveforms generated electronically fit the narrowband model because they are usually transmitted by tuned transmitters. Some wide-band signals do not fit the model and do not appear sinusoidal at all. Examples include transient signals generated by mechanical or explosive methods that are highly impulsive.

The signals that are discussed in this book are predominantly associated with electronically generated, modulated carrier, and narrowband waveforms. That does not mean that they can always be processed by what are known as narrowband methods. It means only that the real signal is derivable from the complex envelope without fold-over. The issue of wide vs narrowband signal representation concerns modeling of the Doppler effect that occurs when a signal is reflected from a moving object. It is discussed extensively in Chapter 2.

1.5 LINEAR SYSTEMS

1.5.1 Definition

In the field of signal processing, a system can be defined as a mathematical object that "operates" on a signal to produce another signal. Symbolically, denoting $x(t)$ as the *input* signal to a system $\mathcal{S}$, and $y(t)$ as the *output* signal, the relationship is written as

$$y(t) = \mathcal{S}[x(t)].$$

Suppose $x_1(t)$, $x_2(t)$ represent input signals with associated outputs $y_1(t)$, $y_2(t)$. $\mathcal{S}$ is said to be a *linear system* if, for constants a, b

$$\begin{aligned}\mathcal{S}[ax_1(t) + bx_2(t)] &= a\mathcal{S}[x_1(t)] + b\mathcal{S}[x_2(t)] \\ &= ay_1(t) + by_2(t).\end{aligned}$$

Furthermore, systems for which $y(t-t_0) = \mathcal{S}[x(t-t_0)]$ and t_0 is any fixed time, are said to be *time-invariant.* Linear time-invariant systems are a widely used model for the design and analysis of signal processing and other physical systems. They are most generally described in terms of homogeneous differential equations that are easily analyzed by Fourier or closely related techniques. Linear time-varying systems represent an important generalization that is discussed in Chapter 6.

1.5.2 The Impulse Response and Convolution Integral

The *impulse response* is the system output corresponding to an input of a *unit impulse function* $\delta(t)$ and is a commonly used time domain

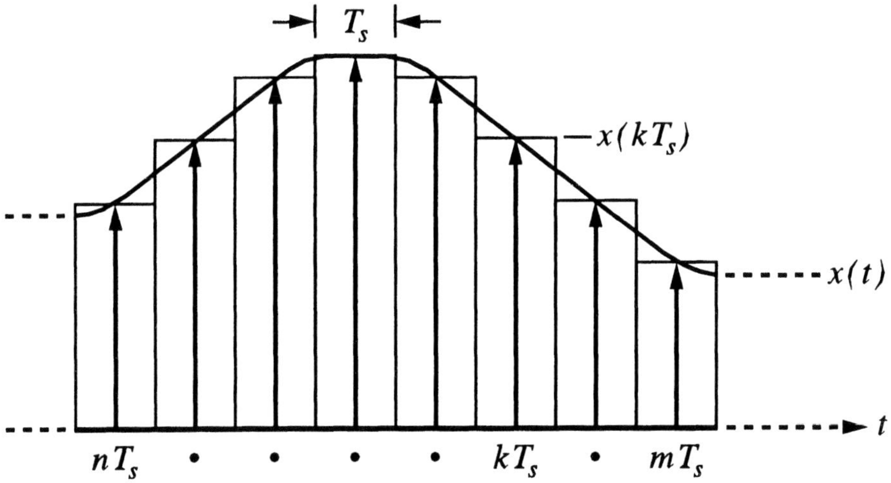

Figure 1.3: Approximation by an impulse sequence

characterization of a linear system. $\delta(t)$ has the *sifting* or reproducing property

$$x(t) = \int_{-\infty}^{\infty} x(u)\delta(u-t)du$$

and

$$h(t) = \mathcal{S}[\delta(t)]$$

defines the impulse response. $h(t)$ is also called the *weighting function* and defines the response of a linear system to a general input $x(t)$. Figure 1.3 shows a portion of a function $x(t)$ that weights a sampling sequence of impulses located at discrete times kT_s with interval T_s and k an integer. This generates the discrete sequence $x(kT_s)$ that approximates the continuous input $x(t)$ the sum of which over a finite interval $(m-n)T_s$ approximates the integral of the continuous function

$$\int_{nT_s}^{mT_s} x(t)dt \approx \sum_{k=n}^{m} T_s x(kT_s).$$

The output corresponding to the sequence is the sum of the responses to the individual weighted impulses with

$$y(t) = \mathcal{S}[x(t)] \approx \sum_{k=-\infty}^{\infty} x(kT_s)\mathcal{S}[\delta(t-kT_s)]$$

$$= \sum_{k=-\infty}^{\infty} T_s x(kT_s) h(t - kT_s).$$

As the sampling interval T_s becomes vanishingly small, the sums reduce to integrals yielding

$$y(t) = \mathcal{S}[x(t)] = \int_{-\infty}^{\infty} x(\tau) h(t - \tau) d\tau. \tag{1.10}$$

This is called the *convolution integral* and it is assumed that the input and impulse responses are absolutely integrable. This is generally the case for physical systems since they are usually described by piecewise, continuous, bounded functions.

1.5.3 The Transfer Function

Taking the FT of both sides of the convolution equation (1.10) yields

$$\begin{aligned}
\mathcal{F}y(t) &= Y(\omega) \\
&= \int_{-\infty}^{\infty} \left[\int_{-\infty}^{\infty} x(\tau) h(t - \tau) d\tau \right] \exp(-j\omega t)\, dt \\
&= \int_{-\infty}^{\infty} x(\tau) \left[\int_{-\infty}^{\infty} h(t - \tau) \exp(-j\omega t)\, dt \right] d\tau.
\end{aligned}$$

Changing variables ($u = t - \tau$) in the bracketed integral produces

$$\begin{aligned}
Y(\omega) &= \int_{-\infty}^{\infty} x(\tau) \left[\int_{-\infty}^{\infty} h(u) \exp -j\omega(u + \tau)\, du \right] d\tau \\
&= \int_{-\infty}^{\infty} x(\tau) H(\omega) \exp -j\omega\tau\, d\tau \\
&= X(\omega) H(\omega) \qquad (1.11)
\end{aligned}$$

where $H(\omega)$ is the FT of the impulse response and is called the *system transfer function.* This suggests that computing the inverse FT of the product of the transfer function and the FT of the input is an alternative way to compute system time response. This is an efficient approach and is commonly done because of the availability of fast Fourier transform (FFT) techniques for computers.

The convolution integral and transfer function represent time and frequency domain approaches to linear systems analysis, each providing particular insight into system behavior. Frequency analysis provides information regarding system response to signals with particular spectral characteristics. For example, systems designed to pass or stop (filter) signals with particular frequencies are best analyzed in the frequency domain. On the other hand the time domain representation provides insight into system transient behavior.

A delta function $\delta(\omega - \omega_0)$ in the frequency domain is transformed to the time domain as the monochromatic complex sinusoid

$$\mathcal{F}^{-1}[\delta(\omega - \omega_0)] = \frac{1}{2\pi}\int_{-\infty}^{\infty} \delta(\omega - \omega_0)\exp(j\omega t) = \exp j\omega_0 t.$$

If this is input to a linear system with transfer function $H(\omega)$, the time domain output is the inverse transform of $Y(\omega)$ (1.11) given by

$$y(t) = \frac{1}{2\pi}\int_{-\infty}^{\infty} \delta(\omega - \omega_0)H(\omega)\exp(j\omega t)\,d\omega = H(\omega_0)\exp j\omega_0 t.$$

This is called the *sinusoidal steady state response* and is the response of the system assuming that the sinusoidal input has been applied for an infinite interval prior to the time of observation and all startup transients have decayed away. A more complete description of system response must take into account these transients that occur when the input is switched on or off. Some transients such as the unit step function

$$\mathcal{U}(t-\tau) = \begin{cases} 0 & \text{if } t \le \tau, \\ 1 & \text{if } t \ge \tau \end{cases}$$

that is used to model switching, are not absolutely integrable and hence do not possess Fourier transforms. To deal with these, the more general *Laplace transform* having the convergent kernel $\exp -(\sigma + j\omega)t$ is applied. Details about it are found in the references.

1.6 STOCHASTIC PROCESSES

All signals are fundamentally random or have a random component if for no other reason than the temperature of our environment is not at absolute zero and there is always a thermal noise component. The interrogation energy reflection process inherent in echo-location is often regarded as random because of scatter complexity and interference due to environmental backscatter. Random signals in the time domain are modeled as functions $s(t;\eta)$ where η is a member of a *sample space* $\mathcal{H}$ whose members are randomly chosen and in some manner parameterize all possible signal time series that could be observed. This set is called an *ensemble* [13].

For example, consider a finite collection of identical, independently operating radio receivers each with a grounded antenna that represent a sample collection of an infinite ensemble of such receivers. The output of each is a noise voltage $n(t;\eta)$ where η is a parameter representing a particular receiver, i.e., $\eta = 1, 2, 3, \cdots$ for receiver number $1, 2, 3$, etc. The noise voltages of all of the receivers in the collection represent a sampling of an ensemble consisting of the underlying space of all possible noise voltages. An experiment is performed whereby at any given time t, η is the result of a random draw and the noise voltage of the corresponding receiver is measured. The measured noise voltage at that particular time is a random number because it depends upon the randomly chosen value of η and could have taken on any value existing within the collection at that time. On the other hand the choice of a particular value of η defines a particular receiver and hence a single non-random time function $n(t;\eta)$ for all time. This implies that every conceivable time series may be associated with some collection of receivers, a value of η, and the output of the associated receiver. Once η and its associated receiver is chosen, the time function $n(t;\eta)$ is determined for all time although it may be highly complex and unpredictable. This is because the noise output of any one of the receivers in the collection is due to the thermal motion of electrons and knowledge of $n(t;\eta)$ for $t \leq t_1$ cannot be used to predict its value for $t > t_1$.

The value of the noise process $n(t;\eta)$ for fixed time is a random variable and the family of time functions is called a *stochastic or random process*. Random variables are denoted by sans-serif fonts to clearly identify them in the subsequent text. For example, the random variables associated with the stochastic noise and signal processes $n(t;\eta)$ and $s(t;\eta)$ at any fixed time t are denoted as $\mathsf{n}(t)$, $\mathsf{s}(t)$ respectively. Likewise those associated with spectra $N(\omega;\eta)$ and $S(\omega;\eta)$ at fixed ω are denoted by $\mathsf{N}(\omega)$ and $\mathsf{S}(\omega)$.

The analysis of stochastic processes is most easily carried out using the probability density function (pdf) or the cumulative distribution function (cdf) for the value of a stochastic process $x(t;\eta)$ at any time t. These describe the overall behavior of the ensemble of functions at a given time. The cdf for the ensemble is defined as

$$\mathsf{F}(x;t) = \wp(\mathsf{x} \in (-\infty, x)) \text{ at time } t$$

where $\wp\{\bullet\}$ denotes probability and x is the random variable representing all possible values of $x(t)$. [2] The associated pdf for the random variable x at time t is

$$\mathsf{f}(x;t) = \frac{d}{dx}\mathsf{F}(x;t)$$

Similarly, define the vector

$$\mathbf{x}(t_1, t_2, \cdots, t_N) = \begin{pmatrix} x(t_1) & \vdots & x(t_N) \end{pmatrix}$$

and its associated random vector

$$\mathsf{x} = \begin{pmatrix} \mathsf{x}_1 & \vdots & \mathsf{x}_N \end{pmatrix}$$

whose i-th element represents all possible values of $x(t_i)$. The joint cdf for the elements of the vector is

$$\begin{aligned} \mathsf{F}(\mathbf{x}; t_1, \cdots, t_N) = \wp(&\mathsf{x}_1 \in (-\infty, x_1)) \text{ at time } t_1, \\ &\cdots, \mathsf{x}_N \in (-\infty, x_N) \text{ at time } t_N \end{aligned} \tag{1.12}$$

and the joint ensemble pdf is

$$\mathsf{f}(\mathbf{x}; t_1, \cdots, t_N) = \frac{d^N}{dx_1 \cdots dx_N}\mathsf{F}(\mathbf{x}; t_1, \cdots, t_N). \tag{1.13}$$

The joint pdf depends explicitly on the time variables but a special case arises when $\mathsf{F}(\mathbf{x}; t_1, \cdots, t_N) = \mathsf{F}(\mathbf{x}; (t_1 + \epsilon), \cdots, (t_N + \epsilon)) \, \forall \epsilon$. Such processes are said to be *stationary*.

Unless required for clarity, the time references in equations (1.12,1.13) are henceforth dropped for notational convenience and the joint cdf and pdf are denoted as

$$\mathsf{F}(\mathbf{x}) = \mathsf{F}(x_1, x_2, \cdots, x_N)$$

and

$$\mathsf{f}(\mathbf{x}) = \mathsf{f}(x_1, x_2, \cdots, x_N)$$

with the understanding that x_i represents $x(t_i)$.

---

[2] The density and distribution functions of a random variable x are denoted by $\mathsf{f}(x)$ and $\mathsf{F}(x)$. If it is necessary to differentiate between the pdf's and distributions of random quantities x and y the notation $\mathsf{f_x}(x)$, $\mathsf{F_x}(x)$ and $\mathsf{f_y}(y)$, $\mathsf{F_y}(y)$ is used. Otherwise the subscripts are dropped. For convenience the ensemble parameter notation $x(t;\eta)$ is also dropped and $x(t)$ is used when the meaning is clear.

The mean and variance of a stochastic process are defined as expectations

$$\bar{x}(t) \equiv \mathcal{E}\{\mathsf{x}(t)\} = \int_{-\infty}^{\infty} x\mathsf{f}(x,t)dx$$
$$\sigma_x^2 \equiv \mathcal{E}\{[\mathsf{x}(t) - \bar{x}(t)]^2\} = \int_{-\infty}^{\infty} (x(t) - \bar{x}(t))^2\mathsf{f}(x,t)dx \qquad (1.14)$$

respectively. The *autocorrelation* between two random variables x_1 and x_2 derived from the process $x(t;\eta)$ at times t_1, t_2 is

$$\mathcal{E}\{\mathsf{x}_1\mathsf{x}_2\} \equiv R_{xx}(t_1,t_2) = \int_{-\infty}^{\infty}\int_{-\infty}^{\infty} x_1x_2\mathsf{f}(x_1,x_2)dx_1dx_2. \qquad (1.15)$$

A process for which $R_{ss} = 0$ for $t_1 \neq t_2$ is called a *white* process. This term is most often associated with broadband (white) noise interference. Conversely noise processes that have non-zero correlation over a finite interval are often called correlated or *colored noise.*

The quantity $\mathcal{E}\{\mathsf{x}_1\mathsf{y}_2\} = R_{xy}(t_1,t_2)$ is the *cross-correlation* of the random variables $\mathsf{x}(t_1)$, $\mathsf{y}(t_2)$ associated with the different processes $x(t;\eta)$ and $y(t;\eta\prime)$ at times t_1, t_2. It is defined similarly to (1.15) in terms of the joint density function $\mathsf{f}(x_1,y_2)$. The correlation between complex vector quantities $\mathbf{x}(t_1)$ and $\mathbf{y}(t_2)$ is expressed as the matrix

$$\mathbf{R}_{xy}(t_1,t_2) = \mathcal{E}\{\mathbf{x}(t_1)\mathbf{y}^{\mathrm{H}}(t_2)\}$$

with elements

$$R_{x_iy_j}(t_1,t_2) = \mathcal{E}\{\mathsf{x}_i(t_1)\mathsf{y}_j^*(t_2)\} = \mathcal{E}\{(\mathsf{y}_j(t_2)\mathsf{x}_i^*(t_1))^*\} = R_{y_jx_i}(t_2,t_1).$$

The latter expression corresponds to the elements of the conjugate or *Hermitian* transpose $\mathbf{R}_{xy}^{\mathrm{H}}(t_2,t_1)$. The relationship is simply written as $\mathbf{R}_{xy} = \mathbf{R}_{xy}^{\mathrm{H}}$ when the time reference is dropped.

Processes for which $R_{xx}(t_1,\, t_2) = R_{xx}(t_1 - t_2) = R_{xx}(\tau)$ are called *wide sense stationary* (WSS) which is less restrictive than stationarity of the joint pdf. A process is said to be *ergodic* if the expectations over the ensemble (1.14,1.15) are equal to time averages where for example

$$\bar{x}(t) = \lim_{T\to\infty}\frac{1}{T}\int_{-T/2}^{T/2} x(t)dt.$$

1.6.1 Fourier Transforms and Power Spectra

Fourier transforms and the concept of the frequency domain spectrum are applied to stochastic processes with some modification. Consider the single sample function $x(t;\eta)$, and define the function

$$x(t;\eta,T)=\begin{cases} x(t;\eta) & \text{for } t\in[-T/2,T/2], \\ 0 & \text{for } t\notin[-T/2,T/2] \end{cases}. \tag{1.16}$$

The FT of this is

$$X(\omega;\eta,T)=\int_{-T/2}^{T/2} x(t;\eta)\exp(-j\omega t)dt$$

and the energy spectral density for this single time limited realization is $|X(\omega;\eta,T)|^2$ (Joules per Hz). For this to be a meaningful description of the entire ensemble, the average on the interval T is defined by

$$S_{xx}(\omega;T)\equiv\mathcal{E}\{|\mathsf{X}(\omega,T|^2\}\quad\text{(Joules/Hz)}.$$

The spectral contribution for t beyond the finite interval $[-T/2,T/2]$ is taken into account by defining the average power of the spectrum at ω. This is obtained by dividing the expected energy density by T and taking the limit as $T\to\infty$. The energy density has units of Joule-sec., therefore division by T produces the *power spectral density* (psd)

$$S_{xx}(\omega)=\lim_{T\to\infty}\frac{S_{xx}(\omega;T)}{T}\quad\text{(Watts/Hz=Joules)}. \tag{1.17}$$

The psd of a constant white noise process for which $S_{xx}(\omega)=S_{xx}\forall\omega$ is usually denoted as $2\pi N_0$ or merely N_0 if the psd is expressed as a function of frequency in Hz.

1.6.2 The Weiner-Khintchine Theorem

Just as there exists a Fourier relationship between a time function and its spectrum, there also exists a Fourier relationship $S_{xx}=\mathcal{F}R_{xx}(\tau)$ between the autocorrelation function and the power spectral density of a WSS stochastic process. This important result in signal processing and communication theory is known as the *Weiner-Khintchine* (W-K) theorem [14] and is proven by first computing the energy spectral density

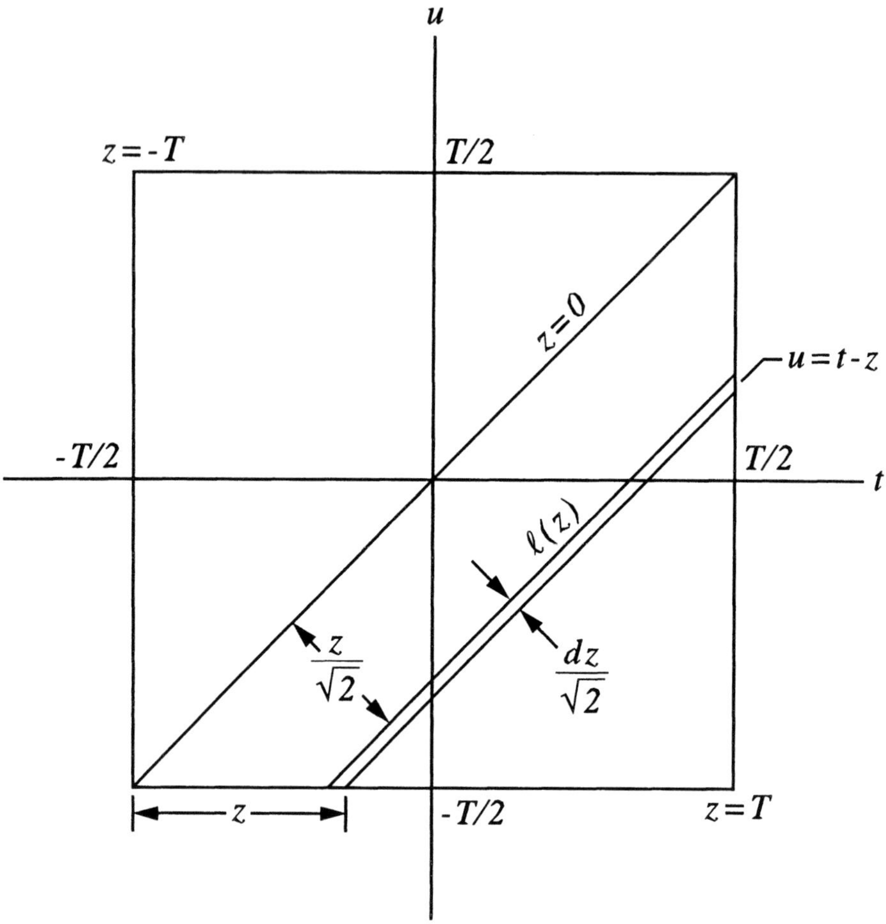

Figure 1.4: Iterated integration of $R(t-u)\exp -j\omega(t-u)$

of the truncated time segment (1.16)

$$S_{xx}(\omega;T) = \int_{-T/2}^{T/2}\int_{-T/2}^{T/2} \mathcal{E}\{\mathsf{x}(t)\mathsf{x}^*(u)\} \exp\ -j\omega(t-u)dtdu$$
$$\int_{-T/2}^{T/2}\int_{-T/2}^{T/2} R_{xx}(t-u)\exp\ -j\omega(t-u)dtdu.$$

The support region for the double integral in the (t,u) plane is shown in Figure 1.4 where $z = t - u$. The integration is conveniently carried out as an iterated integral by first integrating along the strip defined by the line $u = t - z$ and then integrating the result with respect to z. The strip has length $\ell(z)$ and differential area $\ell(z)dz$, and from geometry, $\ell(z) = \sqrt{2}(T - z)$. The differential area is $\ell(z)dz/\sqrt{2} = (T - z)dz$ and $\ell(z)$ is symmetric in z so that the spectral density of the truncated sample reduces to the single integration in z

$$\begin{aligned} S_{xx}(\omega; T) &= \frac{1}{\sqrt{2}} \int_{-T/2}^{T/2} R_{xx}(z)\ell(z) \exp -j\omega z dz \\ &= T \int_{-T/2}^{T/2} \left(1 - \frac{|z|}{T}\right) R_{xx}(z) \exp -j\omega z dz. \end{aligned}$$

The power spectral density is found by substituting this into (1.17)

$$S_{xx}(\omega) = \lim_{T\to\infty} \int_{-T/2}^{T/2} \left(1 - \frac{|z|}{T}\right) R_{xx}(z) \exp -j\omega z dz$$

and in the limit, $\left(1 - \frac{|z|}{T}\right) \approx (1 - \delta(|z| - T))$ so that

$$S_{xx}(\omega) = \int_{-\infty}^{\infty} R_{xx}(z) \exp -j\omega z dz \quad \text{(Watts/Hz)} \tag{1.18}$$

proving the theorem.

$S_{xx}(\omega)$ has the units of Joules or Watts/Hz and represents the average power at each frequency. It is non-negative, real and the corresponding autocorrelation is conjugate-even i.e., $R_{xx}(\tau) = R^*_{xx}(-\tau)$. The inverse FT (1.5) for a time delay τ is $R_{xx}(\tau) = \mathcal{F}^{-1}S_{xx}(\omega)$ and for $\tau = 0$,

$$\frac{1}{2\pi} \int_{-\infty}^{\infty} S_{xx}(\omega)d\omega = R_{xx}(0) = \mathcal{E}\{|\mathsf{x}|^2\} \quad \text{(Watts)}$$

which is the aggregate power over all frequency and the average power of the process $x(t)$. The psd is also defined for the cross correlation between two WSS processes as

$$\begin{aligned} S_{xy}(\omega) &= \lim_{T\to\infty} \frac{\mathcal{E}\{\mathsf{X}(\omega, T)\mathsf{Y}^*(\omega, T)\}}{T} \\ &= \int_{-\infty}^{\infty} R_{xy}(z) \exp -j\omega z dz \quad \text{(Watts/Hz=Joules)} \end{aligned} \tag{1.19}$$

and is called the *cross spectrum* or cross spectral density.

The Fourier series components (1.4)

$$\mathsf{X}_n = \frac{1}{T} \int_0^T \mathsf{x}(t) \exp -jn\omega t dt$$

of a WSS random process $\mathsf{x}(t)$ with $\omega = 2\pi/T$ are asymptotically uncorrelated for long integration times [7, 15]. The spectral correlation is

$$\mathcal{E}\{\mathsf{X}_n\mathsf{X}_m^*\} = \frac{1}{T^2}\int_0^T\int_0^T R_{xx}(t-s)\exp j\omega(ms-nt)dsdt$$

and by changing variables $u = t - s$ to replace t, the correlation becomes

$$\frac{1}{T^2}\int_0^T\int_s^{T-s} R_{xx}(u)\exp j\omega(ms-n(u+s))dsdu.$$

Again by changing variables and letting $v = s/T$, the correlation is

$$T\mathcal{E}\{\mathsf{X}_n\mathsf{X}_m^*\} = \int_0^1 \exp j\omega(m-n)vT\left(\int_{-vT}^{(1-v)T} R_{xx}(u)\exp j\omega nudu\right)dv$$

and the inner integral approaches the psd

$$\lim_{T\to\infty}\int_{-vT}^{(1-v)T} R_{xx}(u)\exp j\omega nudu = \int_{-\infty}^{\infty} R_{xx}(u)\exp j\omega nudu = S_{xx}(n\omega)$$

as the integration interval T increases. The outer integration in v denoted by i_o is equivalent to

$$\begin{aligned} i_o &= \exp\left(\frac{j\omega T(m-n)}{2}\right)\int_{1/2}^{1/2} \exp j\omega T(m-n)udu \\ &= \exp\left(\frac{j\omega T(m-n)}{2}\right)\operatorname{sinc}\left(\frac{\omega T(m-n)}{2}\right) \end{aligned}$$

where $u = v - 1/2$, $\operatorname{sinc}(x) = \sin(x)/x$, and

$$\lim_{T\to\infty} i_o = \begin{cases} 0 & \text{when } m \neq n, \\ 1 & \text{when } m = n. \end{cases}$$

Consequently,

$$\lim_{T\to\infty} T\mathcal{E}\{\mathsf{X}_n\mathsf{X}_m^*\} = \begin{cases} 0 & \text{when } m \neq n, \\ S_{xx}(n\omega) & \text{when } m = n \end{cases} \tag{1.20}$$

and $\mathcal{E}\{\mathsf{X}_n\mathsf{X}_m^*\}$ is asymptotically uncorrelated for $m \neq n$.

1.6.3 Linear System Response to Stochastic Signals

A stochastic linear system input represented by the sample function $x(t;\eta)$ results in the stochastic output

$$\mathsf{y}(t) = \int_{-\infty}^{\infty} \mathsf{x}(\tau)h(t-\tau)d\tau$$
$$y(t;\eta) = \int_{-\infty}^{\infty} x(\tau;\eta)h(t-\tau)d\tau$$

where the random variable y is the output counterpart to the input random variable x spanning the ensemble at any fixed time τ. At any time t, y represents the impulse response weighted contribution of all possible values of $x(\tau;\eta)$ integrated over the time support of the impulse response. The power spectral density of the output is obtained by the same limiting process used to derive the psd of the stochastic input. The output sample function corresponding to the finite duration input (1.16) is given by

$$y(t;\eta,T) = \int_{-\infty}^{\infty} x(\tau;\eta,T)h(t-\tau)d\tau$$

and the corresponding spectrum is

$$Y(\omega;\eta,T) = H(\omega)X(\omega;\eta,T).$$

The energy density is

$$\begin{aligned} S_{yy}(\omega;T) &= \mathcal{E}\{|\mathsf{Y}(\omega,T|^2\} \\ &= |H(\omega)|^2 S_{xx}(\omega;T) \quad \text{(Joules/Hz)} \end{aligned}$$

and the psd is obtained by dividing by the time interval and passing to the limit

$$\begin{aligned} S_{yy}(\omega) &= \lim_{T\to\infty} \frac{S_{yy}(\omega;T)}{T} \\ &= |H(\omega)|^2 S_{xx}(\omega). \end{aligned}$$

The autocorrelation of the output is then obtained by invoking the W-K theorem

$$R_{yy}(\tau) = \mathcal{F}^{-1} S_{yy}(\omega).$$

1.7 SUMMARY

This chapter is designed to provide a brief review of signal theory to supplant but not replace a detailed treatment of the subject that would be obtained in a formal graduate course. It introduces complex signal notation and the concept of the analytic signal as these are used extensively throughout the book and must be understood. The assumption that the signals and waveforms discussed are analytic is realistic because electronically generated sonar waveforms are generally spectrally narrowband. The bandwidth to carrier ratio is small, the real and imaginary components are a Hilbert transform pair, and hence the associated complex bandpass waveforms are analytic. The definition of narrow vs wideband has been broadened in recent years to include consideration of the Doppler effect as a time dilation rather than a frequency shift. This is explored in detail in later chapters.

Bibliography

[1] T.M. Apostol. *Mathematical Analysis* . Addison-Wesley, New York NY, 1974.

[2] H. Dym and H.P. McKean. *Fourier Series and Integrals* . Academic Press, New York NY, 1972.

[3] B.P. Lathi. *Signals, Systems, and Communications* . John Wiley & Sons, New York NY, 1965.

[4] N.S. Tzannes. *Communication and Radar Systems* . Prentice-Hall, Englewood Cliffs NJ, 1985.

[5] R.E. Ziemer and W.H. Tranter. *Principles of Communications: Systems, Modulation, and Noise* . Houghton Mifflin, New York NY, 1976.

[6] H.L. Van Trees. *Detection, Estimation, and Modulation Theory-Part III* . Wiley, 1971.

[7] A.D. Whalen. *Detection of Signals in Noise* . Academic Press, 1971.

[8] W.C. Knight, G.G. Pridham, and S.M. Kay. Digital Signal Processing for Sonar . *Proceedings of the IEEE*, 69(11):1451–1507, Nov. 1981.

[9] J.V. DiFranco and W.L. Rubin. *Radar Detection* . Artech House, Dedham, MA, 1980.

[10] A.W. Rihaczek. *Principles of High Resolution Radar* . McGraw-Hill, New York NY, 1969.

[11] A. Papoulis. *The Fourier Integral and its Applications* . McGraw-Hill, New York NY, 1987.

[12] E. Bedrosian. The Analytical Signal Representation of Modulated Waveforms . *Proceedings of the IEEE*, 50:2071–2076, Oct. 1962.

[13] A. Papoulis. *Probability, Random Variables, and Stochastic Processes* . McGraw-Hill, New York NY, 1965.

[14] Y.W. Lee. *Statistical Theory of Communication.* John Wiley & Sons, New York NY, 1960.

[15] W.B. Davenport and W.L. Root. *An Introduction to the Theory of Random Signals and Noise* . McGraw-Hill, 1958.

Chapter 2

ECHO ENERGY AND TIME BASE

2.1 INTRODUCTION

The concept of echo-location, that is the transmission of acoustic or electromagnetic energy and the subsequent reception of energy reflected from physical objects or scatterers, is primarily motivated by (1.) the desire to detect their presence, and (2.) to estimate associated parameters such as size, range, speed, and bearing. Secondary objectives which relate to these may involve identification, discrimination, and tracking of specific scattering objects in the presence of competing backscatter sources and noise.

Perhaps one of the earliest nautical applications of acoustic echo-location was the determination of the proximity of icebergs by the firing of a cannon or the sounding of a bell or fog horn. Equipment was rudimentary: the sound source, a stopwatch, and a good pair of ears, consequently, the information obtained was minimal consisting of rough estimates of size, range, and possibly direction. With the development of improved hardware in the form of transducer arrays and electronics motivated by military requirements during the two world wars, it became possible not only to more precisely estimate range by timing echos, but also to measure velocity (Doppler) and bearing [1]. After the second war, the fundamental contributions of Shannon, Wiener, and others to the science of information theory, and those of Woodward [2], North [3], Marcum [4], Swerling [5], and others to the theory of radar detection and signal processing, paved the way toward the design of modern acoustic echo-location systems with applications in such diverse areas as medical imaging, seismology, non-destructive testing, and sonar. Through the use of modern echo-location techniques, it is possible not only to measure rudimentary attributes such as distance, bearing, and velocity, but also generate high resolution seismic images for the identification of geological features in the earth, non-invasively probe the human body via ultrasound, and to map and survey the ocean floor and identify scattering sources via sonar.

Much of modern sonar system performance is attributable better understanding of the physics of echo formation and scattering processes, new concepts in signal processing, and increased computer power permitting their real time implementation. A phenomenological approach is taken in this and subsequent chapters. The effects of the propagation medium, signal reflection, and scattering upon the transmitted signal are treated in terms of simple linear system models. In this way, the essential signal processing concepts are emphasized.

2.2 THE ECHO ENERGY CYCLE

In its most rudimentary form, the echo location process may be considered as an echo energy cycle or interrogation cycle modeled as a simple block diagram as illustrated in Fig. (2.1). The cycle begins with the specification of a normalized, complex waveform $f(t)$ defined on a finite time interval T with

$$\int_{-T/2}^{T/2} f(t)f^*(t)dt = \int_{-T/2}^{T/2} \mid f(t) \mid^2 dt = 1.$$

The transmitter generates the electrical signal $\sqrt{E_0}f(t)$ with energy E_0 that is transmitted by a projector array into the medium. The energy projected in a particular direction is a function of the array gain or directivity index D_I of the array [1, 6] and is expressed as the energy intensity or flux at a distance of 1 meter E_x by the decibel equation [1]

$$E_x = E_0 + D_I - 11 \text{ (dB re 1 Joule/m}^2 \text{ @ 1 meter)}.$$

The factor of 11 dB represents the area of the unit sphere, i.e., $10 \log 4\pi = 11$. An additional constant may be incorporated to account for transduction loss if the projector is not 100% efficient. In order to simplify the

[1]The notation E_0 is used interchangeably where noted to represent energy as dB re 1 Joule.

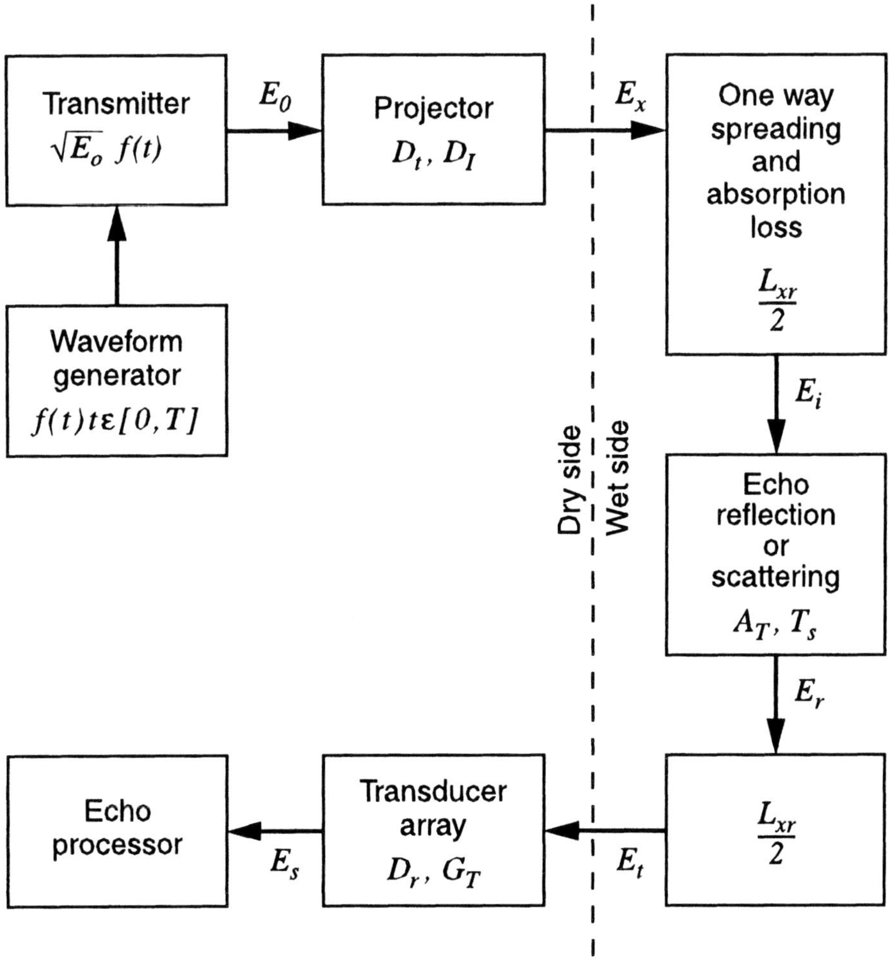

Figure 2.1: The echo energy cycle

discussion and stress the fundamental concepts of echo formation, the transmitted signal is treated as a scalar quantity and the projector and receive arrays are described by their directivity indices. A detailed discussion of beam formation and array processing is provided in Chapter 7.

Sonar transmit power is usually defined in terms of the sound pressure level p_r 1 meter from the source which is called the source level S_l and is expressed in dB re 1μPa (micro-Pascal) as $S_l = 20 \log p_r$, where $1\mu\text{Pa} = 10^{-6}\text{Newtons/m}^2$. The average decibel power intensity P_x is related to the average source level and the propagation impedance of the medium z by

$$P_x = S_l - 10 \log z \quad (\text{dB re 1 Watt/m}^2 \text{ @ 1 meter}). \tag{2.1}$$

The energy intensity one meter from the source is

$$E_x = S_l + T_x - 10 \log z \quad (\text{dB re 1 Joule/m}^2 \text{ @ 1 meter})$$

where T_x is 10 log T. In water, z is 1.5×10^{18} $(\mu\ P_a)^2/(\text{Watts/m}^2)$ and consequently

$$\begin{aligned} E_x(\text{in water}) &= S_l + T_x - 182 \quad (\text{dB re 1 Joule/m}^2 \text{ @ 1 meter}) \\ P_0(\text{transmitter power}) &= S_l - 171 - D_I \quad (\text{dB re 1 Watt}). \end{aligned}$$

The acoustic wavefront expands spherically in a homogeneous medium and the incident energy intensity at range r is

$$\begin{aligned} E_i &= E_x - L_{xr}/2 \\ &= E_x - (20 \log(r/1) + \alpha r) \quad (\text{dB re 1 Joule/m}^2) \end{aligned}$$

where α is the frequency dependent acoustic absorption coefficient (dB/m) [1]. L_{xr} is the two way spreading and absorption (propagation) loss and is equivalent to $(40 \log\ r + 2\alpha r)$. The scattering mechanism is considered to be a point reflector of effective area $A_t(\text{m}^2)$ that re-radiates the incident energy spherically. The resulting reflected energy intensity E_r at one meter is

$$E_r = E_i + 10 \log\ A_t - 11 \quad (\text{dB re 1 Joule/m}^2).$$

The target strength T_s expressed as

$$\begin{aligned} T_s &= E_r - E_i \\ &= 10 \log\ A_t - 11 \quad (\text{dB}) \end{aligned}$$

is the decibel equivalent of the ratio of the reflected intensity at one meter to the incident intensity and is a common measure of the reflective strength of a scatterer. For example, a zero dB target is equivalent to an object with an effective reflecting area of 4π square meters.

If the receive transducer is collocated with the transmit projector, the echo energy intensity at the transducer face E_t, is the reflected energy E_r subject to the same one way loss $(L_{xr}/2)$ and is expressed in terms of the source level and target strength as

$$E_t = S_l + T_x + T_s - L_{xr} - 182 \quad (\text{dB re 1 Joule/m}^2). \tag{2.2}$$

The transducer converts mechanical pressure to voltage and is characterized by the conversion factor $G_t = 20 \log(\text{ volts}/\mu \text{ Pa})$ or equivalently as $10 \log \ (\text{Watts}/(\mu \text{ Pa})^2)$. The electrical echo power P_s and energy E_s expressed in dB are related by $E_s = P_s + T_x$, and in turn, the power is related to the sound pressure level (spl) at the transducer by the factor G_t. Denoting the received signal spl by p_{sp} with units of μPa, the echo energy expressed in dB is

$$E_s = G_t + T_x + 20 \log \, p_{sp} \quad (\text{dB re 1 Joule})$$

but the pressure is related to the propagation impedance and the received acoustic power intensity P_t as in (2.1) so that

$$20 \log \, p_{sp} = 10 \log \, z + P_t \quad (\text{dB re } 1\mu \text{ Pa}^2) \tag{2.3}$$

and consequently,

$$\begin{aligned} E_s &= G_t + 10 \log \, z + P_t + T_x \\ &= G_t + 182 + E_t \quad (\text{dB re 1 Joule}). \end{aligned}$$

Combining equations (2.3) and (2.2) results in

$$\begin{aligned} E_s &= S_l + T_x - L_{xr} + G_t + T_s \quad (\text{dB re 1 Joule}) \\ P_s &= S_l - L_{xr} + G_t + T_s \quad (\text{dB re 1 Watt}) \end{aligned} \tag{2.4}$$

which are the electrical echo energy and power at the transducer output. This is the echo *sonar equation* and is augmented by adding the terms D_t, D_r to account for the transmit and receive beam attenuation when the beams are not aligned with the scatterer [1, 7]. The input signal will invariably be embedded in additive interference that consists of extraneous noise and reverberation. Noise arises from a multitude of

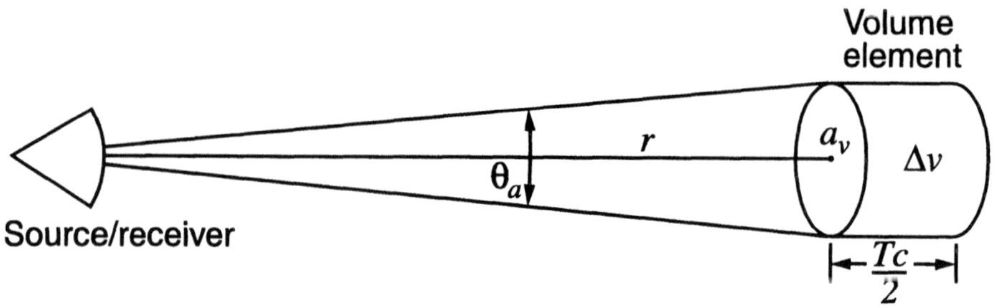

Figure 2.2: Volume reverberation backscatter element

sources such as: shipping, wave breaking, flow over a rigid body, biological sources, rain, and electronic noise. Noise originating from distant sources is called *ambient noise* and is reduced in proportion to the directivity index of the receive array. The aggregate contribution from these is lumped into a single noise parameter and if it is spectrally flat over the system bandwidth W_s, it is represented by its power spectral density N_0 with units of (Watt/Hz=Joules). The noise power that appears at the input to the signal processor is then $p_n = N_0 W_s$. The dB equivalent $10 \log N_0$ is usually given the shorthand notation (dBs).

Reverberation is backscattered signal energy from dispersed material in the water column itself (*volume reverberation*) or from reflection from the surface and/or bottom boundaries (*boundary reverberation*). The reverberation contribution to the sonar equation is computed by defining equivalent scattering and target strengths for volume and boundary backscatter and using (2.4) [1,6]. The effective volume target strength is

$$T_{sv} = 10 \log(s_v \triangle v) \quad \text{(dB)}$$

where s_v is called the volume scattering coefficient and represents the incremental backscatter strength per unit volume. Δv is the incremental volume illuminated at range r by a pulse of duration T as shown in Figure 2.2. If the transmitted energy is confined to a circular beam of effective solid angle θ_a,

$$\Delta v = c \frac{T}{2} r^2 \theta_a$$

where c is the sound speed and $\theta_a = 5.9(\lambda/2\pi a)$ for a circular array of radius a. [1]. In dB notation,

$$T_{sv} = S_v + 20 \log r + 10 \log(cT\theta_a/2) \quad \text{(dB)}.$$

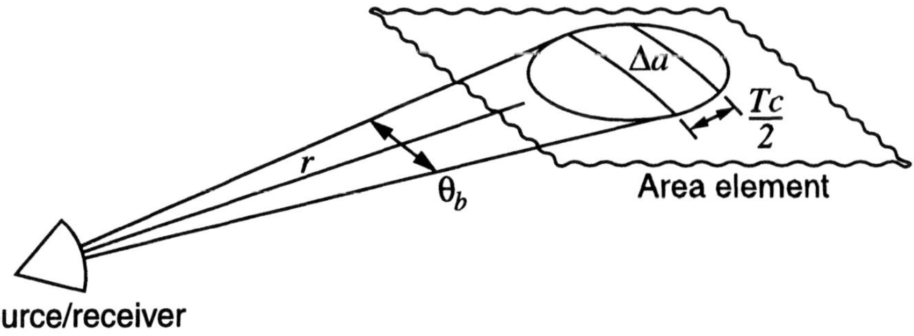

Figure 2.3: Boundary reverberation backscatter element

Substituting this for the target strength in (2.4) yields the volume reverberation input power

$$\begin{aligned} P_v &= S_l - L_{xr} + G_t + S_v + 20\log r + 10\log(cT\theta_a/2) \\ &= S_l + S_v + G_t - 20\log r - 2\alpha r + T_x + 10\log(c\theta_a/2) \\ &\quad \text{(dB re 1 Watt)} \end{aligned} \tag{2.5}$$

or equivalently the energy $P_v + T_x$. The dB difference

$$\mathrm{SIR}_v = P_s - P_v = T_s - S_v - T_x - 20\log r - C_v \quad \text{(vol. reverb. SIR (dB))}$$

is the volume reverberation SIR where $C_v = 10\log(c\theta_a) - 3$. A similar calculation is made for boundary reverberation only the effective target strength is proportional to the scattering coefficient s_b and the illuminated area element Δa of the beam intersection with the boundary as shown in Figure 2.3. The beamwidth is $\theta_b = 4.9\lambda/2\pi a$ for the circular array [1]. In dB notation this is

$$T_{sb} = S_b + 10\log r + 10\log(cT\theta_b/2) \quad \text{(dB)}$$

and the input boundary reverberation power is

$$\begin{aligned} P_b = {} & S_l + S_b + G_t - 30\log r - 2\alpha r \\ & + T_x + 10\log(c\theta_b/2) \quad \text{(dB re 1 Watt).} \end{aligned} \tag{2.6}$$

The resulting SIR for boundary reverberation with $C_b = 10\log(c\theta_b) - 3$ is then

$$\mathrm{SIR}_b = P_s - P_b = T_s - S_b - T_x - 10\log r - C_b \quad \text{(SIR (dB))}$$

The combination of the uncorrelated interference sources is power additive and

$$\mathrm{SIR}_{dB} = 10\log\left(\frac{P_s}{P_v + P_b + N_0 W_s}\right) \quad \text{(total input SIR (dB))} \quad (2.7)$$

where P_s, P_v, P_b are expressed in Watts (not dB). These relationships are the sonar equations for echos embedded in interference and only indicate the general SIR trends with range in volume and boundary reverberation under idealized isotropic conditions. The S_b for example depends upon the boundary type (surface or bottom), grazing angle, depth offset from the boundary, and associated physical parameters such as wave height and bottom roughness. Boundary reverberation is often the most severe interference especially in shallow water. Moreover its level can vary significantly from the idealized $30\log r$ dB predicted under idealized conditions when the bottom is not flat, propagation is not isotropic, or the scattering coefficient varies with range or position of the platform. More details regarding reverberation models are found in [1,6,8–10].

When the statistical characteristics of the reflected signal and the interference are known or may be assumed, the detection performance of a particular sonar system can be expressed in terms of what are known as Receiver Operating Characteristics (ROC's) that are plots of detection and false alarm probability plotted in terms of the SIR at the output of the signal processor. The input and output SIR are related by the *processing gain* which is the average (dB) gain in SIR achieved by the processor and is a function of the waveform, processing algorithm, beam structure, and the second order statistical characteristics of the interference. The performance characteristics of alternative system implementations are readily assessed by employing the sonar equation, models for the interference, the processing gain, and knowledge of the underlying statistics.

2.3 GEOMETRY AND KINEMATICS

The utilization of echo-location to estimate geometric and kinematic (G-K) parameters such as range, relative velocity, and acceleration involves an analysis of the time base of the echo in relation to that of the transmitted signal. The sounding of a fog horn and the timing of the echo is merely a measure of the time base shift or delay where range is estimated using the known speed of sound. Carrying this one step further, a mariner with a good sense of pitch might have been able to estimate the rate at which he was approaching an iceberg by listening for a change in the echo frequency, i.e., the Doppler effect which is actually a compression of the echo time base.

In order to describe time base effects and their relationship to G-K parameters, consider a single point reflector, a transmitter, and a receiver operating in a homogeneous medium characterized by a sound speed c. Each of these may have an arbitrary position and trajectory, and it is assumed that they are isotropic; the transmit, scattering, and reception processes are not affected by their relative geometry. The transmitter and receiver may at any instant of time be either separate (bistatic) or collocated (monostatic) and all three may be stationary or moving with arbitrary trajectories.

First consider the stationary geometry of Figure 2.4a where the positions of the transmitter, the scatterer, and the receiver are defined by the vectors $\mathbf{r}_t, \mathbf{r}_s, \mathbf{r}_r$, respectively. The scalar distances between the three positions are

$$d_{ts} = |\mathbf{r}_t - \mathbf{r}_s|$$
$$d_{sr} = |\mathbf{r}_s - \mathbf{r}_r|$$

and the corresponding differential time delays are

$$\tau_{ts} = d_{ts}/c$$
$$\tau_{sr} = d_{sr}/c.$$

If during the interval $t \in [0, T]$, the normalized waveform $f(t)$ is transmitted, it will be reflected and later received as indicated in Figure 2.4b. Assume that at time t_3, an output $f_r(t_3)$ is observed at the receiver. Neglecting energy considerations, this is identical to the function $f_s(t_2)$ that was observed at the scatterer at time t_2, and was transmitted as $f(t_1)$. The instants of reception, scattering, and transmission are related by the delays τ_{ts}, τ_{sr} and consequently, $f_r(t_3) = f_s(t_3 - \tau_{sr}) = f(t_3 - \tau_{sr} - \tau_{ts})$. The waveform $f(t)$ is defined only for $t \in [0, T]$. In order to define the echo waveform observed at the receiver at some arbitrary time t_3, it is necessary to find the corresponding time t_1 defined during the transmit interval. In other words, a relationship or mapping of the time of reception to that of transmission is required in order to define the received signal.

The time mapping relationships for the static case are simple delays, but are more complex when the transmitter, scatterer, and/or the receiver have arbitrary relative motion. The positions are then dynamic

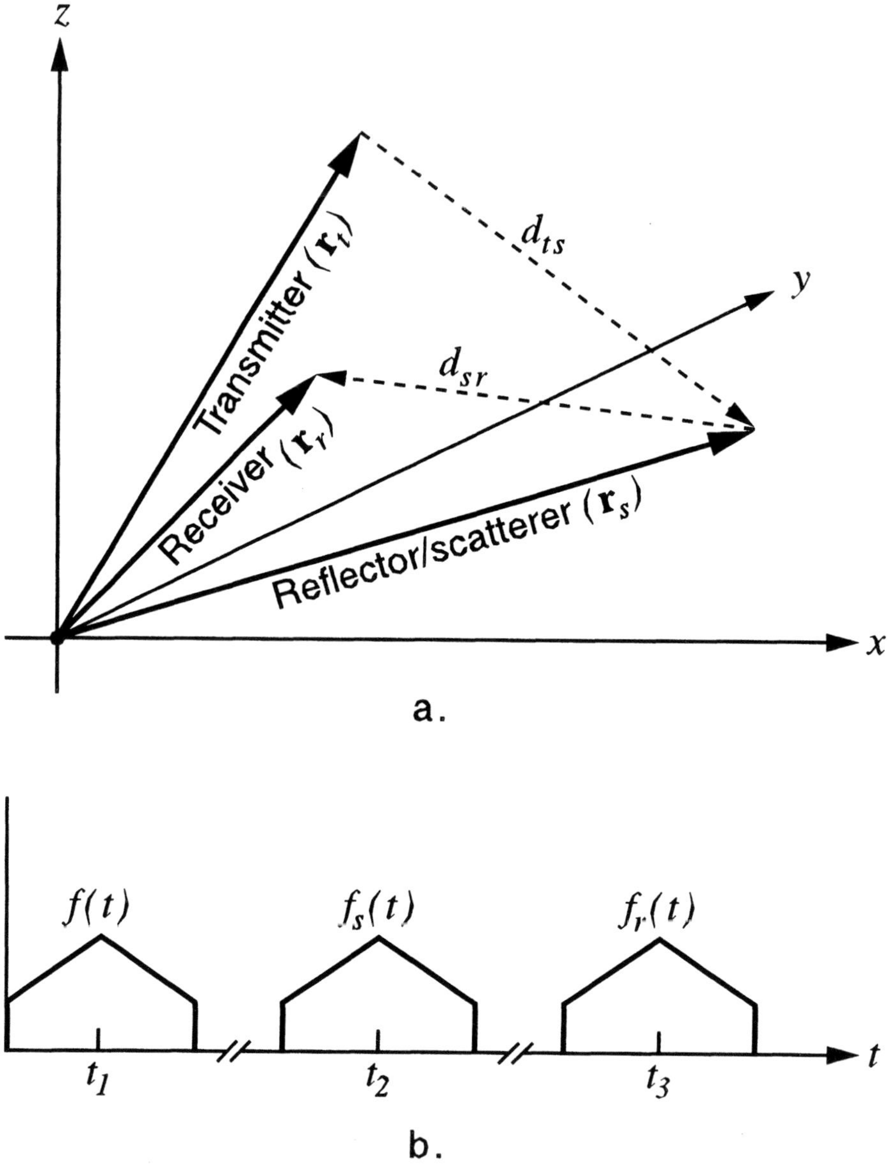

Figure 2.4: Stationary multistatic geometry: (a.) geometry defined by constant transmitter, reflector, and receiver position vectors, (b.) corresponding waveform time functions

and the vectors are defined as the functions of time $\mathbf{r}_t(t), \mathbf{r}_s(t)$, $\mathbf{r}_r(t)$ each following an arbitrary trajectory in 3-space as indicated in Figure 2.5. The scalar distances between the positions are

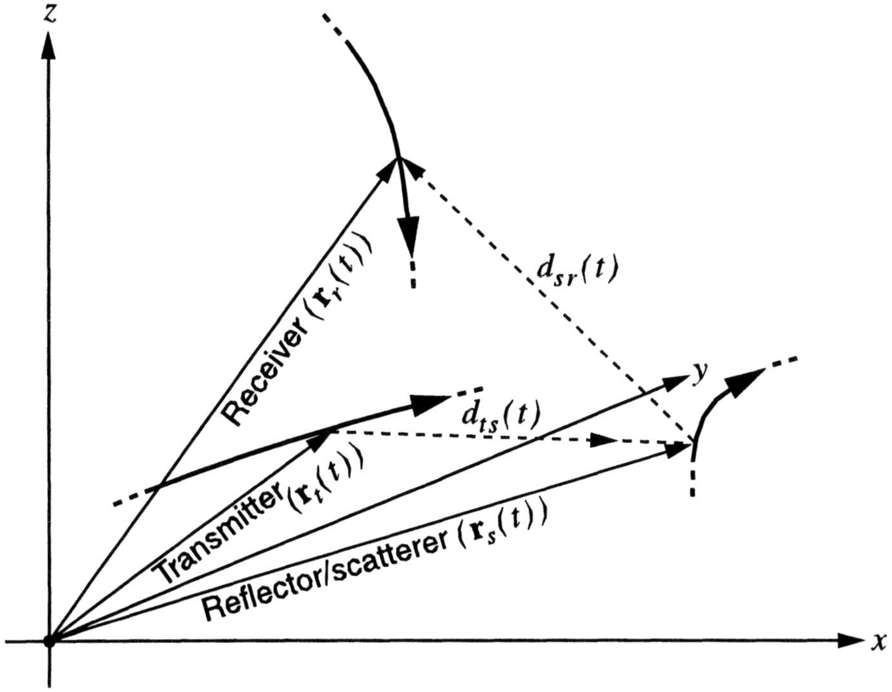

Figure 2.5: Kinematic multistatic geometry defined by time varying transmitter, reflector, and receiver position vectors

$$d_{ts}(t) = |\mathbf{r}_t(t) - \mathbf{r}_s(t)|$$
$$d_{sr}(t) = |\mathbf{r}_s(t) - \mathbf{r}_r(t)|$$

at any instant. The distances over which the sound energy must travel and hence the delays cannot be simply related to the scalar distances at any one specific time since the speed of sound is finite. Information encoded in the transmit function $f(t_1)$ is observed at the scattering site as $f_s(t_2)$ but t_1 and t_2 cannot be related by the delay corresponding to d_{ts} defined for any particular time instant since the transmitter and scatterer are moving in relation to each other. The range over which the sound energy must travel is the distance between the vector position of the transmitter at time t_1 and the vector position of the scatterer at time

t_2. The range between the scatterer at t_2 and the receiver at t_3 likewise defines the range delay for that leg of the cycle.

The relationship between the transmit, reflection, and reception times and the kinematic parameters is called a *time mapping* and is illustrated geometrically by a space-time *trajectory diagram.* Each position vector is written in terms of its Cartesian components. Using the transmit position as an example

$$\mathbf{r}_t(t) = [r_{tx}(t), r_{ty}(t), r_{tz}(t)]$$

and likewise for the scatterer and receiver vectors $\mathbf{r}_s(t)$ and $\mathbf{r}_r(t)$. If each component is divided by the scalar sound speed c, the resulting delay vector $\boldsymbol{\tau}_t(t) = \mathbf{r}_t(t)/c$ has units of time i.e. seconds. Furthermore, if it is augmented by the scalar time variable, the positions in space-time are defined by 4-d vectors

$$\mathbf{t}_t(t) = [\boldsymbol{\tau}_t(t), t] = \left[\frac{r_{tx}(t)}{c}, \frac{r_{ty}(t)}{c}, \frac{r_{tz}(t)}{c}, t\right].$$

The scatterer and receiver positions are similarly defined by $\mathbf{t}_s(t)$ and $\mathbf{t}_r(t)$. The paths of these vectors are called the space-time trajectories and are plotted as the trajectory diagram illustrated in Figure 2.6 for planar motion. Trajectory diagrams are characterized by a time axis and one or more (but no more than three) spatial axis corresponding to the Cartesian distance components divided by the propagation speed c [11]. Consequently, all of the axes have consistent time units.

Acoustic energy propagates according to what is called the *Galilean* model that is, the speed of propagation is defined in relation to the medium. Electromagnetic energy propagates according to the *relativistic* model [12] and diagrams such as (2.6) are used for this situation as well but care must be taken to properly account for relative velocity and the frames of reference used to define time and distance. Propagation in the two regimes is compared later in this chapter.

At time t_1, the signal $f(t_1)$ is generated at the space-time position $\mathbf{t}_t(t_1)$. The corresponding wavefront in a homogeneous medium propagates spherically (a circle in 2-space) about the transmit spatial position defined by the space vector $\mathbf{r}_t(t_1)$. The wavefront has a particular space-time trajectory that is a straight line on the diagram with an orientation that depends upon the evolving geometry. The locus of all possible wavefront trajectories is a right circular hypercone (sometimes called a *ray cone*) with its axis aligned with the time axis and its apex at $\mathbf{t}_t(t_1)$. A

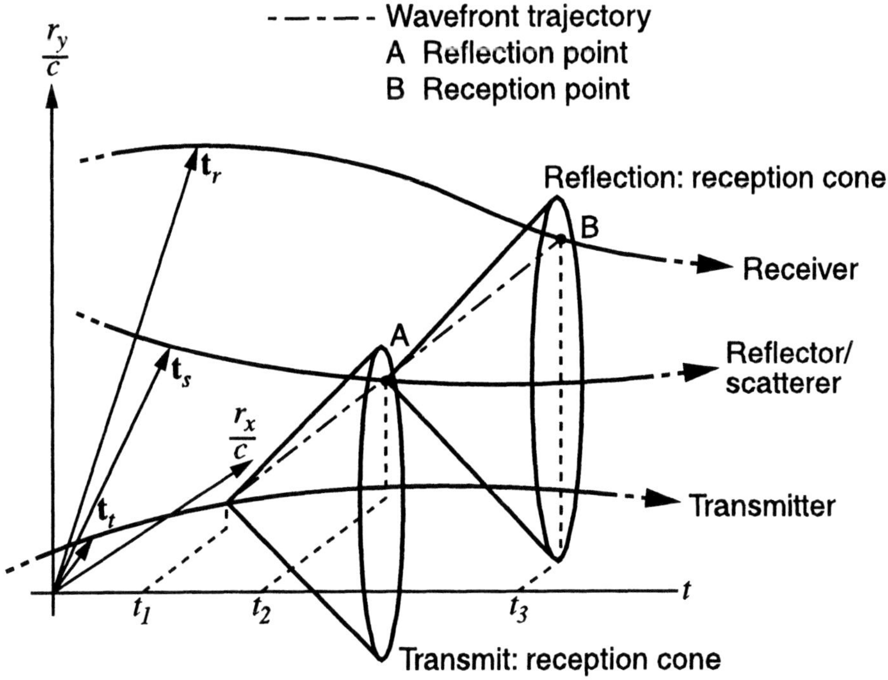

Figure 2.6: Trajectory (ray cone) diagram for planar motion

point on the surface of this cone represents a possible space-time position of the wavefront corresponding to the information encoded by $f(t_1)$ at the apex of the cone at time t_1 and position $\mathbf{r}_t(t_1)$. The wavefront trajectory corresponding to this point is a line or vector connecting the point with the apex of the cone.

At any time $t > t_1$, a trajectory of the wavefront denoted by $\mathbf{t}_w(t) = [\tau_w(t), t]$ must lie on the transmission cone defined by the relationship

$$|\tau_w(t) - \tau_t(t_1)|^2 = (t - t_1)^2.$$

At time t_2, the wavefront locus intersects the scatterer trajectory so that $\mathbf{t}_w(t_2) = \mathbf{t}_s(t_2)$ and

$$|\tau_s(t_2) - \tau_t(t_1)|^2 = (t_2 - t_1)^2. \tag{2.8}$$

The point of intersection defines the particular straight line trajectory traversed by the wavefront from the point of transmission $\mathbf{t}_t(t_1)$ to that of reflection at $\mathbf{t}_s(t_2)$. The transmit-reflection delay for the information encoded as $f(t_1)$ is the trajectory projected on the time axis or $t_2 - t_1$.

The wavefront propagates from the space-time position $\mathbf{t}_s(t_2)$ upon reflection at time t_2 and its trajectory locus satisfies

$$|\tau_w(t) - \tau_s(t_2)|^2 = (t - t_2)^2$$

for $t > t_2$ which defines the reception cone. At time t_3, the reception cone intersects the receiver trajectory such that $\mathbf{t}_w(t_3) = \mathbf{t}_r(t_3)$. This defines the position of reception hence the wavefront reception trajectory is defined by the conical relationship

$$|\tau_r(t_3) - \tau_s(t_2)|^2 = (t_3 - t_2)^2. \tag{2.9}$$

The reflection-reception delay is determined by the projection of the reception trajectory on the time axis or $t_3 - t_2$ as was the case for the transmit-reflection leg. Equations (2.8,2.9) are called the time mapping equations. They make it possible to precisely determine the waveform f_r observed at the receiver at time t_3 as that originally transmitted as $f(t_1)$ given knowledge of the time history of the position vectors. In other words,

$$f_r(t_3) = f(t_1)$$

where the mapping is denoted by

$$(t_1) \overset{\mathcal{M}_t}{\longmapsto} (t_3)$$

and $\mathcal{M}_t$ denotes the implicit mapping of (2.8,2.9). If the transmitted function is defined on the interval $[0, T]$ and a mapping of t_3 produces a negative value for t_1, it indicates that the leading edge of the signal pulse has yet to arrive at the receiver. Likewise, a mapping producing a value greater than T indicates that the pulse has passed. The equations as shown represent the more general bistatic geometry. The monostatic case may easily be treated by defining a common platform delay vector $\tau_p(t) = \tau_t(t) = \tau_r(t)$ for the transmitter and receiver and substituting it into the mapping equations so that the monostatic mappings are

$$\begin{aligned} |\tau_s(t_2) - \tau_p(t_1)|^2 &= (t_2 - t_1)^2 \\ |\tau_p(t_3) - \tau_s(t_2)|^2 &= (t_3 - t_2)^2. \end{aligned} \tag{2.10}$$

The mapping $\mathcal{M}_t$ is generally nonlinear and consequently the delay $t_3 - t_1$ may vary over the duration of a pulse. This introduces a distortion of the waveform time base and possible loss of signal coherence unless the effect can be compensated via prior knowledge or estimates of the kinematics involved.

The mono and bistatic time mapping equations may be used for signal synthesis when simulating kinematic environments but direct synthesis is computationally intense for motion in 3-space described by an arbitrary set of position vectors. A closed form solution is usually unavailable and the mapping to determine the waveform for a specific observation time t_3 requires that the reflection time t_2 be first computed by the iterative solution of (2.9) or (2.10). t_1 is then computed by the iterative solution of (2.8) or (2.10) using the value of t_2 obtained by the first iteration. This sequence is repeated for each time increment or sample of the received signal making it a computationally expensive proposition [13].

An iterative solution can be avoided if the transmitter, scatterer, and receiver all have constant vector velocity over the duration of an interrogation cycle. In this case, the 3-space positions are linear in time and the process is reduced to the solution of two quadratic equations for each mapping of $t_3 \rightarrow t_1$. This significantly reduces the time required to synthesize an echo; an important consideration for real time simulation [11].

2.4 COLLINEAR MOTION

In many cases, the direction in which echo energy is propagated is close to or coincides with the the direction of platform motion. The geometry may be modeled as collinear wherein a monostatic [2] platform-scatterer pair have velocity vectors that are directed along a common line of sight or if the scatterer is slowly moving or moving in nearly the same direction as the platform. The trajectory diagram is consequently two dimensional [14–16] and the wavefront trajectories or ray cones reduce to triangles as shown in Figure 2.7. When the velocities are constant, the mapping equations can be solved in closed form merely by exploiting the geometry of the diagram. The origin of the diagram is defined as the platform position at the time of transmit initiation (pulse leading edge). The initial position of the scatterer is at a range r_0 along an arbitrary line of sight (los). The platform and scatterer velocities are assumed to be constant scalars v_p, v_s directed along the los with positive velocity indicating closing range. Their respective delays relative to the origin are

$$\tau_p(t) = v_p t / c$$

[2] While a bistatic collinear geometry is possible it is somewhat unrealistic and will not be dealt with here.

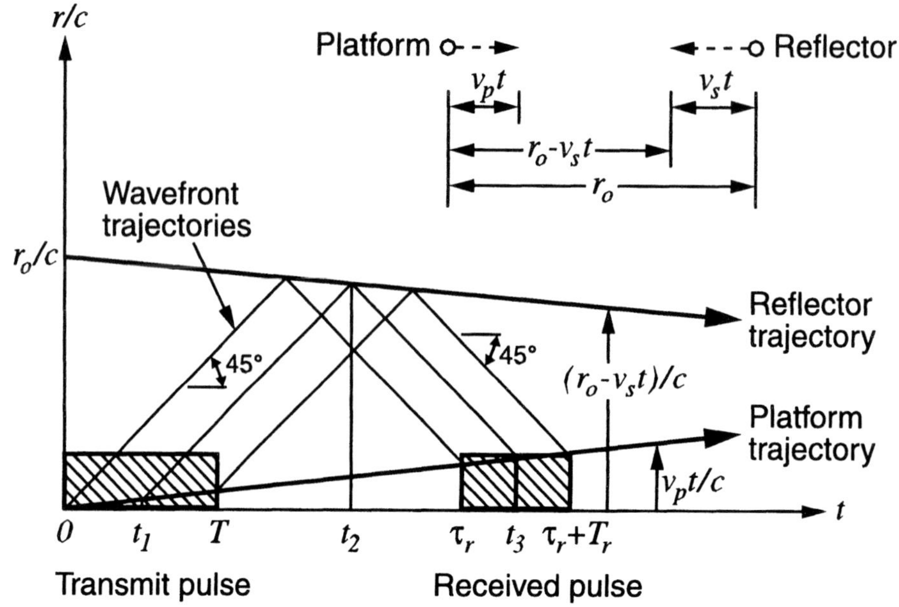

Figure 2.7: Collinear, constant velocity trajectory diagram

$$\tau_s(t) = (r_o - v_s t)/c. \tag{2.11}$$

The speed of propagation in the medium is c and the distance that a wavefront will travel in time t is merely tc. The abscissa of the diagram represents distance divided by c, thus the wavefront trajectories are triangles with legs oriented at 45 deg. with respect to the axis. These are the 2-dimension equivalent of the ray cones illustrated in Figure 2.6. Because of this simple geometry, the times of transmission, scattering, and reception are related to the delays by the isosceles relationships

$$\begin{aligned} t_3 - t_2 &= \tau_s(t_2) - \tau_p(t_3) \\ t_2 - t_1 &= \tau_s(t_2) - \tau_p(t_1) \end{aligned} \tag{2.12}$$

and substituting (2.11) and solving for t_1 in terms of t_3 yields the closed form linear mapping equation

$$\boxed{\begin{aligned} t_1 &= \left[\frac{c + v_s}{c - v_s}\right]\left[\frac{c + v_p}{c - v_p}\right] t_3 - \frac{2r_0 c}{(c - v_s)(c - v_p)} \\ &= \beta t_3 - \tau_0. \end{aligned}} \tag{2.13}$$

The parameter β is called the dilation parameter and is approximated by carrying out the algebraic divisions within the brackets and retaining the first order terms under the assumptions that $v_s \ll c$ and $v_p \ll c$. This results in

$$\begin{aligned} \beta &\approx \left[1 + \frac{2v_s}{c}\right]\left[1 + \frac{2v_p}{c}\right] \\ &\approx \left[1 + \frac{2v_\ell}{c}\right] = \tilde{\beta} \end{aligned} \tag{2.14}$$

where $v_\ell = v_s + v_p$ is the relative velocity along the line of sight and is defined as positive for <u>decreasing</u> range. Likewise, the delay parameter $\tau_0 \approx 2r_0/c$. A pulse defined by the transmit waveform function $f(t_1)$ for $t_1 \in [0, T]$ as denoted by the crosshatched region in the figure, will have a received duration of $T_r = T/\beta$. The pulse is consequently shortened if the los range is decreasing and vice versa.

A constant speed collinear geometry always results in a linear time mapping such that the transmitted and received waveforms are exactly related by the equivalent affine relationships

$$\begin{aligned} f_r(t_3) &= f(\beta(t_3 - \tau)) \\ &= f(\beta t_3 - \tau_0) \end{aligned} \tag{2.15}$$

and the received waveform is merely a delayed and time dilated version of the transmit function. A piecewise linear mapping results if the platform and scatterer velocities remain constant during the whole interrogation cycle and but change between interrogations.

When the motion is collinear and the velocities are not constant but arbitrary time functions, or if the velocities are constant but the motion is not collinear, the mapping equations (2.12) become nonlinear and do not yield to closed form solution. An iterative approach may be required for exact waveform synthesis as for the case of non-collinear motion with time varying velocities [11, 13]. An approximate solution has been obtained by Kelly and Wishner [17] who defined the mapping in terms of a time varying delay that models the received waveform as

$$f_r(t) = f(t - \tau(t)) \tag{2.16}$$

where for convenience the time subscript is dropped with the understanding that $t = t_3$ the time of reception. The delay is approximated by twice the relative los range delay at the instant of reflection $r_\ell(t_2)/c$ as shown in Figure 2.8. This is

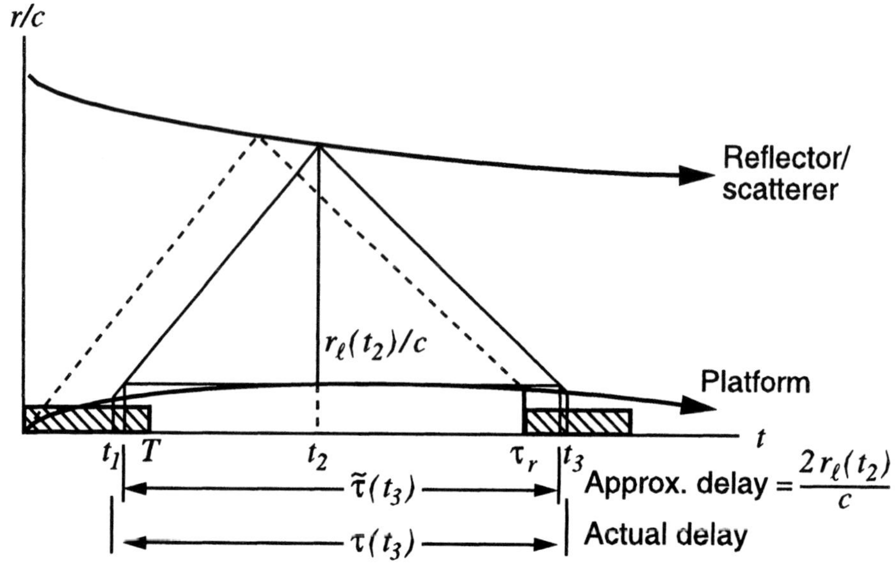

Figure 2.8: Collinear, time varying velocity trajectory diagram

$$
\begin{aligned}
\tau(t) &\approx \tilde{\tau}(t) \\
&= (2/c) r_\ell(u(t)), \\
u(t) &= (t - \tilde{\tau}(t)/2) \\
&= t_2
\end{aligned}
\tag{2.17}
$$

and $\tilde{\tau}$ is approximated by the second order Taylor expansion

$$
\tilde{\tau}(t) \approx \tau_r + (1 - \tilde{\beta})(t - \tau_r) + \tilde{\alpha}(t - \tau_r)^2 \tag{2.18}
$$

about τ_r, the time that the pulse leading edge is received. In order to determine the parameters $\tilde{\beta}, \tilde{\alpha}$, (2.18) is differentiated with respect to t and evaluated at $t = \tau_r$ yielding $(1 - \tilde{\beta}) = \dot{\tilde{\tau}}(\tau_r)$ and $\tilde{\alpha} = \ddot{\tilde{\tau}}(\tau_r)/2$. The time derivatives of $\tilde{\tau}$ are evaluated by differentiating (2.17) and evaluating at τ_r where

$$
\begin{aligned}
\dot{\tilde{\tau}}(\tau_r) &= (2/c) \left(\frac{dr_\ell}{du} \right) \left(\frac{du}{dt} \right)_{t=\tau_r} \\
&= -(2/c) v_\ell (1 - \dot{\tilde{\tau}}(\tau_r)/2).
\end{aligned}
$$

Solving for the delay derivative results in

$$
\begin{aligned}
\dot{\tilde{\tau}}(\tau_r) &= -\frac{2v_\ell}{c - v_\ell} \\
&= (1 - \tilde{\beta})
\end{aligned}
\tag{2.19}
$$

where $v_\ell = -\left(\frac{dr_\ell}{du}\right)_{u=(\tau_r/2)}$ is the relative closing los velocity between the platform and the scatterer evaluated at the instant of leading edge reflection. Differentiating $\dot{\tilde{\tau}}$ yields

$$\begin{aligned}\ddot{\tilde{\tau}} &= (2/c)\left(\frac{dr_\ell}{du}\frac{d^2u}{dt^2}\right) + (2/c)\frac{du}{dt}\left\{\frac{d}{dt}\left(\frac{dr_\ell}{du}\right)\right\} \\ &= (2/c)\left[\frac{dr_\ell}{du}\frac{d^2u}{dt^2} + \frac{du}{dt}\left\{\frac{d^2r_\ell}{du^2}\frac{du}{dt}\right\}\right] \\ &= (2/c)\left[\left(\frac{dr_\ell}{du}\right)\left(\frac{d^2u}{dt^2}\right) + \left(\frac{du}{dt}\right)^2\left(\frac{d^2r_\ell}{du^2}\right)\right].\end{aligned} \tag{2.20}$$

The time derivatives of u are obtained from (2.17) as

$$\begin{aligned}\frac{du}{dt} &= (1 - \dot{\tilde{\tau}}/2) \\ \frac{d^2u}{dt^2} &= -\ddot{\tilde{\tau}}/2\end{aligned}$$

and substituting these into (2.20), using (2.19), and solving the resulting relationship for $\ddot{\tilde{\tau}}$ results in

$$\ddot{\tilde{\tau}}(\tau_r) = -\frac{2a_\ell c^2}{(c - v_\ell)^3}$$

where $a_\ell = -\left(\frac{d^2r_\ell}{du^2}\right)_{u=\tau_r/2}$ is the relative closing acceleration between the platform and scatterer. The parameters of the expansion (2.18) are now evaluated as

$$\begin{aligned}(1 - \tilde{\beta}) &= \frac{-2v_\ell}{c - v_\ell} \approx \frac{-2v_\ell}{c} \\ \tilde{\alpha} &= \frac{-c^2 a_\ell}{(c - v_\ell)^3} \approx \frac{-a_\ell}{c}\end{aligned}$$

and the time mapping is thus equivalent to the evaluation of the argument $(t - \tau(t))$ in (2.16) using the parameters in the approximations (2.17,2.18) with the result that [16, 17]

$$\begin{aligned}f_r(t) &\approx f((1 + \frac{2v_\ell}{c})(t - \tau_r) + \frac{a_\ell}{c}(t - \tau_r)^2) \\ &= f(\tilde{\beta}(t - \tau_r) - \tilde{\alpha}(t - \tau_r)^2).\end{aligned} \tag{2.21}$$

While the linear dilation term is the same as that derived as an approximation to the exact linear mapping problem in (2.13), the exact evaluation of the delay τ_r is a more difficult problem that would require a more precise description of the relative los range as a function of time. If the acceleration component is small however, the delay relationship in (2.13) may be used as an approximation where

$$\tau_r \approx \frac{2r_0}{(c + v_\ell)}. \tag{2.22}$$

The approximate mapping in terms of los velocity and acceleration is particularly useful when it is desirable to evaluate the effect of constant velocity non-collinear motion when the platform and scatterer velocities are resolved into down and cross range components v_ℓ and v_x as shown in Figure 2.9. If $\mathbf{v}_p$ and $\mathbf{v}_s$ are constant velocity vectors, the los component

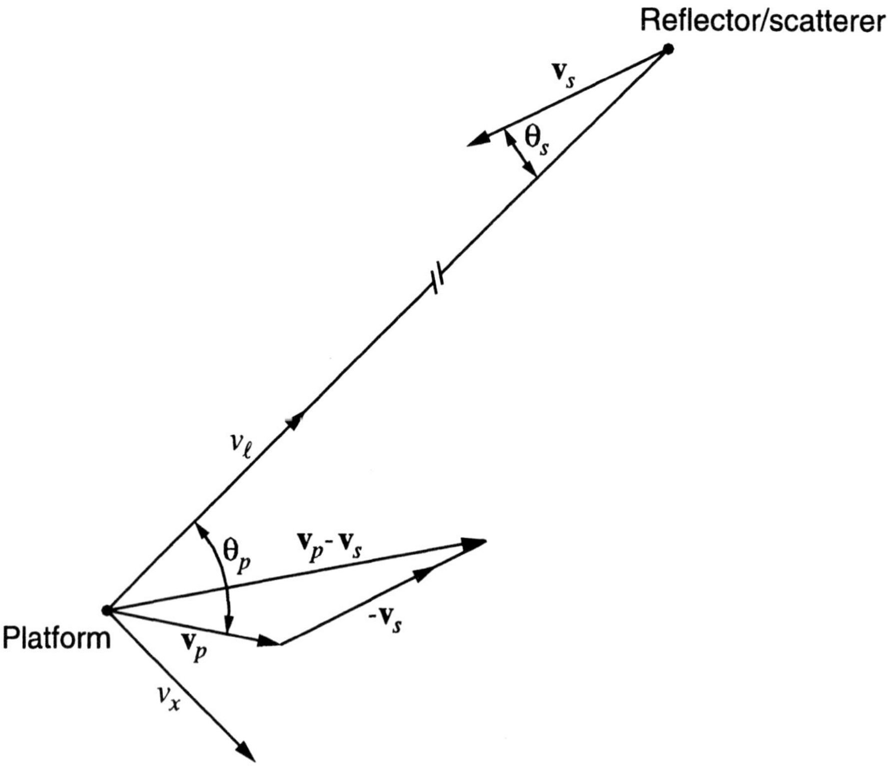

Figure 2.9: Down and cross range velocity components

is defined as

$$\begin{aligned} v_\ell &= (\mathbf{v}_p - \mathbf{v}_s) \cdot \mathbf{u}_r \\ &= v_p \cos\theta_p + v_s \cos\theta_s \end{aligned} \tag{2.23}$$

where θ_p and θ_s are the angles between the platform and scatterer velocities and the unit vector $\mathbf{u}_r$ defining the los direction. The acceleration is obtained by differentiating v_ℓ with respect to time where

$$\begin{aligned} a_\ell &= \dot{v}_\ell \\ &= -(v_p \sin\theta_p + v_s \sin\theta_s)\omega \\ &= -v_x \omega \end{aligned}$$

and ω is angular rotation rate of the los. This is simply v_x/r_ℓ and consequently,

$$a_\ell = -\left(\frac{{v_x}^2}{r}\right). \tag{2.24}$$

The received waveform may now be expressed in terms of both linear and quadratic distortion components related to down and cross range los velocities using (2.21) and (2.23, 2.24). The approximation (2.22) may be used if the los range r_ℓ is the initial range r_0.

The quadratic term clearly becomes small at long range and the dominant effects are delay and dilation as indicated by (2.15). Dilation gives rise to what is commonly known as Doppler frequency shift in sinusoidal waveforms which is generally not considered a fundamental distortion of the waveform. On the other hand, the quadratic (and higher order) terms introduce nonlinearities into the time base that have the effect of modulating or changing the fundamental form of the waveform function. These effects if not compensated or accounted for, can seriously reduce receiver performance [18, 19] and are discussed in Chapter 4.

2.5 SIGNAL MODELS

Having discussed the mechanisms by which kinematic parameters affect the time base of a waveform, it is now possible to develop models for the echo based upon a first order time mapping of the form (2.15). This is justified because many problems involving the use of echo-location for detection and estimation purposes involve relatively benign kinematics, and the interrogation times are short in relation to the time scales associated with changes in the relative velocity. However while restricting the discussion to predominantly linear models, it is necessary to keep in mind that the general time mapping for arbitrary motion is non-linear and higher order kinematics such as acceleration (2.21) can introduce modeling errors and ultimately reduce receiver performance. Fortunately kinematic situations requiring the inclusion of higher order terms to compensate the echo model can usually be anticipated.

The transmit waveform is the complex bandpass function [20, 21]

$$
\begin{aligned}
f(t) &= [a(t) \exp j\psi(t)] \exp j\omega_0 t = \tilde{f}(t) \exp j\omega_0 t \qquad t \in [0, T] \\
\int_{-T/2}^{T/2} |\tilde{f}(t)|^2 &= \int_{-T/2}^{T/2} |a(t)|^2 = 1 \qquad (2.25)
\end{aligned}
$$

where $\tilde{f}(t)$ is the complex envelope, $a(t)$ is the real energy normalized amplitude modulation, $\psi(t)$ is the phase modulation, and ω_0 is the radian carrier frequency. The point scatterer echo return is modeled as

$$s(t) = \sqrt{\beta E_s} f_r(t)$$

where the delayed and dilated received waveform $f_r(t)$ is given by (2.15). The energy is by convention multiplied by the dilation parameter to maintain the waveform normalization over the dilated time interval T/β [22, 23].

Using (2.15,2.25), this takes the form

$$
\begin{aligned}
s(t) &= \sqrt{\beta E_s} f(\beta(t-\tau)) \\
&= \sqrt{\beta E_s} \tilde{f}(\beta(t-\tau)) \exp j\omega_0(\beta(t-\tau)) \\
&= \sqrt{\beta E_s} \left[\tilde{f}(\beta(t-\tau)) \exp j2\pi\phi t\right] \exp j(\omega_0 t - \theta_\tau) \qquad (2.26)
\end{aligned}
$$

where $\omega_0 = 2\pi f_0$, $\theta_\tau = \beta\omega_0\tau$ is the phase delay and

$$\phi = (\beta - 1) f_0 \ \text{(Hz.)}$$

is called the Doppler frequency shift which is approximately related to the relative los velocity by (2.14) with

$$\phi \approx \frac{2v_\ell}{c} f_0.$$

The echo expressed in the form (2.26) is interpreted as the time dilated, delayed, and Doppler shifted envelope function (in square brackets) multiplied by a complex carrier term. The notation

$$
\begin{aligned}
\mathbf{p} &\equiv [\beta, \tau] \quad \text{dilation (wideband) model} \\
&\equiv [\phi, \tau] \quad \text{Doppler frequency (narrowband) model} \qquad (2.27)
\end{aligned}
$$

is used for the delay-Doppler parameters so that the shifted waveform function is expressed as

$$\begin{aligned} f(\mathbf{p},t) &= \sqrt{\beta} f(\beta(t-\tau)),\ t \in [0, T/\beta] \ \ \text{(dilation model)} \\ &= f(t-\tau)\exp j(2\pi\phi t + \theta\tau),\ t \in [0,T] \ \ \text{(frequency model)}. \ (2.28) \end{aligned}$$

The parameters τ and β or ϕ represent delay and Doppler but are more fundamentally a range-velocity or *phase plane* description of the echo. A plot of the ranges and los velocities associated with a sequential series of echos is called a phase plane trajectory and is directly related to the sequentially measured delays and Dopplers that might be plotted in a delay-Doppler map by $r_\ell \approx \tau c/2$ and $v_\ell \approx (\beta - 1)c/2 = \phi c/2f_0$. The terms delay-Doppler map and "phase plane" are used interchangeably for this reason,.

A complex envelope and Doppler shift defined in terms of a specific carrier or frequency centroid are inherently narrowband concepts where in this context, "narrowband" loosely defines those signals that are fundamentally sinusoidal in nature with a modulation bandwidth that is small with respect to the carrier and an envelope that is relatively unaffected by the dilation effect. As the percentage bandwidth of the modulation increases, the signal begins to lose its sinusoidal characteristics and these concepts begin to lose their significance. It then becomes necessary to view the signal as a whole as being affected by the dilation. The model (2.26) is generally regarded as a wideband model even though a carrier and Doppler shift are defined. This is because the dilation affects the whole signal as expressed by the dilation parameter in the envelope as well as the Doppler shift.

An alternative view of the definitions wide vs narrowband is obtained by examining the properties of the dilated echo spectrum. $\tau = 0$ without loss of generality and the bandpass echo is

$$s(t) = \sqrt{\beta E_s}\tilde{f}(\beta t)\exp j(\omega_0 + 2\pi\phi)t.$$

Its spectrum is the Fourier transform

$$\begin{aligned} S(\omega) &= \int_{-\infty}^{\infty} s(t)\exp(-j\omega t)dt \\ &= \sqrt{\frac{E_s}{\beta}}\tilde{F}\left(\frac{\omega - (\omega_0 + 2\pi\phi)}{\beta}\right) \end{aligned}$$

and the envelope spectrum $\tilde{F}(\omega_b)$ is assumed to be band limited in the baseband $\omega_b \in [-\pi W, \pi W]$. The bandpass echo would be confined to the band $[\omega_0 \pm \pi W]$ in the absence of dilation since $F(\omega) = \tilde{F}(\omega_b + \omega_0)$ by definition. When the echo is dilated, the bandpass spectrum is confined to the band defined by the radian frequencies $[\omega_1, \omega_2]$ that are determined by solving for the baseband limits

$$\pi W = \frac{\omega_2 - (\omega_0 + 2\pi\phi)}{\beta}$$

and

$$-\pi W = \frac{\omega_1 - (\omega_0 + 2\pi\phi)}{\beta}$$

yielding

$$\begin{aligned} \omega_2 &= \pi\beta W + \omega_0 + 2\pi\phi \\ \omega_1 &= -\pi\beta W + \omega_0 + 2\pi\phi. \end{aligned}$$

The echo spectrum is not only shifted by the factor ϕ, but is also expanded by the dilation β or in other words, the echo bandwidth is changed by the factor $\Delta W = (\beta - 1)W$. This should not be surprising because of the dual nature of the time and frequency domains where a foreshortened signal duration results in a wider spectrum and vice versa.

Consider the simpler echo model

$$s(t) = \sqrt{E_s}\left[\tilde{f}(t-\tau)\exp\, j2\pi\phi t\right]\exp\, j(\omega_0 t - \theta_\tau) \tag{2.29}$$

which is called the narrowband echo model and is characterized by the assumption that the envelope is unaffected by the dilation, and the effect of relative velocity is manifested only as a Doppler frequency shift of the carrier term. The spectrum corresponding to this model consequently exhibits no expansion of the frequency scale but merely a shift of ϕ (Hz.) up or down depending on the relative velocity. This is clearly an approximation since the dilation effect cannot affect just one part of a waveform however the assumption of the narrowband model can simplify the analysis and implementation of detection and estimation algorithms. The issue of receiver mismatch error and subsequent performance loss introduced by assuming a narrowband model in the presence of dilated echos is reserved for the discussion of ambiguity functions in Chapter 4. For now, narrowband waveforms are defined as those with bandwidths sufficiently small such that the bandwidth expansion ΔW is negligible with respect to the resolution with which the spectrum may be measured.

This is proportional to the inverse of the integration time of the Fourier transform $1/T$ [24] or in terms of the relative los velocity

$$\frac{2v_\ell}{c}W = \Delta W \ll \frac{1}{T}$$

which is usually written as [17, 23]

$$(\beta - 1) \approx \frac{2v_\ell}{c} \ll \frac{1}{TW}.$$

This definition of narrowband depends upon the range of velocities expected as well as the fractional bandwidth of the envelope and indeed, if the velocity range becomes vanishingly small, signals with extremely large TW products could by this definition, be considered as narrowband. On the other hand, it is shown in Chapter 4 that fractional bandwidth by itself is an important indicator of "wideband" vs "narrowband" because of its effect upon ambiguity function volume.

The spectral resolution may also be used to define the magnitude of the acceleration induced second order time mapping requiring compensation as mentioned at the beginning of this section. When acceleration is included in the time mapping, the waveform has the form (2.21) and consequently the narrowband echo model (2.29) with $\tau = 0$ is

$$s(t) = \sqrt{E_s}\left[\tilde{f}(t) \exp\, j2\pi\phi t\right]\left[\exp\left(j\frac{\omega_0 a_\ell}{c}t^2\right)\right] \exp\, j\omega_0 t.$$

The acceleration phase term is quadratic, and consequently the associated frequency is linear. The total acceleration induced carrier frequency shift Δf_a is found by differentiating the phase term, using (2.24), and evaluating the result for the total signal duration T. The result is

$$\begin{aligned}\Delta f_a = \frac{2a_\ell f_0 T}{c} &= \frac{2a_\ell T}{\lambda_0}\\ &= \frac{2v_x^2 T}{r\lambda_0}.\end{aligned}$$

If this is required to be negligible in comparison to the spectral resolution $1/T$, $\Delta f_a \ll 1/T$ or, [23]

$$v_x^2 \ll \frac{r\lambda_0}{2T^2}.$$

If this is not true, the model should include the quadratic term as in (2.21) whether or not it is assumed to be wideband.

2.6 RELATIVISTIC TIME MAPPING

While the emphasis is primarily acoustic echo-location, it is illuminating to relate phenomena that occur in an acoustic context to those that may occur in the electromagnetic realm, i.e. radar. In this spirit, the difference between acoustic and electromagnetic time mapping is presented even though the concept as used in sonar receiver implementation is little used in radar. The reason for this is simply that the speed of physical objects such as a transmitter, scatterer, and receiver in relation to the speed of propagation is vastly smaller in radar than it is in sonar. Consequently, the effects of time dilation are in most cases considered in terms of Doppler shift rather than as a fundamental expansion or distortion of the waveform time base.

In order to understand the difference between time mapping in the two realms, it is necessary to view the propagation of energy from different reference frames and to compare observations made in these frames in both realms. Two such frames are illustrated in Figure 2.10a where two bullet-like objects are shown moving apart with a collinear geometry. They are denoted as objects (1) and (2) that have constant velocities v_1, v_2 respectively. Each carries an observer with the ability to measure time and distance. Both objects are at point (A) which is the origin of the trajectory diagram shown in Figure 2.10b at time zero. At any time $t_1 > 0$, the observer on (1) who is stationary with respect to reference frame (1) observes that his distance r_1 from an arbitrary object at (B) is related to r_2 the distance from (B) to object (2) by

$$r_2 = r_1 + (v_1 + v_2)t_1. \tag{2.30}$$

Likewise observer (2) sees that

$$r_1 = r_2 - (v_1 + v_2)t_2 \tag{2.31}$$

where t_1 and t_2 represent time as measured by each observer. These two equations are called the Galilean transformations relating space-time events between different frames of reference [12,25]. At $t = 0$, observer (1) transmits a wavefront at the origin (A) that propagates to the right at speed $\dot{r_1} = c_1$ with respect to his frame of reference. By taking the time derivative of either (2.30) or (2.31) it is apparent that if observer (1) measures a propagation speed c_1, observer (2) will measure $c_2 = c_1 + v_t$ where $v_t = v_1 + v_2$. Consequently the observed speed of propagation depends upon the reference frame in the Galilean model. The trajectory

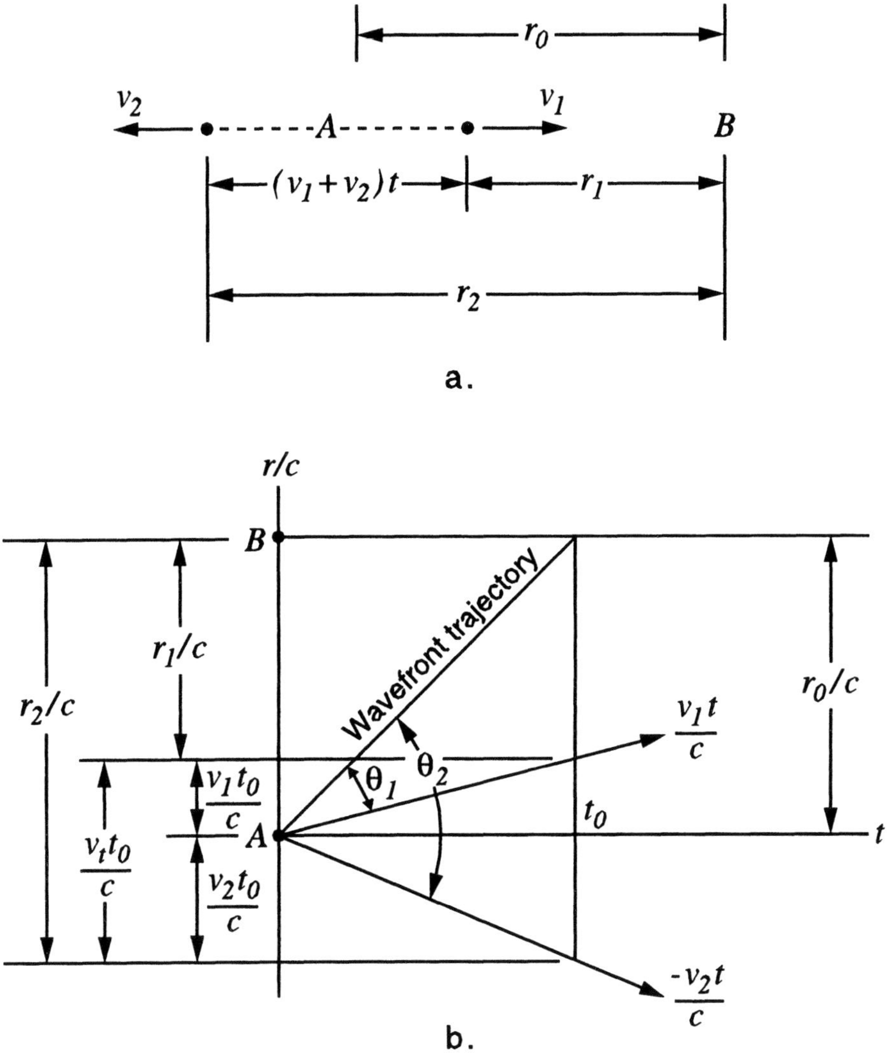

Figure 2.10: Relativistic time mapping reference frames: (a.) collinear geometry, (b.) trajectory diagram

diagram of Figure 2.10b provides another way to view this. A wavefront propagates from point (A) and arrives at (B) at time $t_0 = r_0/c$ where the wavefront trajectory is a 45 deg. line on the diagram. Observer (1) sees that the wave has traveled the distance

$$r_1 = r_0 - v_1 t_0$$

in the time t_0 and concludes that the propagation speed is

$$\begin{aligned} c_1 &= \frac{r_0 - v_1 t_0}{t_0} \\ &= c - v_1. \end{aligned}$$

Likewise, observer (2) measures the travel distance

$$r_2 = r_0 + v_2 t_0$$

and concludes that the propagation speed is $c_2 = c + v_2$ where clearly, $c_2 = c_1 + v_t$.

Trajectory diagrams are consistent with the Galilean model since the apparent propagation speed depends upon the reference frame. This is illustrated geometrically by the fact that the angles θ_1, θ_2 between the wavefront and reference frame trajectories are not the same. The inconsistent observations of propagation speed are a consequence of the Galilean assumption that a physical medium is involved in the propagation of energy through space which is indeed true in the acoustic realm. This is not true of electromagnetic propagation in a vacuum for which the measured round trip speed of propagation (light speed) is *constant in all inertial frames*. The coordinant transformations consistent with relativistic propagation are

$$\begin{aligned} r_1 &= k_l(r_2 - v_t t_2) \\ r_2 &= k_l(r_1 + v_t t_1) \\ v_t &= \frac{v_1 + v_2}{1 + \frac{v_1 v_2}{c^2}}. \end{aligned} \tag{2.32}$$

k_l is called the Lorentz constant and does not depend upon the reference frame [25]. It is evaluated by reconsidering the propagation experiment discussed above with the requirement that the speed of propagation in both frames be the same. Consequently, the distance from each observer to a photon or wavefront propagating at light speed along the los increases at the same rate in each reference frame so that $r_1 = ct_1$ and

$r_2 = ct_2$. Substituting these relationships into (2.32) yields

$$ct_1 = k_l(ct_2 - v_t t_2)$$
$$ct_2 = k_l(ct_1 + v_t t_1)$$

and solving for k_l results in

$$k_l = \frac{1}{\sqrt{1 - (\frac{v_t}{c})^2}}. \tag{2.33}$$

The time transformations are derived by substituting r_1 from the first equation of (2.32) into the second and solving for t_2 yielding

$$r_1 = k_l(k_l(r_1 + v_t t_1) - v_t t_2)$$

and

$$t_2 = \frac{v_t {k_l}^2 t_1 - r_1(1 - {k_l}^2)}{k_l v_t}.$$

Substituting (2.33) for k_l yields the time transformation

$$t_2 = k_l(t_1 + \frac{v_t}{c^2} r_1) \tag{2.34}$$

and carrying out the same procedure for t_1 by substituting the first equation of (2.32) into the second yields

$$t_1 = k_l(t_2 - \frac{v_t}{c^2} r_2). \tag{2.35}$$

In most physics texts [12,25,26], equation (2.34) and the second relationship in (2.32) are defined as the *Lorentz transformations* while (2.35) and the first relationship of (2.32) are defined as their inverses. The transformations may also be expressed in terms of delay variables $\tau_1 = r_1/c$ and $\tau_2 = r_2/c$ where

$$\tau_2 = k_l(\tau_1 + \frac{v_t}{c} t_1)$$
$$t_2 = k_l(t_1 + \frac{v_t}{c} \tau_1)$$

and

$$\tau_1 = k_l(\tau_2 - \frac{v_t}{c} t_2)$$
$$t_1 = k_l(t_2 - \frac{v_t}{c} \tau_2). \tag{2.36}$$

Written in this way, the relationships are useful for the derivation of the time mapping relationships in the electromagnetic realm. To do this requires the establishment of separate reference frames for a platform-scatterer pair for which the subscripts (1) and (3) denote quantities associated with the platform during the transmit and receive legs respectively and the subscript (2) associated with the scatterer and reflection process. At time $t_1 = t_2 = 0$ the platform and scatterer are r_0 meters apart measured in the frame of the platform and moving toward each other at a relative speed of v_t meters per second. At that time, a photon or quantum of information is transmitted from the platform and travels toward the scatterer at light speed taking Δt_1 seconds to make the traverse. Since the scatterer is approaching at speed v_t, the photon travels a distance $r_1 = c \, \Delta \, t_1 = r_0 - v_t \, \Delta \, t_1$, and consequently, the travel time and distance as measured in the platform frame are

$$\Delta t_1 = \frac{r_o}{c + v_t}$$

and

$$r_1 = c \, \Delta \, t_1 .$$

The corresponding transit time as seen in the scatterer frame denoted by Δt_2 is obtained from the Lorentz relationship (2.34) where

$$\begin{aligned} \Delta t_2 &= k_l(\Delta t_1 + \frac{{v_t}^2}{c} r_1) \\ &= k_l(\Delta t_1 + \frac{v_t}{c} \, \Delta \, t_1) \\ &= \frac{k_l r_0}{c} \end{aligned} \tag{2.37}$$

and the corresponding photon transit distance is

$$\begin{aligned} r_2 &= c \, \Delta \, t_2 \\ &= k_l r_0 . \end{aligned}$$

Now let a waveform $f(t_1)$ be transmitted at the platform and observed at some distant point in the platform frame as $f(t_1 - \tau_1)$. In the scatterer frame at an arbitrary point corresponding to a delay τ_2, the signal waveform is observed to be the function $f_s(t_2 - \tau_2)$. In order to relate the functions observed in the two frames it is assumed that the time bases are linearly related by a constant $\sqrt{\beta_e}$ such that

$$\begin{aligned} f_s\,(t_2 - \tau_2) &= f\left(\sqrt{\beta_e}(t_2 - \tau_2)\right) \\ &= f\,(t_1 - \tau_1) \end{aligned} \tag{2.38}$$

implying a mapping from the time of transmit to that of scattering by a time base dilation and shift given by

$$t_1 - \tau_1 = \sqrt{\beta_e}\,(t_2 - \tau_2)\,.$$

Using the transformations (2.36), this is

$$\begin{aligned} t_1 - \tau_1 &= k_l\left(t_2 - \frac{v_t}{c}\tau_2\right) - k_l\left(\tau_2 - \frac{v_t}{c}t_2\right) \\ &= k_l\left(1 + \frac{v_t}{c}\right)(t_2 - \tau_2) \\ &= \sqrt{\beta_e}\,(t_2 - \tau_2) \end{aligned}$$

and

$$\sqrt{\beta_e} = k_l\left(1 + \frac{v_t}{c}\right). \tag{2.39}$$

The delay τ_1 may be arbitrarily chosen and the initiation of transmission or the pulse leading edge is defined as $f(t_1)$ at the origin when $t_1 = 0$ by letting $\tau_1 = 0$. The leading edge arrives at the scatterer after the time interval $r_0/(c + v_t)$ as measured in the platform frame and after the equivalent interval $k_l r_0/c$ as observed in the scatterer frame. Since it is assumed that the event defined by pulse initiation occurs at the time origin in both frames, the information or photon corresponding to $f(0)$ will arrive at the scatterer at the time $t_2 = \Delta t_2$. At this instant

$$\sqrt{\beta_e}\,(t_2 - \tau_2) = 0.$$

Thus, $\tau_2 = \Delta t_2$, and the waveform as measured in the scatterer frame of reference expressed in terms of that transmitted by the platform is

$$f\left(\sqrt{\beta_e}\left(t_2 - \frac{k_l r_0}{c}\right)\right) = f(t_1) \quad \{t_1 \in 0, T\}$$

where T is the transmit pulse duration. The one way time mapping is expressed in terms of the relative velocity v_t (2.32) by using (2.33, 2.37, 2.39) with the result that

$$t_1 = \sqrt{\frac{c + v_t}{c - v_t}}\,t_2 - \frac{r_0}{(c - v_t)} \quad \text{(electromagnetic)}.$$

At the end of the pulse

$$\sqrt{\beta_e}\left(t_2 - \frac{k_l r_o}{c}\right) = T$$

and consequently the pulse duration measured by an observer on the scatterer is

$$\begin{aligned} T_2 &= \frac{T}{\sqrt{\beta_e}} \\ &= \left(\sqrt{\frac{c - v_t}{c + v_t}}\right) T \quad \text{(electromagnetic).} \end{aligned}$$

The electromagnetic results are compared with those obtained from the acoustic relationships (2.11, 2.12) by solving for t_1 in terms of t_2 yielding the acoustic mapping

$$t_1 = \left(\frac{c + v_s}{c - v_p}\right) t_2 - \frac{r_0}{c - v_p} \quad \text{(acoustic)}$$

and the pulse duration as measured at the scatterer

$$T_2 = \left(\frac{c - v_p}{c + v_s}\right) T \quad \text{(acoustic).}$$

It is clear that propagation in the acoustic realm is distinguished from that in the electromagnetic realm by the fact that the velocities of the platform and scatterer contribute individually to the acoustic mapping relationships. This is not the case for electromagnetic propagation where only the relative velocity v_t (2.32) is significant. There is no stationary frame of reference provided by a propagation medium for the definition of individual velocities.

The receive leg mapping required for the complete solution is obtained by applying the same principles. The echo return is considered as a transmission from the scatterer back to the platform initiated at the time $t_1 = r_0/(c + v_t)$ measured in the platform reference frame when the pulse leading edge arrives at the scatterer. Since the platform is stationary in its own frame of reference, the return pulse leading edge must traverse the same range and consequently the total delay of the leading edge as seen by the platform is

$$\tau_3 = \frac{2r_0}{(c + v_t)}. \tag{2.40}$$

In order to relate the waveform $f_r(t_3 - \tau_3)$ received at the platform corresponding to $f_s(t_2 - \tau_2)$ that was reflected at the scatterer, a linear relationship as in equation (2.38) is assumed. Moreover since the relative velocity is unchanged, the same constant is used thus

$$f_r(t_3 - \tau_3) = f_s\left(\sqrt{\beta_e}(t_3 - \tau_3)\right).$$

The scattered waveform $f_s(t)$ is in turn expressed in terms of the original transmit waveform by (2.38). Consequently,

$$\begin{aligned} f_s\left(\sqrt{\beta_e}(t_3 - \tau_3)\right) &= f_s\left(\sqrt{\beta_e}t_3 - \sqrt{\beta_e}\tau_3\right) \\ &= f\left(\sqrt{\beta_e}\left(\sqrt{\beta_e}t_3 - \sqrt{\beta_e}\tau_3\right)\right) \\ &= f\left(\beta_e(t_3 - \tau_3)\right) \end{aligned}$$

and the received waveform is

$$\begin{aligned} f_r\left(t_3 - \tau_3\right) &= f\left(\beta_e(t_3 - \tau_3)\right) \\ &= f\left(t_1\right). \end{aligned}$$

The complete time mapping from time of reception back to time of transmission is written in terms of the fundamental parameters of range and relative velocity using (2.33, 2.39, 2.40) where

$$\boxed{t_1 = \left(\frac{c + v_t}{c - v_t}\right) t_3 - \frac{2r_0}{c - v_t}} \quad \text{(electromagnetic time map)}$$

and the received pulse duration is

$$\begin{aligned} T_3 &= \left(\frac{c - v_t}{c + v_t}\right) T \quad \text{(electromagnetic pulse duration)} \\ &= \frac{T}{\beta_e}. \end{aligned}$$

The corresponding acoustic mapping (2.13) is

$$\boxed{\begin{aligned} t_1 = &\left[\frac{c + v_s}{c - v_s}\right]\left[\frac{c + v_p}{c - v_p}\right] t_3 \\ &- \frac{2r_0 c}{(c - v_s)(c - v_p)} \qquad (2.41) \end{aligned}} \quad \text{(acoustic time map)}$$

and the received acoustic pulse duration is

$$T_3 = \frac{T}{\beta} \quad \text{(acoustic pulse duration)}$$

where β is the acoustic dilation parameter multiplying t_3 in (2.6). Note that the individual velocities appear in β but not in β_e however $\beta \approx \beta_e$

if $c \gg v_s$ and v_p since

$$\begin{aligned}\beta &= \frac{c^2 + (v_s + v_p)c + v_s v_p}{c^2 - (v_s + v_p)c + v_s v_p} \\ &\approx \left(\frac{c + v_t}{c - v_t}\right) = \beta_e\end{aligned}$$

and $v_t \simeq (v_s + v_p)$.

2.7 SUMMARY

This chapter sets the stage for later discussions of detection, estimation, and scattering models by introducing the concept of the echo energy or interrogation cycle. Sonar receivers often include an integrator followed by a square law detector and the signal and interference responses are proportional to the received energy. Detection performance is gauged by the signal to interference energy ratio at the receiver output and it is natural to describe signals in terms of their energy at various points in the cycle. Consideration of signal and interference energy is consistent with the linear scattering models and scattering functions that are introduced in Chapter 6. The energy cycle is denoted in decibel form to conform to traditional treatments of the sonar equation by Urick [1], Burdick [6], and others.

An in-depth discussion of the signal time base is justified because sonar signals are more readily subject to time base distortion than are radar signals propagating at the speed of light. The concept of time base distortion especially that of the first order dilation effect is central to discussions of the wideband (wavelet) description of Doppler that is discussed in the section on signal models. The graphical approach using trajectory diagrams for the calculation of Doppler and delay aids in the visualization of the mapping of time from reception to transmission and the derivation of expressions for the distortion and dilation effects. The time mapping relationships derived from the ray cone relationships can be directly used for simulation studies in order to synthesize echos in kinematic situations. All of the time base distortion effects caused by non-linear trajectories are automatically included in the echo. The section on relativistic time dilation illustrates the fundamental differences between the Galilean (sonar) and relativistic (radar) models of time mapping. Most engineers have had a modern physics course that includes relativity but a comparison between the models is usually not included.

Bibliography

[1] R.J. Urick. *Principles of Underwater Sound* . McGraw-Hill, 1975.

[2] P.M. Woodward. *Probability and Information Theory with Applications to Radar* . Pergamon Press, London, 1953.

[3] D.O. North. Analysis of the Factors Which Determine Signal/Noise Discrimination in Radar . Technical report, RCA Laboratories, June 1943. TR PTR-6C.

[4] J.I. Marcum. A Statistical Theory of Target Detection by Pulsed Radar. *IRE Trans. Info. Theory*, IT-6(2):59–267, April 1960.

[5] P. Swerling. Probability of Detection for Fluctuating Targets . Technical report, Rand Corp., April 1954.

[6] W.S. Burdic. *Underwater Acoustic System Analysis* . Prentice-Hall, 1984.

[7] W.C. Knight, G.G. Pridham, and S.M. Kay. Digital Signal Processing for Sonar . *Proceedings of the IEEE*, 69(11):1451–1507, Nov. 1981.

[8] S.C. Clay and H. Medwin. *Acoustical Oceanography* . Wiley Interscience, 1977.

[9] R.L. Swarts and C.J. Eggen. Simplified Model of the Spectral Characteristics of High Frequency Surface Scatter . *Journ. of the Acoust. Soc. Amer.*, 59(4):846–851, April 1976.

[10] W.S. Hodgkiss. An Oceanic Reverberation Model . *IEEE Journ. of Oceanic Engineering*, OE-9(2), April 1984.

[11] D.W. Ricker. A Non-Iterative Time Mapping Algorithm for Linear Motion In 3-Space . *Journ. of the Acoust. Soc. Amer.*, 89(3):1198–1200, Mar. 1991.

[12] R.T. Weidner and R.L. Sells. *Elementary Modern Physics* . Allyn and Bacon, Boston MA, 1968.

[13] A.R. Gondeck. Doppler Time Mapping . *Journ. of the Acoust. Soc. Amer.*, 73(5), May 1983.

[14] A.W. Rihaczek. *Principles of High Resolution Radar* . McGraw-Hill, New York NY, 1969.

[15] R.A. Altes. *Methods of Wideband Signal Design for Radar and Sonar Systems* . PhD Thesis in Electrical Engineering, University of Rochester, 1970. ASTIA doc. no. AD732494.

[16] E.L. Titlebaum. Echolocation Systems in Nature; A Theory of Optimum Insensitivity to Motion for Bats . In *Proc. 1978 Ultrasonics Symposium*, volume IEEE cat. no. 78CH1344-1SU, 1978.

[17] E.J. Kelly and R.P. Wishner. Matched Filter Theory for High Velocity Accelerating Targets . *IEEE Trans. on Military Electronics*, MIL-9(1):56–69, 1965.

[18] R.O. Harger. *Synthetic Aperture Radar Systems* . Academic Press, 1970.

[19] W.M. Brown and C.J. Palermo. Effects of Phase Errors on Resolution . *IEEE Trans. on Military Electronics*, MIL-9:4–9, Jan. 1965.

[20] W.L. Rubin and J.V. DiFranco. Analytic Representation of Wide Band Radio Frequency Signals . *Journal of the Franklin Inst.*, 275(3):197–204, Mar. 1963.

[21] E. Bedrosian. The Analytical Signal Representation of Modulated Waveforms . *Proceedings of the IEEE*, 50:2071–2076, Oct. 1962.

[22] R.A. Altes and E.L. Titlebaum. Bat Signals as Optimally Doppler Tolerant Waveforms. *Journ. of the Acoust. Soc. Amer.*, 48((4)2):1014–1020, 1970.

[23] L.G. Weiss. Wavelets and Wideband Correlation Processing . *IEEE Sig. Proc. Magazine*, Jan. 1994.

[24] A. Papoulis. *The Fourier Integral and its Applications* . McGraw-Hill, New York NY, 1987.

[25] P.A. Tipler. *Physics* . Worth Pub. Inc., 1982.

[26] C.H. Blanchard. *Introduction to Modern Physics* . Prentice-Hall, 1958.

Chapter 3

DETECTION AND ESTIMATION

3.1 INTRODUCTION

The concepts of echo detection and parameter estimation are introduced in this chapter by considering point scattering which is the simplest scattering model. It results from signal propagation in a homogeneous non-dispersive medium and reflection from a single distant point that may be moving at constant speed with a velocity component along the line of sight. The resulting echo is a Doppler dilated and delayed signal with the same fundamental form as the transmitted waveform (2.26). An understanding of approaches to its detection and the estimation of its constituent parameters is fundamental to an understanding of the signal processing required for the more complex echos treated in later chapters. Common examples are those that are delay and/or Doppler spread because of multiple propagation paths (multipath) especially in shallow water, boundary reverberation, and scattering from fish schools.

Assuming that transmission has occurred, the resulting echo would ideally be the only acoustic energy received and as was the case for the mariners of old, all that would be required would be a power detector (an ear) and a means to measure delay, Doppler shift, and possibly direction (two ears). Unfortunately, nature tends to be perverse and the echo invariably will be only a portion (often small) of the total received acoustic field. Collectively, the extraneous acoustic energy represents *interference* because it interferes with and masks the desired echo energy. Interference associated with echo-location consists of backscatter of the transmitted signal from scattering mechanisms that are not of interest e.g. radar clutter and acoustic reverberation, and as non-backscattered ambient or internal electronic noise. Whatever the source, interference complicates the detection and estimation problem and forces the designer to confront the issue of receiver performance and its optimization according to reasonable measures.

Interference is usually random and consequently, performance is based upon probabilistic measures. For example, in the case of echo detection, it is necessary to consider the probabilities that the echo is or is not detected given that a transmission has occurred or conversely, the probability that the receiver indicates a detection when indeed no transmission has taken place. In the case of parameter estimation, performance measures may be based on the probability that the estimated parameters are "close" to the actual parameter values by some suitable criterion.

3.2 BINARY DETECTION

Binary echo detection is the process by which a decision is made between two hypotheses regarding the presence or lack of an echo in the received signal. The signal may be a random scalar or vector time series or a random vector $\mathbf{r}$ of length n derived from it that represents direct measurement (e.g. samples) or functions of direct measurements of the received acoustic field. It may or may not contain an echo but will always consist of some amount of random noise or backscattered interference. The echo component may or may not be random depending upon the backscattering and propagation mechanisms involved but the total signal is considered to be random regardless of the nature of the echo because of the interference. The received signal data in vector form is

$$\mathbf{r} = \mathbf{s} + \mathbf{n} \tag{3.1}$$

indicating the presence of a random echo with additive random interference. The vector

$$\mathbf{r} = \mathbf{n}. \tag{3.2}$$

consists of interference only in the absence of the echo. A non-random signal vector in (3.1) would be denoted as s.

The use of vector notation is a generality that permits the consideration of both scalar and multidimensional data. Scalar data, for example could be a scalar function of time or frequency such as discussed in

Chapter 2 while data that is sampled, received from multiple sources such as an acoustic array, or represented by a generalized Fourier series would be an example of multidimensional data. Whatever the form of the data, the detection problem remains the same. The receiver must decide whether or not the data set $\mathbf{r}$ originated from a source that reflected echo energy or more generally, if the data originated from measurements whose value depends upon the presence of a scatterer that we wish to detect. The 'bottom line' is that given these measurements, and consequently the data vector $\mathbf{r}$, one and only one of the following statements must be declared: decide that a signal is present (3.1) which is called hypothesis H_1, or decide that no signal is present (3.2) which is called hypothesis H_0.

The detection problem is to decide H_0 or H_1, or stated alternately, consider how to formulate the policy or strategy by which a decision is made given the data vector $\mathbf{r}$. Since H_0 and H_1 are mutually exclusive (neither can be true at the same time), the set of all data vectors $\mathbf{r}$, the n-dimensional Euclidean space R^n, is divided into two distinct subsets. One in which each data vector would cause the receiver to declare H_0, and the other in which each data vector would cause it to declare H_1. These subsets of R_n are described by the functions $D_0(\mathbf{r})$ and $D_1(\mathbf{r})$ that in mathematical parlance are called set membership functions and are defined as

$$D_0(\mathbf{r}) = \begin{cases} 1, & \text{if } \mathbf{r} \Rightarrow H_0, \\ 0, & \text{if } \mathbf{r} \Rightarrow H_1. \end{cases}$$

and

$$D_1(\mathbf{r}) = \begin{cases} 0, & \text{if } \mathbf{r} \Rightarrow H_0, \\ 1, & \text{if } \mathbf{r} \Rightarrow H_1. \end{cases}$$

Note that the defining of two separate functions is redundant since

$$D_1(\mathbf{r}) = 1 - D_0(\mathbf{r}) \tag{3.3}$$

however, this clarifies later derivations.

These functions are mathematical representations of decisions however there must be some basis upon which the decisions are made. A reasonable approach is the evaluation of the cost of making a particular decision. Naturally, a good decision policy will have a low cost, and bad policy will have a high cost. To be more specific, define $C_{i,j}$ as a measure of the cost or benefit resulting from deciding H_i when in fact the correct decision should be H_j where i, j are either 0 or 1. If $C_{i,j}$ is positive, it is interpreted as a cost or penalty and if it is negative, it represents a benefit.

The values $C_{i,j}$ are real numbers representing perhaps some unit of payment and, in order to evaluate an optimal decision strategy defined by the costs and the functions D_0 and D_1, a single meaningful measure of overall effectiveness is required. One such a measure is the *Bayes risk* [1–3] which is the overall cost averaged over all possible decisions

$$\begin{aligned} C(D_0, D_1) = &C_{0,1}\, \wp(D_0 H_1) + C_{1,0}\, \wp(D_1 H_0) \\ &+ C_{0,0}\, \wp(D_0 H_0) + C_{1,1}\, \wp(D_1 H_1). \end{aligned} \tag{3.4}$$

$\wp(AB)$ denotes the joint probability of the events AB. For example, $\wp(D_0 H_1)$ is interpreted as the probability that the echo is actually present <u>and</u> $D_0 = 1$ i.e. the incorrect decision is made that no echo is present. The cost is equivalently expressed in terms of conditional probabilities

$$\begin{aligned} C(D_0, D_1) = &C_{0,1}\wp(D_0|H_1)\wp(H_1) + C_{1,0}\wp(D_1|H_0)\wp(H_0) \\ &+ C_{0,0}\wp(D_0|H_0)\wp(H_0) + C_{1,1}\wp(D_1|H_1)\wp(H_1) \end{aligned} \tag{3.5}$$

where $\wp(H_i)$ is the prior probability that hypothesis H_i actually true.

The cost is easily evaluated if the data density functions $\mathsf{f}(\mathbf{r}|H_0)$, $\mathsf{f}(\mathbf{r}|H_1)$, and the prior probabilities are known. For example, $\wp(D_0|H_1)$ is the probability that $D_0 = 1$ given the information that the echo is present. Mathematically, this is

$$\wp(D_0|H_1) = \int_{R^n} D_0(\mathbf{r})\mathsf{f}(\mathbf{r}|H_1)d\mathbf{r} \tag{3.6}$$

where the integral is taken over the entire Euclidean space R^n, but the integrand make a contribution only when $D_0(\mathbf{r}) = 1$. The density function $\mathsf{f}(\mathbf{r}|H_1)$ thus contributes to the integral only when $\mathbf{r}$ is in the portion of the Euclidean space where vectors are associated with data indicating no presence of an echo. The other conditional probabilities are similarly evaluated and the average cost is written as

$$\begin{aligned} C(D_0, D_1) = &\int_{R^n} [C_{0,1}\, \wp(H_1)\mathsf{f}(\mathbf{r}|H_1) + C_{0,0}\, \wp(H_0)\mathsf{f}(\mathbf{r}|H_0)]\, D_0(\mathbf{r}) \\ &+ [C_{1,0}\, \wp(H_0)\mathsf{f}(\mathbf{r}|H_0) + C_{1,1}\, \wp(H_1)\mathsf{f}(\mathbf{r}|H_1)]\, D_1(\mathbf{r})d\mathbf{r}. \end{aligned} \tag{3.7}$$

The quantities within the square brackets are functions of the data $\mathbf{r}$ and are denoted by $C_0(\mathbf{r})$ and $C_1(\mathbf{r})$ and

$$C(D_0, D_1) = \int_{R^n} \{C_0(\mathbf{r})D_0(\mathbf{r}) + C_1(\mathbf{r})D_1(\mathbf{r})\}\, d\mathbf{r}. \tag{3.8}$$

To optimize the decision strategy by minimizing overall cost it is only necessary to choose when $D_0(\mathbf{r}) = 1$ or when $D_1(\mathbf{r}) = 1$. Because of (3.3), only one term in the sum composing the integrand will contribute to the integral for any given value of $\mathbf{r}$. Thus, for any $\mathbf{r}$ there are two choices for the value of the integrand reflecting the choices for $D_0(\mathbf{r})$ and $D_1(\mathbf{r})$. They are $C_0(\mathbf{r})$ and $C_1(\mathbf{r})$ and to minimize the cost, choose the functions D_0 and D_1 such that the only the smaller of these contributes to the integral in (3.8) for any particular value of the data vector $\mathbf{r}$. That is,

$$C_0(\mathbf{r}) < C_1(\mathbf{r}) \Leftrightarrow D_0(\mathbf{r}) = 1,$$

and

$$C_1(\mathbf{r}) < C_0(\mathbf{r}) \Leftrightarrow D_1(\mathbf{r}) = 1.$$

These relationships define $D_0(\mathbf{r})$ and $D_1(\mathbf{r})$ and hence the optimal detection strategy is expressed more succinctly as

$$C_0(\mathbf{r}) \underset{H_0}{\overset{H_1}{\gtrless}} C_1(\mathbf{r}).$$

This notation indicates that H_1 is decided when the left side is larger than the right side and vice versa. Substituting the terms in the integrand of (3.7) for C_0 and C_1 and rearranging, this becomes

$$L_r(\mathbf{r}) = \frac{\mathsf{f}(\mathbf{r}|H_1)}{\mathsf{f}(\mathbf{r}|H_0)} \underset{H_0}{\overset{H_1}{\gtrless}} \frac{(C_{1,0} - C_{0,0})\wp(H_0)}{(C_{0,1} - C_{1,1})\wp(H_1)} \doteq \gamma. \tag{3.9}$$

The probabilities $\mathsf{f}(\mathbf{r}|H_0)$ and $\mathsf{f}(\mathbf{r}|H_1)$ are often referred to as *likelihood functions* since they represent the conditional probability or likelihood of the data $\mathbf{r}$ given the hypothesis H_0 (3.2) or H_1 (3.1). Consequently, $L_r(\mathbf{r})$ is called the *likelihood ratio*, and γ is referred to as the *likelihood threshold.*

Equation (3.9) reveals that the likelihood threshold depends upon the prior probabilities $\wp(H_0)$ and $\wp(H_1)$ and the cost/benefit weights $C_{0,1}$, $C_{1,0}$, $C_{0,0}$, $C_{1,1}$. In particular, consider a simple example of the decision strategy wherein the benefit is the same for either correct decision. The penalty is the same for either incorrect decision and no knowledge about the presence or absence of the echo is assumed. First, the overall cost (3.4) is a function of two costs and two benefits. Of the costs, one is associated with deciding H_0 when H_1 is true and the other with deciding H_1 when H_0 is true. These are proportional to $C_{0,1}$ and $C_{1,0}$. To assume both errors are equally grave implies

$$C_{0,1} = C_{1,0}. \tag{3.10}$$

Likewise, if the benefits accruing from correctly deciding that the echo is or is not present are the same,

$$C_{1,1} = C_{0,0}. \tag{3.11}$$

Moreover, to say that there exists absolutely no knowledge as to whether or not the echo is actually present implies that

$$\wp(H_0) = \wp(H_1) = 1/2. \tag{3.12}$$

Substituting (3.10,3.11,3.12) into the definition of the likelihood ratio (3.9) yields

$$L_r(\mathbf{r}) = \frac{\mathsf{f}(\mathbf{r}|H_1)}{\mathsf{f}(\mathbf{r}|H_0)} \underset{H_0}{\overset{H_1}{\gtrless}} 1.$$

Clearly , assuming as little as possible about costs and the presence of an echo yields a simple decision policy; given a particular set of data or measurements $\mathbf{r}$, choose the hypothesis corresponding to which probability density, $\mathsf{f}(\mathbf{r}|H_0)$ or $\mathsf{f}(\mathbf{r}|H_1)$, is larger. That is,

$$\begin{aligned} \mathsf{f}(\mathbf{r}|H_0) > \mathsf{f}(\mathbf{r}|H_1) &\Rightarrow \text{ choose } H_0, \\ \mathsf{f}(\mathbf{r}|H_0) < \mathsf{f}(\mathbf{r}|H_1) &\Rightarrow \text{ choose } H_1. \end{aligned}$$

Once the decision policy or strategy is determined, it is a simple matter to evaluate its overall cost given a particular set of cost weights $C_{i,j}$ and prior probabilities.

Example 3.1 A Dice Game Decision Strategy

A simple example involving a dice game illustrates how a decision strategy based upon the likelihood ratio test is affected by the prior probabilities and the cost/benefit weights. Consider a game that involves two types of dice one of which is a regular 6-sided die (H_0). The other

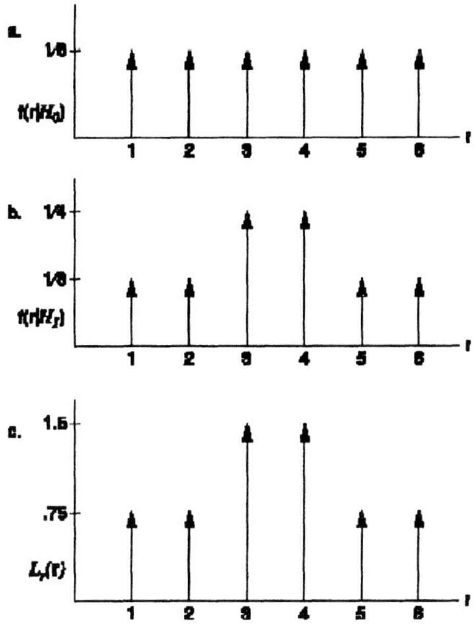

Figure 3.1: Data probabilities for the dice game example: (a) 6-sided die, (b) 8-sided die, (c) likelihood ratio

is an eight sided die (octahedron) (H_1) with the numbers 3 and 4 each imprinted on <u>two</u> of the faces and the numbers 1,2,5,6 each appearing once on one of the remaining 4 faces. The dice are fair such that the probability of rolling a particular number (1-6) is 1/6 for the regular die and the probability of rolling 1,2,5,6 is 1/8 and rolling 3 or 4 is 1/4 for the eight sided one. These represent the probabilities of the "data" which is the number appearing when a die is thrown conditioned upon which is used. These are shown in Figure 3.1.

A quantity n_0 of the regular dice and n_1 of the others is put into a bin and uniformly mixed. A die is chosen randomly from the bin and thrown. The only information provided is the number that appeared. This is the data. Assuming that the numbers of each type of die in the bin and the cost weights are known, the decision to be made is which die, H_0 or H_1 was chosen. A strategy must be adopted that will minimize the losses or maximize the benefits if a wager is made on the outcome. To do this by using a likelihood ratio test, it is necessary to determine the threshold γ in (3.9) based upon knowledge of the quantities of each type in the bin and the costs/benefits of the decisions. The data in this example is a scalar r taking an integer value on [1,6] and the likelihood ratio $L_r(r)$ (3.9) is the ratio of the conditional probabilities evaluated at each integer value as shown in the figure.

$L_r(r)$ has only two discrete values; $L_r(r) = .75$ when $r = (1, 2, 5, 6)$ and $L_r(r) = 1.5$ when $r = (3, 4)$. This means that there are only three possible strategies for making a decision;

$$S_1: \ -\infty < \gamma < .75 \Rightarrow \text{decide } H_1 \ \forall \ r,$$
$$S_2: \ .75 < \gamma < 1.5 \Rightarrow \text{decide } H_1 \text{ for } r = (3, 4), \text{ and } H_0 \text{ otherwise},$$
$$S_3: \ 1.5 < \gamma < \infty \Rightarrow \text{decide } H_0 \ \forall \ r.$$

The overall cost or Bayes risk for each of these strategies is evaluated by using (3.5) and a discrete and generalized version of (3.6) given by

$$\wp(D_i|H_j) = \sum_k D_i \ \wp(r_k|H_j) \tag{3.13}$$

where $\wp(r_k|H_j)$ represents the probability that $r = r_k$ given hypothesis H_j. The first strategy denoted by S_1, is adopted when the likelihood ratio is greater than the threshold for all possible values of the data. Consequently, $D_1 = 1$, $D_0 = 0$, and using (3.13) and the density functions shown in Figure 3.1, $\wp(D_1|H_0) = \wp(D_1|H_1) = 1$ and $\wp(D_0|H_0) = \wp(D_0|H_1) = 0$. The cost associated with strategy S_1 is evaluated from (3.5) as

$$C_{s1} = C_{1,0}\wp(H_0) + C_{1,1}\wp(H_1)$$

and using the same procedure, the costs for the other strategies are evaluated as;

$$C_{s2} = .50\wp(H_1)(C_{0,1} + C_{1,1}) + .33\wp(H_0)(C_{1,0} + 2C_{0,0})$$
$$C_{s3} = C_{0,1}\wp(H_1) + C_{0,0}\wp(H_0).$$

Assume that each time the game is played, $n_0 = 200$, $n_1 = 150$, and the penalty for deciding H_1 when H_0 is picked is $C_{1,0} = 5$ units of payment. On the other hand, the correct decision of H_0 when H_0 is picked is $C_{0,0} = -5$ units recalling that a benefit or payment to you is a negative cost. Likewise let $C_{0,1} = 10$ and $C_{1,1} = -10$. The prior probabilities are computed from n_0, n_1 and using (3.9), $\gamma = .67$. Clearly, the likelihood ratio $L_r(r)$ shown in the figure is greater than the threshold for all values of the data and consequently the optimum strategy is S_1;

always choose H_1, the eight sided die. The costs for each of the strategies are:

$$C_{s1} = -1.43$$
$$C_{s2} = -.95$$
$$C_{s3} = +1.43$$

and indeed the optimum strategy is S_1 that on the average, returns 1.43 units of payment . The worst strategy is be S_3 that, on the average, would cost 1.43 units.

If costs can be assigned to specific decisions and the prior probabilities are known, it is a simple matter to compute the likelihood threshold γ in (3.9) that minimizes the overall risk and hence devise the optimal decision strategy. In many situations including those often involving echo-location, costs cannot be assigned in a meaningful way and the prior probabilities are not known. Consequently it is impossible to compute the likelihood threshold γ and the corresponding strategy by using (3.9). Moreover, the density $\mathsf{f}(\mathbf{r}|H_1)$ will not be precisely known since it will in part, depend upon the relative power levels of the signal and interference. On the other hand the interference-only density $\mathsf{f}(\mathbf{r}|H_0)$ is often known and may be used as the basis for an alternate decision strategy which does not require the assessment of costs or prior information. In order to do this some new terminology is required.

The volume under the density $\mathsf{f}(\mathbf{r}|H_1)$ for data $\mathbf{r}$ such that $D_1 = 1$ is the conditional probability $\wp(D_1|H_1)$. In an echo-location context this is called the *detection probability.*

$$\begin{aligned}\wp_d &= \wp(D_1|H_1)\\ &= \int_{R_n} D_1(\mathbf{r})\mathsf{f}(\mathbf{r}|H_1)d\mathbf{r}\end{aligned}$$

and simply stated, $\wp_d$ is the probability that the echo is detected given that a signal was transmitted. Conversely, the volume defined for data where $D_0 = 1$ is $\wp(D_0|H_1)$ and is called the *probability of false rest*

$$\begin{aligned}\wp_{fr} &= \wp(D_0|H_1) = 1 - \wp_d\\ &= \int_{R_n} D_0(\mathbf{r})\mathsf{f}(\mathbf{r}|H_1)d\mathbf{r}\end{aligned}$$

and represents the probability that the echo is not detected given that a transmission has occurred. When no transmission occurs but the receiver indicates a detection, the situation is called a *false alarm* and is represented by the volume under $\mathsf{f}(\mathbf{r}|H_0)$ defined by $D_1 = 1$ which is

$$\begin{aligned}\wp_f &= \wp(D_1|H_0) \\ &= \int_{R_n} D_1 \mathsf{f}(\mathbf{r}|H_0) d\mathbf{r}.\end{aligned}$$

Rather than establishing γ by using (3.9), an alternate approach called the *Nehman-Pearson criterion* [1,3–5] seeks an optimum test maximizing $\wp_d$ (minimize $\wp_{fr}$) with a fixed $\wp_f$ as a constraint. This is equivalent to the minimization of the expression

$$\begin{aligned}\mathcal{P} &= \wp_{fr} + \gamma_{np}\left(\int_{R_n} D_1(\mathbf{r})\mathsf{f}(\mathbf{r}|H_0)d\mathbf{r} - \wp_f\right) \\ &= \int_{R_n} (1 - D_1(\mathbf{r}))\mathsf{f}(\mathbf{r}|H_1)d\mathbf{r} + \gamma_{np}\left(\int_{R_n} D_1(\mathbf{r}))\mathsf{f}(\mathbf{r}|H_0) - \wp_f\right) \\ &= \int_{R_n} D_1(\mathbf{r})[\gamma_{np}\mathsf{f}(\mathbf{r}|H_0 - \mathsf{f}(\mathbf{r}|H_1)]d\mathbf{r} + (1 - \gamma_{np}\wp_f)\end{aligned}$$

where γ_{np} is a Lagrange multiplier [1]. For fixed $\wp_f$ and $\gamma_{np} \geq 0$, $\mathcal{P}$ is minimized by setting $D_1(\mathbf{r}) = 1$ only for data $\mathbf{r}$ for which the integrand is negative. In other words the minimization of $\mathcal{P}$ is equivalent to the likelihood ratio test (3.9)

$$L_r(\mathbf{r}) = \frac{\mathsf{f}(\mathbf{r}|H_1)}{\mathsf{f}(\mathbf{r}|H_0)} \underset{H_0}{\overset{H_1}{\gtrless}} \gamma_{np} \tag{3.14}$$

where $\gamma = \gamma_{np}$ is the likelihood threshold (3.9) that must be specified to satisfy the constraint.

The detection and false alarm probabilities are the probabilities that L_r exceeds the threshold under the conditions H_1, H_0 hence,

$$\begin{aligned}\wp_d &= \int_{\gamma_{np}}^{\infty} \mathsf{f}(L_r(\mathbf{r})|H_1)dL_r \\ \wp_f &= \int_{\gamma_{np}}^{\infty} \mathsf{f}(L_r(\mathbf{r})|H_0)dL_r.\end{aligned}$$

If the data $\mathbf{r}$ is scalar ($\mathbf{r} \to \mathsf{r}$) and the conditional probabilities are continuous, γ_{np} establishes intervals on the real line (r axis) which define D_1

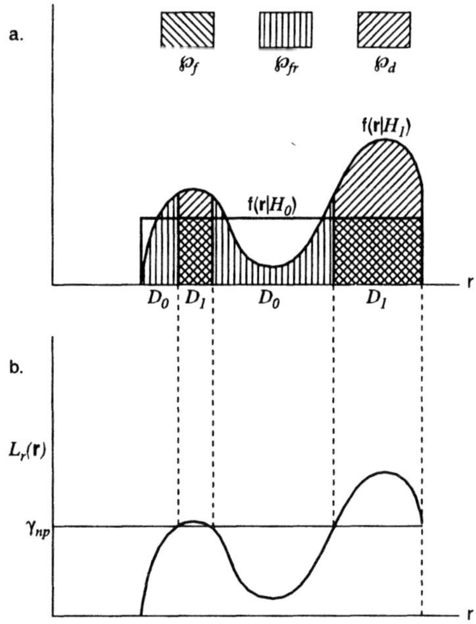

Figure 3.2: Bimodal likelihood example: (a) conditional densities, (b) likelihood ratio

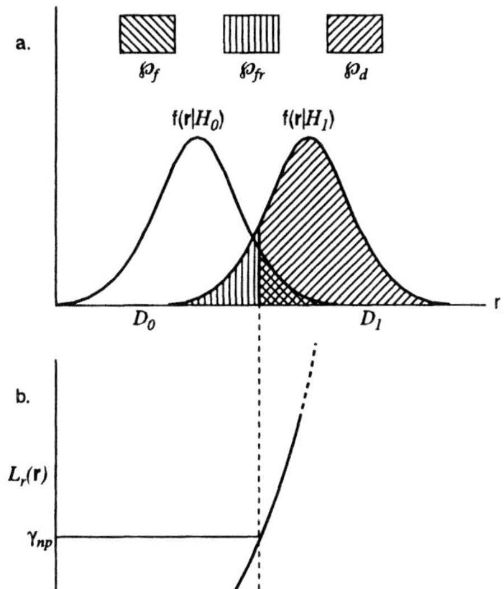

Figure 3.3: Monotonic likelihood example: (a) conditional densities, (b) likelihood ratio

and D_0 as illustrated for the rather pathological example shown in Figure 3.2. Since $\mathsf{f}(\mathsf{r}|H_1)$ is bimodal and $\mathsf{f}(\mathsf{r}|H_0)$ is uniform in this example, $L_r(\mathsf{r})$ will be a linearly scaled version of $\mathsf{f}(r|H_1)$ and the region associated with $D_1(\mathsf{r}) = 1$ may exist as either a single connected interval or as two disconnected intervals depending upon the value of γ_{np}. The threshold is determined when the area under $\mathsf{f}(\mathsf{r}|H_0)$ is equal to the specified $\wp_f$ indicated by the shaded area in the figure.

If the densities are such that $L_r(\mathsf{r})$ is monotonic as illustrated in Figure 3.3, the likelihood threshold always divides the real line into at most two regions: $\mathsf{r} \geq \gamma_r \Rightarrow D_1 = 1$ and, $\mathsf{r} < \gamma_r \Rightarrow D_0 = 1$ where γ_r is defined as the <u>data</u> threshold. Consequently,

$$\begin{aligned}
\wp_d &= \int_{\gamma_r}^{\infty} \mathsf{f}(\mathsf{r}|H_1)d\mathsf{r} \\
\wp_{fr} &= \int_{-\infty}^{\gamma_r} \mathsf{f}(\mathsf{r}|H_1)d\mathsf{r} \\
\wp_f &= \int_{\gamma_r}^{\infty} \mathsf{f}(\mathsf{r}|H_0)d\mathsf{r}.
\end{aligned} \tag{3.15}$$

The probabilities are uniquely determined for a particular value of γ_r and a plot of $\wp_d$ vs $\wp_f$ called a *Receiver Operating Characteristic* or ROC curve [1, 3] is often used to provide a comparative assessment of receiver performance. ROC curves are usually parameterized by the signal to interference ratio (SIR) which is a measure of the power or energy of the echo signal relative to that of the interference.

ROC curves have two important attributes. All ROC's resulting from continuous likelihood ratio tests have negative second derivatives meaning that $\wp_d \geq \wp_f \forall \wp_f$. Were this not so, a chance test (coin toss) resulting in a ROC curve with unity slope would be better. Moreover, the slope of a ROC curve is equal to the value of the threshold associated with $\wp_d$ and $\wp_f$ at that point. Proof of this is found in [1].

Example 3.2 Square Law Detection of Gaussian Processes

Let the input echo and noise be independent scalar complex Gaussian processes $\mathsf{z} = \mathsf{x} + j\mathsf{y}$ with independent identically distributed (iid) real and imaginary parts and variances $\sigma_x^2 = \sigma_y^2$. The input is passed through a square law energy detector [4, 6] that forms the scalar statistic $\mathsf{r} = |\mathsf{z}|^2$ under the hypotheses $\mathsf{r}|H_1$ and $\mathsf{r}|H_0$. The corresponding density functions of r are

$$\begin{aligned} f(r|H_1) &= \frac{1}{{\sigma_1}^2} \exp -\frac{r}{{\sigma_1}^2} \\ f(r|H_0) &= \frac{1}{{\sigma_0}^2} \exp -\frac{r}{{\sigma_0}^2} \end{aligned} \tag{3.16}$$

with variances

$$\begin{aligned} {\sigma_1}^2 &= \mathcal{E}\{|z|^2\}|H_1 = {\sigma_s}^2 + {\sigma_n}^2 \\ {\sigma_0}^2 &= \mathcal{E}\{|z|^2\}|H_0 = {\sigma_n}^2. \end{aligned} \tag{3.17}$$

The likelihood ratio (3.9) evaluated with (3.16) is

$$L_r(r) = \left(\frac{1}{1+\text{SNR}}\right) \exp\left\{\frac{r}{{\sigma_n}^2}\left[\frac{\text{SNR}}{1+\text{SNR}}\right]\right\} \underset{H_0}{\overset{H_1}{\gtrless}} \gamma$$

where $\text{SNR} = \sigma_s^2/\sigma_n^2$ is the *Signal to Noise Ratio*. An equivalent test is obtained by taking the natural logarithm of $L_r(r)$ to form

$$\begin{aligned} \ell_r(r) &= \ln\, L_r(r) \\ &= \ln\left(\frac{1}{1+\text{SNR}}\right) + \frac{r}{{\sigma_n}^2}\left[\frac{\text{SNR}}{1+\text{SNR}}\right] \underset{H_0}{\overset{H_1}{\gtrless}} \gamma_l \end{aligned}$$

and $\gamma_l = \ln\,\gamma$. This is called the *log likelihood ratio* (llr) and its use is justified because the logarithm is monotonic in its argument. Solving for the data results in the test

$$r \underset{H_0}{\overset{H_1}{\gtrless}} {\sigma_n}^2\left(\frac{1+\text{SNR}}{\text{SNR}}\right)\left\{\gamma_l - \ln\left[\frac{1}{1+\text{SNR}}\right]\right\} = \gamma_r \tag{3.18}$$

where the noise variance is assumed to be known and γ_r is the data threshold which may be set either by considering Bayes risk or by the Nehman-Pearson criterion. If the prior probabilities are known and costs can be assigned, γ may be evaluated from (3.9). Its logarithm is computed to form γ_l, and γ_r which minimizes the Bayes risk is computed as indicated for a particular SNR. The Nehman-Pearson criterion results in a threshold that is independent of SNR because it only fixes the false alarm probability. $\wp_f$ is computed from (3.15) using the appropriate relationships in (3.16,3.17) resulting in

$$\gamma_r = -{\sigma_n}^2 \ln \wp_f.$$

The Nehman-Pearson likelihood threshold γ_{np} corresponding to this value of γ_r could be obtained from (3.18) by solving for γ_l and hence for γ.

The ROC curves for this example are easily obtained by solving (3.15) for $\wp_f, \wp_d$ using (3.16) resulting in $\wp_d = \exp(-\gamma/\sigma_1^2)$, $\wp_f = \exp(-\gamma/\sigma_0^2)$ and

$$\wp_d = \wp_f^{\frac{1}{1+\mathrm{SNR}}}.$$

The curves are plotted in Figure 3.4 with SNR as a parameter and clearly, performance improves with increasing SNR. When SNR $= 0$, $\wp_d = \wp_f$

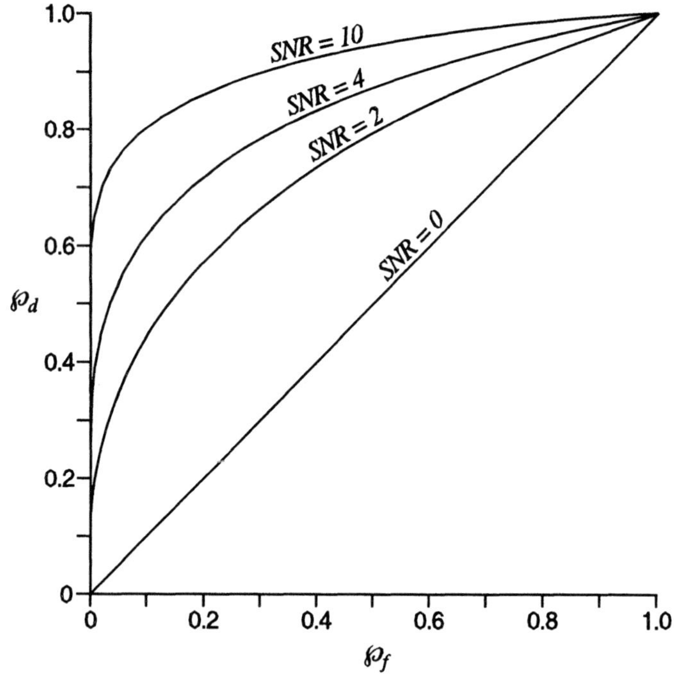

Figure 3.4: ROC curves for the square law detector

and the decision becomes a coin toss no matter what threshold is chosen. As SNR increases, $\wp_d$ increases for constant $\wp_f$ as the curves crowd into the upper left quadrant of the plot. The derivation of the receiver operating characteristics is a simple calculation yielding a closed form solution for this particular example. The derivation of ROC curves for more complex receiver structures and data models is often considerably more involved. [4,7,8].

3.3 MULTIPLE HYPOTHESES

The solution to the binary detection problem in terms of likelihood ratios illustrates a general approach to the making of decisions based upon probabilistic models of the data and the decisions to be made. The binary problem involved only a decision between two hypotheses namely whether or not the echo was present. Usually, the echo-location problem is more complex because in addition to a decision regarding the presence or absence of an echo, it may be necessary, given that a detection has occurred, to decide which of many possible echoes has been received. The fact that multiple signals and hypotheses regarding those signals exist is a consequence of the Doppler (dilation) and delay parameters $[\beta, \tau]$ or $[\phi, \tau]$ which give rise to echo signals of the form (2.26). Indeed, if the parameters can take on a continuum of values over specific ranges, there are an infinite number of possible signals. The problem of deciding which has been received is actually a problem of parameter estimation. This can be considered as a multiple hypothesis testing problem if it is assumed that the parameters can only take on a countably finite set of values. This is not an overly restrictive assumption since with some knowledge of the possible ranges of parameter values, as fine a grid of hypothesis values as desired can be chosen. The tacit assumption is that when the parameters are continuous over their range, it is possible by some measure, to choose the grid point that is closest to the actual value and hence the best estimate. As shall be later shown in the discussion of parameter estimation and ambiguity functions, the choice of signal waveform largely determines the required resolution of the grid.

The notation is simplified by denoting the set of Doppler and delay (dilation) parameters by the vector $\mathbf{p}$ (2.27) and it is assumed that they can take on $M = M_\tau M_D$ discrete values denoted by $\mathbf{p}_k \quad k = 1, M$ where M_τ and M_D represent the number of discrete delay and Doppler values respectively. The received echo signal function corresponding to a particular parameter set $\mathbf{p}_k = [\phi_k, \tau_k]$ or equivalently $[\beta_k, \tau_k]$ has the form given by (2.28) and is denoted by $s(\mathbf{p}_k, t)$. The combined detection and estimation problem may be considered as a multiple hypothesis testing problem with hypotheses $H_0, H_1, ..., H_M$ corresponding to the possible parameter values. As before, H_0 is defined as the null hypothesis indicating no detection. It is known that the optimum test for the presence or absence of a signal is the binary test which computes the likelihood ratio. Consequently, a binary test is constructed for each hypothesis that may be envisioned as a bank of M receivers each receiving the same data $\mathbf{r}$ and performing the test

$$L_{rk}(\mathbf{r}) = \frac{f(\mathbf{r}|H_k)}{f(\mathbf{r}|H_0)} \underset{H_0}{\overset{H_k}{\gtrless}} \gamma.$$

When the problem is binary, the threshold is determined based upon Bayes cost or by the Nehman-Pearson criterion. In the latter case an acceptable false alarm probability is specified and the threshold set accordingly. When multiple hypotheses are possible the definition of what constitutes a detection and a false alarm must be considered. For example it could be required that a detection be declared only when a threshold crossing occurs in more than one receiver.

If one or more threshold crossings occur in the receiver bank, it is interpreted as a rejection of the null hypothesis and a detection is declared. If just one crossing occurs say, in the $k-th$ receiver, the interpretation is that hypothesis H_k is true and the corresponding echo signal with parameters $\mathbf{p}_k$ was received. However if more than one detection occurs, a criterion for choosing the correct hypothesis and hence the best estimate of the signal parameters is required.

When the parameters are random and multiple hypothesis are possible, the probability that the $k-th$ has occurred given data $\mathbf{r}$ is denoted by $\wp(H_k|\mathbf{r})$ and is called the *a-posteriori* probability and is related to the *a-priori* probability $\wp(H_k)$ by Bayes rule

$$\wp(H_k|\mathbf{r}) = \frac{f(\mathbf{r}|H_k)\wp(H_k)}{f(\mathbf{r})}$$

where

$$f(\mathbf{r}) = \sum_{k=1}^{M} f(\mathbf{r}|H_k)\wp(H_k)$$

is the total probability density of the data. The a-priori probability represents the prior knowledge concerning the relative chances for the occurrence of the signals. For example, if they are equally likely, and there is no other information (data), $\wp(H_k) = 1/M \ \forall k$. The a-posteriori probabilities reflect the re-assessment of the relative chances of occurrence given that some information (data) perhaps in the form of measurements $\mathbf{r}$ has been received. The ratio of the a-posteriori and a-priori probabilities is a measure of the information gained from the data. A more detailed discussion of this may be found in [1, 4]. The most likely hypothesis regarding which signal is received based upon the data corresponds to the maximum a-posteriori probability and detectors based upon this measure are called MAPP detectors. Assuming that more than one threshold crossing occurs in the bank of detectors, the MAPP estimate of the parameters $\mathbf{p}_{mp}$ can be found by using the likelihood ratios generated by each binary detector along with the a-priori probabilities. The ratio of any two a-posteriori probabilities indexed by k, j is

$$\begin{aligned}
\frac{\wp(H_k|\mathbf{r})}{\wp(H_j|\mathbf{r})} &= \frac{\mathsf{f}(\mathbf{r}|H_k)\wp(H_k)}{\mathsf{f}(\mathbf{r}|H_j)\wp(H_j)} \\
&= \frac{L_k(\mathbf{r})\wp(H_k)}{L_j(\mathbf{r})\wp(H_j)}
\end{aligned}$$

where

$$\begin{aligned}
L_k(\mathbf{r}) &= \frac{\mathsf{f}(\mathbf{r}|H_k)}{\mathsf{f}(\mathbf{r}|H_0)} \\
L_j(\mathbf{r}) &= \frac{\mathsf{f}(\mathbf{r}|H_j)}{\mathsf{f}(\mathbf{r}|H_0)}
\end{aligned}$$

are the likelihood ratios computed by binary detectors k and j that indicated threshold crossings. Clearly, when multiple detections occur, a repeated application of this ratio test or equivalently, the determination of $\max_k L_k(\mathbf{r})\wp(H_k)$ yields the MAPP parameter estimate $\mathbf{p}_{mp}$ within the resolution of the grid chosen. The maximization of $L_k(\mathbf{r})$ is equivalent to the maximization of the likelihood function $\mathsf{f}(\mathbf{r}|H_k)$ since $\mathsf{f}(\mathbf{r}|H_0)$ does not depend upon the parameters of the echo. Both functions yield the maximum likelihood (ML) parameter estimate $\mathbf{p}_{ml}$ and in the absence of prior knowledge, or for the case where the parameters are non-random but unknown, $\mathbf{p}_{mp} = \mathbf{p}_{ml}$.

3.4 NON-RANDOM POINT ECHO DETECTION

Interference whether it be backscattered energy such as reverberation or extraneous noise is modeled as a random process that is additive to the echo return. The complex random data vector $\mathbf{r}$ of arbitrary order n is composed of either an interference vector denoted by $\mathbf{n}$ or a signal $\mathbf{s}$ plus interference vector where,

$$\mathbf{r} = \begin{cases} \mathbf{n}, & \text{under } H_0, \\ \mathbf{s} + \mathbf{n}, & \text{under } H_1. \end{cases} \tag{3.19}$$

The noise and interference phenomena experienced in an echo-location context often result from the combined effects of a multitude of individual random sources. For this reason and for analytical convenience a Gaussian model is appropriate for interference comprised of reverberation and ambient noise processes in the absence of rough reflecting boundaries as in a high sea state or over a rocky bottom.[1] $\mathbf{n}$ is therefore modeled as Gaussian with

$$\begin{aligned} \mathcal{E}\{\mathbf{n}\} &= 0 \\ \mathcal{E}\{\mathbf{n}\mathbf{n}^{\mathrm{H}}\} &= \mathbf{R}_n \end{aligned} \tag{3.20}$$

where both signal and interference vectors are assumed to be complex, $()^{\mathrm{H}}$ indicates the Hermitian transpose, and $\mathbf{R}_n$ is the $n \times n$ interference covariance matrix with $\mathbf{R}_n^{\mathrm{H}} = \mathbf{R}_n$. This latter relationship defines a Hermitian matrix. It is assumed that $\mathbf{s}$ is a known non-random (deterministic) echo signal reflected from a highly localized (point like) scattering site. The reflector is assumed to be at some arbitrary range and may be moving with uniform velocity. This is the simplest possible scattering model and while not realistic in all applications, serves as a starting point for a discussion of detection in Gaussian interference. The general approach taken to develop detection criteria for this simple model will later be applied to more complex random scatter models.

With these assumptions, the pdf's under the detection hypotheses H_1 and H_0 are

$$\mathsf{f}(\mathbf{r}|H_k) = \left[\pi^n \left|\mathbf{R}_n\right|\right]^{-1} \exp -(\mathbf{r} - \bar{\mathbf{r}})^{\mathrm{H}} \mathbf{R}_n{}^{-1} (\mathbf{r} - \bar{\mathbf{r}}) \tag{3.21}$$

[1] Reverberation from rough boundaries in the ocean may not be Gaussian but "heavy tailed" meaning that there is the possibility of increased false alarm probability for detectors designed under a Gaussian assumption. This is discussed in [9–11] and their associated references.

with

$$\bar{\mathbf{r}} = \mathcal{E}\{\mathbf{r}\} = \begin{cases} 0, & \text{if } k = 0, \\ \mathbf{s}, & \text{if } k = 1. \end{cases}$$

The likelihood ratio (3.14) is

$$\begin{aligned} L_r(\mathbf{r}) &= \exp - \left[(\mathbf{r} - \mathbf{s})^{\mathrm{H}} \mathbf{R}_n^{-1} (\mathbf{r} - \mathbf{s}) - \mathbf{r}^{\mathrm{H}} \mathbf{R}_n^{-1} \mathbf{r} \right] \\ &= \exp \left[\mathbf{s}^{\mathrm{H}} \mathbf{R}_n^{-1} \mathbf{r} + \mathbf{r}^{\mathrm{H}} \mathbf{R}_n^{-1} \mathbf{s} - \mathbf{s}^{\mathrm{H}} \mathbf{R}_n^{-1} \mathbf{s} \right] \end{aligned}$$

and the log likelihood ratio (llr)

$$\ell_r(\mathbf{r}) = \left[\mathbf{s}^{\mathrm{H}} \mathbf{R}_n^{-1} \mathbf{r} + \mathbf{r}^{\mathrm{H}} \mathbf{R}_n^{-1} \mathbf{s} - \mathbf{s}^{\mathrm{H}} \mathbf{R}_n^{-1} \mathbf{s} \right] \underset{H_0}{\overset{H_1}{\gtrless}} \gamma_l. \tag{3.22}$$

is formed as an equivalent test as was done in Example 3.2

The first two terms on the right hand side are complex conjugate scalars [2] thus their sum is equivalent to twice the real part of either and the log likelihood detector is equivalent to the test

$$\begin{aligned} \Re\left(\mathbf{s}_{\mathrm{H}} \mathbf{R}_n^{-1} \mathbf{r} \right) = \Re\left(\mathbf{r}^{\mathrm{H}} \mathbf{g} \right) &\underset{H_0}{\overset{H_1}{\gtrless}} \frac{1}{2} \left(\gamma_l + \mathbf{s}^{\mathrm{H}} \mathbf{g} \right) = \gamma_l' \\ \mathbf{g} &= \mathbf{R}_n^{-1} \mathbf{s} \end{aligned} \tag{3.23}$$

where $\Re$ represents the real part of the complex scalar quantity and $\mathbf{g}$ is called the processing vector. The detector is thus implemented by forming the real inner product of the data vector with the known signal vector weighted by the inverse of the interference covariance matrix. This is compared to a modified threshold that is in part determined from probabilistic considerations such as the Nehman-Pearson criterion plus a term representing a generalized signal to noise ratio (SIR). If the interference components are uncorrelated and of equal energy N_0, $\mathbf{R}_n$ is diagonal and the test statistic reduces to the replica correlator $\Re(\mathbf{r}^{\mathrm{H}}\mathbf{s})$. The weighting of the data or equivalently the signal by $\mathbf{R}_n^{-1}$ is called a *prewhitening* operation [1].

[2] This can be shown by making use of the fact that the covariance matrix is Hermitian and thus diagonalizable by a unitary matrix. Consequently its inverse is also Hermitian. See Strang [12].

The remaining task is to relate the development of the detection problem describing the signal and interference processes as generalized complex vectors to the time continuous signals that are actually transmitted and received. Again, assuming an additive Gaussian interference model, the received scalar data is expressed as a time continuous complex random process

$$\mathsf{r}(t) = \begin{cases} \mathsf{n}(t), & \text{under } H_0, \\ s(t) + \mathsf{n}(t), & \text{under } H_1, \quad 0 \le t \le T \end{cases}$$

where $\mathsf{n}(t)$ is a Gaussian process, and $s(t) = \sqrt{E_s} f(\mathbf{p}, t)$ is a deterministic echo signal of the form (2.26) with a particular parameter set $\mathbf{p}$ engendered by the delay and Doppler of the scatterer. $\mathbf{p}$ may represent a continuum of parameters or in the context of a multiple hypothesis problem, may be considered as one of a set of discrete parameter subsets comprising the totality of possible echoes as discussed in the previous section. In an echo-location context, the interference is usually considered to consist of two components, a backscattered component generated by the reflection of the transmitted energy from interfering scattering mechanisms, and noise that is not transmission dependent such as electronic noise or noise generated by interfering acoustic (sonar) or electromagnetic (radar) sources. In sonar, backscattered interference is called reverberation and includes reflections from medium boundaries such as the bottom or surface in the ocean, bubbles, and biological sources [13, 14]. In radar it is called clutter and often results from reflections from land or sea surfaces, atmospheric phenomena such as precipitation, and from anomalous sources dubbed angels that may be birds or inhomogeneities in the refractive index of the atmosphere [15]. Regardless of the physical sources, it is assumed that the interference $\mathsf{n}(t)$ is the sum of the independent stochastic processes $\mathsf{n}_c(t)$ representing a correlated interference process that may in general be non-stationary, and $\mathsf{n}_0(t)$ representing white noise with spectral density N_0 (Joules). Both processes are assumed to be zero mean and the covariance of the combination is

$$\mathcal{E}\{\mathsf{n}(t)^*\mathsf{n}(u)\} = R_n(t,u) = R_c(t,u) + N_0\delta(t-u) \tag{3.24}$$

with finite power such that

$$\int_{\Omega_u}\int_{\Omega_t} |R_n(t,u)|^2 dt du \le \left[\int_{\Omega_t} \mathcal{E}\{|\mathsf{n}(t)|^2\} dt\right]^2 < \infty$$

where Ω_t and Ω_u represent the nonzero support regions of their respective integrands. Most generally $\mathsf{n}_c(t)$ is considered as a non-stationary stochastic process as implied by the two variable dependence of its covariance. However, an important subclass of processes are those that are wide sense stationary for which the covariance is $R_c(t-u)$ and depends only upon the time difference $(t-u)$ not the absolute time.

The combined covariance is Hermitian since

$$R_n{}^*(t,u) = \mathcal{E}\{\mathsf{n}(t)\mathsf{n}^*(u)\} = R_n(u,t)$$

and is positive definite which is shown by considering any deterministic square integrable complex function $x(t)$ such that

$$\int_{\Omega_t} |x(t)|^2 \, dt = E_x \quad 0 < E_x < \infty.$$

A new random variable y is formed from

$$\mathsf{y} = \int_{\Omega_t} x(t)\mathsf{n}_c(t)dt$$

that has variance

$$\begin{aligned} \sigma_y^2 &= \mathcal{E}\left\{\int_{\Omega_t}\int_{\Omega_u} x^*(t)\mathsf{n}_c{}^*(t)\mathsf{n}_c(u)x(u)dudt\right\} \\ &= \int_{\Omega_t}\int_{\Omega_u} x^*(t)R_c(t,u)x(u)dudt \geq 0 \end{aligned} \tag{3.25}$$

which is a positive semi-definite condition. If y is formed using the combined interference process $\mathsf{n}(t)$, its variance will be the sum of (3.25) and the strictly positive term due to the white noise

$$N_0 \int_{\Omega_t} |x(t)|^2 \, dt > 0.$$

Consequently,

$$\int_{\Omega_t}\int_{\Omega_u} x^*(t)R_n(t,u)x(u)dudt > 0$$

which is the positive definite property for covariance functions. It is noteworthy that positive definiteness can only be guaranteed in the presence of a white noise process.

In order to obtain a countable discrete data representation consistent with the vector model (3.19), the continuous data is expanded in terms of a generalized Fourier series such that

$$\mathsf{r}(t) = \mathbf{r}^{\mathrm{T}}\boldsymbol{\xi}(t) \tag{3.26}$$

where $()^{\mathrm{T}}$ represents the vector-matrix transpose, r is the data vector (3.19), and the components of $\boldsymbol{\xi}(t)$ are a countable set of functions $\xi_i(t)$, $1 \leq i \leq n$ comprising the first n terms of a complete orthonormal (CON) set. Orthonormality implies that

$$\int_{\Omega_t} \xi_i^*(t)\xi_j(t)dt = \delta_{ij}$$

or equivalently,

$$\int_{\Omega_t} \boldsymbol{\xi}(t)\boldsymbol{\xi}^{\mathrm{H}}(t)dt = \mathbf{I}_n \tag{3.27}$$

where $\mathbf{I}_n$ is the identity matrix of order n. Completeness of an orthonormal set for a finite energy and deterministic function $x(t)$ that is defined on an interval $[0, T]$ is defined by the convergence of the integral square error

$$\lim_{n\to\infty} \int_{\Omega_t} \left(x(t) - \sum_{i=1}^{n} x_i\xi_i(t) \right)^2 dt = 0$$

and assuming that $x(t)$ and the series $\sum_{i=1}^{n} x_i\xi_i(t)$ are continuous [16]

$$\Rightarrow \; x(t) = \lim_{n\to\infty} \sum_{i=1}^{n} x_i\xi_i(t). \tag{3.28}$$

The expansion (3.26) will only make sense as a characterization of the continuous random data $\mathsf{r}(t)$ if conditions may be specified guaranteeing the equivalence in some form. A reasonable condition is that the series converge in mean square

$$\lim_{n\to\infty} \mathcal{E}\left\{ \left|\mathsf{r}(t) - \mathsf{r}^{\mathrm{T}}\boldsymbol{\xi}(t)\right|^2 \right\} = 0 \quad t \in \Omega_t \tag{3.29}$$

and it is assumed that the orthonormal set of functions may be so chosen. The coefficient vector r is computed from

$$\begin{aligned}
\int_{\Omega_t} \mathsf{r}(t)\boldsymbol{\xi}^{\mathrm{H}}(t)dt &= \int_{\Omega_t} \left[\mathsf{r}^{\mathrm{T}}\boldsymbol{\xi}(t)\right]\boldsymbol{\xi}^{\mathrm{H}}(t)dt \\
&= \mathsf{r}^{\mathrm{T}} \int_{\Omega_t} \boldsymbol{\xi}(t)\boldsymbol{\xi}^{\mathrm{H}}(t)dt \\
&= \mathsf{r}^{\mathrm{T}}\mathbf{I}_n = \mathsf{r}^{\mathrm{T}}
\end{aligned}$$

and therefore,

$$\begin{aligned}
\mathsf{r} &= \int_{\Omega_t} \mathsf{r}(t)\boldsymbol{\xi}^*(t)dt = \mathbf{s}(\mathbf{p}) + \mathsf{n} \\
\mathbf{s} &= \int_{\Omega_t} s(t)\boldsymbol{\xi}^*(t)dt = \sqrt{E_s}\mathbf{f}(\mathbf{p}) = \sqrt{E_s}\int_{\Omega_t} f(\mathbf{p},t)\boldsymbol{\xi}^*(t)dt \\
\mathsf{n} &= \int_{\Omega_t} \mathsf{n}(t)\boldsymbol{\xi}^*(t)dt. \qquad (3.30)
\end{aligned}$$

The individual data coefficients are computed from the scalar relationship

$$\mathsf{r}_i = \int_{\Omega_t} \mathsf{r}(t)\xi_i^*(t)dt \qquad (3.31)$$

as likewise are the signal s_i and noise n_i coefficients. Since they are linearly related to the random data, the coefficients are Gaussian with $\mathcal{E}\{\mathsf{r}_i|H_1\} = s_i$, $\mathcal{E}\{\mathsf{r}_i|H_0\} = 0$, covariance (3.20), and pdf (3.21).

The continuous data has been reduced to a countable set of random variables without any particular specification for the orthonormal set of functions. The choice for example, could be based upon some measure of representational efficiency such that fewer terms would be required to achieve a particular average error (3.29). Another option would be to choose the functions so as to expedite the implementation of the likelihood ratio test. Under a Gaussian assumption, the test is given by (3.23) and involves the inversion of the covariance matrix (or an estimate of it) which is a numerically intense operation. This computation would be considerably simplified if the functions $\boldsymbol{\xi}(t)$ could be specified such that the random coefficients in the expansion are uncorrelated and hence independent rendering the covariance matrix (3.20) diagonal. Since under both H_1 and H_0, the covariance is that of the interference alone, the condition sought is that $\mathbf{R}_n = \boldsymbol{\lambda}$ which is a diagonal matrix of real constants such that

$$R_{nij} = \mathcal{E}\{\mathsf{n}_i^*\mathsf{n}_j\} = \lambda_i\delta_{ij}, \quad i,j = 1,\cdots n.$$

Using (3.20,3.30), the covariance is expressed as

$$\begin{aligned}
\mathbf{R}_n = \mathcal{E}\{\mathsf{n}\mathsf{n}^{\mathrm{H}}\} &= \mathcal{E}\left\{\int_{\Omega_t}\int_{\Omega_u} \mathsf{n}(u)\boldsymbol{\xi}^*(u)\mathsf{n}^*(t)\boldsymbol{\xi}^{\mathrm{T}}(t)dudt\right\} \\
&= \int_{\Omega_u}\left[\int_{\Omega_t} R_n(t,u)\boldsymbol{\xi}(t)dt\right]\boldsymbol{\xi}^{\mathrm{H}}(u)du = \boldsymbol{\lambda} \qquad (3.32)
\end{aligned}$$

and invoking the orthonormality condition (3.27) results in

$$\boldsymbol{\lambda}\int_{\Omega_u}\boldsymbol{\xi}(u)\boldsymbol{\xi}^{\mathrm{H}}(u)du=\int_{\Omega_u}\left[\boldsymbol{\lambda}\boldsymbol{\xi}(u)\right]\boldsymbol{\xi}^{\mathrm{H}}(u)du=\boldsymbol{\lambda}.$$

The required relationship specifying the function set is obtained by equating the terms within the square brackets yielding

$$\int_{\Omega_t}R_n(t,u)\boldsymbol{\xi}(t)dt=\boldsymbol{\lambda}\boldsymbol{\xi}(u).$$

Each component function is thus related to the covariance by a scalar integral equation

$$\int_{\Omega_t}R_n(t,u)\xi_i(t)dt=\lambda_i\xi_i(u)\quad\forall i$$

and if the $\xi(t)$ can be chosen satisfying this, it is assured that the Gaussian interference components n_i will be uncorrelated and hence independent. The inverse of the covariance required to evaluate the log likelihood is now much more easily computed since

$$\mathbf{R}_n^{-1}=\boldsymbol{\lambda}^{-1}=\begin{bmatrix}1/\lambda_1 & \cdots & 0\\ \vdots & \ddots & \vdots\\ 0 & \cdots & 1/\lambda_n\end{bmatrix}.$$

This is called a *Karhunen-Loéve* (K-L) expansion and has been widely applied in image processing [17], pattern recognition [18], and signal and array processing [1, 19 21]. The constants λ_i and the functions $\xi_i(t)$ are respectively called the eigenvalues and corresponding eigenfunctions of the interference covariance $R_n(t,u)$. An alternative procedure constructs a K-L expansion directly from an arbitrary CON set by utilizing the fact that the covariance is diagonalizable [12].

The covariance function is shown to be expressible in terms of the eigenvalues and eigenfunctions of a K-L expansion by approximating the continuous interference process by the first $n<\infty$ terms of a K-L expansion to form $\tilde{\mathsf{n}}(t)=\mathbf{n}^{\mathrm{T}}\boldsymbol{\xi}(t)$. The covariance function of the truncated expansion is

$$\begin{aligned}\tilde{R}_n(t,u)&=\mathcal{E}\left\{\boldsymbol{\xi}^{\mathrm{H}}(t)\mathbf{n}^*\mathbf{n}^{\mathrm{T}}\boldsymbol{\xi}(u)\right\}\\&=\boldsymbol{\xi}^{\mathrm{H}}(t)\left(\mathcal{E}\left\{\mathbf{n}^*\mathbf{n}^{\mathrm{T}}\right\}\right)\boldsymbol{\xi}(u).\end{aligned}$$

The covariance matrix of the truncated series is Hermitian with real eigenvalues hence $\mathcal{E}\{\mathbf{n}^*\mathbf{n}^{\mathrm{T}}\} = \mathcal{E}\{(\mathbf{n}^*\mathbf{n}^{\mathrm{T}})^*\} = \mathcal{E}\{\mathbf{n}\mathbf{n}^{\mathrm{H}}\}$ and

$$\tilde{R}_n(t,u) = \boldsymbol{\xi}^{\mathrm{H}}(t)\boldsymbol{\lambda}\boldsymbol{\xi}(u) \quad (n < \infty)$$

by (3.32). *Mercer's theorem* [1,22] states that convergence is uniform as $n \to \infty$ for a positive semi-definite (or definite) covariance kernel and

$$\lim_{n\to\infty} \tilde{R}_n(t,u) = R_n(t,u) = \boldsymbol{\xi}^{\mathrm{H}}(t)\boldsymbol{\lambda}\boldsymbol{\xi}(u) \tag{3.33}$$

where $\boldsymbol{\lambda}$ becomes a countably infinite diagonal eigenvalue matrix.

This result is used to show that the K-L expansion itself is mean square convergent (3.29) by defining the error for finite n as

$$\begin{aligned} e_n &= \mathcal{E}\left\{|\mathsf{r}(t) - \mathbf{r}^{\mathrm{T}}\boldsymbol{\xi}(t)|^2\right\} \\ &= \mathcal{E}\left\{|\mathsf{r}(t)|^2 - 2\Re\left[\mathsf{r}(t)\mathbf{r}^{\mathrm{H}}\boldsymbol{\xi}^*(t)\right] + \boldsymbol{\xi}^{\mathrm{H}}(t)\mathbf{r}\mathbf{r}^{\mathrm{H}}\boldsymbol{\xi}(t)\right\} \quad n < \infty. \end{aligned}$$

Substituting (3.26) for $\mathsf{r}(t)$ and using (3.19) assuming that the echo is present (H_1), yields

$$e_n = \mathbf{R}_n(t,t) - 2\Re\left[\boldsymbol{\xi}^{\mathrm{H}}(t)\boldsymbol{\lambda}\boldsymbol{\xi}(t)\right] + \left[\boldsymbol{\xi}^{\mathrm{H}}(t)\boldsymbol{\lambda}\boldsymbol{\xi}(t)\right] + |s(t)|^2 - |\mathbf{s}^{\mathrm{T}}\boldsymbol{\xi}(t)|^2.$$

The quadratic form is real and completeness of the expansion (3.28) is assumed so that in the limit, the signal terms drop out. The equivalence of the covariance terms is guaranteed by Mercer's theorem (3.33) thus

$$\lim_{n\to\infty} e_n = 0$$

and the K-L expansion is mean square convergent. Note that this result does not depend upon the conditions H_1 or H_0.

Under H_1, the total energy in the data is

$$\begin{aligned} E_r &= \mathcal{E}\left\{\int_{\Omega_t} \mathsf{r}^*(t)\mathsf{r}(t)dt\right\} \\ &= E_s + \int_{\Omega_t} R_n(t,t)dt \end{aligned}$$

where the interference energy is equivalently written as

$$E_n = \int_{\Omega_t} R_n(t,t)dt = \int_{\Omega_t} \boldsymbol{\xi}^{\mathrm{H}}(t)\boldsymbol{\lambda}\boldsymbol{\xi}(t)dt = \sum_{i=1}^{\infty}\lambda_i \int_{\Omega_t} \boldsymbol{\xi}_i^{\mathrm{H}}(t)\boldsymbol{\xi}_i(t)dt.$$

The $\xi_i(t)$ are orthonormal (3.27) consequently,

$$E_n = \sum_{i=1}^{\infty} \lambda_i = T_r[\boldsymbol{\lambda}] = \mathcal{E}\left\{\sum_{i=1}^{\infty} |\mathsf{n}_i|^2\right\} \tag{3.34}$$

where $T_r[\,]$ represents the trace or sum of the diagonal terms of a matrix. The λ_i represent the interference energy associated with the corresponding $\xi_i(t)$ $0 \le t \le T$ and likewise, since the function set is assumed to be complete,

$$E_s = \sum_{i=1}^{\infty} |s_i|^2 . \tag{3.35}$$

The equivalence of the energy computed via a time integration and that obtained from the summation of the expansion coefficients is merely a general statement of *Parseval's Theorem* [1, 16] that requires only that the expansion be complete, not necessarily a K-L expansion. The concept is perhaps most generally familiar in spectral analysis where the signal energies obtained by integrating in the time and frequency domains are equivalent [23]. The K-L expansion results in a simplification of the likelihood ratio (3.22) expressed using (3.32) as

$$\begin{aligned} \ell_r(\mathbf{r}) &= 2\Re\left[\mathbf{r}^{\mathrm{H}}\boldsymbol{\lambda}^{-1}\mathbf{s}\right] - \mathbf{s}^{\mathrm{H}}\boldsymbol{\lambda}^{-1}\mathbf{s} \\ &= 2\sqrt{E_s}\Re\left[\mathbf{r}^{\mathrm{H}}\mathbf{g}\right] - E_s\mathbf{f}^{\mathrm{H}}\mathbf{g} \\ \mathbf{g} &= \boldsymbol{\lambda}^{-1}\mathbf{f} \end{aligned} \tag{3.36}$$

where $\mathbf{g}$ is called the processing vector for correlated interference. From (3.19, 3.30, 3.34, 3.35) this is equivalent to

$$\ell_r(\mathsf{r}) = \begin{cases} 2\Re\sum_{i=1}^{\infty}\frac{\mathsf{n}_i^* s_i}{\lambda_i} + \sum_{i=1}^{\infty}\frac{|s_i|^2}{\lambda_i}, & \text{under } H_1, \\ 2\Re\sum_{i=1}^{\infty}\frac{\mathsf{n}_i^* s_i}{\lambda_i} - \sum_{i=1}^{\infty}\frac{|s_i|^2}{\lambda_i}, & \text{under } H_0. \end{cases}$$

This is a Gaussian random quantity with an expected value equal to plus or minus the sum of the component SIRs depending whether or not the signal is present. The implementation of the likelihood receiver is straightforward since the signal and the noise covariance are assumed to be known beforehand. It is only necessary to generate a replica of the signal function with the known parameter set $\mathbf{p}$, compute the K-L coefficients of the replica and the data, form the requisite inner products to form (3.22), and compare with a suitable threshold γ_l.

The completely known signal problem is the simplest situation and corresponds to the transmission and subsequent detection of a known message over a non-dispersive channel with known delay and Doppler. In an echo-location context, the fundamental form of the signal may be known but usually, the delay and Doppler parameters are not and the problem involves both detection and parameter estimation. An approach to the solution of this joint problem is the implementation of the detector as a bank of likelihood receivers performing multiple parameter hypotheses as discussed in Section 3.2. The performance of the receiver bank as a detector of the correct echo will depend upon the relative response of a likelihood receiver when mismatched to the data parameters whether or not the parameters take on a continuous or discrete set of values. The maximization of the likelihood as a function of the parameter hypothesis is often called the *generalized likelihood problem* [24, 25].

To examine this concept, it is convenient to emphasize the parameter dependence by denoting the K-L coefficients of the signal time series $s(\mathbf{p}, t)$ as the vector $\mathbf{s}(\mathbf{p})$ and denote by $\hat{\mathbf{p}}$ the parameters hypothesized by a particular receiver in the receiver bank. As indicated in (3.36), the expected response of the receiver making the correct parameter hypothesis under H_1 is merely the sum of the component signal to interference ratios. This receiver will have the maximum expected response, all others hypothesizing incorrect parameters will, on the average, have a lesser response. This is shown by considering the response in the receiver hypothesizing $\hat{\mathbf{p}}$ when the data is parameterized by $\mathbf{p}$. The expected log likelihood ratio under H_1 dependent upon signal parameters $\mathbf{p}$ and replica parameters $\hat{\mathbf{p}}$ is

$$\begin{aligned}\bar{\ell}_r(\mathbf{p}, \hat{\mathbf{p}}) &= 2\Re\left[\mathbf{s}^{\mathrm{H}}(\hat{\mathbf{p}})\boldsymbol{\lambda}^{-1}\mathbf{s}(\mathbf{p})\right] - \mathbf{s}^{\mathrm{H}}(\hat{\mathbf{p}})\boldsymbol{\lambda}^{-1}\mathbf{s}(\hat{\mathbf{p}}) \\ &= 2\left[\Re\left\{\mathbf{s}^{\mathrm{H}}(\mathbf{p})\mathbf{g}(\hat{\mathbf{p}})\right\} - \frac{1}{2}\mathbf{s}^{\mathrm{H}}(\hat{\mathbf{p}})\mathbf{g}(\hat{\mathbf{p}})\right]\end{aligned} \tag{3.37}$$

which reduces to

$$\bar{\ell}_r(\mathbf{p}, \mathbf{p}) = \mathbf{s}^{\mathrm{H}}(\mathbf{p})\boldsymbol{\lambda}^{-1}\mathbf{s}(\mathbf{p}) = \mathbf{s}^{\mathrm{H}}(\mathbf{p})\mathbf{g}(\mathbf{p}). \tag{3.38}$$

when $\hat{\mathbf{p}} = \mathbf{p}$. The weighted square difference between $\mathbf{s}(\hat{\mathbf{p}})$ and $\mathbf{s}(\mathbf{p})$ is the positive semi-definite quadratic form

$$0 \leq \left(\mathbf{s}(\hat{\mathbf{p}}) - \mathbf{s}(\mathbf{p})\right)^{\mathrm{H}} \boldsymbol{\lambda}^{-1} \left(\mathbf{s}(\hat{\mathbf{p}}) - \mathbf{s}(\mathbf{p})\right)$$

which expands to

$$0 \leq \left[\mathbf{s}^{\mathrm{H}}(\hat{\mathbf{p}})\boldsymbol{\lambda}^{-1}\mathbf{s}(\hat{\mathbf{p}}) + \mathbf{s}^{\mathrm{H}}(\mathbf{p})\boldsymbol{\lambda}^{-1}\mathbf{s}(\mathbf{p})\right] - 2\Re\left[\mathbf{s}^{\mathrm{H}}(\hat{\mathbf{p}})\boldsymbol{\lambda}^{-1}\mathbf{s}(\mathbf{p})\right].$$

Rearranging,

$$\begin{aligned} 2\Re\left[\mathbf{s}^{\mathrm{H}}(\hat{\mathbf{p}})\boldsymbol{\lambda}^{-1}\mathbf{s}(\mathbf{p})\right] - \mathbf{s}^{\mathrm{H}}(\hat{\mathbf{p}})\boldsymbol{\lambda}^{-1}\mathbf{s}(\hat{\mathbf{p}}) &\leq \mathbf{s}^{\mathrm{H}}(\mathbf{p})\boldsymbol{\lambda}^{-1}\mathbf{s}(\mathbf{p}) \\ \Rightarrow \bar{\ell}_r(\mathbf{p},\hat{\mathbf{p}}) &\leq \bar{\ell}_r(\mathbf{p},\mathbf{p}) \end{aligned}$$

with equality only for $\hat{\mathbf{p}} = \mathbf{p}$. This proves that the expected response is maximized in the receiver hypothesizing the correct parameters. Note that (3.38) may be written as

$$\bar{\ell}_r(\mathbf{p},\mathbf{p}) = \sum_{i=1}^{\infty} \frac{|s_i(\mathbf{p})|^2}{\lambda_i}$$

where the signal vector components differ depending upon the particular parameters engendered by the range and velocity of the scatterer. Consequently, the expected peak response will generally vary with the parameters. In other words $\bar{\ell}_r(\mathbf{p},\mathbf{p}) \neq \bar{\ell}_r(\hat{\mathbf{p}},\hat{\mathbf{p}})$ even if the total signal and interference energies are the same. It is not necessarily true that a receiver hypothesizing $\hat{\mathbf{p}}$ will have a greater expected response when presented with data parameterized by $\hat{\mathbf{p}}$ than it would when presented with data parameterized by $\mathbf{p}$. In other words its expected response to the wrong parameters $\bar{\ell}_r(\mathbf{p},\hat{\mathbf{p}})$ is not guaranteed to be less than its response to the right ones $\bar{\ell}_r(\hat{\mathbf{p}},\hat{\mathbf{p}})$. On the other hand, since $\bar{\ell}_r(\hat{\mathbf{p}},\mathbf{p}) \leq \bar{\ell}_r(\hat{\mathbf{p}},\hat{\mathbf{p}})$, its expected response to the correct parameters $\hat{\mathbf{p}}$ is guaranteed to be greater than that of any other receiver presented with $\hat{\mathbf{p}}$. This means that a single receiver "tuned" to particular fixed set of parameters could not be used to sift incoming data with the object of identifying an echo or message with the correct parameters. To do so would imply that its response would necessarily be maximized by the correct parameters in comparison to its response to all other data. Conversely, it is possible to sift incoming data with a tunable receiver or bank of receivers with fixed tuning and to identify the correct data by a comparison of the responses of each receiver much like the tuning of a radio. This is consistent with the concept of ML estimation discussed in Section 3.2.

All of this is a consequence of there being a correlated interference component as indicated in (3.24). If only uncorrelated (white) noise is present, the peak expected response will be independent of parameter variations for constant energy signals since

$$\bar{\ell}_r(\mathbf{p},\mathbf{p}) = \frac{1}{N_0} \sum_{i=1}^{\infty} |s_i(\mathbf{p})|^2$$

which is simply the signal to interference ratio E_s/E_n. The first term of the expected value of the likelihood $\bar{\ell}_r(\hat{\mathbf{p}},\mathbf{p})$ (3.37) is equivalent to $\sqrt{E_s}\,[2\Re\{\mathbf{f}^{\mathrm{H}}(\mathbf{p})\mathbf{g}(\hat{\mathbf{p}})\}]$. The square bracketed quantity is one form of a normalized *ambiguity function* (AF) [6,26,27] so called because the transmit function is assumed normalized to unit energy $\mathbf{f}(\mathbf{p})^{\mathrm{H}}\mathbf{f}(\mathbf{p}) = 1\ \forall\mathbf{p}$. As will be discussed later, in one form or another, the AF plays a central role in the design and analysis of echo-location systems. It is crucial to the design of waveforms with respect to the specification of their resolution properties for parameter estimation, and for the minimization of interference.

With this in mind, it is illuminating to derive representations of the likelihood and AF in terms of the continuous data and waveform time functions rather than in terms of the K-L coefficients. Substituting the K-L vector definitions (3.30) into the vector form of the likelihood function (3.36) results in

$$\ell_r(\mathbf{r}) = 2\Re\left[\int_{\Omega_u} \mathsf{r}(u)g^*(\hat{\mathbf{p}},u)du\right] - \int_{\Omega_u} s(\hat{\mathbf{p}},u)g^*(\hat{\mathbf{p}},u)du \qquad (3.39)$$

where the processing function

$$g(\hat{\mathbf{p}},u) = \int_{\Omega_t} s(\hat{\mathbf{p}},t)Q_n(t,u)dt = \mathbf{g}^{\mathrm{T}}(\hat{\mathbf{p}})\boldsymbol{\xi}(u)$$

is the continuous equivalent of the processing vector and

$$Q_n(t,u) = \boldsymbol{\xi}^{\mathrm{H}}(t)\lambda^{-1}\boldsymbol{\xi}(u) = Q_n^*(u,t) \qquad (3.40)$$

is called the inverse kernel. It is so called because it bears an inverse relationship to the covariance

$$\int_{\Omega_t} R_n(z,u)Q_n(u,t)du = \delta(z-t) = \text{(Dirac delta)}$$

that is shown by rewriting it using (3.33, 3.40) and invoking the K-L orthonormality conditions. The equivalent likelihood ratio test (3.23) is

$$\Re\int_{\Omega_u} \mathsf{r}(u)g^*(\hat{\mathbf{p}},u)du \underset{H_0}{\overset{H_1}{\gtrless}} \frac{1}{2}\left(\gamma_l + \int_{\Omega_u} s(\hat{\mathbf{p}},u)g^*(\hat{\mathbf{p}},u)du\right) = \gamma_l'$$

and the modified threshold is a function of the parameter hypothesis. This is not surprising since the interference contains a correlated component n_c that may vary with both time and frequency. Consequently the receiver corresponding to a particular parameter hypothesis must be adjusted to reflect the local signal to interference ratio. This is done by modifying the threshold.

The expected likelihood receiver response $\bar{\ell}_r$ equivalent to (3.37) and the associated wideband AF is expressed in terms of the normalized transmitted waveform $f(t)$ and the signal model (2.26) by defining

$$s(\mathbf{p},t) = \sqrt{E_s}f(\mathbf{p},t) = \sqrt{E_s}\sqrt{\beta}f(\beta(t-\tau))$$

and evaluating the signal only portion of the likelihood. This results in

$$\begin{aligned} \bar{\ell}_r(\mathbf{p},\hat{\mathbf{p}}) = 2\sqrt{E_s} & \left[\Re \int_{\Omega_t} f(\mathbf{p},t)g^*(\hat{\mathbf{p}},t)dt\right] \\ & -\sqrt{E_s}\int_{\Omega_t} f(\hat{\mathbf{p}},t)g^*(\hat{\mathbf{p}},t)dt \end{aligned} \tag{3.41}$$

where the bracketed term is the normalized AF and the processing function is related to the transmitted waveform by

$$g(\hat{\mathbf{p}},u) = \sqrt{E_s}\int_{\Omega_t} f(\hat{\mathbf{p}},t)Q_n(t,u)dt. \tag{3.42}$$

It is possible to derive an alternate expression for the processing function by using the inverse relation between Q_n and R_n and writing

$$\begin{aligned} \int_{\Omega_u} R_n(z,u)g^*(\hat{\mathbf{p}},u)du &= \int_{\Omega_t}\int_{\Omega_u} R_n(z,u)Q_n(u,t)s^*(\hat{\mathbf{p}},t)dudt \\ &= \int_{\Omega_t} s^*(\hat{\mathbf{p}},t)\int_{\Omega_u} R_n(z,u)Q_n(u,t)dudt \\ &= \int_{\Omega_t} s^*(\hat{\mathbf{p}},t)\delta(z-t)dt \\ &= s^*(\hat{\mathbf{p}},z) = \sqrt{E_s}f^*(\hat{\mathbf{p}},z). \end{aligned}$$

$g(\hat{\mathbf{p}},u)$ is thus the solution to the integral equation

$$\sqrt{E_s}f(\hat{\mathbf{p}},t) = \int_{\Omega_u} R_n(u,t)g(\hat{\mathbf{p}},u)du \tag{3.43}$$

and if there is no correlated component in the interference process such that $R_c(t,u) = 0$ in (3.24), the process is white Gaussian noise (WGN) and $R_n(u,t) = N_0\delta(u-t)$. Substituting this into the integral equation results in

$$g(\hat{\mathbf{p}},t) = \frac{\sqrt{E_s}}{N_0}f(\hat{\mathbf{p}},t).$$

This is a significant result as it means that when the interference is WGN, the maximum likelihood receiver (3.39) for a deterministic signal forms the statistic

$$\ell_r(\mathsf{r}) = \frac{\sqrt{E_s}}{N_0}\left[2\Re\int_{\Omega_u} \mathsf{r}(u) f^*(\hat{\mathbf{p}}, u) du\right] - \frac{E_s}{N_0} \tag{3.44}$$

which is a form of what is often called a *replica correlator* or *matched filter* because it processes the data time series with a parameterized replica or match of the known signal waveform [3, 4, 6, 28]. Conversely, the more general receiver (3.39) is often called the optimal *mismatched filter*. The equivalent test in WGN is

$$\Re\int_{\Omega_u} \mathsf{r}(u) f^*(\hat{\mathbf{p}}, u) du \mathop{\gtrless}_{H_0}^{H_1} \frac{1}{2}\left(\frac{\gamma_l + \frac{E_s}{N_0}}{\frac{\sqrt{E_s}}{N_0}}\right) = \gamma_l'$$

and the threshold is only a function of the signal and noise energies, not of the particular parameter hypothesis. Consequently, the same threshold may be applied in each receiver. This is a consequence of the white noise assumption which implies that the interference is uniform in time and frequency and the signal will be competing with the same noise energy for all hypotheses. The expected matched receiver response is

$$\begin{aligned}\bar{\ell}_r(\mathbf{p}, \hat{\mathbf{p}}) &= \frac{E_s}{N_0}\left[2\Re\int_{\Omega_t} f(\mathbf{p}, t) f^*(\hat{\mathbf{p}}, t) dt - 1\right] \\ &= \frac{E_s}{N_0}, \text{ when } \hat{\mathbf{p}} = \mathbf{p}\end{aligned}$$

which is the white noise equivalent of (3.37) and the same as was obtained from the discrete K-L formulation.

Under the condition of deterministic scattering in WGN, the normalized ambiguity function takes the particular form of the first term within the brackets. Since both functions in the integrand are the same, it is usually more precisely defined as an *auto* ambiguity function (AAF) denoted by $\Phi_{ff}(\mathbf{p}, \hat{\mathbf{p}})$. An ambiguity function involving different functions as in the discrete form (3.37) or continuous form (3.41) is called a *cross* ambiguity functions (CAF) and denoted by $\Phi_{fg}(\mathbf{p}, \hat{\mathbf{p}})$. The behavior of an AF as a function of its parameters is examined by expressing it in terms of the actual normalized bandpass waveforms (2.26) where, assuming the wideband model

$$\Phi_{ff}(\mathbf{p},\hat{\mathbf{p}}) = \Re(\chi_{ff}(\mathbf{p},\hat{\mathbf{p}}))$$
$$\chi_{ff}(\mathbf{p},\hat{\mathbf{p}}) = \int_{\Omega_t} f(\mathbf{p},t)f^*(\hat{\mathbf{p}},t)dt \quad (3.45)$$

and

$$f(\mathbf{p},t) = \sqrt{\beta}f(\beta(t-\tau)).$$

The inner product $\chi_{ff}(\mathbf{p},\hat{\mathbf{p}})$ is called the *uncertainty function* (UF) and is defined separately because the ambiguity function while always functionally dependent upon χ, can take different forms. Clearly when $\hat{\mathbf{p}} = \mathbf{p}$, $\Phi_{ff}(\mathbf{p},\mathbf{p}) = 1$ because the waveforms are normalized. $f(t)$ is defined for $t \in [0,T]$ and is zero elsewhere thus the echo will only exist for $t \in [\tau, \tau + T/\beta]$ and if for example the delay hypothesis is $\hat{\tau} < \tau$ the integration support $\Omega_t = [\tau, \hat{\tau} + T/\hat{\beta}]$. If $(\hat{\tau} + T/\hat{\beta}) < \tau$, the echo and the processor will be disjoint and no signal output will be observed.

Example 3.3 The AF of a CW Waveform

The sinusoidal (CW) waveform function $f(t) = a(t)\exp j\omega_0 t$ with $a(t) = \sqrt{1/T}$ and pulse length T provides an example of AF behavior and consequently that of the expected likelihood ratio. Its auto-AF is

$$\Phi_{ff}(\mathbf{p},\hat{\mathbf{p}}) =$$
$$\Re\left[\exp -j\omega_0(\beta\tau - \hat{\beta}\hat{\tau})\frac{\sqrt{\beta\hat{\beta}}}{T}\int_{\tau}^{\hat{\tau}+T/\hat{\beta}} \exp\ j\omega_0 \triangle \beta t dt\right] \quad (3.46)$$

with $\Delta\beta = \beta - \hat{\beta}$. Assume that the correct Doppler hypothesis is made such that $\Delta\beta = \beta - \hat{\beta} = 0$ and that the echo and processor overlap in time with $\hat{\tau} \leq \tau$. The response along the delay axis as $\hat{\tau}$ is varied is

$$\Phi_{ff}(0,\Delta\tau) = \left(1 - \frac{\beta(\Delta\tau)}{T}\right)\cos\omega_0\beta\Delta\tau \quad (3.47)$$

where the notation $\Phi_{ff}(\Delta\beta,\Delta\tau)$ is adopted when the actual parameters are specified. The bracketed term is a linear attenuation that is unity at the origin and approaches zero as $\Delta\tau = \tau - \hat{\tau}$ approaches the compressed pulse duration at which point the echo and processor become disjoint. The oscillatory term is also unity at the origin but has local maxima at delays corresponding to integral multiples of the wavelength. Clearly only the maximum occurring at the origin is the true global maximum owing to the effect of the linear attenuation term. Similarly, the behavior of Φ in the Doppler direction ($\triangle\tau = 0$) exhibits the same oscillatory effect behavior with a period $\triangle\beta \approx \pi/\omega_0 T$. The implication of this is that in considering the bank of receivers each hypothesizing a different set of parameters, it would be necessary to use a much finer grid than might otherwise be deemed necessary to adequately localize the scatterer. If in the process of testing multiple parameter hypotheses a maximum were found, it would be necessary to examine adjacent maxima to ascertain the true or global maximum and hence the ML estimate. This reveals a fundamental shortcoming of the deterministic scattering model. It results in a form of the likelihood receiver that is phase sensitive and thus sensitive to parameter variations occurring on a wavelength scale.

Such precision is not required for most echo-location applications and in any case nearly impossible to achieve except under the most highly controlled conditions. This is because; 1. true point like scattering can be physically justified only if echo reflection occurs from a region with delay extent that is small with respect to the inverse waveform bandwidth, and 2. echoes often propagate through a random non-homogeneous medium in which the refractive index varies with time and location and there are multiple propagation paths sometimes involving secondary scattering. Such conditions are typically observed in sonar and air acoustics and in radar propagation through the ionosphere. The combination of non point-like scattering and random propagation introduces random time varying amplitude and phase variations in the echo signal. Consequently, the phase sensitive likelihood detector formulated under a deterministic signal hypothesis cannot be used for most echo-location applications. It is thus necessary to consider the more realistic random scattering model and its associated detection and estimation strategy.

3.5 SLOWLY FLUCTUATING POINT SCATTERING

3.5.1 White Gaussian Noise (WGN)

Deterministic scattering and echo propagation is an idealistic view of nature that may not be supported by prevailing echo formation and propagation conditions. If the echo signal is regarded as a random Gaussian process it is conveniently expressed as a vector $\mathbf{r}$ with expectation $\mathcal{E}\{\mathbf{r}\} = 0$ and covariance $\mathcal{E}\{\mathbf{r}\mathbf{r}^{\mathrm{H}}\} = \mathbf{R}_s$. The data pdf's under the binary hypotheses are

$$f(\mathbf{r}|H_k) = [\pi^n |\mathbf{R}_k|]^{-1} \exp -(\mathbf{r})^{\mathrm{H}}\mathbf{R}_k{}^{-1}(\mathbf{r}), \quad k = 0, 1$$

with

$$\mathcal{E}\{\mathbf{r}\mathbf{r}^{\mathrm{H}}\} = \begin{cases} \mathbf{R}_1 = \mathbf{R}_s + \mathbf{R}_n & \text{for } H_1, \\ \mathbf{R}_0 = \mathbf{R}_n & \text{for } H_0. \end{cases}$$

The llr corresponding to that of the non-random echo (3.22) is

$$\begin{aligned} \ell_r(\mathbf{r}) &= \mathbf{r}^{\mathrm{H}}(\mathbf{R}_0^{-1} - \mathbf{R}_1^{-1})\mathbf{r} - (\ln|\mathbf{R}_1| - \ln|\mathbf{R}_0|) \\ &= \ell_d + \eta \end{aligned}$$

and the data dependent term is

$$\begin{aligned} \ell_d(\mathbf{r}) &= (\mathbf{R}_s(\mathbf{R}_s + \mathbf{R}_0)^{-1}\mathbf{r})^{\mathrm{H}}\mathbf{R}_0^{-1}\mathbf{r} \\ &= \hat{\mathbf{s}}^{\mathrm{H}}\mathbf{R}_0^{-1}\mathbf{r}, \quad (\mathbf{R}_0 = \mathbf{R}_n). \end{aligned} \tag{3.48}$$

$\hat{\mathbf{s}}$ is an estimate of the signal vector that is derived from the random data and prior knowledge of the covariances. Except for the taking of the real part, the implementation of ℓ_d is of the same form as that for the known non-random echo (3.23). It is known as the *estimator-correlator* (EC) detector [29] because the random echo signal is estimated from the received data and then correlated the whitened data. The EC is applicable to the detection of echos that are spread in delay and/or Doppler and is discussed in detail in Chapter 6.

The vector implementation is reduced to a scalar process when the scattering is highly localized and involves multiple reflections occurring over a delay spread that is small in comparison to the inverse waveform bandwidth. It is consistent with specular scattering from large structures and the spreading effect often generated by inhomogeneities and multiple propagation paths in the ocean medium. Moreover because of relative motion, changes in geometry, and temporal variations in the medium boundaries, the contributions of individual scatterers and propagation paths may be considered to be random.

When a large number of random slightly delayed echoes are received, the signal is modeled as having been generated by a slowly fluctuating point source that preserves the fundamental form of the transmitted waveform but introduces a random component as a multiplier [2,6]. The data time series is modeled as

$$\mathsf{r}(t) = \sqrt{E_0}\mathcal{S}_{sp}f(\mathbf{p},t) + \mathsf{n}(t) = \mathsf{s}(t) + \mathsf{n}(t) \tag{3.49}$$

where $\mathcal{S}_{sp}$ is called the signal point *spreading function.* It is a complex zero mean Gaussian variable with $\mathcal{E}\{|\mathcal{S}_{sp}|^2\} = \mathcal{R}_{sp}$ that is called the signal point *scattering function.* $\mathcal{S}_{sp}$ is assumed to be independent of the interference, constant over an illumination interval, and independent between interrogations (pings). In other words it does not change within a time period commensurate with the interrogation interval but changes randomly with each transmission. This is called the slowly fluctuating point scatter model [6]. $\mathcal{R}_{sp}$ represents the combined point echo scattering and channel attenuation such that the received signal and transmitted energies E_s and E_0 are related by $E_s = \mathcal{R}_{sp}E_0$. Random fluctuation that occurs on a shorter time scale or scattering from more than one point or a continuum of points results in an echo that is spread in delay and/or Doppler. In other words, the echo duration and bandwidth may be greater than those of the transmitted waveform. Spread scattering is described by a more general model represented by parameter dependent spreading and scattering functions denoted as $\mathcal{S}_s(\mathbf{p})$ and $\mathcal{R}_s(\mathbf{p})$ respectively.

It will be initially assumed that the interference is additive WGN with spectral density N_0 and the time series is expanded using a CON set of functions as indicated in (3.30). This time a slightly different approach is taken noting that when WGN is assumed for the deterministic model, the likelihood receiver reduces to the matched filter (3.44). The integral term is of the same form as the defining relation for a data coefficient (3.31) using the normalized waveform as one member of the orthonormal set. Therefore the CON set is chosen such that the first of the functions $\xi_i(t)$ is the normalized waveform

$$\xi_1(t) = f(\mathbf{p},t) \tag{3.50}$$

with the rest chosen to be orthonormal to it such that

$$\int_{\Omega_t} f(\mathbf{p},t)\xi_i^*(t)dt = 0 \quad for\ i \neq 1.$$

The first data coefficient is

$$\mathsf{r}_1 = \int_{\Omega_t} \mathsf{r}(t)\xi_1^*(t)dt = \int_{\Omega_t} (\mathsf{s}(t) + \mathsf{n}(t))f^*(\mathbf{p},t)dt \tag{3.51}$$

and

$$\mathsf{r}_1 = \begin{cases} \sqrt{E_0}\mathcal{S}_{sp} + \mathsf{n}_1 & \text{under } H_1, \\ \mathsf{n}_1 & \text{under } H_0. \end{cases}$$

$\mathsf{r}_i = \mathsf{n}_i$ under H_1 and H_0 for all of the other data coefficients with $(i > 1)$ because of the assumed orthonormality of the waveform with all other members of the CON set. Only the first term reflects the presence or absence of the echo in the time series when the expansion is formed in this way. All of the other terms are independent of its existence implying that r_1 alone is required for detection. Note that the data time series may be written as

$$\begin{aligned} \mathsf{r}(t) &= \mathsf{r}_1 f(\mathbf{p}, t) + \sum_i \mathsf{r}_i \xi_i(t) \quad i > 1 \\ &= \mathsf{r}_1 f(\mathbf{p}, t) + \mathsf{u}(t) \end{aligned} \tag{3.52}$$

where $\mathsf{u}(t)$ contains all of the other expansion terms. It is also uncorrelated with r_1 since

$$\begin{aligned} \mathcal{E}\{\mathsf{r}_1\mathsf{r}_i^*\} &= \mathcal{E}\{\mathsf{r}_1\mathsf{n}_i^*\} \\ &= \mathcal{E}\{(\mathsf{s}_1 + \mathsf{n}_1)\mathsf{n}_i^*\} \\ &= \mathcal{E}\{\mathsf{n}_1\mathsf{n}_i^*\} \\ &= \int_{\Omega_t}\int_{\Omega_u} \mathcal{E}\{\mathsf{n}(t)\mathsf{n}^*(u)\} f^*(\mathbf{p}, t)\xi_i(u) du dt \\ &= N_0 \int_{\Omega_t} f^*(\mathbf{p}, t)\xi_i(t) dt = 0. \end{aligned}$$

r_1 alone reflects the presence of the signal and is statistically uncorrelated with $\mathsf{u}(t)$. Hence, it is sufficient for detection and is defined as a *sufficient statistic* [1,3].

The Gaussian pdf for a vector process is given by (3.21) but the vector is reduced to a scalar as only one member of the CON set (r_1) is necessary to characterize the data. In addition, both the interference and signal are random and $\mathcal{E}\{\mathsf{r}_1\} = 0$ under both H_1 and H_0. The scalar covariance of r_1 is

$$R_r|H_1 = \mathcal{E}\{|\mathsf{r}_1|^2|H_1\} = \mathcal{E}\{|\sqrt{E_0}\mathcal{S}_{sp} + \mathsf{n}_1|^2\} = E_s + N_0 = \sigma_1^2$$

in the presence of the signal with $E_s = E_0\mathcal{R}_{sp}$ and

$$R_r|H_0 = \mathcal{E}\{|\mathsf{r}_1|^2|H_0\} = N_0 = \sigma_0^2$$

in its absence. These variances are energies and denoting them as σ_1^2 and σ_0^2 respectively, the Gaussian pdfs under hypotheses H_i, $i = 0, 1$ are

$$\mathsf{f}(\mathsf{r}_1|H_i) = \frac{1}{\pi\sigma_i^2} \exp - \left(\frac{|\mathsf{r}_1|^2}{\sigma_i^2}\right) \quad i = 0, 1. \tag{3.53}$$

The likelihood ratio is

$$\begin{aligned} L_r(\mathsf{r}_1) &= \frac{\mathsf{f}(\mathsf{r}_1|H_1)}{\mathsf{f}(\mathsf{r}_1|H_0)} \\ &= \frac{\sigma_0^2}{\sigma_1^2} \exp\left(|\mathsf{r}_1|^2\left(\frac{1}{\sigma_0^2} - \frac{1}{\sigma_1^2}\right)\right) \end{aligned} \tag{3.54}$$

and the equivalent log-likelihood ratio test (llrt) becomes

$$\ell_r(\mathsf{r}_1) = \ln\left(\frac{\sigma_0^2}{\sigma_1^2}\right) + |\mathsf{r}_1|^2\left(\frac{1}{\sigma_0^2} - \frac{1}{\sigma_1^2}\right) \underset{H_0}{\overset{H_1}{\gtrless}} \gamma_l. \tag{3.55}$$

This is equivalent to $\ell_r(\mathbf{r})$ (3.48) for $\mathbf{r}$ scalar with $\mathbf{R}_0 \to \sigma_0^2$ and $\mathbf{R}_1 = \mathbf{R}_s + \mathbf{R}_0 \to \sigma_1^2$. Solving for the statistic $|\mathsf{r}_1|^2$ and gathering the other terms in the modified threshold γ_l', the llrt becomes

$$\begin{aligned} |\mathsf{r}_1|^2 = |\mathsf{s}_1 + \mathsf{n}_1|^2 &= \left|\int_{\Omega_t} \mathsf{r}(t) f^*(\mathbf{p}, t)dt\right|^2 \\ &= \left|\int_{\Omega_t} (\mathsf{s}(t) + \mathsf{n}(t)) f^*(\mathbf{p}, t)dt\right|^2 \underset{H_0}{\overset{H_1}{\gtrless}} \gamma_l' \end{aligned} \tag{3.56}$$

where

$$\begin{aligned} \gamma_l' &= \frac{\gamma_l - \ln\frac{\sigma_0^2}{\sigma_1^2}}{\frac{1}{\sigma_0^2} - \frac{1}{\sigma_1^2}} \\ &= \frac{(1 + \mathrm{SNR})(\gamma_l + \ln(1 + \mathrm{SNR}))N_0}{\mathrm{SNR}} \end{aligned} \tag{3.57}$$

and SNR is the input signal to noise ratio E_s/N_0. As was the case for deterministic scattering in WGN, the threshold level depends upon the SNR and is not a function of the parameters. The likelihood ratio test for detection is derived without regard to the parameters with the assumption that the normalized waveform comprising the first term in the CON set (3.50) has parameters matched to the underlying data parameters. In other words, the receiver hypothesizing $\mathbf{p}$ is optimum in the likelihood sense only for data of the form (3.49) with the same parameters. When the parameters are unknown, multiple receivers (matched filters) of the form

$$|\mathsf{r}_1|^2 = \left| \int_{\Omega_t} \mathsf{r}(t) f^*(\hat{\mathbf{p}}, t) dt \right|^2 . \tag{3.58}$$

are synthesized each hypothesizing a particular parameter set $\hat{\mathbf{p}}$.

The receiver structure resulting from the slowly fluctuating scatter model in WGN is similar yet fundamentally different from that resulting from a purely deterministic scatter assumption. Both structures rely on a form of replica correlation in which the input data time series is integrated with the hypothesized waveform however phase information is suppressed in the fluctuating model. The operation of finding the maximum of the real part of the complex correlator output in the deterministic case is equivalent to finding the zero phase point but phase conveys no information in the fluctuating model since the echo is assumed to be a complex random variable. Consequently, phase information is suppressed and only magnitude information is used. In both cases, the expected value of the likelihood function is maximized when $\hat{\mathbf{p}} = \mathbf{p}$.

The corresponding auto-AF is obtained from (3.58) by taking the expectation of the signal term

$$\begin{aligned}\mathcal{E}\{\left\{ \left| \int_{\Omega_t} \sqrt{E_0} \mathcal{S}_{sp} f(\mathbf{p}, t) f^*(\hat{\mathbf{p}}, t) dt \right|^2 \right\} &= E_0 \mathcal{R}_{sp} \left| \int_{\Omega_t} f(\mathbf{p}, t) f^*(\hat{\mathbf{p}}, t) dt \right|^2 \\ &= E_0 \mathcal{R}_{sp} \Phi_{ff}(\mathbf{p}, \hat{\mathbf{p}}) = E_s \Phi_{ff}(\mathbf{p}, \hat{\mathbf{p}})\end{aligned}$$

where E_s is the received echo energy and the AF $\Phi_{ff}(\mathbf{p}, \hat{\mathbf{p}})$ is now defined as the magnitude square rather than the real part of the uncertainty function χ_{ff} (3.45. It is also bounded such that

$$\begin{aligned}\Phi_{ff}(\mathbf{p}, \hat{\mathbf{p}}) &= \left| \int_{\Omega_t} f(\mathbf{p}, t) f^*(\hat{\mathbf{p}}, t) dt \right|^2 \\ &\leq \left(\int_{\Omega_t} |f(t, \mathbf{p})|^2 \, dt \right) \left(\int_{\Omega_t} |f(t, \hat{\mathbf{p}})|^2 \, dt \right) = 1 \end{aligned} \tag{3.59}$$

by the Schwarz inequality and the waveform normalization. This is equivalent to the maximization of the deterministic model AF (3.45) where the integral term becomes real and is maximized by the correct parameter hypothesis. The expected value of $|\mathrm{r}_1|^2$ which is equivalent to the variance of r_1 is $E_s\Phi(\mathbf{p},\hat{\mathbf{p}}) + N_0$ and N_0 under H_1 and H_0 respectively. The expectation under H_1 achieves its maximum of $E_s + N_0$ only in the receiver making the correct hypothesis.

Example 3.4 Ambiguity Function Comparison

A comparison of the two forms of the AF corresponding to the deterministic and fluctuating models is made by evaluating (3.59) using the same function as was used in example (3.3). This is simply the magnitude square rather than the real part of the bracketed quantity in (3.46) which is

$$\Phi_{ff}(\mathbf{p},\hat{\mathbf{p}}) = \frac{\beta}{\hat{\beta}}\left(1-\frac{\hat{\beta}\,\Delta\,\tau}{T}\right)^2 \operatorname{sinc}^2\left[\frac{T\omega_0}{2}\left(\frac{\beta}{\hat{\beta}}-1\right)\left(\frac{\Delta\tau\hat{\beta}}{T}-1\right)\right]$$

where $\operatorname{sinc}(x) = \sin(x)/x$. This reduces to

$$\Phi_{ff}(1,\Delta\tau) = \left(1-\frac{\beta\,\Delta\,\tau}{T}\right)^2$$

when $\hat{\beta} = \beta$ where unity in the argument represents the condition $\beta/\hat{\beta} = 1$. Except for squaring, this the same term appearing in (3.47) but there is no oscillatory term and the AF is monotonic on the delay axis. For example, if delay resolution is defined as the minimum value of $\Delta\tau$ for which the AF is zero on the delay axis, the resolution of the deterministic AF is $\Delta\tau = 1/4\beta f_0$ or comparable to the period of the carrier. The resolution for the fluctuating model is $\Delta\tau = T/\beta$ and comparable to the pulse duration. The resolutions of both models are comparable on the Doppler axis since the first null in the sinc function occurs when

$$\frac{\beta}{\hat{\beta}} = 1 + \frac{1}{f_0 T}.$$

The operation of taking the magnitude square of the output of a replica correlator as a test statistic is often called square law detection and because it is a nonlinear operation, the Gaussianity of the underlying input data is not preserved. r_1 is a complex Gaussian random variable and $|r_1|^2$ is composed of the sum of the squares of two real Gaussian scalars (the real and imaginary parts) under both H_1 and H_0 where,

$$|r_1|^2 = r_{1r}^2 + r_{1i}^2.$$

$|r_1|^2$ has an exponential density as shown in Example 3.2 and

$$\begin{aligned} \wp_f &= \exp -\frac{\gamma_l}{\sigma_0^2} = \exp -\frac{\gamma_l'}{N_0} \\ \wp_d &= \exp -\frac{\gamma_l}{\sigma_1^2} = \exp -\frac{\gamma_l'}{E_s\Phi_{ff}(\mathbf{p},\hat{\mathbf{p}}) + N_0}. \end{aligned} \tag{3.60}$$

The threshold may be eliminated between these equations generating the ROC relationship

$$\wp_d = \wp_f^{\frac{1}{1+D_{\mathbf{p}\hat{\mathbf{p}}}}} \tag{3.61}$$

where $D_{\mathbf{p}\hat{\mathbf{p}}}$ is called the *detection index* (DI) [2,6,30,31] and is a function of the parameters for replica correlation. It is defined in terms of the modified likelihood ratio test statistic (3.56) as

$$\begin{aligned} D_{\mathbf{p}\hat{\mathbf{p}}} &= \frac{\mathcal{E}\{|r_1|^2|H_1\} - \mathcal{E}\{|r_1|^2|H_0\}}{\mathcal{E}\{|r_1|^2|H_0\}} = \frac{\sigma_1^2 - \sigma_0^2}{\sigma_0^2} \\ &= \frac{E_s\Phi_{ff}(\mathbf{p},\hat{\mathbf{p}})}{N_0} \leq \frac{E_s}{N_0}. \end{aligned} \tag{3.62}$$

This represents the average signal to noise ratio realized at the output of a receiver hypothesizing $\hat{\mathbf{p}}$ when the data is parameterized by $\mathbf{p}$. $D_{\mathbf{pp}} = E_s/N_0$ is the received or input signal to noise energy ratio and reflects the totality of available signal energy. The ML parameter estimate $\mathbf{p}_{ml}$ discussed in Section 3.2 is associated with the receiver having the greatest average value of the detection statistic $|r_1|^2$ or equivalently, the greatest probability of detection for a given threshold (3.60). Both occur in the receiver for which $\mathbf{p} = \hat{\mathbf{p}}$ maximizing Φ and hence $D_{\mathbf{p}\hat{\mathbf{p}}}$.[3]

Detection performance of a particular receiver as measured by (3.61) is directly proportional to the detection index which is the signal to noise energy ratio realized at the receiver output. If for some reason, perhaps because of equipment limitations the statistic (3.58) could not be used,

[3]This definition of the DI differs from that of some other authors because of differing models for the likelihood function. It is related to another performance measure called the *deflection* [31,32].

a simple recourse might be to attempt to detect echoes by measuring the instantaneous power in the input time series $r(t)$. A detection would occur whenever the power $|r(t)|^2$ exceeded a specified threshold. The data time series is a Gaussian stochastic process and the power measured at a particular time has an exponential density. Detection performance would still be described by the ROC relationship (3.61) but the detection index would be the signal to noise power rather than energy ratio and would be parameter independent. In the absence of an echo (H_0), the input power is the noise power spectral density (psd) multiplied by the effective noise bandwidth W_s of the receiver $N_0 W_s$ (Joules)(sec.)$^{-1}$ = Watts.[4] The signal power is the received energy divided by the effective pulse duration E_s/T and thus the power SNR is

$$\text{SNR (input power)} = \frac{E_s}{N_0 T W_s}.$$

The performance improvement of energy vs power detection is proportional to the ratio of the detection indices which is equivalent to the ratio of the output to input signal to noise ratios. This quantity is commonly called the *processing gain* denoted by (G_p) [13] and is

$$G_p = \Phi_{ff}(\mathbf{p}, \hat{\mathbf{p}}) T W_s. \tag{3.63}$$

which is the output energy SNR (3.62) divided by the input power ratio. This is an important result because it indicates that the detection performance achieved by a matched filter in WGN is directly proportional to the signal energy and not the specific waveform transmitted. Since most practical systems are peak power limited, performance is improved by increasing system integration time or equivalently, the pulse duration. There are tradeoffs of course, one for example being imposed by a minimum search range requirement and another from the specification of desired Doppler resolution both of which are functions of T.

3.6 THE RICIAN MODEL

The deterministic and slowly fluctuating models are the extremes of pure determinism and randomness for point scattering but there is an inter-

[4]It was assumed in the derivation of the receiver that the noise was white implying a flat spectrum of infinite bandwidth. This is not a realizable situation in practice as all real systems have a finite bandwidth imposed by the hardware. White noise in the context of echo-location generally means that the noise spectrum is flat within the system bandwidth W_s which is designed to be sufficiently wide to accommodate the expected Doppler range and bandwidth of the signal waveform used.

mediate model that incorporates elements of both. It treats the signal as having a deterministic envelope but random phase under the assumption that for most echo-location scenarios it is unrealistic to consider range delay within a wavelength as determinable. The phase delay $\theta_\tau = \beta\omega_0\tau$ of (2.26) is assumed to have an added component $\theta = \omega_0\delta\tau \in [0, 2\pi]$. $f_0 = \omega_0/2\pi$ is the carrier frequency and the delay increment $\delta\tau$ is uniformly distributed on $0, 1/f_0$ [4,33].

Let the Doppler dilation $\beta = 1$ and the nominal delay $\tau = 0$. The data model for WGN is

$$\mathsf{r}(t) = \sqrt{E_s} f(t) \exp j\omega_0\delta\tau + \mathsf{n}(t)$$

where $\mathsf{n}(t)$ is complex Gaussian and its real and imaginary components have variance $\sigma^2 = N_0/2$. The sufficient statistic (3.51) is

$$\begin{aligned}\mathsf{r}_1 = \mathsf{r}_{1r} + j\mathsf{r}_{1i} &= \sqrt{E_s}\exp j\omega_0\delta\tau + \mathsf{n}_1 \\ &= \sqrt{E_s}(\cos(\omega_0\delta\tau) + \mathsf{n}_{1r} + j(\sin(\omega_0\delta\tau) + \mathsf{n}_{1i}))\end{aligned}$$

and the joint density of the data components conditioned on a realization of the phase $\theta = \omega_0\delta\tau$ is

$$\mathsf{f}(r_{1r}r_{1i}|\theta) = \frac{1}{\sqrt{2\pi\sigma^2}}\exp -\left[\frac{(r_{1r} - \sqrt{E_s}\cos\theta)^2 + (r_{1i} - \sqrt{E_s}\sin\theta)^2}{2\sigma^2}\right].$$

Converting to polar coordinants yields the magnitude $\tilde{r} = \sqrt{r_{1r}^2 + r_{1i}^2}$ and angle $\xi = \tan^{-1}(r_{1r}/r_{1i})$ with Jacobian $\tilde{r}$ so that the density becomes

$$\begin{aligned}\mathsf{f}(\tilde{r},\xi|\theta) &= \frac{\tilde{r}}{2\pi\sigma^2}\exp -\left[\frac{(\tilde{r}\cos\xi - \sqrt{E_s}\cos\theta)^2 + (\tilde{r}\sin\xi - \sqrt{E_s}\sin\theta)^2}{2\sigma^2}\right] \\ &= \frac{\tilde{r}}{2\pi\sigma^2}\exp -\left[\frac{\tilde{r}^2 + E_s - 2\tilde{r}\sqrt{E_s}\cos(\theta - \xi)}{2\sigma^2}\right].\end{aligned}$$

Integrating over ξ yields the conditional density of the magnitude

$$\mathsf{f}(\tilde{r}|\theta) = \frac{\tilde{r}}{2\pi\sigma^2}\exp -\frac{\tilde{r}^2 + E_s}{2\sigma^2}\int_0^{2\pi}\exp\left(\frac{\sqrt{E_s}\tilde{r}\cos(\theta - \xi)}{\sigma^2}\right)d\xi.$$

The integral is equivalent to $2\pi I_0(\sqrt{E_s}\tilde{r}/\sigma^2)$ which is independent of θ, I_0 is the modified Bessel function of zero order, and

$$I_0(x) = \sum_{n=0}^{\infty}\frac{x^{2n}}{2^{2n}(n!)^2} = 1 + \frac{x^2}{4} + \frac{x^4}{64} + \cdots. \tag{3.64}$$

The pdf of the magnitude when the signal is present (H_1) is the Rician density

$$\mathsf{f}(\tilde{r}|H_1) = \frac{\tilde{r}}{\sigma^2} I_0\left(\frac{\sqrt{E_s}\tilde{r}}{\sigma^2}\right) \exp - \left[\frac{\tilde{r}^2 + E_s}{2\sigma^2}\right]$$

that reverts to the Rayleigh density

$$\mathsf{f}(\tilde{r}|H_0) = \frac{\tilde{r}}{\sigma^2} \exp - \frac{\tilde{r}^2}{2\sigma^2}.$$

when the signal is absent (H_0). The ratio of the pdfs under H_1 and H_0

$$\begin{aligned} L_{\tilde{r}} &= \frac{\mathsf{f}(\tilde{r}|H_1)}{\mathsf{f}(\tilde{r}|H_0)} \\ &= \exp - \frac{E_s}{2\sigma^2} I_0\left(\frac{\sqrt{E_s}\tilde{r}}{\sigma^2}\right) \end{aligned}$$

is the likelihood ratio and

$$I_0\left(\frac{\sqrt{E_s}\tilde{\mathsf{r}}}{N_0}\right) \underset{H_0}{\overset{H_1}{\gtrless}} \gamma \exp - \frac{E_s}{N_0}$$

is the lrt in terms of the random magnitude $\tilde{\mathsf{r}}$ with $\sigma^2 = N_0/2$. Note that, unlike the llrt for the slowly fluctuating model (3.56), the random magnitude $\tilde{\mathsf{r}} = |\mathsf{r}_1|$ rather than magnitude square appears in the argument of the lrt. The pdf's can be expressed in terms of normalized variables

$$r_n = \frac{\tilde{r}}{\sigma}, \quad s_n = \frac{\sqrt{E_s}}{\sigma}$$

where $d\tilde{r} = \sigma d r_n$ and

$$\begin{aligned} \mathsf{f}(r_n|H_1) &= r_n \exp - \left[\frac{r_n^2 + s_n^2}{2}\right] I_0(r_n s_n) \\ \mathsf{f}(r_n|H_0) &= r_n \exp - \frac{r_n^2}{2}. \end{aligned}$$

The ROC's for the Rician model are obtained by integrating these from a threshold to infinity.

$$\begin{aligned} \wp_d &= \int_\gamma^\infty r_n \exp - \left[\frac{r_n^2 + s_n^2}{2}\right] I_0(r_n s_n) dr_n \\ \wp_f &= \int_\gamma^\infty r_n \exp - \frac{r_n^2}{2} dr_n. \end{aligned}$$

The detection probability has been tabulated as the Marcum Q-function [34] and the false alarm probability is one minus the cumulative Rayleigh probability. Plots of the Q-function are provided in [33] and ROC curves are provided in [4].

The Bessel function is nearly quadratic and approximated by the first two terms of the expansion (3.64) for small values of its argument. This yields the test

$$\tilde{r}^2 = |r_1|^2 \underset{H_0}{\overset{H_1}{\gtrless}} \left(\gamma \exp \frac{E_s}{N_0} - 1\right)\left(\frac{4N_0^2}{E_s}\right) = \gamma'$$

in terms of the modified threshold γ'. This is called the small signal or low SNR approximation and is equivalent to the square law form of the llrt for the fluctuating model (3.56). This simpler form is often employed even when the scattering is modeled as Rician because the ROC performance is nearly the same as that achieved by synthesizing the Bessel function [4]. Suboptimal performance is often acceptable as a tradeoff for receiver complexity at high SNR and the approximation is appropriate for most sonar situations.

3.7 CORRELATED GAUSSIAN INTERFERENCE

In echo-location, correlated interference arises from extraneous noise sources or from backscatter of the transmit signal in the form of reverberation or clutter. While extraneous sources may or may not have flat spectra, backscatter mechanisms usually are characterized by spectra which are on the average, similar to that of the transmitted waveform. Moreover, interference may be non-stationary because its level and to a lesser extent, spectra may change with time due to receiver motion and reflection from fixed boundaries. When Gaussian interference is correlated, its second order properties are characterized by the covariance $R_n(t, u)$ and as shown in Section 3.3, for deterministic scattering, the likelihood detector takes the form of a phase sensitive mismatched filter. On the other hand, under the random slowly fluctuating scatter model in WGN, the detector suppresses phase information and computes the magnitude of a matched filter. It would therefore stand to reason that the optimum detector for the fluctuating scatterer in correlated Gaussian interference (CGI) would compute the magnitude of the output of a mismatched filter. This is indeed the case and is shown by again choosing a CON set ξ such that the first term is (3.50) and deriving a sufficient statistic r_1. Rather than compute r_1 as was done in the WGN case it is assumed that a mismatched receiver of the form (3.39) is required and hypothesize that

$$\mathsf{r}_1 = \int_{\Omega_t} \mathsf{r}(t) g^*(\mathbf{p}, t) dt. \tag{3.65}$$

Having chosen the CON set as indicated, only r_1 reflects the presence of the signal in the time series. It therefore only remains to show that r_1 and $\mathsf{u}(t)$ (3.52) are statistically uncorrelated for r_1 to be a sufficient statistic. The requirement is

$$\mathcal{E}\{\mathsf{r}_1 \mathsf{u}^*(t)\} = 0 \tag{3.66}$$

and since the CON set is chosen such that $\mathsf{r}_i = \mathsf{n}_i \; i > 1$ and the signal and interference are assumed to be uncorrelated,

$$\mathcal{E}\{\mathsf{n}_1 \mathsf{u}^*(t)\} = 0 \;\Rightarrow\; \mathcal{E}\{\mathsf{r}_1 \mathsf{u}^*(t)\} = 0.$$

Using (3.52),

$$\begin{aligned} \mathsf{u}(t) &= \mathsf{r}(t) - \mathsf{r}_1 f(\mathbf{p}, t) \\ &= (\mathsf{s}(t) - \mathsf{s}_1 f(\mathbf{p}, t)) + (\mathsf{n}(t) - \mathsf{n}_1 f(\mathbf{p}, t)) \end{aligned}$$

and again invoking uncorrelated signal and interference, (3.66) reduces to the requirement that

$$\mathcal{E}\{\mathsf{n}_1 (\mathsf{n}(t) - \mathsf{n}_1 f(\mathbf{p}, t))^*\} = 0.$$

or equivalently

$$\mathcal{E}\{\mathsf{n}_1 \mathsf{n}^*(t)\} = \mathcal{E}\{|\mathsf{n}_1|^2\} f^*(\mathbf{p}, t)\}. \tag{3.67}$$

Since $\mathsf{r}_1 = \mathsf{s}_1 + \mathsf{n}_1$, (3.65) is used to write

$$\mathsf{n}_1 = \int_{\Omega_u} \mathsf{n}(u) g^*(\mathbf{p}, u) du$$

and substituting this into (3.67) yields

$$\int_{\Omega_u} R_n(t,u)\ g^*(\mathbf{p},u)du =$$

$$f^*(\mathbf{p},t)\int_{\Omega_t} g(t)\left[\int_{\Omega_u} R_n(t,u)g^*(\mathbf{p},u)du\right]dt. \quad (3.68)$$

The relationship

$$\int_{\Omega_u} R_n(t,u)g^*(\mathbf{p},u)du = kf^*(\mathbf{p},t) \quad (3.69)$$

is a solution to this where k is a constant and the parameter notation in the argument of g denotes the parameter dependence implied by the relationship. Substituting into (3.68) results in

$$kf^*(\mathbf{p},t)\int_{\Omega_t} g(\mathbf{p},t)f^*(\mathbf{p},t)dt = kf^*(\mathbf{p},t)$$

that requires the constraint $\int_{\Omega_t} gf^*dt = 1$. The conjugate of (3.69)

$$\int_{\Omega_u} R_n(u,t)g(\mathbf{p},u)du = kf(\mathbf{p},t) \quad (3.70)$$

is similar to the relationship (3.43) derived for deterministic scattering. When $g(\mathbf{p},t)$ is computed for arbitrary k, there is no guarantee that the constraint will be satisfied. However the properly scaled processing function can be easily computed after the fact.

Assume that an arbitrary constant k is chosen and is used to calculate the function $\tilde{g}(\mathbf{p},t)$ by solving the integral equation (3.70). The quantity

$$\chi_{\tilde{g}f}(\mathbf{p},\mathbf{p}) = \int_{\Omega_t} \tilde{g}(\mathbf{p},t)f^*(\mathbf{p},t)dt$$

is computed and the scaled processing function $g = \tilde{g}/\chi_{\tilde{g}f}$ is formed. This is substituted into (3.68) with the desired result. A direct solution for the proper constant k is obtained by multiplying both sides of (3.69) by $g(\mathbf{p},t)$, integrating, and applying the constraint with the result that k is equivalent to the interference energy at the output of the mismatched receiver $\mathcal{E}\{|\mathsf{n}_1|^2\} = \sigma_0^2(\mathbf{p})$ that is in turn a function of the parameter $\mathbf{p}$ where

$$k = \sigma_0^2(\mathbf{p}) = \int_{\Omega_t}\int_{\Omega_u} g(\mathbf{p},t)g^*(\mathbf{p},u)R_n(t,u)dudt. \quad (3.71)$$

Alternatively, the inverse kernel (3.42) may be used where

$$g(\mathbf{p}, u) = k \int_{\Omega_t} f(\mathbf{p}, t) Q_n(t, u) dt. \tag{3.72}$$

Applying the constraint and solving for k, yields

$$k = \sigma_0^2(\mathbf{p}) = \frac{1}{\int_{\Omega_t} \int_{\Omega_u} f(\mathbf{p}, t) f^*(\mathbf{p}, u) Q_n(t, u) du dt} = \mathcal{E}\{|\mathsf{n}_1|^2\}. \tag{3.73}$$

$g(\mathbf{p}, u)$ obtained in this manner is the same as that obtained by the normalizing of $\tilde{g}$ by $\chi_{\tilde{g}f}$. It is equivalent within a constant to the processing function (3.42) derived for the deterministic signal and is the continuous version of the discrete processing vector (3.23). The parameter notation $\sigma_0^2(\mathbf{p})$ is introduced to emphasize the fact that when the constraint is satisfied for each hypothesis, a different constant that is a function of the parameters must be computed. The requirement that $\chi_{gf}(\mathbf{p}, \mathbf{p}) = 1$ in order that r_1 be a sufficient statistic for the slowly fluctuating scatter problem is equivalent to the required modification of the likelihood threshold as a function of the parameter hypothesis discussed for the deterministic case. It guarantees that when the parameter hypothesis matches the parameters of the echo, the signal response of a likelihood receiver using the processing function (3.72) depends only upon the signal energy and not the parameters. The optimum test is given by (3.55) however the statistic is now formed from (3.65,3.70). As before in order to test the parameters, a bank of receivers is synthesized each forming

$$\begin{aligned}
|\mathsf{r}_1|^2 &= \left| \int_{\Omega_t} \mathsf{r}(t) g^*(\hat{\mathbf{p}}, t) dt \right|^2 \\
&= \left| \int_{\Omega_t} \mathsf{n}(t) g^*(\hat{\mathbf{p}}, t) dt \right|^2 \underset{H_0}{\overset{H_1}{\gtrless}} \gamma_1' \quad (H_0) \\
&= \left| \int_{\Omega_t} (\mathsf{s}(\mathbf{p}, t) + \mathsf{n}(t)) g^*(\hat{\mathbf{p}}, t) dt \right|^2 \underset{H_0}{\overset{H_1}{\gtrless}} \gamma_1' \quad (H_1).
\end{aligned} \tag{3.74}$$

The threshold is given by (3.57) but the variances σ_0^2, σ_1^2 now become functions of the parameters. The expected interference response of the receiver hypothesizing $\hat{\mathbf{p}}$ is

$$\begin{aligned}\mathcal{E}\{|\mathsf{r}_1|^2|H_0\} &= \sigma_0^2(\hat{\mathbf{p}}) \\ &= \int_{\Omega_t} g(\hat{\mathbf{p}},t)\int_{\Omega_u} R_n(t,u)g^*(\hat{\mathbf{p}},u)dudt \end{aligned} \tag{3.75}$$

which is equivalent to (3.71) with $\mathbf{p}$ replaced by $\hat{\mathbf{p}}$. In the presence of the echo (H_1) with parameters $\mathbf{p}$, the expected receiver response is

$$\begin{aligned}\mathcal{E}\{|\mathsf{r}_1|^2|H_1\} &= \sigma_1^2(\mathbf{p},\hat{\mathbf{p}}) \\ &= E_s\Phi_{fg}(\mathbf{p},\hat{\mathbf{p}}) + \sigma_0^2(\hat{\mathbf{p}}) \end{aligned} \tag{3.76}$$

where

$$\begin{aligned}\Phi_{fg}(\mathbf{p},\hat{\mathbf{p}}) &= \left|\int_{\Omega_t} f(\mathbf{p},t)g^*(\hat{\mathbf{p}},t)dt\right|^2 \\ &= |\chi_{fg}(\mathbf{p},\hat{\mathbf{p}})|^2 \end{aligned} \tag{3.77}$$

is the cross-ambiguity function between f and g. χ_{fg} is the cross uncertainty function analogous to the auto-UF defined in (3.45).[5] Each receiver in the bank acts as a "stand alone" structure performing a binary hypothesis test for the detection of a signal with specific parameters denoted by $\hat{\mathbf{p}}$. Consequently, the threshold γ_l' is computed based upon the assumption that the echo is parameterized by $\hat{\mathbf{p}}$. The expressions for σ_0^2, σ_1^2 with $\hat{\mathbf{p}} = \mathbf{p}$ may be substituted into (3.57) yielding

$$\begin{aligned}\gamma_l'(\mathbf{p}) &= \frac{\sigma_0^2(\mathbf{p})(1+D_{\mathbf{pp}})(\gamma_l + \ln(1+D_{\mathbf{pp}}))}{D_{\mathbf{pp}}} \\ D_{\mathbf{pp}} &= \frac{E_s}{\sigma_0^2(\mathbf{p})} \quad \forall\, \mathbf{p}. \end{aligned} \tag{3.78}$$

Unlike the case for WGN, the interference and consequently, the matched parameter detection index (3.62) vary with the parameter hypothesis. The resulting effect on receiver performance is illustrated by the detection and false alarm probabilities (3.60).

[5]It is important to note that the definition of the ambiguity function depends upon the scattering assumption under which it is derived. In the discussion of deterministic scattering in section (3.3), the ambiguity function was defined as the real part of χ. The magnitude square definition arising from the slowly fluctuating model is that which is most often defined in the classic radar literature [6, 26, 27].

The false alarm probability in the receiver hypothesizing $\hat{\mathbf{p}}$ with threshold γ_l is

$$\wp_f = \exp\ -\frac{\tilde{\gamma}_l}{\sigma_0^2(\hat{\mathbf{p}})} \tag{3.79}$$

in the absence of the echo (H_0). The corresponding detection probability H_1 for an echo with parameters $\mathbf{p}$ is

$$\begin{aligned}\wp_d &= \exp\ -\frac{\tilde{\gamma}_l}{\sigma_1^2} \\ &= \exp\ -\frac{\tilde{\gamma}_l}{E_s\Phi_{fg}(\mathbf{p},\hat{\mathbf{p}}) + \sigma_0^2(\hat{\mathbf{p}})}.\end{aligned} \tag{3.80}$$

If it is desired to maintain a constant false alarm rate (CFAR) in all receivers [3], the threshold must be adjusted as a function of $\hat{\mathbf{p}}$ such that $\tilde{\gamma}_l/\sigma_0^2(\hat{\mathbf{p}})$ is constant. In other words, the threshold must vary across the receiver bank. $\wp_f$ and $\wp_d$ are related by the detection index (3.61) and when the receiver is mismatched,

$$D_{\mathbf{p}\hat{\mathbf{p}}} = \frac{E_s\Phi_{fg}(\mathbf{p},\hat{\mathbf{p}})}{\sigma_0^2(\hat{\mathbf{p}})} \tag{3.81}$$

where $D_{\mathbf{p}\hat{\mathbf{p}}} \rightarrow D_{\mathbf{pp}}$ (3.78) as $\hat{\mathbf{p}} \rightarrow \mathbf{p}$. When a CFAR threshold is used, $\wp_d$ is a function of both the local SIR$=E_s/\sigma_0^2(\hat{\mathbf{p}})$ and the parameter mismatch as described by the cross-ambiguity function. $D_{\mathbf{p}\hat{\mathbf{p}}}$ as derived for the correlated interference case is analogous to that derived in WGN (3.62). It is similar because it represents an SIR weighted by the degree of parameter mismatch. It differs because the SIR also depends upon the parameter hypothesis.

In practice, the threshold relationship (3.78) is not very useful for the actual setting of thresholds as it requires knowledge of the echo energy that like the parameters, is not usually known a-priori. Indeed this is the case as well for WGN and deterministic scattering (3.44). It is however an indicator of how the likelihood test threshold γ_l (3.56) must be altered as a function of the energy and the parameters in order to directly utilize the statistic (3.65). Generally a CFAR threshold for each receiver is set using (3.79) and the Nehman-Pearson criterion. The detection probability is then a function of the echo energy and the input parameters resulting in ROC curves such as shown in Figure 3.4.

$\wp_d$ is proportional to $D_{\mathbf{p}\hat{\mathbf{p}}}$ for fixed $\wp_f$ and $D_{\mathbf{p}\hat{\mathbf{p}}}$ is maximized in the receiver making the correct hypothesis corresponding to the ML parameter estimate. This requires that

$$D_{\mathbf{p}\hat{\mathbf{p}}} \leq D_{\mathbf{pp}} \Leftrightarrow \Phi_{fg}(\mathbf{p},\hat{\mathbf{p}}) \leq \frac{\sigma_0^2(\hat{\mathbf{p}})}{\sigma_0^2(\mathbf{p})}$$

$$\Rightarrow \Phi_{fg}(\mathbf{p},\hat{\mathbf{p}})\frac{\sigma_0^2(\mathbf{p})}{\sigma_0^2(\hat{\mathbf{p}})} \leq 1. \tag{3.82}$$

The implication is that, in contrast to the auto-AF (3.59), the cross-AF formed by the normalized waveform and the processing function (3.72), is not necessarily bounded by unity. It is however constrained to be unity when $\hat{\mathbf{p}} = \mathbf{p}$ in order to satisfy the condition for a sufficient statistic.

In order to demonstrate (3.82) it is convenient to define the linear operator $\mathcal{Q}$

$$\mathcal{Q}f_{\hat{\mathbf{p}}} = \int_{\Omega_u} Q_n(u,t) f(\hat{\mathbf{p}},u)du \tag{3.83}$$

and $f_{\hat{\mathbf{p}}}$, $f_{\mathbf{p}}$ as the parameterized waveform functions $f(\hat{\mathbf{p}},t)$, $f(\mathbf{p},t)$ respectively. Let

$$\rho_O(\mathbf{p},\hat{\mathbf{p}}) = \int_{\Omega_t}\int_{\Omega_u} f(\mathbf{p},t)f^*(\hat{\mathbf{p}},u)Q_n(t,u)dudt$$

then using (3.83) and the short hand notation $< x,y >= \int_{\Omega_t} x(t)y^*(t)dt$ for the inner product yields

$$\begin{aligned}\rho_O(\mathbf{p},\hat{\mathbf{p}}) &= < f_{\mathbf{p}}, \mathcal{Q}f_{\hat{\mathbf{p}}} > \\ &= \frac{1}{\sigma_0^2(\mathbf{p})} \quad (\text{for } \hat{\mathbf{p}} = \mathbf{p}).\end{aligned} \tag{3.84}$$

The cross-AF defined by (3.72, 3.73, 3.77) is

$$\Phi_{fg}(\mathbf{p},\hat{\mathbf{p}}) = \left|\sigma_0^2(\hat{\mathbf{p}})\rho_O(\mathbf{p},\hat{\mathbf{p}})\right|^2 \tag{3.85}$$

and the inequality (3.82) becomes

$$\Phi_{fg}(\mathbf{p},\hat{\mathbf{p}})\left(\frac{\sigma_0^2(\hat{\mathbf{p}})}{\sigma_0^2(\mathbf{p})}\right) = |\rho_O(\mathbf{p},\hat{\mathbf{p}})|^2\,\sigma_0^2(\mathbf{p})\sigma_0^2(\hat{\mathbf{p}}) \leq 1.$$

which by (3.84) is equivalent to

$$\Phi_{fg}(\mathbf{p},\hat{\mathbf{p}})\left(\frac{\sigma_0^2(\mathbf{p})}{\sigma_0^2(\hat{\mathbf{p}})}\right) = \frac{|< f_{\mathbf{p}}, \mathcal{Q}f_{\hat{\mathbf{p}}} >|^2}{< f_{\mathbf{p}}, \mathcal{Q}f_{\mathbf{p}} >< f_{\hat{\mathbf{p}}}, \mathcal{Q}f_{\hat{\mathbf{p}}} >} \leq 1 \tag{3.86}$$

with equality for $\hat{\mathbf{p}} = \mathbf{p}$. A necessary condition for a maximum of the detection index at $\hat{\mathbf{p}} = \mathbf{p}$ is that there be an extremal solution of (3.86) indicated by a zero of its gradient with respect to the parameter set $\hat{\mathbf{p}}$ at $\mathbf{p}$. By defining the gradient operator for N_p parameters as

$$\nabla_{\hat{\mathbf{p}}} = \left[\frac{\partial}{\partial \hat{p}_1}, \dots, \frac{\partial}{\partial \hat{p}_{N_p}}\right]^{\mathrm{T}},$$

the constituent derivatives of the numerator and denominator are written as

$$\begin{aligned} \nabla_{\hat{\mathbf{p}}}(< f_{\hat{\mathbf{p}}}, \mathcal{Q} f_{\hat{\mathbf{p}}} >) &= 2\Re\{< f_{\hat{\mathbf{p}}}, \mathcal{Q} \nabla_{\hat{\mathbf{p}}} f_{\hat{\mathbf{p}}} >\} \\ \nabla_{\hat{\mathbf{p}}} |< f_{\mathbf{p}}, \mathcal{Q} f_{\hat{\mathbf{p}}} >|^2 &= 2\Re\left\{< f_{\mathbf{p}}, \mathcal{Q} f_{\hat{\mathbf{p}}} >< f_{\mathbf{p}}, \mathcal{Q} \nabla_{\hat{\mathbf{p}}} f_{\hat{\mathbf{p}}} >\right\}. \end{aligned}$$

Using these in the gradient of the quotient (3.86) and evaluating at $\hat{\mathbf{p}} = \mathbf{p}$ results in

$$\begin{aligned} &\nabla_{\hat{\mathbf{p}}} \left\{ \Phi_{fg}(\mathbf{p}, \hat{\mathbf{p}}) \frac{\sigma_0^2(\mathbf{p})}{\sigma_0^2(\hat{\mathbf{p}})} \right\}_{\hat{\mathbf{p}}=\mathbf{p}} = \\ &\frac{< f_{\mathbf{p}}, \mathcal{Q} f_{\mathbf{p}} > \Re\left\{< f_{\mathbf{p}}, \mathcal{Q} f_{\mathbf{p}} >< f_{\mathbf{p}}, \mathcal{Q} \nabla_{\hat{\mathbf{p}}} f_{\mathbf{p}} >\right\}}{< f_{\mathbf{p}}, \mathcal{Q} f_{\mathbf{p}} >^3} \\ &- \frac{|< f_{\mathbf{p}}, \mathcal{Q} f_{\mathbf{p}} >|^2 \Re\{< f_{\mathbf{p}}, \mathcal{Q} \nabla_{\hat{\mathbf{p}}} f_{\mathbf{p}} >\}}{< f_{\mathbf{p}}, \mathcal{Q} f_{\mathbf{p}} >^3}. \end{aligned}$$

Since $< f_{\mathbf{p}}, \mathcal{Q} f_{\mathbf{p}} >$ is a real quantity for all parameters, this reduces to

$$(< f_{\mathbf{p}}, \mathcal{Q} f_{\mathbf{p}} >)^2 \Re\left\{< f_{\mathbf{p}}, \mathcal{Q} \nabla_{\hat{\mathbf{p}}} f_{\mathbf{p}} > - < f_{\mathbf{p}}, \mathcal{Q} \nabla_{\hat{\mathbf{p}}} f_{\mathbf{p}} >\right\} = 0$$

and by (3.81) the extremal solution is

$$\nabla_{\hat{\mathbf{p}}} \left\{ \frac{D_{\mathbf{p}\hat{\mathbf{p}}}}{D_{\mathbf{p}\mathbf{p}}} \right\}_{\hat{\mathbf{p}}=\mathbf{p}} = 0.$$

That the extremal is indeed a maximum and $D_{\mathbf{p}\hat{\mathbf{p}}} \leq D_{\mathbf{p}\mathbf{p}}$ may be shown by examination of the second derivative, an exercise left to the curious reader.

This indicates that the receiver making the correct hypothesis provides the maximum likelihood estimate of the parameters since both the detection index and hence the detection probability are locally maximized. Unlike the situation when the interference is WGN, the parameter corresponding to the absolute maximum of the square law response

$|\mathsf{r}_1|^2$ is biased by $\sigma_0^2(\hat{\mathbf{p}})$. It is necessary to compensate each receiver for the local interference level in order to realize an *unbiased estimator.* This results in each receiver's output $|\mathsf{r}_1|^2$ being compared to a threshold relative to $\sigma_0^2(\hat{\mathbf{p}})$ for CFAR. The decision that an echo is characterized by a particular parameter set $\mathbf{p}$ results from the receiver hypothesizing $\mathbf{p}$ having the greatest response relative to that threshold. On the average this results in the maximization of the detection index.

3.8 WIDE SENSE STATIONARITY

The derivation of the processing function for continuous time series requires the solution of integral equations such as (3.70). The covariance is a function only of the time difference $(t-u)$ if the interference is wide sense stationary (WSS) which simplifies the evaluation of the processing function. The defining relationships (3.70,3.72) become the convolutions

$$\int_{\Omega_u} R_n(u-t)g(\mathbf{p},u)du = kf(\mathbf{p},t)$$

$$\int_{\Omega_t} f(\mathbf{p},t)Q_n(t-u)dt = g(\mathbf{p},u)$$

that are readily solved in the frequency domain by using Fourier transforms. The frequency domain equivalents are

$$G(\mathbf{p},\omega) = \frac{kF(\mathbf{p},\omega)}{N(\omega)} \tag{3.87}$$

where $\mathcal{F}$ is the Fourier transform operator,

$$\mathcal{F}\sqrt{\beta}x(\beta(t-\tau)) = (1/\sqrt{\beta})X(\omega/\beta)\exp -\jmath\omega t,$$

and

$$\begin{aligned} G(\mathbf{p},\omega) &= \mathcal{F}\,g(\mathbf{p},t) \\ F(\mathbf{p},\omega) &= \mathcal{F}\,f(\mathbf{p},t) \\ \mathsf{R}(\mathbf{p},\omega) &= \mathsf{S}(\mathbf{p},\omega) + \mathsf{N}(\omega) = \mathcal{F}\,\mathsf{r}(\mathbf{p},t) \\ N(\omega) &= \lim_{T\to\infty}(\mathcal{E}\{|\mathsf{N}(\omega)|^2\}/T) = \mathcal{F}\,R_n(t-u). \end{aligned}$$

$N(\omega)$ is the power spectral density (psd) $S_{nn}(\omega)$ (1.17) of the interference process and is a real scalar. The inverse kernel and the covariance are simple inverses in the frequency domain so that

$$\mathcal{F}\,Q_n(t-u) = \frac{1}{N(\omega)}$$

and it is assumed that the interference process has a white component that guarantees that $N(\omega) \neq 0$ over the frequency band of interest. The constraint $\chi_{fg}(\hat{\mathbf{p}}, \hat{\mathbf{p}}) = 1$ for each receiver carries over by Parseval's rule to the frequency domain as

$$\chi_{fg}(\hat{\mathbf{p}}, \hat{\mathbf{p}}) = \frac{1}{2\pi} \int_{-\infty}^{\infty} G(\hat{\mathbf{p}}, \omega) F^*(\hat{\mathbf{p}}, \omega) d\omega = 1. \tag{3.88}$$

The normalized processing function for each receiver is computed by applying this to (3.87) to form

$$\frac{k}{2\pi} \int_{-\infty}^{\infty} \frac{|F(\hat{\mathbf{p}}, \omega|^2}{N(\omega)} = 1$$

and

$$k = \sigma_0^2(\hat{\mathbf{p}}) = \frac{1}{\frac{1}{2\pi} \int_{-\infty}^{\infty} \frac{|F(\hat{\mathbf{p}}, \omega|^2}{N(\omega)} d\omega} \tag{3.89}$$

which is the frequency domain equivalent of (3.73). The normalized processing function

$$G(\hat{\mathbf{p}}, \omega) = \frac{\frac{F(\hat{\mathbf{p}}, \omega)}{N(\omega)}}{\frac{1}{2\pi} \int_{-\infty}^{\infty} \frac{|F(\hat{\mathbf{p}}, \omega)|^2}{N(\omega)} d\omega} = \frac{F(\hat{\mathbf{p}}, \omega)}{N(\omega)} \sigma_0^2(\hat{\mathbf{p}}) \tag{3.90}$$

is the wide sense stationary equivalent to (3.72, 3.73) and is obtained by using this result in (3.87) Under WSS conditions, the nature of σ_0^2 as a measure of interference becomes more apparent because it is expressed explicitly in terms of a normalized signal to interference ratio. If the received echo signal has energy E_s, (3.89) indicates that the ratio $E_s/\sigma_0^2(\hat{\mathbf{p}})$ is the integration of the signal to interference energies of each frequency component and represents the maximum possible SIR and detection index available to a receiver making the hypothesis $\hat{\mathbf{p}}$. The detection statistic (3.74) is expressed in the frequency domain with

$$\mathsf{r}_1 = \frac{1}{2\pi} \int_{-\infty}^{\infty} \mathsf{R}(\omega) G^*(\hat{\mathbf{p}}, \omega) d\omega \tag{3.91}$$

and its variance under H_1 or H_0 is given by (3.75, 3.76) using (3.89).

ML estimation requires the condition (3.82) as is the case for non-stationary interference processes. The ambiguity function analogous to

(3.77, 3.84, 3.85) is

$$\begin{aligned}\Phi_{fg}(\mathbf{p},\hat{\mathbf{p}}) &= \left|\frac{1}{2\pi}\int_{-\infty}^{\infty} F(\mathbf{p},\omega)G^*(\hat{\mathbf{p}},\omega)d\omega\right|^2 \\ &= \left|\frac{\sigma_0^2(\hat{\mathbf{p}})}{2\pi}\int_{-\infty}^{\infty}\frac{F(\mathbf{p},\omega)F^*(\hat{\mathbf{p}},\omega)}{N(\omega)}d\omega\right|^2 \\ &= \left|\sigma_0^2(\hat{\mathbf{p}})\rho_0(\mathbf{p},\hat{\mathbf{p}})\right|^2 \end{aligned} \tag{3.92}$$

in the frequency domain. $N(\omega)$ is positive real and is written in terms of its constituent complex spectrum $N_c(\omega)$ as

$$N(\omega) = N_c(\omega)N_c^*(\omega)$$

thus the right hand side of (3.92) is bounded by

$$\begin{aligned}\left|\frac{\sigma_0^2(\hat{\mathbf{p}})}{2\pi}\int_{-\infty}^{\infty}\left(\frac{F(\mathbf{p},\omega)}{N_c(\omega)}\right)\left(\frac{F(\hat{\mathbf{p}},\omega)}{N_c(\omega)}\right)^* d\omega\right|^2 &\leq \left(\frac{\sigma_0^2(\hat{\mathbf{p}})}{2\pi}\int_{-\infty}^{\infty}\frac{|F(\mathbf{p},\omega)|^2}{N(\omega)}d\omega\right) \\ &\quad\cdot\left(\frac{\sigma_0^2(\hat{\mathbf{p}})}{2\pi}\int_{-\infty}^{\infty}\frac{|F(\hat{\mathbf{p}},\omega)|^2}{N(\omega)}d\omega\right) \\ &= \chi_{fg}(\mathbf{p},\mathbf{p})\chi_{fg}(\hat{\mathbf{p}},\hat{\mathbf{p}}).\end{aligned}$$

by the Schwarz inequality. The inequality becomes

$$\Phi_{fg}(\mathbf{p},\hat{\mathbf{p}}) \leq \frac{\sigma_0^2(\hat{\mathbf{p}})}{2\pi}\int_{-\infty}^{\infty}\frac{|F(\mathbf{p},\omega)|^2}{N(\omega)}d\omega = \frac{\sigma_0^2(\hat{\mathbf{p}})}{\sigma_0^2(\mathbf{p})}$$

from the constraint relationship (3.89) and bounds the CAF by (3.82) establishing that $D_{\mathbf{p}\hat{\mathbf{p}}} \leq D_{\mathbf{pp}}$ under WSS conditions as well. For WGN, $N(\omega) = N_0 \;\forall\; \omega$ and $G(\hat{\mathbf{p}},\omega) = F(\hat{\mathbf{p}},\omega)$ as indicated by (3.82, 3.90). The detection statistic is

$$\mathsf{r}_1 = \frac{1}{2\pi}\int_{-\infty}^{\infty} \mathsf{R}(\omega)F^*(\hat{\mathbf{p}},\omega)d\omega$$

which is the frequency domain equivalent of r_1 as derived for WGN (3.56).

Whether or not correlated interference is WSS, the basic receiver structure remains the same except for details of implementation. In either case, prior knowledge of the interference covariance or spectrum and the transmit waveform is required. The statistic r_1 is formed from (3.65) or (3.91) and compared with a suitable threshold for detection. When the echo is characterized by parameters that are to be estimated, a bank of receivers each having its own threshold is employed. Each threshold is determined with regard to the desired detection criteria and the hypothesized parameters and their relationship to the interference process. A detection is declared when one or more of the receiver outputs exceed its threshold and the ML estimate of the echo parameters corresponds to the receiver having the greatest output magnitude in relation to the local interference level $\sigma_0^2(\hat{\mathbf{p}})$. CFAR processing wherein each receiver has the same $\wp_f$, requires that $\tilde{\gamma}_l = c_d\sigma_0^2(\hat{\mathbf{p}})$ where c_d is a constant. The receivers are modified so that a constant threshold can be applied to all by linearly transforming each output to form the statistic

$$\mathsf{d}_1 = \frac{|\mathsf{r}_1|^2 - \sigma_0^2(\hat{\mathbf{p}})}{\sigma_0^2(\hat{\mathbf{p}})} = \frac{|\mathsf{r}_1|^2}{\sigma_0^2(\hat{\mathbf{p}})} - 1 \tag{3.93}$$

where

$$\mathcal{E}\{\mathsf{d}_1\} = D_{\mathbf{p}\hat{\mathbf{p}}}. \tag{3.94}$$

CFAR requires that under H_0,

$$\begin{aligned} |\mathsf{r}_1|^2 &\leq \tilde{\gamma}_l = c_d\sigma_0^2 \\ \Rightarrow \mathsf{d}_1 &\leq c_d - 1 \end{aligned}$$

and consequently, the same threshold may be used throughout. If multiple threshold crossings occur, the receiver with the greatest output in relation to $c_d - 1$ provides the ML parameter estimate. Receivers for operation in correlated non-stationary and WSS interference forming the modified statistic d_1 are shown in Figure 3.5. A multi-hypothesis receiver bank using these is shown in Figure 3.6.

Example 3.5 CFAR Thresholding

A simple signal and WSS interference model illustrates CFAR thresholding in a receiver bank. CFAR requires that a threshold be assigned to each receiver such that $\tilde{\gamma}_l/\sigma_0^2(\hat{\mathbf{p}})$ be a constant. Equations (3.79, 3.80) are the probabilities with $\sigma_0^2(\hat{\mathbf{p}})$ given by (3.89) under WSS conditions and (3.76) is

$$\begin{aligned} \sigma_1^2 &= E_s\Phi_{fg}(\mathbf{p}, \hat{\mathbf{p}}) + \sigma_0^2(\hat{\mathbf{p}}) \\ &= E_s + \sigma_0^2(\hat{\mathbf{p}}) \quad (\hat{\mathbf{p}} = \mathbf{p}). \end{aligned}$$

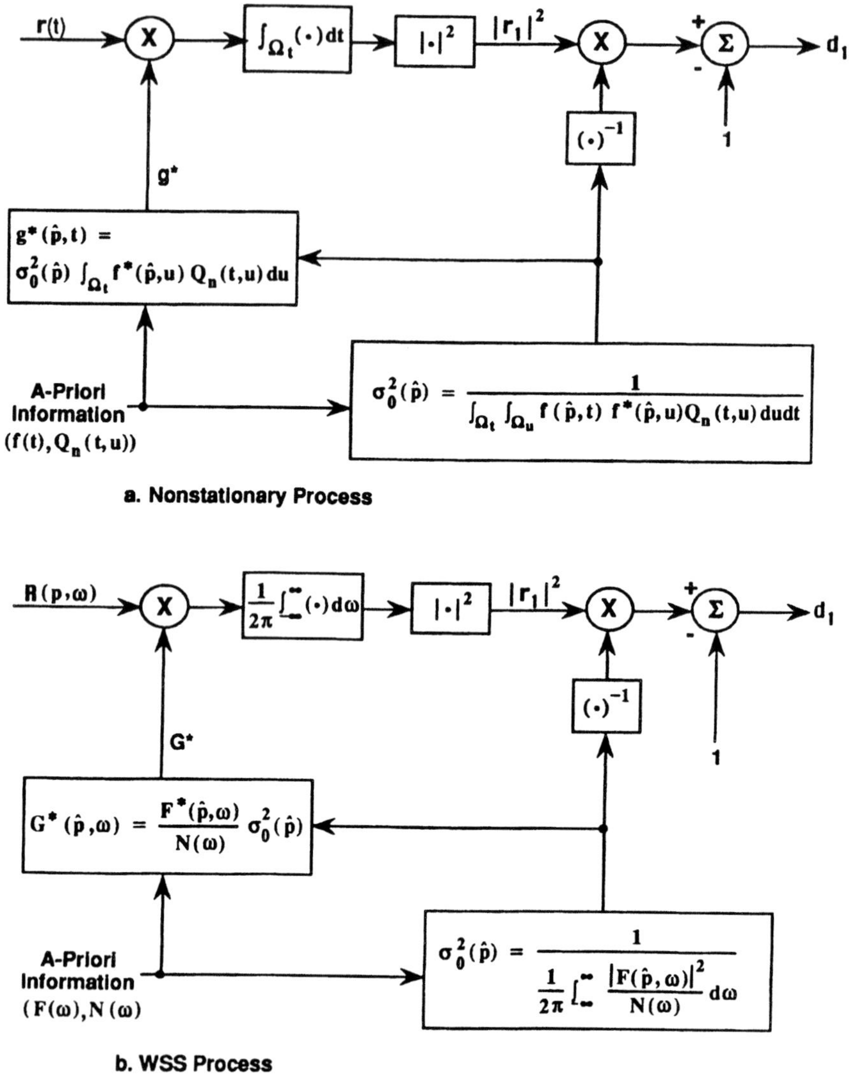

Figure 3.5: Receiver structures for correlated interference

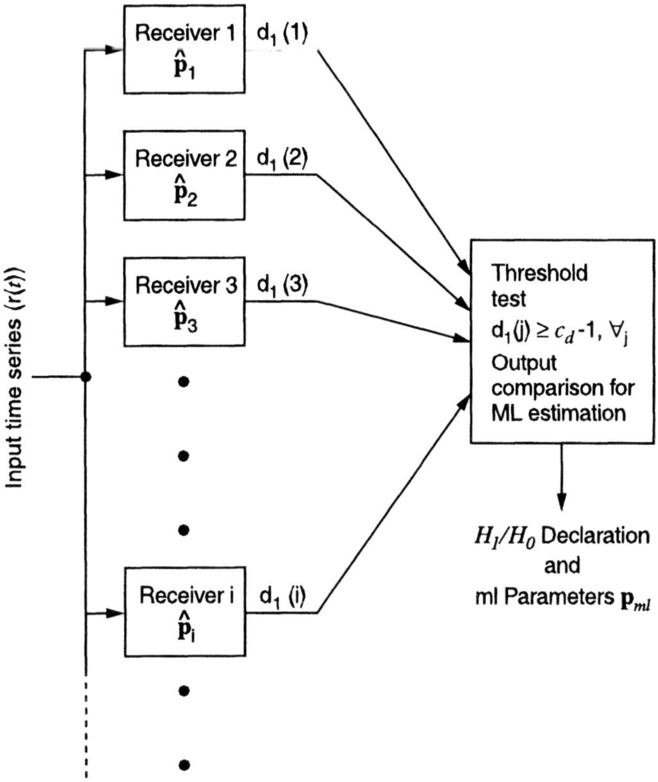

Figure 3.6: Multi-hypothesis receiver bank

For the purpose of illustration, let the transmitted waveform be the simple continuous wave (CW) function

$$f(t) = a(t) \exp j\omega_0 t \quad t \in [0, T]$$

where $a(t)$ is a real window function.[6] Furthermore assume that Doppler ϕ is the only parameter of interest and that range delay τ is zero. with bandwidth $W \sim 1/T << \omega_0/2\pi$.

Consequently, the parameter set is $\mathbf{p} = [0, \phi]$ and the echo return is modeled as narrowband frequency shifted signal (2.26) with $\theta_\tau = 0$.

[6]Window functions are usually real, non-negative, symmetric, normalized, and unimodal (single maximum) time functions defined over the pulse duration and are most often used to taper spectral sidebands [35–38]. The simplest is the rectangular window $a(t) = 1/T$, $t \in [0, T]$ and zero elsewhere.

When the Doppler band of interest is large with respect to the echo bandwidth, the echo power spectrum at a particular Doppler is approximated by a delta function

$$|S(\mathbf{p},\omega)|^2 = E_s\,|F(\mathbf{p},\omega)|^2 \approx E_s\delta(\omega - 2\pi\phi). \tag{3.95}$$

A bank of receivers hypothesizing discrete Dopplers $\hat{\phi}_i$ form signal replicas with spectra $\delta(\omega - 2\pi\hat{\phi}_i)$ at increments commensurate with the desired Doppler resolution which should be no larger than approximately the inverse time duration $1/T$. The interference is assumed to possess a non-white power spectrum $N(\omega)$ that is illustrated along with the receiver hypothesis spectra in Figure 3.7a.

Substituting the waveform spectrum into the expression for $\sigma_0^2(\hat{\mathbf{p}})$ given by (3.89) and expressing the result in terms of the parameters yields the relationship

$$\sigma_0^2(\hat{\mathbf{p}}) = N(2\pi\hat{\phi}) \tag{3.96}$$

where $\hat{\phi}$ assumes the discrete values $\hat{\phi}_i$ corresponding to the receiver hypotheses. The discrete values of $N(2\pi\hat{\phi}_i)$ at each frequency hypothesis are illustrated by the solid dots in the figure. Figure 3.7b illustrates the constant echo energy, matched receiver H_1 response across the receiver bank along with a CFAR threshold $\tilde{\gamma}_l$ set such that the ratio $\tilde{\gamma}_l/\sigma_0^2$ is constant for all Dopplers. The detection index is given by (3.81) using (3.89, 3.92) under WSS conditions. When $\hat{\mathbf{p}} = \mathbf{p}$,

$$D_{\mathbf{pp}} = \frac{E_s}{2\pi}\int_{-\infty}^{\infty}\frac{|F(\mathbf{p},\omega)|^2}{N(\omega)}d\omega \tag{3.97}$$

and substituting (3.95), the detection index at each matched hypothesis $\hat{\phi}_i = \phi_i$ becomes

$$\begin{aligned} D_{\mathbf{pp}} = D_{\phi_i\phi_i} &= \frac{E_s}{\sigma_0^2(\phi_i)} \\ &= \frac{E_s}{N(2\pi\phi_i)}. \end{aligned}$$

Since $D_{\mathbf{pp}}$ is inversely proportional to $\sigma_0^2(\mathbf{p})$ and $\wp_d$ and $\wp_f$ are related by (3.61), CFAR thresholding causes $\wp_d$ to increase monotonically as Doppler deviates from the band center and the interference rolls off. This of course assumes matched parameter situations, that is, the comparison is between receivers matched at each hypothesis to signals of identical

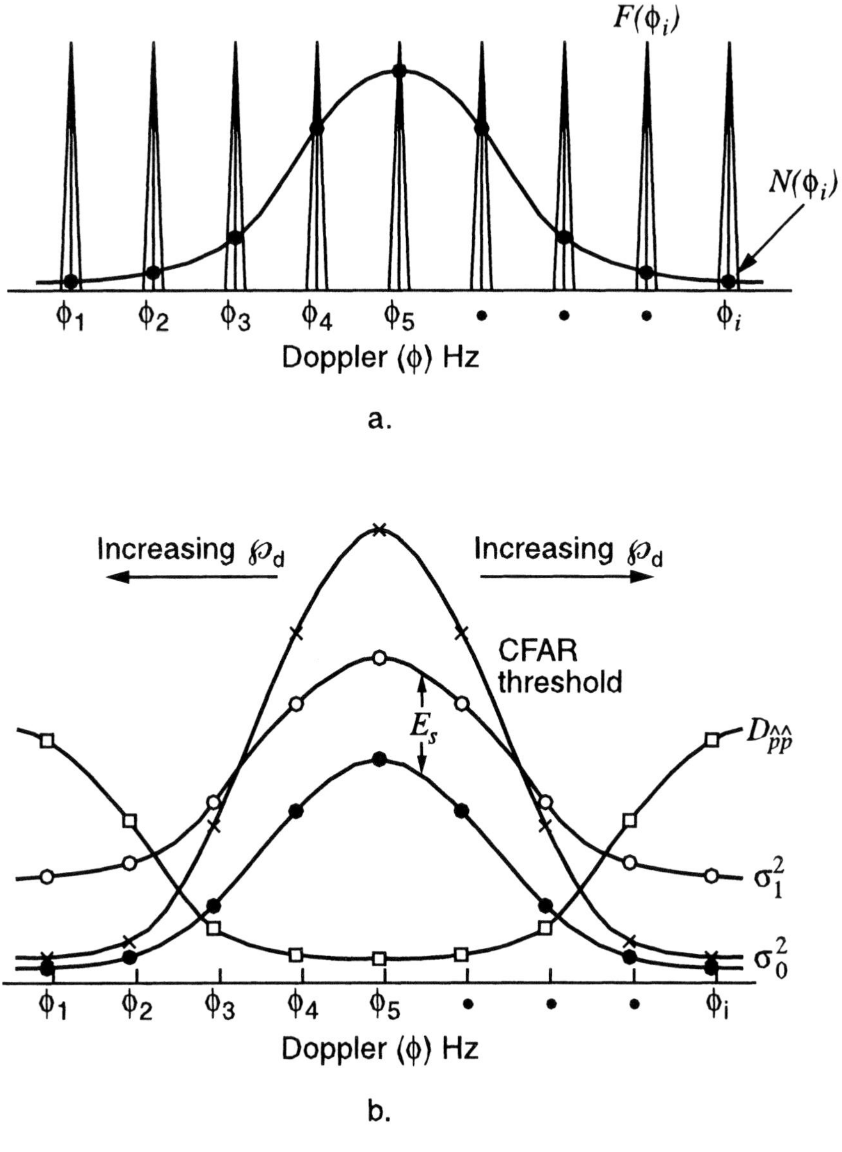

Figure 3.7: CFAR threshold example: (a) signal and interference spectra, (b) receiver outputs with the ambiguity constraint imposed

energy. For example, a receiver in the middle of the band at Doppler ϕ_5 would have a much lower detection probability given a signal with Doppler ϕ_5 than would a receiver at ϕ_1 given a signal with Doppler ϕ_1 and the same energy. As shown earlier, the detection index of the receiver matched to the input signal will always be greater than that of any other receiver.

The optimal mismatched receiver requires information about the interference covariance or spectrum in order to form the processing waveform. If this is lacking or not readily predictable the conventional matched filter receiver formed under the WGN assumption (3.58) may be used. This may result in reduced performance, the amount depending upon the interference characteristics.

The ratio of the detection indices of the optimal mismatched and conventional receivers is a convenient measure of relative performance. It is related to the spectral characteristics of the interference under WSS conditions where $D_{\mathbf{pp}}$ of the optimal mismatched receiver is given by (3.97). The WSS detection index for the conventional receiver is determined by using WSS versions of (3.75,3.76) with $f(\mathbf{p},t)$ substituted for $g(\mathbf{p},t)$ to evaluate the variances under H_0, H_1. The interference energy is

$$\sigma_0^2(\mathbf{p}) = \frac{1}{2\pi}\int_{-\infty}^{\infty} |F(\mathbf{p},\omega)|^2 N(\omega) d\omega \tag{3.98}$$

and σ_1^2 for matched parameters is given by (3.76). The detection index is defined by (3.81) with Φ_{ff} substituted for Φ_{fg} resulting in

$$D_{\mathbf{p}\hat{\mathbf{p}}} = \frac{E_s \Phi_{ff}(\mathbf{p},\hat{\mathbf{p}})}{\frac{1}{2\pi}\int_{-\infty}^{\infty} |F(\mathbf{p},\omega)|^2 N(\omega) d\omega} \quad \text{(conventional receiver)}$$

where $\Phi_{ff}(\mathbf{p},\hat{\mathbf{p}}) = 1$ when $\hat{\mathbf{p}} = \mathbf{p}$. The ratio of the optimal mismatched index (3.97) to that of the conventional receiver when $\hat{\mathbf{p}} = \mathbf{p}$ is

$$r_{oc} = \left(\frac{1}{2\pi}\int_{-\infty}^{\infty} |F(\mathbf{p},\omega)|^2 N(\omega) d\omega\right) \cdot \left(\frac{1}{2\pi}\int_{-\infty}^{\infty} \frac{|F(\mathbf{p},\omega)|^2}{N(\omega)} d\omega\right). \tag{3.99}$$

The spectral density $N(\omega)$ is a non-negative real quantity so that

$\sqrt{N(\omega)}$ is defined and

$$\left[\frac{1}{2\pi}\int_{-\infty}^{\infty}(|F(\mathbf{p},\omega)|\sqrt{N(\omega)})\left(\frac{|F(\mathbf{p},\omega)|}{\sqrt{N(\omega)}}d\omega\right)\right]^2$$
$$\leq\left(\frac{1}{2\pi}\int_{-\infty}^{\infty}|F(\mathbf{p},\omega)|^2N(\omega)d\omega\right)\left(\frac{1}{2\pi}\int_{-\infty}^{\infty}\frac{|F(\mathbf{p},\omega)|^2}{N(\omega)}d\omega\right)$$

by the Schwarz inequality. The transmitted waveform is energy normalized so that the left hand term is unity and $r_{oc} \geq 1$ for the matched parameter hypothesis $\hat{\mathbf{p}} = \mathbf{p}$. Equality occurs when $N(\omega)$ is constant in frequency which is the WGN condition for which the matched receiver is optimum. It is suboptimum for non-white interference spectra.

The WSS relationship (3.97) suggests that it is possible to maximize $D_{\mathbf{pp}}$ by proper signal design. If the interference spectral density $N(\omega)$ is known, it is possible to vary the transmit and hence echo spectrum $F(\mathbf{p},\omega)$ so as to seek a maximum of $D_{\mathbf{pp}}$. Let Doppler be the parameter of interest as in Example 3.5 and seek a signal such that $D_{\mathbf{pp}}$ is maximized for a particular $F(\omega - 2\pi\phi)$. This would be the echo-location problem of designing a signal that provides maximum performance for a slowly fluctuating point scatterer embedded in reverberation or clutter confined to a spectral band about ϕ.

Let $F_m(\omega) = |F(\omega - 2\pi\phi)|^2$ and consider the maximization of the integral (3.97)

$$\frac{1}{2\pi}\int_{-\infty}^{\infty}\frac{F_m(\omega)}{N(\omega)}d\omega$$

the argument of which is a bounded positive definite function over the band of interest. Maximization of the square of a positive definite quantity is equivalent to maximization of the quantity itself and by the Schwarz inequality,

$$\left(\int_{-\infty}^{\infty}\frac{F_m(\omega)}{N(\omega)}d\omega\right)^2 \leq \int_{-\infty}^{\infty}(F_m(\omega))^2\,d\omega\int_{-\infty}^{\infty}\frac{1}{(N(\omega))^2}d\omega$$

with equality for $F_m(\omega) = |F(\omega - 2\pi\phi)|^2 = \frac{\alpha}{N(\omega)}$. α is a constant that is determined by applying the energy normalization

$$\frac{1}{2\pi}\int_{-\infty}^{\infty}F_m(\omega)d\omega = 1$$

yielding

$$\alpha = \frac{2\pi}{\int_{-\infty}^{\infty}\frac{1}{N(\omega)}d\omega}$$

and the optimal waveform spectral density

$$|F(\omega - 2\pi\phi)|^2 = \frac{2\pi}{N(\omega)\int_{-\infty}^{\infty}\frac{1}{N(\omega)}d\omega}.$$

This is optimum only for the specified ϕ where for example, $\phi = 0$ when it is desired to specify a waveform for the detection of a stationary object in wide sense stationary clutter/reverberation. It should be noted that only the power spectrum of the waveform is optimized by this procedure not the actual spectrum $F(\omega)$. This is because a second order characterization of the interference process has been considered either via the covariance or the associated power spectrum, both of which are sufficient when considering Gaussian processes. The designer is free to choose the underlying waveform modulation function as long as the power spectrum remains as specified. Indeed in most situations, the waveform is the only function fully under the designer's control [39] and the choice of the modulation depends in part upon the desired resolution properties of the signal. This and the general problem of signal design for operation in correlated interference has received extensive treatment in the literature [40–45] and is discussed in Chapter 6.

3.9 PARAMETER ESTIMATION

3.9.1 Maximum Likelihood Estimation

We now turn to a more detailed examination of the estimation problem in order to consider the accuracy and bias of parameter estimates and the role played by the transmitted waveform and its ambiguity function. In the discussion of the detection problem, parameter estimation was considered as an adjunct to echo detection implemented in terms of a multi-hypothesis decision test. Echos were modeled as discrete messages to be detected by a bank of receivers each tuned for a specific message characterized by a parameter hypothesis $\hat{\mathbf{p}}$. This is equivalent to a repeated application of a likelihood ratio test and the maximum receiver response indicates the maximum likelihood parameter estimate.

In an echo-location context, parameters such as range delay and Doppler have continuous values over some finite range and are usually regarded as non-random but unknown. If the detection and estimation process is implemented as a discrete hypothesis testing procedure (receiver bank), it is necessary to specify the desired search grid resolution and consequently the number of receiver realizations required to cover the parameter space of interest. There is clearly a cost vs benefit trade-off between estimation accuracy and the number of receiver realizations required. Each increment in the size of the receiver bank incurs a cost in processing time and/or complexity and it is necessary to analyze the effects of signal waveform, SIR, and the range and number of parameters to be estimated as a part of the design process.

The slowly fluctuating point scatterer discussed in Section 3.4 is a simple model suitable for the discussion of both detection and parameter estimation in an echo-location context. It is therefore assumed that the data time series is reduced to the random statistic r_1 the form of which depends upon the assumptions regarding the interference process (3.58, 3.65, 3.91). In the absence of an echo (H_0) the mean square response σ_0^2 of a receiver forming r_1 reflects only the interference. This is constant as a function of the parameters in WGN but more generally, in correlated interference it is a function of one or more of the parameters and is denoted by $\sigma_0^2(\hat{\mathbf{p}})$ (3.75, 3.89, 3.96). Under H_1, the random data time series $\mathsf{r}(t)$ contains the slowly fluctuating echo signal $\mathsf{s}(\mathbf{p}, t)$ where $\mathbf{p}$ is a unknown but non-random parameter set taking on a continuum of possible values over some finite range. Since the input data is random, the response of receivers forming realizations of r_1 and parameter estimates based upon the responses will be random. The accuracy of the estimate and the role played by the receiver structure, signal waveform, and interference level are therefore items for consideration. It is preferable to adopt a CFAR thresholding scheme in the presence of correlated interference when the false alarm rate is of paramount concern. This especially true for echo-location in the presence of non-uniform clutter or reverberation. The detection probability is then maximized for the parameter hypothesis corresponding to the maximum of the detection index $D_{\mathbf{p}\hat{\mathbf{p}}}$. The ML parameter estimate $\mathbf{p}_{ml}$ is equal to that $\hat{\mathbf{p}}$ maximizing the statistic d_1 in comparison to a constant threshold for any single realization of the data. This is because $\mathcal{E}\{\mathsf{d}_1\} = D_{\mathbf{p}\hat{\mathbf{p}}} \leq D_{\mathbf{pp}}$ and $\mathbf{p}_{ml} = \mathbf{p}$ maximizes the detection index in comparison to any other value of $\hat{\mathbf{p}}$. Both the log likelihood ratio and the detection index may be used as performance measures as they are related by

$$\mathcal{E}\{\ell_r(\mathsf{r}_1)\} = \ln\left(\frac{1}{1 + D_{\mathbf{p}\hat{\mathbf{p}}}}\right) + D_{\mathbf{p}\hat{\mathbf{p}}}$$

and $D_{\mathbf{p}\hat{\mathbf{p}}} \geq 0 \; \forall \; \mathbf{p}, \hat{\mathbf{p}}$. The relationship is monotonic and both are maximized by $\hat{\mathbf{p}} = \mathbf{p}_{ml}$ when the same values of σ_0^2, σ_1^2 are used.

White Gaussian noise (WGN) energy $\sigma_0^2 = N_0$ is not a function of the parameters. $D_{\mathbf{p}\hat{\mathbf{p}}}$ is given by (3.62),

$$\mathcal{E}\{|\mathsf{r}_1|^2|H_1\} = \sigma_1^2 = E_s\Phi_{ff}(\mathbf{p},\hat{\mathbf{p}}) + \sigma_0^2,$$

and the auto-ambiguity function is bounded by unity and maximized for $\hat{\mathbf{p}} = \mathbf{p}$. The ML estimate corresponding to the $\hat{\mathbf{p}}$ maximizing $|\mathsf{r}|^2$ (3.58) is equivalent to that obtained by maximizing d_1 or the log likelihood.

$$\mathcal{E}\{|\mathsf{r}_1|^2|H_1\} = \sigma_1^2 = E_s\Phi_{fg}(\mathbf{p},\hat{\mathbf{p}}) + \sigma_0^2(\hat{\mathbf{p}})$$

when the interference is correlated and $\sigma_0^2(\hat{\mathbf{p}})$ is given by (3.73) and (3.88) for non-stationary and WSS processes respectively for the constrained receiver formulation. The cross-ambiguity function is constrained to be unity when its parameters match but was shown to be bounded by the ratio $\sigma_0^2(\hat{\mathbf{p}})/\sigma_0^2(\mathbf{p})$ (3.82) that may exceed unity. Additionally, $|\mathsf{r}_1|^2$ is a function of the variable interference background and consequently an unbiased estimate is not guaranteed by merely determining the absolute maximum of $|\mathsf{r}_1|^2$.

The parameters affect the signal auto-AF but not the interference background under WGN conditions. The narrowband auto-ambiguity function is "wide sense stationary" with respect to its parameters meaning that $\Phi_{ff}(\mathbf{p},\hat{\mathbf{p}}) = \Phi_{ff}(\mathbf{p}-\hat{\mathbf{p}})$. Consequently, the detection index depends only upon the degree of mismatch between the echo and replica and not upon the absolute value of the hypothesis being made for that model. In other words, the interference energy is independent of the parameters, the auto-AF depends only upon the properties of the waveform employed, and for CFAR, the maximum attainable detection probability is the same for all $\mathbf{p}$. The peak of the detection index coincides with that of the auto-AF thus $\mathbf{p}_{ml}$ corresponds to that $\hat{\mathbf{p}}$ maximizing the statistic $|\mathsf{r}_1|^2$, the detection parameter d_1 (3.93), or the log likelihood.

When the interference is correlated and the optimum receiver structure employed, the detection index becomes in part, a function of the cross-AF between the transmitted waveform $f(\mathbf{p},t)$ and the processing function $g(\mathbf{p},t)$. This in turn depends upon the second order statistical properties of the interference that may vary as a function of the parameter hypothesis $\hat{\mathbf{p}}$. In non-stationary interference, the cross-AF is obtained from (3.72, 3.73, 3.77) as

$$\Phi_{fg}(\mathbf{p},\hat{\mathbf{p}}) = \left|\sigma_0^2(\hat{\mathbf{p}})\int_{\Omega_t}\int_{\Omega_u} f(\mathbf{p},t)f^*(\hat{\mathbf{p}},u)Q_n(u,t)du\,dt\right|^2 \quad (3.100)$$

with $\sigma_0^2(\mathbf{p}, \hat{\mathbf{p}}) = k(\hat{\mathbf{p}})$ as given by (3.73). Under WSS conditions,

$$\Phi_{fg}(\mathbf{p}, \hat{\mathbf{p}}) = \left| \frac{\sigma_0^2(\hat{\mathbf{p}})}{2\pi} \int_{-\infty}^{\infty} \frac{F(\mathbf{p}, \omega) F^*(\hat{\mathbf{p}}, \omega)}{N(\omega)} d\omega \right|^2 \tag{3.101}$$

and $\sigma_0^2(\hat{\mathbf{p}}) = k(\hat{\mathbf{p}})$ as shown by (3.89). Equations (3.81, 3.82) in the previous section indicate that

$$D_{\mathbf{p}\hat{\mathbf{p}}} \leq D_{\mathbf{p}\mathbf{p}} \Rightarrow \frac{E_s \Phi_{fg}(\mathbf{p}, \hat{\mathbf{p}})}{\sigma_0^2(\hat{\mathbf{p}})} \leq \frac{E_s}{\sigma_0^2(\mathbf{p})} \tag{3.102}$$

providing a non-uniform bound for the cross-AF. The bound is uniform and equal to unity in the case of WGN. The quantity $E_s/\sigma_0^2(\mathbf{p})$ represents the maximum available signal to noise ratio (SIR) available for a signal with energy E_s and parameterized by $\mathbf{p}$. It indicates the maximum attainable detection probability for CFAR in non-uniform interference that may vary as illustrated in Figure 3.7. The bounding relationship (3.102) indicates that the maximum is realized only for the correct parameter hypothesis $\hat{\mathbf{p}} = \mathbf{p}$. This is illustrated by defining

$$\text{SIR}(\mathbf{p}) = \frac{E_s}{\sigma_0^2(\mathbf{p})}, \quad \text{SIR}(\hat{\mathbf{p}}) = \frac{E_s}{\sigma_0^2(\hat{\mathbf{p}})}$$

and using (3.102) so that

$$\Phi_{fg}(\mathbf{p}, \hat{\mathbf{p}}) \leq \frac{\sigma_0^2(\mathbf{p})}{\sigma_0^2(\hat{\mathbf{p}})} = \frac{\text{SIR}(\mathbf{p})}{\text{SIR}(\hat{\mathbf{p}})}. \tag{3.103}$$

A non-uniform interference process and associated SIR is illustrated in Figure 3.8a for a single parameter. The abscissa of the plots might represent a clutter or reverberation frequency spectrum when the parameter is Doppler as in Example 3.5. The achievable $\text{SIR}(\hat{\mathbf{p}})$ is the difference (in dB) between the E_s and $\sigma_0^2(\hat{\mathbf{p}})$ as indicated in the figure. The upper bound of the cross-AF is obtained from (3.103) for a specific value of the data parameter $\mathbf{p}$. It is constrained to be unity (0 dB) at $\hat{\mathbf{p}} = \mathbf{p}$ but clearly may exceed this when the maximum attainable SIR at $\mathbf{p}$ exceeds that at $\hat{\mathbf{p}}$. This curve only establishes the upper bound. The actual shape of the cross-AF will depend upon the waveform function $f(t)$ and the interference distribution according to (3.100) or (3.101). The peak of the cross-AF does not necessarily coincide with the correct parameter in the figure but the detection index

$$D_{\mathbf{p}\hat{\mathbf{p}}} = \text{SIR}(\hat{\mathbf{p}}) \Phi_{fg}(\mathbf{p}, \hat{\mathbf{p}})$$

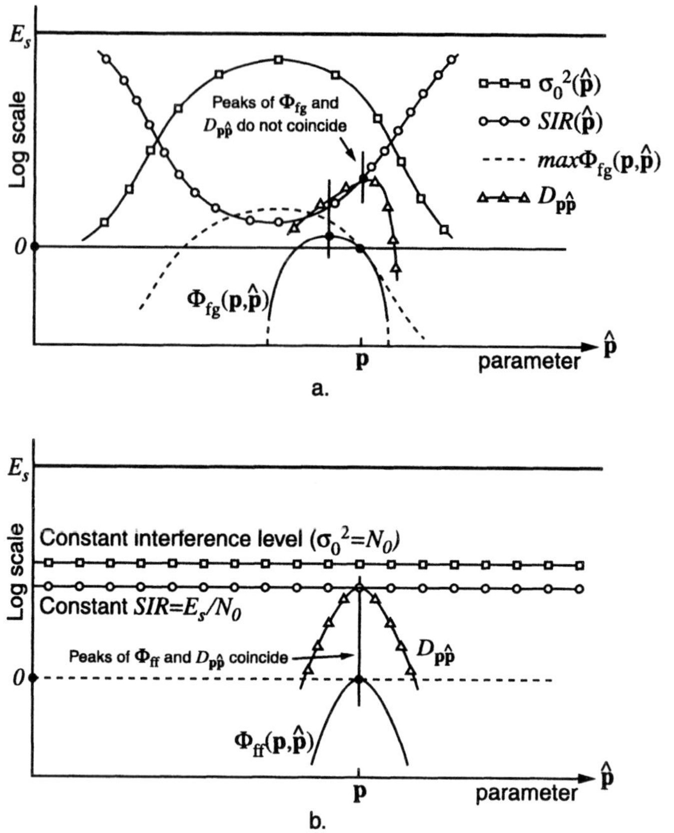

Figure 3.8: Interference effects in estimation: (a) correlated interference (b) white Gaussian noise (WGN)

clearly peaks at the right place. This situation contrasts with that of WGN shown in Figure 3.8b where the auto-AF and detection index are maximized at the same point. This illustrates the potential difficulty of parameter estimation in the presence of non-uniform interference that in many cases is highly non-stationary with properties that are difficult to predict.

The estimation of parameters by using a bank of receivers and seeking the maximum response upon presentation of a realization of the data is an implementation of the maximum likelihood approach [6]. It is not the only way to estimate parameters and in order to develop a generally useful measure of estimator performance, it is necessary to consider the problem in a broader context wherein no particular estimation scheme is assumed.

3.9.2 The Cramér-Rao Bound (CRB)

The maximization of the detector response in a receiver bank is an example of the process by which a realization r_1 of the random statistic (3.58) derived from an input time series (3.49) is used to obtain the maximum likelihood estimate $\hat{\mathbf{p}} = \mathbf{p}_{ml}$ of the actual parameters $\mathbf{p}$. The input is random hence the estimates derived from realizations of it are also random and characterized by a (usually unknown) density function. It is possible however to assess the average performance of an estimator by establishing a lower bound for the estimate error variance. The *Cramér-Rao Bound* (CRB) is widely used and serves as a performance benchmark for the evaluation of different approaches. It is also used to identify those signal waveform attributes that are critical for the estimation of particular parameters.

Consider the generation an estimate $\hat{p}$ of a single non-random parameter p from a realization of the received echo data. The input is the random vector $\mathbf{r}$ or is transformed into vector form from a continuous time series by a suitable expansion and is characterized by the conditional multidimensional pdf $\mathsf{f}(\mathbf{r}|p)$. This is called the *likelihood function* and represents the likelihood of the observation $\mathbf{r}$ when the random data vector $\mathbf{r}$ is parameterized by p [1,3]. It is related to the likelihood ratio used for detection by the pdf under H_0 as in (3.54). The expected value of the random estimates $\hat{\mathrm{p}}$ over the data ensemble is

$$\bar{p} = \mathcal{E}\{\hat{p}(\mathbf{r})\} = \int_{-\infty}^{\infty} \hat{p}(\mathbf{r})\mathsf{f}(\mathbf{r}|p)d\mathbf{r} \tag{3.104}$$

and the error between the expected estimate and the parameter is

$$\bar{e}_b = (\bar{p} - p) = \int_{-\infty}^{\infty} (\hat{p} - p)\mathsf{f}(\mathbf{r}|p)d\mathbf{r} \tag{3.105}$$

and called the estimator *bias* with $\hat{p} \equiv \hat{p}(\mathbf{r})$. This is differentiated with respect to p resulting in

$$\frac{d\bar{e}_b}{dp} = -\int_{-\infty}^{\infty} \mathsf{f}(\mathbf{r}|p)d\mathbf{r} + \int_{-\infty}^{\infty} (\hat{p} - p)\frac{\partial \mathsf{f}(\mathbf{r}|p)}{\partial p}d\mathbf{r}$$

assuming that the partial derivatives of $\mathsf{f}(\mathbf{r}|p)$ with respect to p exist and are integrable. The first term is the density integrated over its range which is unity thus

$$\int_{-\infty}^{\infty} (\hat{p} - p) \frac{\partial \mathsf{f}(\mathbf{r}|p)}{\partial p} d\mathbf{r} = \left(1 + \frac{d\bar{e}_b}{dp}\right). \tag{3.106}$$

The log likelihood function (llf) and its derivatives with respect to the parameter are denoted by

$$\begin{aligned} l(\mathbf{r}, p) &= \ln \mathsf{f}(\mathbf{r}|p) \\ \dot{l}(\mathbf{r}, p) &= \frac{\partial \ln \mathsf{f}(\mathbf{r}|p)}{\partial p} \\ \ddot{l}(\mathbf{r}, p) &= \frac{\partial^2 \ln \mathsf{f}(\mathbf{r}|p)}{\partial p^2}. \end{aligned}$$

The partial derivative in (3.106) is expressed as

$$\frac{\partial \mathsf{f}(\mathbf{r}|p)}{\partial p} = \mathsf{f}(\mathbf{r}|p)\dot{l}(\mathbf{r}, p) \tag{3.107}$$

and (3.106) becomes

$$\int_{-\infty}^{\infty} (\hat{p} - p)\mathsf{f}(\mathbf{r}|p)\dot{l}(\mathbf{r}, p) d\mathbf{r} = \left(1 + \frac{d\bar{e}_b}{dp}\right).$$

The density function is non-negative and the integrand on the left may be split forming

$$\int_{-\infty}^{\infty} \left[(\hat{p} - p)\sqrt{\mathsf{f}(\mathbf{r}|p)}\right] \left[\sqrt{\mathsf{f}(\mathbf{r}|p)}\dot{l}(\mathbf{r}, p)\right] d\mathbf{r} = \left(1 + \frac{d\bar{e}_b}{dp}\right).$$

Squaring both sides and invoking the Schwarz inequality establishes the bound

$$\left[\int_{-\infty}^{\infty} (\hat{p} - p)^2 \mathsf{f}(\mathbf{r}|p) d\mathbf{r}\right] \left[\int_{-\infty}^{\infty} \mathsf{f}(\mathbf{r}|p) \left(\dot{l}(\mathbf{r}, p)\right)^2 d\mathbf{r}\right] \geq \left(1 + \frac{d\bar{e}_b}{dp}\right)^2.$$

The first term on the left is the mean square estimation error $\sigma_e^2 = \mathcal{E}\{(\hat{p} - p)^2\}$ and the second is the expectation of the square of the derivative of the llf. Consequently the mean square error is bounded from below by

$$\sigma_e^2 \geq \frac{\left(1 + \frac{d\bar{e}_b}{dp}\right)^2}{\mathcal{E}\left\{\left(\dot{l}(\mathbf{r}, p)\right)^2\right\}}.$$

The variance of the estimate is expressed in terms of the bias and the mean square error by (3.105), where

$$\begin{aligned}\sigma_{\hat{p}}^2 &= \mathcal{E}\{(\hat{p} - \bar{p})^2\} \\ &= \sigma_e^2 - \bar{e}_b^2\end{aligned}$$

and the variance is bounded by

$$\sigma_{\hat{p}}^2 \geq \frac{\left(1 + \frac{d\bar{e}_b}{dp}\right)^2}{\mathcal{E}\left\{\left(\dot{l}(\mathbf{r}, p)\right)^2\right\}} - \bar{e}_b^2.$$

The estimator is unbiased, $\bar{p} = p$, and the lower bound (CRB) is

$$\boxed{\sigma_{\hat{p}}^2 \geq \frac{1}{\mathcal{E}\left\{\left(\dot{l}(\mathbf{r}, p)\right)^2\right\}} \qquad (3.108)} \qquad \text{(scalar CRB)}$$

when $\bar{e}_b = 0$. An alternative form of the bound is obtained by observing that

$$\int_{-\infty}^{\infty} \mathsf{f}(\mathbf{r}|p) d\mathbf{r} = 1.$$

Differentiating with respect to p and using (3.107) yields

$$\int_{-\infty}^{\infty} \dot{l}(\mathbf{r}, p) \mathsf{f}(\mathbf{r}|p) d\mathbf{r} = 0$$

which indicates that the derivative of the llf for fixed p is a zero mean process when the data realization $\mathbf{r}$ is replaced by the random data $\mathbf{r}$. Taking the second derivative and again using (3.107) results in

$$\int_{-\infty}^{\infty} \ddot{l}(\mathbf{r}, p) \mathsf{f}(\mathbf{r}|p) d\mathbf{r} = -\int_{-\infty}^{\infty} \left(\dot{l}(\mathbf{r}, p)\right)^2 \mathsf{f}(\mathbf{r}|p) d\mathbf{r}.$$

which is a relationship between the expectations

$$\mathcal{E}\left\{\ddot{l}(\mathbf{r}, p)\right\} = -\mathcal{E}\left\{\left(\dot{l}(\mathbf{r}, p)\right)^2\right\}.$$

Consequently the equivalent form for the unbiased estimator is

$$\boxed{\sigma_{\hat{p}}^2 \geq -\frac{1}{\mathcal{E}\left\{\ddot{l}(\mathbf{r}, p)\right\}}.} \qquad \text{(equivalent scalar CRB)}$$

This and (3.108) are alternative forms of the Cramér-Rao Bound (CRB) and establish a lower bound upon the variance of <u>any</u> unbiased estimator [1, 3, 46]. It is satisfied with equality only if the derivative of the log density has the form

$$\dot{l}(\mathbf{r}, p) = c(p)(\hat{p} - p). \tag{3.109}$$

Estimators for which this is true are called *efficient estimators*.

The density $f(\mathbf{r}|p)$ represents the incremental probability of observing non-random realizations of the data vector $\mathbf{r}$ drawn from the random data $\mathbf{r}$ given that the data is parameterized by a particular value of p. Conversely, when it is considered as a function of p for a particular realization, it represents the likelihood or the relative probability of the occurrence of the observation $\mathbf{r}$ for different values of the parameter. In other words, given a realization of the data, the value of p maximizing the likelihood function or equivalently the llf, maximizes the probability of having observed the particular observation $\mathbf{r}$. It is called the *maximum likelihood* (ML) estimate p_{ml} and satisfies the necessary condition

$$\dot{l}(\mathbf{r}, p)\Big|_{p=p_{ml}} = 0$$

which is called the *likelihood equation.* In the case of the slowly fluctuating echo model, the receiver bank generating the statistic d_1 (3.93) parameterized by the trial parameter $\hat{p}$ is an example of a ML estimator where p_{ml} is that value of $\hat{p}$ maximizing the response. If the derivative is linear in the parameters as in (3.109), the non trivial solution $(\hat{p} - p) = 0$ implies that the the estimator is efficient and $\hat{p} = p_{ml}$. In other words, if an efficient estimator exists, it will be a maximum likelihood estimator. The estimate may be equivalently found by maximizing the likelihood function multiplied or divided by any other function that does not depend upon the parameters. Consequently the estimate is also obtained by seeking the maximum of the log likelihood ratio ℓ_r since the data density is independent of the signal parameters under the null hypothesis H_0 [1].

3.9.3 Multiple Parameters

Multiple parameter estimation naturally arises in echo-location when joint estimates of range delay, Doppler (dilation), and in some cases, bearing are desired. The extension of the CRB to the vector parameter case is a straightforward extension of the scalar case but the likelihood function derived from the random data now becomes a multidimensional surface in parameter space denoted by $l(\mathbf{r}, \mathbf{p})$. The solution of the likelihood equation then becomes a search for the null point of the gradient

$$\nabla_{\mathbf{p}} l(\mathbf{r}, \mathbf{p})\big|_{\mathbf{p}=\mathbf{p}_{ml}} = 0 \qquad (3.110)$$

for a realization of the data where $\nabla_{\mathbf{p}}$ is the vector gradient operator

$$\nabla_{\mathbf{p}}^{\mathrm{T}} = \left[\frac{\partial}{\partial p_1}, \cdots \frac{\partial}{\partial p_i}, \cdots \frac{\partial}{\partial p_{N_p}}\right]$$

and N_p is the number of parameters. Regardless of the estimator structure used, the resulting estimates are random and it is necessary to consider not only the individual estimate variances but the possibility of error coupling as expressed by the covariance $\mathbf{R}_{\hat{\mathbf{p}}} = \mathcal{E}\{(\hat{\mathbf{p}} - \bar{\mathbf{p}})(\hat{\mathbf{p}} - \bar{\mathbf{p}})^{\mathrm{T}}\}$.

The CRB derivation parallels that of the scalar case where the first step defined the expected value of the estimate (3.104) which in this case is the vector

$$\bar{\mathbf{p}} = \mathcal{E}\{\hat{\mathbf{p}}(\mathbf{r})\} = \int_{-\infty}^{\infty} \hat{\mathbf{p}}(\mathbf{r}) \mathsf{f}(\mathbf{r}|\mathbf{p}) d\mathbf{r}.$$

The estimator bias analogous to (3.105) becomes

$$\bar{\mathbf{e}}_b = (\bar{\mathbf{p}} - \mathbf{p}) = \int_{-\infty}^{\infty} (\hat{\mathbf{p}} - \mathbf{p}) \mathsf{f}(\mathbf{r}|\mathbf{p}) d\mathbf{r}.$$

and the gradient of the bias error with respect to the parameters is the matrix

$$\begin{aligned}\nabla_{\mathbf{p}} \bar{\mathbf{e}}_b^{\mathrm{T}} &= \int_{-\infty}^{\infty} \nabla_{\mathbf{p}} \left(\mathsf{f}(\mathbf{r}|\mathbf{p})(\hat{\mathbf{p}} - \mathbf{p})^{\mathrm{T}}\right) d\mathbf{r} \\ &= \int_{-\infty}^{\infty} \mathsf{f}(\mathbf{r}|\mathbf{p}) \nabla_{\mathbf{p}} (\hat{\mathbf{p}} - \mathbf{p})^{\mathrm{T}} + \nabla_{\mathbf{p}} \mathsf{f}(\mathbf{r}|\mathbf{p})(\hat{\mathbf{p}} - \mathbf{p})^{\mathrm{T}} d\mathbf{r}.\end{aligned}$$

The parameter vector $\mathbf{p}$ consists of distinct non-random scalar parameters that are assumed to be independent, that is, they are not functionally related. Consequently,

$$\frac{\partial}{\partial p_i}(\hat{p}_j - p_j) = \begin{cases} 0 & \text{if } i \neq j, \\ -1 & \text{if } i = j \end{cases}$$

and the gradient of the bias is

$$\nabla_{\mathbf{p}} \bar{\mathbf{e}}_b^{\mathrm{T}} = -\mathbf{I} + \int_{-\infty}^{\infty} \nabla_{\mathbf{p}} \mathsf{f}(\mathbf{r}|\mathbf{p})(\hat{\mathbf{p}} - \mathbf{p})^{\mathrm{T}} d\mathbf{r}.$$

If the vector equivalent of (3.107) is substituted for the gradient of the pdf, the integral becomes the expectation of the outer product of the error vector $\hat{\mathbf{p}} - \mathbf{p}$ and the gradient of the likelihood function. Rearranging produces

$$\mathcal{E}\left\{\nabla_{\mathbf{p}} l(\mathbf{r},\mathbf{p})(\hat{\mathbf{p}}-\mathbf{p})^{\mathrm{T}}\right\} = \mathbf{I} + \nabla_{\mathbf{p}} \bar{\mathbf{e}}_b^{\mathrm{T}} \tag{3.111}$$

and if the estimator is unbiased, the right hand side reduces to the identity.

The CRB of the unbiased estimator is derived by assuming that $\bar{\mathbf{p}} = \mathbf{p}$ and forming the vector

$$\mathbf{w} = \begin{bmatrix} \hat{\mathbf{p}} - \mathbf{p} \\ \nabla_{\mathbf{p}} l(\mathbf{r},\mathbf{p}) \end{bmatrix}$$

and its associated covariance

$$\mathbf{W} = \mathcal{E}\left\{\mathbf{w}\mathbf{w}^{\mathrm{T}}\right\} = \begin{bmatrix} \mathbf{R}_{\hat{\mathbf{p}}} & \mathbf{I} \\ \mathbf{I} & \mathbf{J} \end{bmatrix}.$$

The off diagonal terms are identities for the unbiased estimator by virtue of (3.111). $\mathbf{J} = \mathcal{E}\{(\nabla_{\mathbf{p}} l)(\nabla_{\mathbf{p}} l)^{\mathrm{T}}\}$ is called the *Fisher Information Matrix* [1,3,46]. It is positive definite with diagonal terms

$$J_{ii} = \mathcal{E}\left\{\left(\frac{\partial l}{\partial p_i}\right)^2\right\} > 0 \quad \forall i$$

and thus its inverse $\mathbf{J}^{-1}$ exists. $\mathbf{R}_{\hat{\mathbf{p}}}$ is symmetric and positive semidefinite (psd) because it is a covariance matrix. Therefore, $\mathbf{W}$ is also psd and the quadratic form $Q = \mathbf{x}^{\mathrm{T}}\mathbf{W}\mathbf{x} \geq 0 \ \forall \mathbf{x}$. The vector $\mathbf{x}$ may be partitioned as $\mathbf{x}^{\mathrm{T}} = [\mathbf{x}_1^{\mathrm{T}}, \mathbf{x}_2^{\mathrm{T}}]$ so that

$$Q = \mathbf{x}_1^{\mathrm{T}}\mathbf{R}_{\hat{\mathbf{p}}}\mathbf{x}_1 + 2\mathbf{x}_2^{\mathrm{T}}\mathbf{x}_1 + \mathbf{x}_2^{\mathrm{T}}\mathbf{J}\mathbf{x}_2 \geq 0.$$

For any $\mathbf{x}_1$, the vector $\mathbf{x}_2$ minimizing Q is determined from the extremal solution

$$\begin{aligned} \frac{\partial Q}{\partial \mathbf{x}_2} &= 2\mathbf{x}_1 + 2\mathbf{J}\mathbf{x}_2 = 0 \\ &\Rightarrow \mathbf{x}_2 = -\mathbf{J}^{-1}\mathbf{x}_1. \end{aligned}$$

Substituting this for $\mathbf{x}_2$, results in the minimum Q for fixed $\mathbf{x}_1$

$$\begin{aligned} Q \geq Q_{min} &= \mathbf{x}_1^{\mathrm{T}}\mathbf{R}_{\hat{\mathbf{p}}}\mathbf{x}_1 - 2\mathbf{x}_1^{\mathrm{T}}\mathbf{J}^{-1}\mathbf{x}_1 + \mathbf{x}_1^{\mathrm{T}}\mathbf{J}^{-1}\mathbf{J}\mathbf{J}^{-1}\mathbf{x}_1 \\ &= \mathbf{x}_1^{\mathrm{T}}\left(\mathbf{R}_{\hat{\mathbf{p}}} - \mathbf{J}^{-1}\right)\mathbf{x}_1 \geq 0 \end{aligned}$$

indicating that $\left(\mathbf{R}_{\hat{\mathbf{p}}} - \mathbf{J}^{-1}\right)$ is also psd. Its diagonal terms are either positive or zero and the error variance for the i-th parameter is bounded from below by

$$\boxed{\sigma_{pi}^2 = \mathcal{E}\{(\hat{p}(\mathbf{r}) - p)_i^2\} \geq \left(\mathbf{J}^{-1}\right)_{ii} \quad (3.112)}$$

(multiple parameter CRB)

which represents the CRB for the vector (multiparameter) case. The bound is satisfied with equality and the estimator is said to be efficient if it is an unbiased ML estimator (3.110) that is linear in the error

$$\nabla_{\mathbf{p}} l(\mathbf{r}, \mathbf{p}) = c(\mathbf{p})(\hat{\mathbf{p}} - \mathbf{p}) \tag{3.113}$$

where $c(\mathbf{p})$ is a scalar function. For this the Fisher matrix is

$$\mathbf{J} = \mathcal{E}\{\nabla_{\mathbf{p}} l \nabla_{\mathbf{p}}^{\mathrm{T}} l\} = c(\mathbf{p})^2 \mathbf{R}_{\hat{\mathbf{p}}} \tag{3.114}$$

and using (3.111) for the unbiased estimator results in

$$\mathcal{E}\{(\nabla_{\mathbf{p}} l)(\hat{\mathbf{p}} - \mathbf{p})^{\mathrm{T}}\} = \mathcal{E}\{c(\mathbf{p})(\hat{\mathbf{p}} - \mathbf{p})(\hat{\mathbf{p}} - \mathbf{p})^{\mathrm{T}}\} = c(\mathbf{p})\mathbf{R}_{\hat{\mathbf{p}}} = \mathbf{I}.$$

Consequently, $\mathbf{J} = c(\mathbf{p})\mathbf{I}$ and $\mathbf{J}^{-1}$ is diagonal with elements

$$\left(\mathbf{J}^{-1}\right)_{ii} = \frac{1}{c(\mathbf{p})}.$$

(3.114) is equivalent to

$$\mathbf{J} = [c(\mathbf{p})\mathbf{I}]\mathbf{R}_{\hat{\mathbf{p}}}[\mathbf{I}c(\mathbf{p})] = \mathbf{J}\mathbf{R}_{\hat{\mathbf{p}}}\mathbf{J}$$

and

$$\mathbf{J}\mathbf{R}_{\hat{\mathbf{p}}} = \mathbf{I} \Rightarrow \mathbf{R}_{\hat{\mathbf{p}}} = \mathbf{J}^{-1}.$$

The ML estimator may also be shown to be efficient if $c(\mathbf{p})$ is replaced by a matrix function $\mathbf{C}(\mathbf{p})$ and $\nabla_{\mathbf{p}} l = \mathbf{C}(\mathbf{p})(\hat{\mathbf{p}} - \mathbf{p})$. In general, if an estimator is efficient it is a ML estimator and linear in the error which is both a necessary and sufficient condition [3]. As for the single parameter case, an alternative form of $\mathbf{J}$ is derived by evaluating the second gradient of the likelihood yielding the relationship

$$\int_{-\infty}^{\infty} \nabla_{\mathbf{p}} \left[\mathsf{f}(\mathbf{r}|\mathbf{p}) \nabla_{\mathbf{p}} l(\mathbf{r}, \mathbf{p})\right]^{\mathrm{T}} d\mathbf{r} = 0.$$

Carrying out the gradient operation on the product and using

$$\nabla_{\mathbf{p}} \mathsf{f}(\mathbf{r}|\mathbf{p}) = \mathsf{f}(\mathbf{r}|\mathbf{p}) \nabla_{\mathbf{p}} l(\mathbf{r}, \mathbf{p})$$

yields

$$\int_{-\infty}^{\infty} \mathsf{f}(\mathbf{r}|\mathbf{p}) \left(\nabla_{\mathbf{p}}[\nabla_{\mathbf{p}} l(\mathbf{r},\mathbf{p})]^{\mathrm{T}}\right) d\mathbf{r} = -\int_{-\infty}^{\infty} \mathsf{f}(\mathbf{r}|\mathbf{p})[\nabla_{\mathbf{p}} l(\mathbf{r},\mathbf{p})][\nabla_{\mathbf{p}} l(\mathbf{r},\mathbf{p})]^{\mathrm{T}} d\mathbf{r}.$$

This is just the relationship between the expectations and

$$\mathbf{J} = -\mathcal{E}\left\{\nabla_{\mathbf{p}}[\nabla_{\mathbf{p}} l(\mathbf{r},\mathbf{p})]^{\mathrm{T}}\right\} = \mathcal{E}\left\{(\nabla_{\mathbf{p}} l(\mathbf{r},\mathbf{p}))(\nabla_{\mathbf{p}} l(\mathbf{r},\mathbf{p}))^{\mathrm{T}}\right\}$$

with components

$$J_{ij} = -\mathcal{E}\left\{\frac{\partial^2 l(\mathbf{r},\mathbf{p})}{\partial p_i \partial p_j}\right\} = \mathcal{E}\left\{\left(\frac{\partial l(\mathbf{r},\mathbf{p})}{\partial p_i}\right)\left(\frac{\partial l(\mathbf{r},\mathbf{p})}{\partial p_j}\right)\right\}.$$

The Fisher matrix is not necessarily diagonal and error coupling in the evaluation of the bound arises because the diagonal terms of the inverse reflect the effects of the cross terms J_{ij}. If it were diagonal, $J_{ij} = 0 \quad i \neq j$ and the bound would depend only the scalar inverses of the diagonal terms J_{ii}.

Example 3.6; The CRB for two parameters

If there are only two parameters such as delay and Doppler,

$$\mathbf{J} = \begin{bmatrix} J_{11} & J_{12} \\ J_{21} & J_{22} \end{bmatrix}$$

and

$$\mathbf{J}^{-1} = \frac{1}{J_{11}J_{22} - J_{12}J_{21}} \begin{bmatrix} J_{22} & -J_{12} \\ -J_{21} & J_{11} \end{bmatrix}.$$

The error variances σ_{pi}^2, σ_{p2}^2 for unbiased estimation of the parameters (p_1, p_2) are thus bounded by

$$\begin{array}{l} \sigma_{p1}^2 \geq \frac{J_{22}}{J_{11}J_{22}-J_{12}J_{21}} \\ \sigma_{p2}^2 \geq \frac{J_{11}}{J_{11}J_{22}-J_{12}J_{21}}. \end{array} \quad (3.115)$$

Note that when there is no error coupling, the cross terms $J_{12} = J_{21} = 0$ and

$$\sigma_{pi}^2 \geq \frac{1}{J_{ii}}$$

as was the case for the scalar parameter.

3.9.4 The Role of the Ambiguity function

The derivation of the explicit CRB relationships for the slowly fluctuating white noise Gaussian and the deterministic signal models illustrates the roles played by the signal waveform and the "quality " of the data as measured by the SIR. The CRB for the slowly fluctuating model follows the development of Van Trees ([6] chpt 10) where the CRB is derived in terms of the gradient of the data dependent term

$$\ell_d = |\mathrm{r}_1|^2 \left(\frac{1}{\sigma_0^2} - \frac{1}{\sigma_1^2} \right)$$

of the log-likelihood ratio (3.55). This is justified because the pdf under H_0 is not a function of the parameters. $\sigma_1^2 = E_s + N_0$, $\sigma_0^2 = N_0$, and SNR$=E_s/N_0$ so that

$$\ell_d = \frac{\mathrm{SNR}}{1+\mathrm{SNR}} \left(\frac{1}{N_0} \right) |\mathrm{r}_1|^2.$$

The ML parameter estimate is obtained from the solution

$$\nabla_{\hat{\mathbf{p}}} \ell_d |_{\hat{\mathbf{p}}=\mathbf{p}_{ml}} = 0$$

to the likelihood equation which in this case is the gradient of ℓ_d and depends only upon the realization $|r_1|^2$. The CRB (3.112) is obtained from the Fisher matrix

$$\mathbf{J} = -\mathcal{E}\{\nabla_{\hat{\mathbf{p}}} \left(\nabla_{\hat{\mathbf{p}}} \ell_d(\hat{\mathbf{p}})\right)^{\mathrm{T}}\}$$

evaluated at $\hat{\mathbf{p}} = \mathbf{p}$ corresponding to the extremal $\nabla_{\hat{\mathbf{p}}} \ell_d = 0$. Adopting the short hand notation for the outer product operator $\nabla_{\hat{\mathbf{p}}}^2 l = \nabla_{\hat{\mathbf{p}}} \left(\nabla_{\hat{\mathbf{p}}} l\right)^{\mathrm{T}}$ (which is not to be confused with the scalar Laplacian which is an inner product), the second gradient of the llr is the matrix

$$\nabla_{\hat{\mathbf{p}}}^2 \ell_d(\hat{\mathbf{p}}) = \frac{\mathrm{SNR}}{1+\mathrm{SNR}} \left(\frac{1}{N_0} \right) \nabla_{\hat{\mathbf{p}}}^2 |\mathrm{r}_1|^2.$$

The statistic evaluated for point scattering and WGN is

$$\begin{aligned} |\mathrm{r}_1|^2 &= \left| \sqrt{E_0} \mathcal{S}_{sp} \chi_{ff}(\mathbf{p}, \hat{\mathbf{p}}) + \mathrm{n}_1 \right|^2 \\ &= E_0 |\mathcal{S}_{sp}|^2 \Phi_{ff}(\mathbf{p}, \hat{\mathbf{p}}) + |\mathrm{n}_1|^2 + \text{uncorrelated cross terms} \end{aligned}$$

and the Fisher matrix is

$$\boxed{\mathbf{J} = -\frac{\mathrm{SNR}^2}{1+\mathrm{SNR}} \nabla^2_{\hat{\mathbf{p}}} \Phi_{ff}(\mathbf{p}, \hat{\mathbf{p}})\big|_{\hat{\mathbf{p}}=\mathbf{p}} \qquad (3.116)}$$

(random point scatter in WGN)

where the auto-ambiguity function takes the equivalent inner product forms

$$\Phi_{ff}(\mathbf{p}, \hat{\mathbf{p}}) = \left|\int_{-\infty}^{\infty} f(t, \mathbf{p}) f^*(t, \hat{\mathbf{p}}) dt\right|^2 = \left|\mathbf{f}^{\mathrm{H}}(\mathbf{p})\mathbf{f}(\hat{\mathbf{p}})\right|^2.$$

This relationship illustrates the roles of the ambiguity function and the signal to interference ratio in establishing estimation accuracy for the slowly fluctuating model. Generally, because of the inverse relationship (3.112) or as illustrated by Example 3.6, high accuracy implied by a lower CRB is associated with a highly resolvent ambiguity function. These are characterized by a highly localized maximum point in parameter space and depend only upon the waveform chosen when the interference is constant as in WGN [6, 47–50] When the interference varies with the parameters and the optimal likelihood ratio receiver using $g(t)$ is constructed, accuracy as reflected by the bound becomes a function of the interference distribution via its effect upon the form of the CAF, the variation of $\sigma_0^2(\hat{\mathbf{p}})$ in the vicinity of $\mathbf{p}$, and the SIR.

It is enlightening to contrast the slowly fluctuating result with that for the deterministic scatter model discussed in section (3.3). Its likelihood function is given by (3.21) where the data is discretized by the K-L expansion (3.30, 3.32). Taking the log and retaining only the terms that are functions of $\hat{\mathbf{p}}$ yields

$$l(\mathbf{r}, \hat{\mathbf{p}}) = \sqrt{E_s}\left(2\Re\left\{\mathbf{r}^{\mathrm{H}}\boldsymbol{\lambda}^{-1}\mathbf{f}(\hat{\mathbf{p}})\right\}\right) - E_s\mathbf{f}(\hat{\mathbf{p}})^{\mathrm{H}}\boldsymbol{\lambda}^{-1}\mathbf{f}(\hat{\mathbf{p}}).$$

The ML estimate is obtained by evaluating the null point of the gradient of l (3.110) and the Fisher matrix is

$$\begin{aligned}\mathbf{J} &= -\mathcal{E}\{\nabla^2_{\hat{\mathbf{p}}} l\} \\ &= -E_s\nabla^2_{\hat{\mathbf{p}}} 2\Re\left\{\mathbf{f}(\mathbf{p})^{\mathrm{H}}\boldsymbol{\lambda}^{-1}\mathbf{f}(\hat{\mathbf{p}})\right\}\big|_{\hat{\mathbf{p}}=\mathbf{p}} + E_s\nabla^2_{\hat{\mathbf{p}}}\left[\mathbf{f}(\hat{\mathbf{p}})^{\mathrm{H}}\boldsymbol{\lambda}^{-1}\mathbf{f}(\hat{\mathbf{p}})\right]\big|_{\hat{\mathbf{p}}=\mathbf{p}}.\end{aligned}$$

Recalling that the processing function for deterministic scattering (3.36) was defined as $\mathbf{g}(\hat{\mathbf{p}}) = \boldsymbol{\lambda}^{-1}\mathbf{f}(\hat{\mathbf{p}})$ and the cross-ambiguity function as $\Phi_{fg}(\mathbf{p}, \hat{\mathbf{p}}) = \Re\{\mathbf{f}^{\mathrm{H}}(\mathbf{p})\mathbf{g}(\hat{\mathbf{p}})\}$, the Fisher matrix is

$$\begin{aligned}\mathbf{J} &= -E_s\left[\nabla^2_{\hat{\mathbf{p}}} 2\Re\left\{\mathbf{f}(\mathbf{p})^{\mathrm{H}}\mathbf{g}(\hat{\mathbf{p}})\right\} + \nabla^2_{\hat{\mathbf{p}}}\left(\mathbf{f}(\hat{\mathbf{p}})^{\mathrm{H}}\mathbf{g}(\hat{\mathbf{p}})\right)\right]_{\hat{\mathbf{p}}=\mathbf{p}} \\ &= -E_s\left[\nabla^2_{\hat{\mathbf{p}}} 2\Phi_{fg}(\mathbf{p}, \hat{\mathbf{p}}) + \nabla^2_{\hat{\mathbf{p}}}\Phi_{fg}(\hat{\mathbf{p}}, \hat{\mathbf{p}})\right]_{\hat{\mathbf{p}}=\mathbf{p}}.\end{aligned}$$

The diagonal matrix $\boldsymbol{\lambda}$ represents the interference energy in each component thus the quantity $E_s\boldsymbol{\lambda}^{-1}$ is the matrix of component signal to interference ratios. Consequently, the Fisher matrix and hence the estimate accuracy as indicated by the CRB increases without limit with the SIR. This is more apparent for the WGN case where $\boldsymbol{\lambda} = N_0 I$, SNR$=E_s/N_0$ and

$$\mathbf{J} = -\mathrm{SNR}\ \nabla^2_{\hat{\mathbf{p}}} 2\Re\left\{\mathbf{f}(\mathbf{p})^{\mathrm{H}}\mathbf{f}(\hat{\mathbf{p}})\right\}\Big|_{\hat{\mathbf{p}}=\mathbf{p}} + \mathrm{SNR}\,\nabla^2_{\hat{\mathbf{p}}}\left(\mathbf{f}(\hat{\mathbf{p}})^{\mathrm{H}}\mathbf{f}(\hat{\mathbf{p}})\right).$$

The transmit function is normalized to unit energy for all parameters thus the second term is zero and

$$\boxed{\mathbf{J} = -\mathrm{SNR}\ \nabla^2_{\hat{\mathbf{p}}} 2\Phi_{ff}(\mathbf{p},\hat{\mathbf{p}})\Big|_{\hat{\mathbf{p}}=\mathbf{p}} \qquad (3.117)}$$

(deterministic point scattering in WGN)

where the auto-ambiguity function has the equivalent forms

$$\Phi_{ff}\mathbf{p}\hat{\mathbf{p}}) = \Re\left\{\int_{-\infty}^{\infty} f(t,\mathbf{p}) f^*(t,\hat{\mathbf{p}})dt\right\} = \Re\left\{\mathbf{f}^{\mathrm{H}}(\mathbf{p})\mathbf{f}(\hat{\mathbf{p}})\right\}.$$

When the parameters are the delay and Doppler parameters τ and β or equivalently ϕ for the narrowband case, the outer product gradient operator is

$$\nabla^2_{\hat{\mathbf{p}}} = \begin{bmatrix} \frac{\partial^2}{\partial\tau^2} & \frac{\partial^2}{\partial\tau\partial\beta} \\ \frac{\partial^2}{\partial\beta\partial\tau} & \frac{\partial^2}{\partial\beta^2} \end{bmatrix}$$

and the elements of the $\mathbf{J}$ matrix for WGN are

$$\begin{aligned} J_{11} &= -S_n \frac{\partial^2\Phi_{ff}}{\partial\hat{\tau}^2} \\ J_{12} &= -S_n \frac{\partial^2\Phi_{ff}}{\partial\hat{\tau}\partial\hat{\beta}} \\ J_{21} &= -S_n \frac{\partial^2\Phi_{ff}}{\partial\hat{\beta}\partial\hat{\tau}} \\ J_{22} &= -S_n \frac{\partial^2\Phi_{ff}}{\partial\hat{\beta}^2} \end{aligned} \qquad (3.118)$$

where S_n=SNR or SNR$^2/(1 + \mathrm{SNR})$ for the deterministic and random cases respectively.

Note that the role of the SNR differs for the two models . Estimation accuracy as indicated by the slowly fluctuating bound is proportional to

SNR^2 when the interference dominates but the marginal performance gain diminishes with increasing SIR and becomes linear at high SIR. This apparent anomaly is explained by the fact that with the random slowly fluctuating scatter model (3.49) no matter how high the average SIR, the random scalar $\mathcal{S}_{sp}$ will occasionally assume small values effectively dropping the signal out of the input time series. Parameter estimates made under these circumstances will naturally be very poor. The deterministic signal does not fluctuate and hence performance will vary linearly with SNR. The estimation performance of the slowly fluctuating model as measured by the CRB will always lag that of the deterministic model by the factor SNR/(1+SNR).

3.10 SUMMARY

Echo detection is fundamentally a problem in statistical decision theory that is complicated by the need to simultaneously estimate echo parameters such as delay, Doppler, and bearing angle. The likelihood criterion for Gaussian signals is derived from the minimization of Bayes risk and it is shown that the implementation of the likelihood ratio results in the replica correlator or matched filter. The ambiguity function is fundamental to the study of both sonar and radar echo-location and is the expectation of the matched filter under signal alone conditions. It is a characteristic of the waveform and ultimately determines parameter estimation accuracy and detection performance. The auto-ambiguity function arises from the likelihood ratio test for point scattering in white noise. When the interference is correlated, the optimum receiver is mismatched and the expected point scatter response is a cross-ambiguity function between the transmitted and optimum processing waveform functions $f(\mathbf{p}, t)$, $g(\hat{\mathbf{p}}, t)$ in the parameter domain. The relationship between these depends only upon the interference covariance as in (3.70, 3.71) or equivalently the inverse kernel (3.72, 3.73). The choice of $f(t)$ depends upon other factors such as the desired delay-Doppler resolution and the optimization of detection performance for spread (non point like) scattering and interference. This is discussed further in Chapter 6.

The treatment of parameter estimation, the CRB, and the Fisher matrix are derived from sources given in the references. The comparison of the Fisher matrix for the deterministic and slowly fluctuating cases is presented in order to illustrate the relative effect of the SNR for each model and the role played by the waveform ambiguity function. The

CRB depends in part upon the resolution properties of the waveform ambiguity function which leads to the consideration of signal design for estimation as well as detection.

Bibliography

[1] H.L. Van Trees. *Detection, Estimation, and Modulation Theory-Part I*. Wilcy, 1968.

[2] W.C. Knight, G.G. Pridham, and S.M. Kay. Digital Signal Processing for Sonar . *Proceedings of the IEEE*, 69(11):1451–1507, Nov. 1981.

[3] L.L. Sharf. *Statistical Signal Processing* . Addison-Wesley, 1991.

[4] J.V. DiFranco and W.L. Rubin. *Radar Detection* . Artech House, Dedham, MA, 1980.

[5] R.V. Hogg and E.A. Tanis. *Probability and Statistical Inference* . Macmillan, ny, 1988.

[6] H.L. Van Trees. *Detection, Estimation, and Modulation Theory-Part III* . Wiley, 1971.

[7] D.K. Barton. *Radars: Frequency Agility and Diversity* , volume 6. Artech House, Dedham, Mass., 1977.

[8] M.A. Weiner. Detection Probability for Partially Correlated Chi-Square Targets . *IEEE Trans. on Aerosp. and Elect. Systs.*, 24(4):411–416, July 1988.

[9] D.A. Abraham. Choosing a non-Rayleigh Reverberation Model. In *Proc. Oceans '99 MTS/IEEE*, volume 1, pages 284–288, Seattle, WA, 1999.

[10] D.A. Abraham. Efficacy Analysis of the Power Law Detector for non-Rayleigh Distributed Reverberation in Active Sonar. In *Proc. 2001 Aerosp. Conf.*, volume 4, pages 1739–48, Big Sky, MT, 2001.

[11] D.M. Drumheller and M.L. Glasser. Detection of Chi-Square Fluctuating Targets in Arbitrary Clutter. *IEEE Trans. on Aerosp. and Elect. Systs.*, 33(3):784–794, July 1997.

[12] G. Strang. *Linear Algebra and its Applications* . Academic Press, 1980.

[13] R.J. Urick. *Principles of Underwater Sound* . McGraw-Hill, 1975.

[14] S.C. Clay and H. Medwin. *Acoustical Oceanography* . Wiley Interscience, 1977.

[15] M.I. Skolnik. *Introduction to Radar Systems* . McGraw-Hill, 1980.

[16] F.B. Hildebrand. *Methods of Applied Mathematics* . Prentice-Hall, 1965.

[17] H.C. Andrews and B.R. Hunt. *Digital Image Restoration* . Prentice-Hall, 1977.

[18] R.O. Duda and P.E. Hart. *Pattern Classification and Scene Analysis* . Wiley Interscience, 1973.

[19] L.H. Sibul and L.J. Ziomek. Generalized Wideband Cross-Ambiguity Function . In *Proc. IEEE Intl. Conf. on Acoust. Spch. and Sig. Proc.*, Atlanta GA, 1981.

[20] E.J. Kelly, E.J. Reed, and W.L. Root. The Detection of Radar Echoes in Noise . *SIAM*, 8:481–507, 1960.

[21] H. Cox. Interrelated Problems in Estimation and Detection II. In *Proc. NATO Advanced Study Inst. on Sig. Proc.*, Enschede, Netherlands, Aug. 1968. NATO.

[22] M. Loève. *Probability Theory* . D. Van Nostrand, Princeton NJ, 1963.

[23] A. Papoulis. *The Fourier Integral and its Applications* . McGraw-Hill, New York NY, 1987.

[24] C.W. Helstrom. *Elements of Signal Detection and Estimation* . Prentice-Hall, Englewood Cliffs N.J., 1995.

[25] R.N. Carpenter and S. Kay. GLRT Detection of Signals in Reverberation. In *Proc. ICASSP-95*, volume 5, pages 3127–3130, Detroit,MI, 1995.

[26] C.E. Cook and M. Bernfeld. *Radar Signals* . Academic Press, New York NY, 1967.

[27] A.W. Rihaczek. *Principles of High Resolution Radar* . McGraw-Hill, New York NY, 1969.

[28] Multiple authors. Special Issue on Matched Filtering . *IRE Trans. Info. Theory*, IT-6(3), June 1960.

[29] D.W. Ricker and A.J. Cutezo. A Model Based Estimator Correlator (EC) Structure. *IEEE Trans. Sig. Proc.*, 48(10):2733–2742, Oct. 2000.

[30] H. Cox. Interrelated Problems in Estimation and Detection I . In *Proc. NATO Advanced Study Inst. on Sig. Proc.*, Enschede, Netherlands, Aug. 1968. NATO.

[31] R.O. Nielsen. *Sonar Signal Processing.* Artech House, Boston/London, 1991.

[32] B. Picinbono and P. Duvaut. Optimal Linear-Quadratic Systems for Detection and Estimation. *IEEE Trans. Info. Theory*, 34(2):304–311, Mar. 1988.

[33] A.D. Whalen. *Detection of Signals in Noise* . Academic Press, 1971.

[34] J.I. Marcum. A Statistical Theory of Target Detection by Pulsed Radar. *IRE Trans. Info. Theory*, IT-6(2):59–267, April 1960.

[35] D.C. Rife and G.A. Vincent. Use of Discrete Fourier Transform in the Measurement of Frequencies and Levels of Tones . *Bell Sytem Tech. Journ.*, 49(2):197–228, Feb. 1970.

[36] F.J. Harris. On the Use of Windows for Harmonic Analysis with the Discrete Fourier Transform . *Proceedings of the IEEE*, 66(1):51–83, Jan. 1978.

[37] N.C. Geckinli and D.Yavuz. Some Novel Windows and a Concise Tutorial Comparison of Window Families . *IEEE Trans. Acoust.,Speech, and Signal Proc.*, 26(6):501–507, Dec. 1978.

[38] H. Rohling and J. Schuerman. Discrete Time Window Functions With Arbitrarily Low Sidelobe Level . *Signal Processing*, 5(2):127–138, Mar. 1983.

[39] L.H. Sibul and E.L. Titlebaum. Signal Design for Detection of Targets in Clutter . *Proceedings of the IEEE*, 69(4):481–482, April 1981.

[40] D.F. DeLong and E.M. Hofstetter. On the Design of Optimum Radar Waveforms for Clutter Rejection . *IEEE Trans. Info. Theory*, IT-13:454–463, July 1967.

[41] L.J. Spafford. Optimum Radar Signal Processing in Clutter . *IEEE Trans. Info. Theory*, IT-14:731–743, Sept. 1968.

[42] H.L. Van Trees. Optimum Signal Design and Processing for Reverberation Limited Environments. *IEEE Trans. on Military Electronics*, MIL-9:212–229, July-Oct. 1965.

[43] T.G. Kincaid. On Optimum Waveforms for Correlation Detection in the Sonar Environment: Reverberation limited Conditions . *Journ. of the Acoust. Soc. Amer.*, 44(3):789–796, Sept. 1968.

[44] L.J. Ziomek and L.H. Sibul. Maximization of the SIR for a Doubly Spread Target: Problems in Nonlinear Programming . *Signal Processing*, 5:355–368, 1983.

[45] D.M. Drumheller and D.W. Ricker. Receiver-Transmitter Optimization for Detection in Doubly Spread Channels . *Journ. of the Acoust. Soc. Amer.*, 89(4):1714–1723, April 1991.

[46] H.V. Poor. *An Introduction to Signal Detection and Estimation* . Springer-Verlag, 1988.

[47] J. Costas. A Study of a Class of Detection Waveforms Having Nearly Ideal Range-Doppler Ambiguity Properties . *Proceedings of the IEEE*, 72:996–1009, Aug. 1984.

[48] E.L. Titlebaum and L.H. Sibul. Time-Frequency Hop Signals Part II: Coding Based Upon the Theory of Quadratic Congruences . *IEEE Trans. on Aerosp. and Elect. Systs.*, 17(4):494–499, July 1981.

[49] E.L. Titlebaum, S.V. Marić, and J.R. Bellegarda. Ambiguity Properties of Quadratic Congruential Coding . *IEEE Trans. on Aerosp. and Elect. Systs.*, 27(1):18–25, Jan. 1991.

[50] D.W. Ricker. Constrained Bandwidth Waveforms with Minimal Dilation Sensitivity . *IEEE Trans. on Aerosp. and Elect. Systs.*, 29(3):666–679, July 1993.

Chapter 4

AMBIGUITY FUNCTIONS

4.1 INTRODUCTION

Ambiguity functions and their associated uncertainty functions, play a central role in echo-location as they relate signal characteristics to system and parameter estimation performance. Estimation performance as measured by the CR bound was shown in the previous chapter to be proportional to the resolution or "peakedness" of the ambiguity function (AF). This in turn depends upon the characteristics of the transmitted signal. As will be discussed in later chapters, detection performance in complex (non point-like) backscatter depends strongly on the distributions of signal and interference energy relative to the AF in the space defined by the parameter set $\mathbf{p}$. When the parameters are the usual delay and Doppler parameters associated with sonar or radar, the parameter space is often referred to as the *phase space/plane* because the parameters are directly related to range (delay) and velocity (Doppler). Range (r), delay (τ), velocity (v), and Doppler (ϕ, β) are related by $r \approx c\tau/2$ and $v \approx \phi c/2f_0$ or $v \approx (\beta - 1)c/2$ for the narrow and wideband representations respectively and a monostatic line of sight geometry. In this context, an uncertainty function (UF) is a complex two dimensional function and its associated real ambiguity function (AF) represents a surface.

Ambiguity functions arise naturally from the maximum likelihood criterion for detection and estimation in a Gaussian environment but are only one of many "time-frequency" representations used for signal analysis. All share a common common characteristic in that that they map a function of one dimension (time) into another two dimensional function of time or time delay and frequency or scale. The Wigner-Ville distribution (WVD), short term Fourier transform, and the wavelet transform are other examples that are often used for signal analysis. All are 2-d representations that express the time vs instantaneous frequency (or dilation) characteristics of a time function [1–3].

This chapter explores the characteristics of wide and narrowband ambiguity functions, alternative signal representations, waveform design parameters and their relationship to estimation accuracy, and ambiguity distortion due to model mismatch. The primary emphasis is on the form of the AF arising from the slowly fluctuating model as it is commonly employed for most high frequency sonar applications. The deterministic or exactly known signal and the random phase (Rician) models are treated as special cases.

4.2 MODELS AND NOTATION

The ambiguity function of waveforms $f(\mathbf{p},t)$, $g(\hat{\mathbf{p}},t)$ under the slowly fluctuating point scatter model (3.49) is written as

$$\Phi_{fg}(\mathbf{p};\hat{\mathbf{p}}) = |\chi_{fg}(\mathbf{p},\hat{\mathbf{p}})|^2 \tag{4.1}$$

where $\chi_{fg}(\mathbf{p},\hat{\mathbf{p}})$ is the uncertainty function (UF) defined as the inner product

$$\chi_{fg}(\mathbf{p},\hat{\mathbf{p}}) =< f_{\mathbf{p}}, g_{\hat{\mathbf{p}}} >= \int_{\Omega_t} f(\mathbf{p},t)g^*(\hat{\mathbf{p}},t)dt.$$

Ω_t represents the total time support of the integrand that depends upon the parameters. These equations represent *cross*-ambiguity and uncertainty functions (CAF's, CUF's) because $f(t)$ and $g(t)$ are different. When they are the same, the term *auto*-ambiguity (uncertainty) function (AAF, AUF) is used. In the interest of brevity, the generic terms ambiguity (uncertainty) function (AF, UF) will be used when the meaning is clear from the context.

$\mathbf{p},\hat{\mathbf{p}}$ represent parameter sets (2.27) that depend upon delay (range) and velocity (Doppler) and modify the signal (2.28) according to the wide or narrowband model. The parameters are identified with the attributes of the received signal and the hypotheses made in forming a receiver bank. Waveforms are assumed to be time limited, zero average, complex analytic waveform functions that are square integrable. Moreover they are assumed to be *admissible* functions [4–6] which requires that

$$c_x = \int_{-\infty}^{\infty} \frac{|X(\omega)|^2}{\omega} d\omega < \infty \tag{4.2}$$

for $x(t)$ with Fourier transform $X(\omega)$. This is equivalent to

$$X(0) = \int_{-\infty}^{\infty} x(t)dt = 0.$$

c_x is called the admissibility constant and has units of energy spectral density or Joules/Hz. Admissibility is required in order that a function be recoverable from its uncertainty function and that its wideband ambiguity volume be bounded [6]. Admissible functions have energy density spectra that approach zero more quickly than ω or in other words, have no DC component. The relationship of admissibility to bandwidth is illustrated by the example of a band limited function with constant spectral density $|X_0|^2$ in the band W Hz and energy $E_x = |X_0|^2 W$ such that

$$|X(\omega)|^2 = |X_0|^2 \quad \omega \in 2\pi \left(f_0 - \frac{W}{2},\ f_0 + \frac{W}{2} \right).$$

Evaluating the admissibility constant for this results in

$$c_x = \frac{E_x}{W} \ln \left(\frac{1 + \frac{W}{2f_0}}{1 - \frac{W}{2f_0}} \right)$$

and

$$\lim_{W \to 0} c_x = E_x.$$

Consequently, for narrowband functions, the admissibility constant is equivalent to the total energy (or unity if the function is normalized), but increases without bound as the bandwidth approaches twice the carrier frequency and the lower band edge approaches 0. In practice, physical limitations preclude the transmission of DC and all of the waveforms and signals discussed in this book are considered to be admissible.

4.2.1 Wide and Narrowband AF Models

The terms wide and narrowband (WB and NB) as applied to functions subjected to delay and Doppler effects in an echo-location context are somewhat misleading as they refer to the means by which delay and velocity effects are modeled and parameterized, not specifically to the spectral bandwidth. However, they have become common usage for the description of time dilation or scale (wideband) and frequency shift (narrowband) models of the Doppler effect. The first order effect of scatterer velocity on the time function describing an echo is a dilation of the time base. This is the fundamental model for collinear scattering from a constant velocity moving object and gives rise to the wideband (dilation) model $\sqrt{\beta} f(\beta(t-\tau))$ for an energy normalized time function. When the dilation factor β is small in relation to the inverse TW product and W is small in relation to the carrier frequency f_0, the signal (2.29) is defined as narrowband for the purpose of modeling the Doppler effect. The dilation of the complex envelope is neglected giving rise to the narrowband (frequency shift) model $f(t-\tau) \exp j2\pi\phi t$. In each case the delay parameter is τ but Doppler or velocity is parameterized by dilation β and frequency ϕ for the wide and narrowband models respectively. Assuming that Doppler is subsonic i.e. $|v| \leq c$, the parameter domain Ω of a waveform $f(\mathbf{p}, t)$ is composed of wide and narrowband subdomains

$$\begin{aligned} \mathbf{p} &= [\beta, \tau], \ \Omega_\tau : \tau \in [-\infty, \infty], \ \Omega_\beta : \beta \in [0, \infty] \\ \mathbf{p} &= [\phi, \tau], \ \Omega_\tau : \tau \in [-\infty, \infty], \ \Omega_\phi : \phi \in [-f_0, \infty]. \end{aligned} \quad (4.3)$$

The ambiguity origin parameters $\beta = 1$, $\tau = \phi = 0$ define zero Doppler and delay and are denoted by $\mathbf{p}_0 = [1, 0]$ (wideband) and $\mathbf{p}_0 = [0, 0]$ (narrowband). The dilation model is the more general model for first order velocity effects (no acceleration) while the frequency shift model represents an approximation to the dilation model for narrowband signals and low velocities.

To be notationally precise, AF's and UF's should be expressed in terms of four parameters or equivalently, two parameter sets even when one of the sets is $\mathbf{p}_0$. The wideband versions are

$$\begin{aligned} \chi_{fg}(\mathbf{p}, \hat{\mathbf{p}}) &= \chi_{fg}(\beta, \tau, \hat{\beta}, \hat{\tau}) = < f_{\mathbf{p}}, g_{\hat{\mathbf{p}}} > \\ &= \sqrt{\beta \hat{\beta}} \int_{\Omega_t} f(\beta(t-\tau)) g^*(\hat{\beta}(t-\hat{\tau})) dt \\ \chi_{fg}(\mathbf{p}_0, \hat{\mathbf{p}}) &= \chi_{fg}(1, 0, \hat{\beta}, \hat{\tau}) = < f, g_{\hat{\mathbf{p}}} > \\ &= \sqrt{\hat{\beta}} \int_{\Omega_t} f(t) g^*(\hat{\beta}(t-\hat{\tau})) dt \\ \chi_{fg}(\mathbf{p}_0, \mathbf{p}_0) &= \chi_{fg}(1, 0, 1, 0) = < f, g > = \int_{\Omega_t} f(t) g^*(t) dt. \end{aligned} \quad (4.4)$$

Likewise, the frequency shift (narrowband) versions of the UF are

$$\begin{aligned}\chi_{fg}(\mathbf{p},\hat{\mathbf{p}}) &= \chi_{fg}(\phi,\tau,\hat{\phi},\hat{\tau}) =< f_{\mathbf{p}}, g_{\hat{\mathbf{p}}} > \\ &= \exp j(\hat{\theta}-\theta)\int_{\Omega_t} f(t-\tau)g^*(t-\hat{\tau})\exp j2\pi(\phi-\hat{\phi})tdt \\ \chi_{fg}(\mathbf{p}_0,\hat{\mathbf{p}}) &= \chi_{fg}(0,0,\hat{\phi},\hat{\tau}) =< f, g_{\hat{\mathbf{p}}} > \\ &= \exp j\hat{\theta}\int_{\Omega_t} f(t)g^*(t-\hat{\tau})\exp -j2\pi(\hat{\phi})tdt. \end{aligned} \tag{4.5}$$

$\chi_{fg}(\mathbf{p}_0,\mathbf{p}_0)$ is the same as for (4.4) and θ and $\hat{\theta}$ are the phase constants $2\pi\phi\tau$ and $2\pi\hat{\phi}\hat{\tau}$ respectively. The corresponding AF's are generated in each case by (4.1).

Frequency domain representations are obtained using Parseval's rule and the Fourier equivalents of the constituent functions. The wideband versions are

$$\begin{aligned}\chi_{fg}(\mathbf{p},\hat{\mathbf{p}}) &= \frac{1}{2\pi\sqrt{\beta\hat{\beta}}}\int_{-\infty}^{\infty} F\left(\frac{\omega}{\beta}\right)G^*\left(\frac{\omega}{\hat{\beta}}\right)\exp j\omega(\hat{\tau}-\tau)d\omega \\ \chi_{fg}(\mathbf{p}_0,\hat{\mathbf{p}}) &= \frac{1}{2\pi\sqrt{\hat{\beta}}}\int_{-\infty}^{\infty} F(\omega)G^*\left(\frac{\omega}{\hat{\beta}}\right)\exp j\omega\hat{\tau}d\omega \end{aligned} \tag{4.6}$$

and the corresponding narrowband versions are

$$\begin{aligned}&\chi_{fg}(\mathbf{p},\hat{\mathbf{p}}) = \\ &(\exp j(\hat{\theta}-\theta))\left(\frac{1}{2\pi}\right)\int_{-\infty}^{\infty} F(\omega-2\pi\phi)G^*(\omega-2\pi\hat{\phi})\exp j\omega(\hat{\tau}-\tau)d\omega \\ &\chi_{fg}(\mathbf{p}_0,\hat{\mathbf{p}}) = (\exp j\hat{\theta})\left(\frac{1}{2\pi}\right)\int_{-\infty}^{\infty} F(\omega)G^*(\omega-2\pi\hat{\phi})\exp j\omega\hat{\tau}d\omega. \end{aligned} \tag{4.7}$$

The four parameter notation is cumbersome and is relaxed when the meaning is clear from the context. For example if $\beta = 1$ and $\tau = 0$, the UF is simply expressed in terms of the non-origin parameters

$$\chi_{fg}(\mathbf{p}_0,\hat{\mathbf{p}}) = \chi_{fg}(1,0,\hat{\beta},\hat{\tau}) \equiv \chi_{fg}(\hat{\beta},\hat{\tau}).$$

The choice of the model to use in designing receivers for echo-location represents a tradeoff between ease of implementation and accuracy. The frequency shift model is inherently simpler because no dilation of the processing signal $g(t)$ is required. The detection statistic realized by the receivers (3.5) is generated by advancing the conjugated processing signal through the data stream (or vice versa) to test for delay and simultaneously testing for Doppler by efficient Fourier (FFT) processing. Dilation processing requires the generation or storage of dilated replicas with which to process the data. This is more involved when the processing waveform does not exist as a closed form mathematical function but must be constructed from a time series by multi-rate sampling and interpolation.

4.2.2 The AF of Real Waveforms

If a real waveform rather than its complex analytic form is used, the AF mainlobe will have the overall shape of the envelope AF but will exhibit periodic nulls on the delay axis occurring at intervals $\Delta\tau = 1/2f_0$. These nulls are called grating lobes because they are caused by constructive and destructive interference at twice the carrier frequency in a manner similar to that occurring when coherent monochrome light is passed through a diffraction grating. This phenomenon presents a problem because a particular matched filter delay hypothesis close to the AF origin can fall into a null and generate an unexpected low response even if it is within the envelope mainlobe.

The frequency domain wideband AUF $\chi_{ff}(\mathbf{p}_0, \mathbf{p})$ evaluated along the delay axis with $(\beta = 1)$ is obtained from (4.6) as

$$\chi_{ff}(\tau) = \frac{1}{2\pi} \int_{-\infty}^{\infty} |F(\omega)|^2 \exp j\omega\tau d\omega. \tag{4.8}$$

The corresponding complex time domain bandpass function $f(t)$ written in terms of its complex envelope is

$$f(t) = \tilde{f}(t) \exp j\omega_0 t$$

and the baseband envelope spectrum $\tilde{F}(\omega)$ with a non-admissible DC offset is illustrated in Figure 4.1a. If the envelope spectrum is confined to a bandwidth W(Hz), it is possible to shift it by the carrier ω_0 sufficiently to eliminate the offset producing the admissible analytic function $f(t)$ with spectrum $F(\omega) = \tilde{F}(\omega - \omega_0)$ shown in Figure 4.1b and its real counterpart shown in Figure 4.1c. Denote the envelope AUF obtained by substituting the baseband spectrum $\tilde{F}(\omega)$ for $F(\omega)$ in (4.8) as $\tilde{\chi}_{ff}(\tau)$. Then, the analytic passband AUF is

$$\chi_{ff}(\tau) = \frac{1}{2\pi} \int_{0}^{\infty} |\tilde{F}(\omega - \omega_0)|^2 \exp j\omega\tau d\omega$$

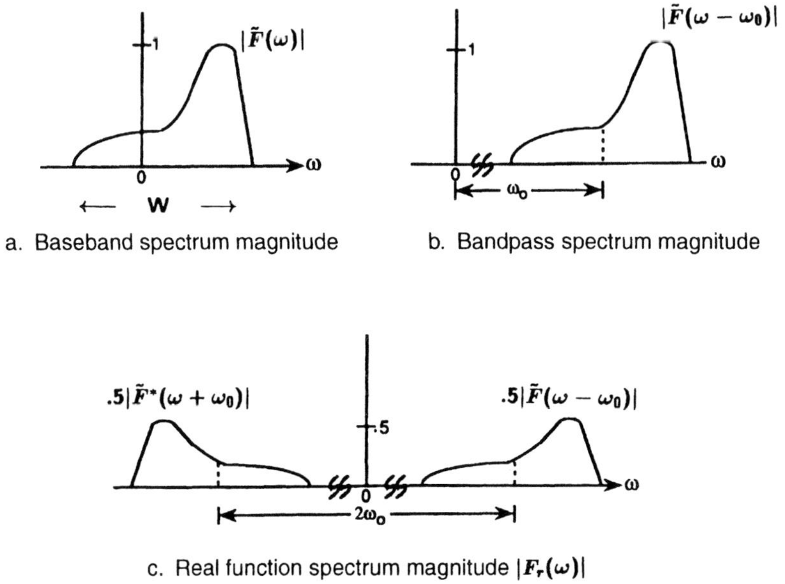

Figure 4.1: Spectral representations: a. baseband spectrum magnitude, b. bandpass spectrum magnitude, c. real function spectrum magnitude $|F_r(\omega)|$

and changing variables by $\nu = \omega - \omega_0$ and assuming that $\omega_0 > \pi W$ yields

$$
\begin{aligned}
\chi_{ff}(\tau) &= \frac{1}{2\pi} \exp j\omega_0 t \int_{-\omega_0}^{\infty} |\tilde{F}(\nu)|^2 \exp j\nu t d\nu \\
&= \frac{1}{2\pi} \exp j\omega_0 t \, \tilde{\chi}_{ff}(\tau).
\end{aligned}
$$

Clearly, the delay axis AF characteristics of the shifted analytic bandpass waveform depend only upon the envelope not the particular carrier shift.

The spectrum of the real shifted waveform is

$$
\begin{aligned}
\tilde{F}_r(\omega) &= \frac{1}{2}\tilde{F}^*(\omega + \omega_0) \quad (\omega < 0) \\
&= \frac{1}{2}\tilde{F}(\omega - \omega_0) \quad (\omega > 0)
\end{aligned}
$$

and its AUF on the delay axis is

$$
\chi_r(\tau) = \frac{1}{8\pi} \int_{-\infty}^{0} |\tilde{F}(\omega + \omega_0)|^2 \exp j\omega\tau d\omega
$$

$$+\frac{1}{8\pi}\int_0^\infty |\tilde{F}(\omega-\omega_0)|^2 \exp j\omega\tau d\omega.$$

Changing variables in the second term as before yields

$$\frac{1}{4}\left(\frac{1}{2\pi}\right)\exp -j\omega_0\tau \int_{-\omega_0}^{\infty} |\tilde{F}(\nu)|^2 \exp j\nu\tau d\nu.$$

If ω_0 is greater than the total envelope radian bandwidth, the integral represents 1/4 of the envelope uncertainty function denoted by $\tilde{\chi}_{ff}$. The contribution to the real waveform AUF by the positive frequency portion of the spectrum is then

$$\frac{1}{4}\tilde{\chi}_{ff}(\tau)\exp -j\omega_0\tau.$$

A similar calculation for the negative frequency term yields the contribution

$$\begin{aligned}
&\frac{1}{4}\left(\frac{1}{2\pi}\right)\exp j\omega_0\tau \int_{\infty}^{\omega_0} |\tilde{F}(\nu)|^2 \exp j\nu\tau d\nu \\
&= \frac{1}{4}\tilde{\chi}_{ff}(\tau)\exp j\omega_0\tau
\end{aligned}$$

and combining the contributions yields the real waveform UF and AF

$$\begin{aligned}
\chi_r(\tau) &= \frac{1}{2}\tilde{\chi}_{ff}(\tau)\cos\omega_0 t \\
\Phi_r(\tau) &= \frac{1}{4}\tilde{\Phi}_{ff}(\tau)\cos^2\omega_0 t.
\end{aligned} \tag{4.9}$$

The conclusion is that the AF of a real band limited admissible signal exhibits ambiguity nulls with period $1/2f_0$ and is attenuated by 1/4 in relation to the corresponding analytic waveform AF. When $f_0 = W/2$ such that the lower band edge of the shifted envelope spectrum occurs at the origin, the nulls occur at $1/2W$.

The AF of a real waveform $f_r(t)$ with a uniform spectrum is easily computed as an example. Let it be energy normalized with magnitude $|F_r(\omega)| = 1/(2\sqrt{W})$ on the intervals $[-\omega_2, -\omega_1]$ and $[\omega_1, \omega_2]$ where $\omega_1 = \omega_0 - \pi W$, $\omega_2 = \omega_0 + \pi W$, and $\omega_0 = 2\pi f_0$. The uncertainty function is

$$\chi_r(\tau) = \frac{1}{8\pi W}\left[\int_{-\omega_2}^{-\omega_1} \exp j\omega\tau d\omega + \int_{\omega_1}^{\omega_2} \exp j\omega\tau d\omega\right]$$

and carrying out the integrations results in the AF

$$\Phi_r(\tau) = |\chi_r(\tau)|^2 = \frac{1}{4}\text{sinc}^2(\pi W\tau)\,\cos^2(\omega_0\tau).$$

where the sinc term represents the AF $\tilde{\Phi}_{ff}$ of the uniform basebanded envelope with nulls at $\pm\tau = k/W$ with $k = 1, 2, \cdots,$. The associated real passband waveform has additional carrier induced nulls at $\tau = (2k-1)/4f_0$. If the waveform is admissible in that it has no DC component the upper and lower spectral sidebands will not overlap at the origin and $2f_0 \geq W$. This means that the carrier nulls will always occur within the AF mainlobe since

$$\tau = \frac{1}{4f_0} \leq \frac{1}{2W}.$$

If the envelope spectrum is expanded to fill the band and $W = 2f_0$, the spectrum becomes uniform over the total width $2W$ and the delay axis AF profile becomes a sinc function with its first null at $\tau = 1/(2W)$ and no carrier nulls occur within the mainlobe. If the one-sided analytic representation is used, the cosine term and the associated nulls disappear for all bandwidths.

The sample spectra of a representative real waveform function and its envelope with f_0=5kHz and W=2kHz are shown in Figures 4.2a,b. It is a BPSK or pseudo-random-noise (PRN) waveform with a chip interval of .5 (msec.)[1] and the τ axis AF slices for the real and analytic counterparts are shown in Figures 4.2c,d. The width of the mainlobe is about $\pm$.5 (msec.) corresponding to the envelope bandwidth but the grating lobes occur at intervals of .1 (msec.) corresponding to $1/2f_0$.

4.2.3 Alternative Signal Representations

Uncertainty and their associated ambiguity functions arise naturally in the echo-location context and are one example of the general class of time-frequency or phase space signal representations that map time functions to a two dimensional time-frequency or time-scale representation. These have arisen as a means for the analysis of those time varying (non-stationary) phenomena such as speech and music for which the

[1] BPSK (bi-phase shift keyed) waveforms are members of the general class of pseudo-random noise (PRN) codes and are generated by switching the phase of a sinusoid by 180 deg. at irregular (pseudo-random) intervals. Their spectra have the shape of a $\sin/x$ function with nulls at the inverse of the minimum switching (chip) interval. These are discussed in Chapter 5.

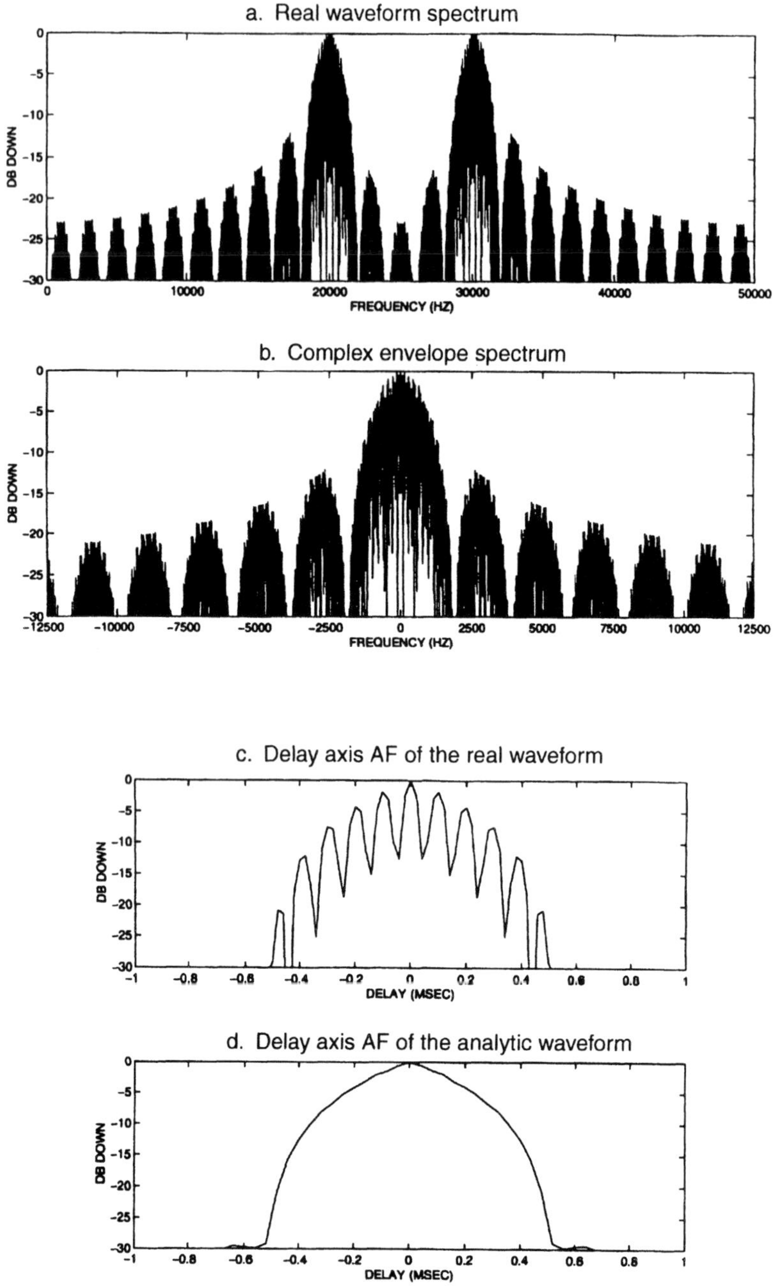

Figure 4.2: BPSK spectra and AF's: a. real waveform spectrum, b. complex envelope spectrum, c. delay axis AF of the real waveform, d. delay axis AF of the analytic waveform

frequency content of the signal varies rapidly within the length of the record [1, 7]. For these, it is important to obtain the instantaneous frequency in order to understand the physical phenomena generating the sounds. While the Fourier transform FT may be used to decompose a signal time series into its frequency components, it does not indicate just when these components occur within the record. The simplest approach is to analyze speech or other signals with rapidly time-varying spectra by Fourier transforming short segments of the overall record (the short term FT). This suffers from the inherent limitation of the FT namely that the shorter the record used, the poorer the resolution of the frequency content since the frequency resolution $\triangle f \approx 1/\triangle t$. Alternative representations such as the the wavelet transformation and various forms of Wigner-Ville distribution have been developed to overcome the inherent limitation of the FT. These are closely related to uncertainty and ambiguity functions and are applicable to the echo-location problem.

The Wavelet Transform (WT)

The WT is a description of functions in terms of the shifting and scaling (dilation) of the independent variable and is applied as a means for multi-resolution analysis in such diverse fields as image processing, signal compression, transient analysis, and non-stationary signal processing [4–6, 8–10]. The wavelet transformation and the dilation model for uncertainty and ambiguity functions are equivalent except for notation in that the dilation (scale) variables are inverted. The WT is defined as

$$W_g f(a,b) = \frac{1}{\sqrt{|a|}} \int_{-\infty}^{\infty} f(u) g^* \left(\frac{u-b}{a} \right) du \tag{4.10}$$

where b represents the shift (delay) parameter that is equivalent to τ, and a is the inverse of the dilation parameter β. This represents the WT of the function $f(u)$ with respect to the *mother wavelet* or *analyzing function* $g(u)$ parameterized by the scale a and shift b. The Gaussian weighted tone kernel $g(u) = \exp(j\omega u - \alpha u^2)$ is a common choice for the analyzing function and the associated WT is called The Morlet wavelet transform [8]. Clearly the WT and the wideband UF are closely related since

$$W_g f(\frac{1}{\beta}, \tau) = \chi_{fg}(1, 0, \beta, \tau).$$

The Wigner-Ville Distribution (WVD)

The WVD is a member of a general class of time-frequency distributions defined by

$$P(t,\xi) = \int_{-\infty}^{\infty}\int_{-\infty}^{\infty}\int_{-\infty}^{\infty} p(\theta,\tau)[\exp -j2\pi\theta(t-u)][\exp -j2\pi\xi\tau] \cdot f(u+\frac{\tau}{2})f^*(u-\frac{\tau}{2})dud\tau d\theta$$

where t is time, ξ, θ are frequency parameters, and specific distributions are defined by the kernel $p(\theta,\tau)$ [1]. In particular the WVD $V_w(t,\xi)$ is defined for $p(\theta,\tau)=1$ and

$$V_w(t,\xi) = P(t,\xi)|_{p=1} = \int_{-\infty}^{\infty} f(t+\frac{\tau}{2})f^*(t-\frac{\tau}{2})\exp -j2\pi\xi\tau d\tau.$$

The WVD is a real but not necessarily positive function and related to the centralized narrowband UF

$$\chi(\tau,\phi) = \int_{-\infty}^{\infty} f(t+\frac{\tau}{2})f^*(t-\frac{\tau}{2})\exp -j2\pi\phi t dt$$

by the 2-d Fourier transform

$$\chi(\tau,\phi) = \int_{-\infty}^{\infty}\int_{-\infty}^{\infty} V_w(t,\xi)\exp j2\pi(\xi\tau-\phi t)dtd\xi.$$

It also has the same time-frequency support as its constituent time function, satisfies the marginals

$$\int_{-\infty}^{\infty} V_w(t,\xi)dt = |F(\xi)|^2$$
$$\int_{-\infty}^{\infty} V_w(t,\xi)d\xi = |f(t)|^2,$$

and preserves the signal energy

$$\int_{-\infty}^{\infty}\int_{-\infty}^{\infty} V_w(t,\xi)dtd\xi = \int_{-\infty}^{\infty}\int_{-\infty}^{\infty} |\chi(\tau,\phi)|d\tau\phi = E_f$$

which is unity for normalized signals.

The WVD represents the instantaneous time-frequency content of the waveform but all portions of the waveform contribute at each value of t

since the τ integration spans the total time support. Artifacts are consequently generated in the t-f plane that do not correspond to the local (in time) instantaneous frequency especially for highly non-stationary signals. This is illustrated by an example similar to one given by Cohen [1] in which a non-stationary waveform is generated by concatenating two subfunctions separated by an interval in which it is zero.

The subfunctions are two rectangularly weighted LFM waveforms each of 100m Hz bandwidth and 100 (msec.) duration and are separated by a 100 (msec.) interval of zero. The WVD is shown in Figure 4.3 where the time-frequency distribution consists of two outer peaks corresponding to the subfunctions but is accompanied by an artifact in the center where the waveform was nulled and no energy is expected. In an effort to

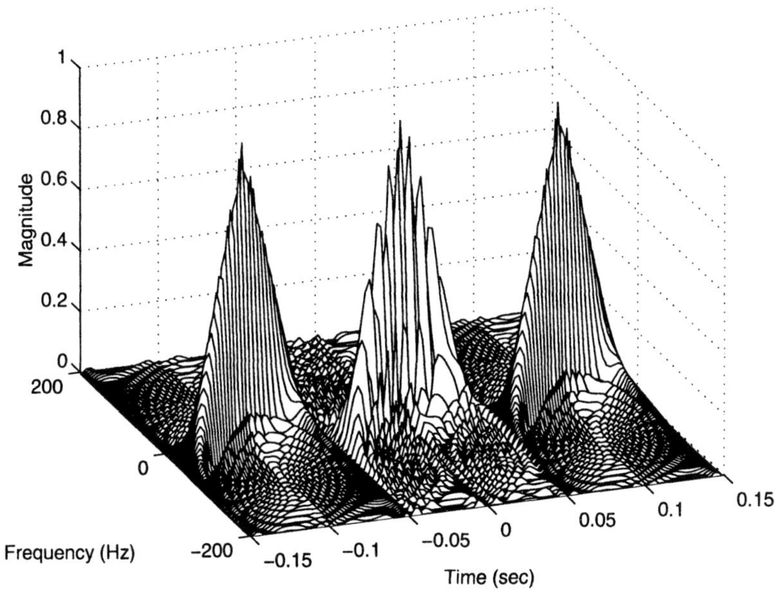

Figure 4.3: WVD of a split LFM waveform

alleviate the artifact problem , Choi and Williams [11] introduced an exponential kernel $p(\theta, \tau) = \exp -\theta^2\tau^2/\sigma$ that weights the contributions of the signal inversely with τ so that the local (with respect to t) signal properties are emphasized. The Choi-Williams distribution (CWD) is

$$C_w(t,\xi) = \int_{-\infty}^{\infty}\int_{-\infty}^{\infty} \sqrt{\frac{\pi\sigma}{\tau^2}} \exp - \left(\frac{\sigma(t-u)^2}{4\tau^2}\right) \exp - j(2\pi\xi\tau) \cdot f(t+\frac{\tau}{2}) f^*(t-\frac{\tau}{2}) d\tau du \tag{4.11}$$

and the relative emphasis placed upon the local signal properties depends upon the parameter σ. Small values of σ emphasize the local t-f properties and the CWD approaches the WVD as $\sigma \to \infty$. Decreasing σ reduces the effective CWD integration interval hence increasing the time localization accuracy. This also reduces the precision of the frequency measurement and smoothes the resulting t-f plot in a manner similar to that which occurs when a smaller interval is used to compute a Fourier transform. WD's corresponding to the above example are shown in Fig-

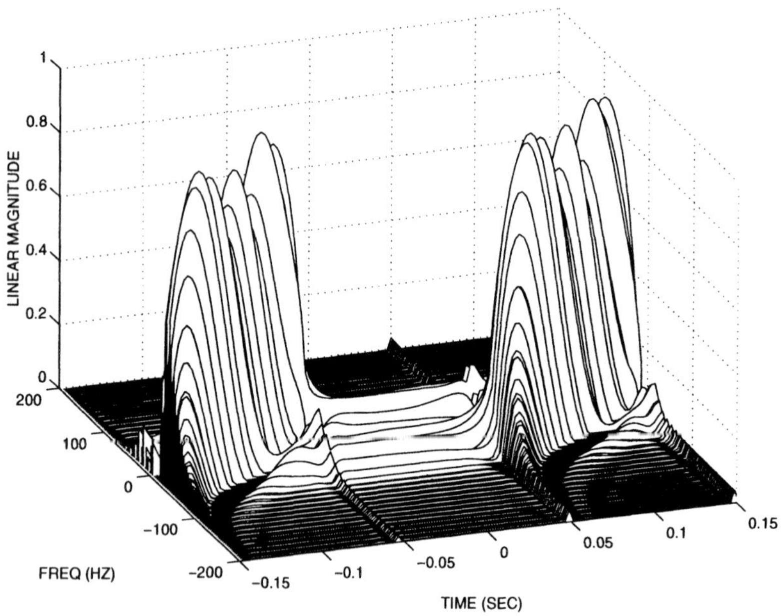

Figure 4.4: CWD of the split LFM waveform with ($\sigma = 1$)

ures 4.4, 4.5 for $\sigma = 1, 20$. Note the attenuation of the WVD artifact and relative smoothness of the CWD when $\sigma = 1$ but the increasing sharpness of the plot and returning presence of the artifact for $\sigma = 20$. Other attributes of the WVD, CWD, and associated distributions are to

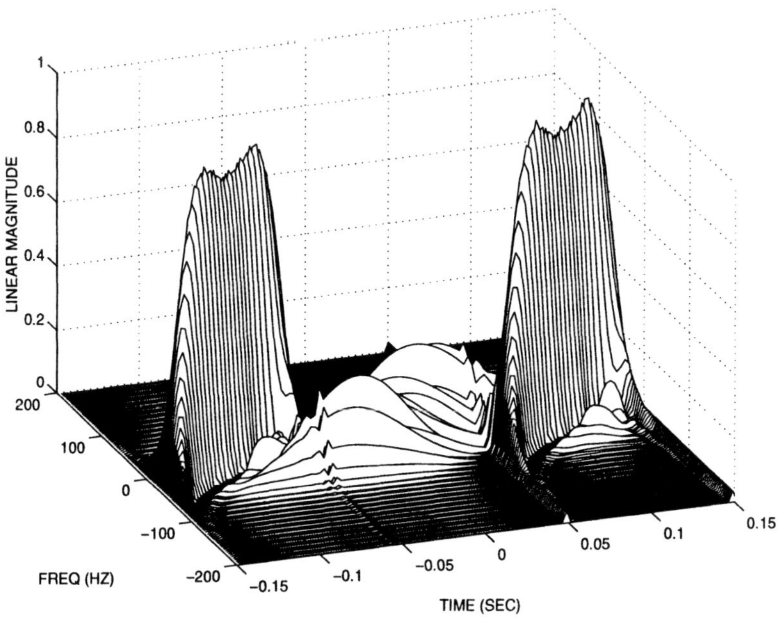

Figure 4.5: CWD of the split LFM with ($\sigma = 20$)

be found in [1–3]. These distributions have important implications for the design of high resolution echo-location waveforms that is discussed in Chapter 5.

4.3 AF AND UF PROPERTIES

There exists an extensive historical literature regarding the properties of ambiguity functions as applied to radar and comprehensive discussions of the subject may be found in books by Woodward [12], Cook and Bernfeld [13], Van Trees [14], and Richaczek [15]. Radar applications are usually regarded as narrowband because the ratio of the signal bandwidth to the carrier is generally very small and the speed of propagation is large with respect to target speed. As a result most of the radar related literature treats only the narrowband case. Subsequent interest in the application of wideband waveforms for acoustic echo-location where the speed of propagation may not greatly exceed reflector speed, led to the development of dilation models for Doppler and hence wideband ambiguity functions.

The properties of the WB-AF as applied to the sonar problem have been explored in the works of Stewart and Westerfield [16], Kelly and Wishner [17], Rihaczec [15], Altes [18, 19], Sibul and Titlebaum [20] among many others. Recent interest in multiresolution analysis and the theory of wavelets has resulted in the application of wideband signal theory to other broad areas of signal processing. The subsequent analysis of wavelet transform properties by people in diverse fields has led to a deeper understanding of not only wavelets but the closely related WB-AF's of echo-location.

4.3.1 General Properties

The properties of the slowly fluctuating uncertainty and ambiguity functions (4.1) that affect echo detection and parameter estimation performance are discussed in this section. Both wide (WB) and narrowband (NB) properties are presented, but the emphasis will be on the WB-AF since NB properties are well represented in the literature and are readily derived as special cases. The discussion of WB-AF properties is drawn from both WB-AF and wavelet literature and some details are left to the references.

Conjugation

$$\begin{aligned} \chi_{fg}(\mathbf{p}, \hat{\mathbf{p}}) &= \chi^*_{gf}(\hat{\mathbf{p}}, \mathbf{p}) \\ \Phi_{fg}(\mathbf{p}, \hat{\mathbf{p}}) &= \Phi_{gf}(\hat{\mathbf{p}}, \mathbf{p}) \end{aligned}$$

This hold for both cases and is easily shown by taking the complex conjugates of (4.4, 4.5).

Change of Variables

$$\begin{aligned} \chi_{fg}(\mathbf{p}, \hat{\mathbf{p}}) &= \chi_{fg}(\mathbf{p}_0, \tilde{\mathbf{p}}) \\ \Phi_{fg}(\mathbf{p}, \hat{\mathbf{p}}) &= \Phi_{fg}(\mathbf{p}_0, \tilde{\mathbf{p}}) \end{aligned} \tag{4.12}$$

with WB parameters

$$\tilde{\mathbf{p}} = [\tilde{\beta}, \tilde{\tau}] = \left[\frac{\hat{\beta}}{\beta}, \beta(\hat{\tau} - \tau)\right] \quad \text{(wideband)}$$

and $\mathbf{p}_0 = [1, 0]$. This is derived by substituting $u = \beta(t - \tau)$ into (4.4) and using (4.5). The corresponding NB parameters are

$$\tilde{\mathbf{p}} = [\hat{\phi} - \phi, \hat{\tau} - \tau] \quad \text{(narrowband)}$$

with $\mathbf{p}_0 = [0, 0]$ and obtained by substituting $u = t - \tau$ in (4.6) and using (4.7). The effect of interchanging the constituent functions is similarly derived and the WB relationships are:

$$\chi_{fg}(\mathbf{p}_0, \mathbf{p}) = \chi^*_{gf}(\mathbf{p}_0, \tilde{\mathbf{p}})$$
$$\Phi_{fg}(\mathbf{p}_0, \mathbf{p}) = \Phi_{gf}(\mathbf{p}_0, \tilde{\mathbf{p}})$$

with

$$\tilde{\mathbf{p}} = [\tilde{\beta}, \tilde{\tau}] = \left[\frac{1}{\beta}, -\beta\tau\right] \quad \text{(wideband)}.$$

The NB relationship for the AF is the same but with $\tilde{\mathbf{p}} = [-\phi, -\tau]$. The narrowband AF symmetry relationship $\Phi_{fg}(0, 0, -\phi, -\tau) = \Phi_{fg}(0, 0, \phi, \tau)$ is derived by direct substitution and

$$\Phi_{fg}(0, 0, -\phi, -\tau) = \Phi_{fg}(0, 0, \phi, \tau) = \Phi_{gf}(0, 0, \phi, \tau).$$

Bounds

If the functions $f(t)$, $g(t)$ have finite energies E_f, E_g,

$$\Phi_{fg}(\mathbf{p}, \hat{\mathbf{p}}) \leq E_f E_g$$

with equality for $f(t) = g(t)$ at the origin defined by $\mathbf{p} = \hat{\mathbf{p}}$. This is true for both cases and is easily shown by invoking the Schwarz inequality. Only auto-ambiguity functions are guaranteed to have their global maximum at the origin and when the functions are associated with waveforms as in the discussions regarding detection and estimation in the previous chapter, they are energy normalized by convention such that $\Phi_{ff}(\mathbf{p}, \mathbf{p}) = E_f = E_g = 1$. This permits the unambiguous definition of signal energy without regard to waveform energy.

Time Function Recovery (Inversion)

Time functions are recovered from the wide and narrowband UF's (4.4, 4.5) by

$$f(t) = \frac{1}{c_g} \int_{\Omega} \chi_{fg}(\mathbf{p}_0, \mathbf{p}) g(\mathbf{p}, t) d\mathbf{p} \quad \text{(wideband)}$$

$$= \frac{1}{E_g} \int_{\Omega} \chi_{fg}(\mathbf{p}_0, \mathbf{p}) g(\mathbf{p}, t) d\mathbf{p} \quad \text{(narrowband)} \tag{4.13}$$

where E_g is the energy of $g(t)$, $c_g < \infty$ is the admissibility constant (4.2), and the integrals represent integration over the space defined by $\mathbf{p} \in \Omega$. When χ_{fg} is expressed as a wavelet transform $W_g f$, the wideband relationship represents the inverse of (4.10) and is called the *inverse wavelet transform* [4, 5]. The WB inverse relationship is proven by expressing it in the frequency domain where the function $g(\mathbf{p}, t)$ is written as the inverse transform

$$g(\mathbf{p}, t) = \sqrt{\beta} g(\beta(t - \tau)) = \frac{1}{2\pi\sqrt{\beta}} \int_{-\infty}^{\infty} G\left(\frac{\omega'}{\beta}\right) \exp j\omega'(t - \tau) d\omega'$$

and χ_{fg} is given by (4.6) with $\mathbf{p}$ substituted for $\hat{\mathbf{p}}$ [6]. The rhs of (4.13) is obtained by combining these relations to form

$$rhs = \frac{1}{2\pi\beta c_g} \int_0^{\infty} \int_{-\infty}^{\infty} \int_{-\infty}^{\infty} F(\omega) G^*\left(\frac{\omega}{\beta}\right) G\left(\frac{\omega'}{\beta}\right) \cdot \frac{\exp j\omega' t}{2\pi} \int_{-\infty}^{\infty} \exp j(\omega - \omega')\tau d\tau d\omega d\omega' d\beta$$

and realizing that the last integral is a delta function. This simplifies things considerably and

$$rhs = \frac{1}{2\pi c_g} \int_0^{\infty} \int_{-\infty}^{\infty} F(\omega) \left| G\left(\frac{\omega}{\beta}\right) \right|^2 \exp j\omega t d\omega \frac{d\beta}{\beta}.$$

Let $z = \omega/\beta$ then $d\beta/\beta = -dz/z$ and

$$rhs = \frac{1}{2\pi c_g} \int_{-\infty}^{\infty} F(\omega) \exp j\omega t dt \int_{-\infty}^{\infty} \frac{|G(z)|^2}{z} dz = f(t)$$

which proves (4.13). The NB proof is obtained by writing

$$g(\mathbf{p}, t) = g(t - \tau) \exp j2\pi\phi t,$$

using (4.5) with u substituted for t , and carrying out the integrations with respect to ϕ, τ, u. Substituting $g(t)$ for $f(t)$ in (4.13) results in

$$g(t) = \frac{1}{c_g} \int_{\Omega} \chi_{gg}(\mathbf{p}_0, \mathbf{p}) g(\mathbf{p}, t) d\mathbf{p} \quad \text{(wideband)}$$
$$= \frac{1}{E_g} \int_{\Omega} \chi_{gg}(\mathbf{p}_0, \mathbf{p}) g(\mathbf{p}, t) d\mathbf{p} \quad \text{(narrowband)}$$

that are examples of what are called reproducing relationships. Integrating the product of a parameterized time function and its AUF over the space of the parameters "reproduces" the time function hence the AUF is called a *reproducing kernel* [4, 5]. The result is the same as would be obtained by substituting the multidimensional delta function $\delta(\mathbf{p}-\mathbf{p}_0)$ for the AUF. Both of these functions are multidimensional as they are defined in the space of the parameters (4.3) and are examples of the non-uniqueness of parameter space representations in reproducing a function of a single variable, in this case time. The proof that a specific time function is recoverable from multiple wide or narrowband representations is beyond the scope of this discussion but details may be found in [4].

Resolution of Identity (ROI)

The ROI is a useful relationship that maps two inner products into a single inner product. The ROI in equivalent notation is

$$
\begin{aligned}
< f, g > &= \frac{1}{c_h} \int_\Omega < f, h_\mathbf{p} > < g, h_\mathbf{p} >^* d\mathbf{p} \\
\chi_{fg}(\mathbf{p}_0, \mathbf{p}_0) &= \frac{1}{c_h} \int_\Omega \chi_{fh}(\mathbf{p}_0, \mathbf{p}) \chi_{gh}^*(\mathbf{p}_0, \mathbf{p}) d\mathbf{p}
\end{aligned} \tag{4.14}
$$

and the proof is an extension of that used for the inverse relationships (4.13). The inner products are expressed as frequency domain integrations using (4.6) and the rhs is integrated over frequency and the space of the parameters. This is

$$
\begin{aligned}
\text{rhs} = \frac{1}{2\pi\beta c_h} \int_0^\infty \int_{-\infty}^\infty \int_{-\infty}^\infty & F(\omega) G^*(\omega') H\left(\frac{\omega'}{\beta}\right) H^*\left(\frac{\omega}{\beta}\right) \\
\cdot \frac{1}{2\pi} \int_{-\infty}^\infty & \exp j(\omega - \omega')\tau d\tau d\omega d\omega' d\beta
\end{aligned}
$$

and results in

$$
\text{rhs} = \frac{1}{2\pi\beta c_h} \int_{-\infty}^\infty \int_0^\infty F(\omega) G^*(\omega) \left| H\left(\frac{\omega}{\beta}\right) \right|^2 d\beta d\omega.
$$

Changing variables and integrating over frequency as before yields the frequency domain inner product

$$
\frac{1}{2\pi} \int_{-\infty}^\infty F(\omega) G^*(\omega) d\omega = < f, g >
$$

completing the proof. The NB version is the same except that the energy E_h replaces the admissibility constant c_h.

Ambiguity Volume

$\Phi_{fg}(\mathbf{p}_0, \mathbf{p})$ is a non-negative function representing a surface defined in the space of the parameters $\mathbf{p}$. It is bounded by the signal energies and if it is an auto-ambiguity function, it achieves its maximum at the origin defined by the point where the parameters of the constituent functions are equal.

The volume under this surface and its distribution are perhaps the most important considerations relating to the specification of waveforms for detection and estimation. It is evaluated from

$$(\text{volume}) = \int_\Omega \Phi_{fg}(\mathbf{p}_0, \mathbf{p}) d\mathbf{p}$$

and is denoted by V_W and V_N for the wide and narrowband models (4.4,4.5) with integration over $[\beta, \tau]$ and $[\phi, \tau]$ respectively. $\mathbf{p}_0$ defines the nominal origin of the AF but the volume may be computed for ambiguity functions centered at any point simply be changing variables as indicated by (4.12).

The narrowband AF is noteworthy because the ambiguity surface bounds a fixed volume that is determined only by the signal energies and not their functional forms. This is shown by using (4.5) to express the narrowband volume V_N as

$$\begin{aligned} V_N &= \int_\Omega \Phi_{fg}(\mathbf{p}_0, \mathbf{p}) d\mathbf{p} \\ &= \int_{-\infty}^{\infty} \int_{-\infty}^{\infty} \left(\int_{-\infty}^{\infty} f(t) g^*(t-\tau) \exp -j2\pi\phi t dt \right) \\ &\qquad \cdot \left(\int_{-\infty}^{\infty} f^*(u) g(u-\tau) \exp j2\pi\phi u du \right) d\tau d\phi \\ &= \int_{-\infty}^{\infty} \int_{-\infty}^{\infty} \int_{-\infty}^{\infty} f(t) f^*(u) g^*(t-\tau) g(u-\tau) \\ &\qquad \cdot \left(\int_{-\infty}^{\infty} \exp -j2\pi\phi(t-u) d\phi \right) dt\, du\, d\tau. \end{aligned}$$

The bracketed integral in ϕ is the delta function $\delta(t-u)$ and the volume becomes

$$V_N = \int_{-\infty}^{\infty} |f(t)|^2 \int_{-\infty}^{\infty} |g(t-\tau)|^2 dt\, d\tau = E_f E_g \quad (\text{Joules}^2) \qquad (4.15)$$

or E_f^2 when $f = g$. V_N is integrated over the time-frequency parameter plane that by itself carries no dimension. The volume reflects only the dimension of the AF itself which is either (energy)^2 or if normalized, unitless. In the latter case, $V_N = 1$.

The narrowband auto-ambiguity function (AAF) has the unique property that its peak value at the origin is equal to its volume and the distribution of a constant amount of ambiguity volume in the τ, ϕ plane is the only attribute affected by choice of waveform. Narrowband ambiguity volume is conserved and cannot be created nor destroyed by choice of waveform resulting in what might be called a "water bed" effect. It can be removed from one portion of the phase plane through changes in waveform but must always appear somewhere else.

The wideband AF is a different matter. While it also is expressed in terms of squared energy, its volume is integrated over the β, τ plane. β is unitless but τ has time dimension and the volume has dimension of squared energy × time (Joules^2 × sec.) or merely time if it is normalized.

The V_W is evaluated by defining the admissible functions $f(t), g(t), h(t)$ and writing the ROI (4.14) as

$$< f, h >= \left(\frac{1}{c_g}\right) \int_\Omega < f, g_{\mathbf{p}} >< h, g_{\mathbf{p}} >^* d\mathbf{p}.$$

Setting $h(t) = f(t)$,

$$\begin{aligned} < f, f >= E_f = \left(\frac{1}{c_g}\right) \int_\Omega | < f, g_{\mathbf{p}} > |^2 d\mathbf{p} &= \left(\frac{1}{c_g}\right) \int_\Omega \Phi_{fg}\mathbf{p}_0, \mathbf{p}) d\mathbf{p} \\ &= \frac{V_W}{c_g} \end{aligned}$$

and

$$V_W = c_g E_f \quad \text{Joules}^2 \times \text{sec.}$$

When $g = f$, the volume of the AAF is $c_f E_f$. The volume is independent of the origin as may be demonstrated by substituting $f_{\hat{\mathbf{p}}}$ for f in the ROI. The energy becomes the inner product $< f_{\hat{\mathbf{p}}}, f_{\hat{\mathbf{p}}} >$ but this is also E_f. Unlike the narrowband volume, the wideband volume depends upon the admissibility constant as well as the energy and moreover, interchanging the constituent functions yields a different result. Interchanging $f(t)$ and $g(t)$ in the ROI yields

$$\text{vol.}\, (\Phi_{gf}(\mathbf{p}_0, \mathbf{p})) = \int_\Omega \Phi_{gf}(\mathbf{p}_0, \mathbf{p}) d\mathbf{p} = c_f E_g$$

which is not necessarily equal to $c_g E_f$.

The wideband volume defined in the $[\beta, \tau]$ plane is fundamentally different from the narrowband volume defined in $[\phi, \tau]$ because it includes time dimension, it is a function of the spectral characteristics of the constituent functions as reflected in the admissibility constants, it is not necessarily bounded, and it depends upon the order of the constituent functions in its definition. That is,

$$\frac{\text{vol.}\,(\Phi_{fg}(\mathbf{p}_0, \mathbf{p}))}{\text{vol.}\,(\Phi_{gf}(\mathbf{p}_0, \mathbf{p}))} = \frac{c_g E_f}{c_f E_g}.$$

Rather than integrating over β, τ, the wideband ambiguity volume may be integrated over time-frequency resulting in a volume measure that is dimensionally consistent with the narrowband volume [17,20]. Assume that the constituent functions are analytic, admissible, and bandpass with bandwidth W such that the finite frequency centroid

$$\omega_0 = \frac{1}{E_g}\int_0^\infty \omega |G(\omega)|^2 d\omega$$

exists. The associated Doppler frequency ϕ is

$$\phi = (\beta - 1)\frac{\omega_0}{2\pi} \tag{4.16}$$

and the alternative WB volume is integrated over ϕ, τ as in the NB case. It is denoted as V_{W-N} and

$$V_{W-N} = \int_{-\infty}^{\infty}\int_{-\infty}^{\infty} \Phi_{fg}(1, 0, \beta, \tau) d\phi d\tau.$$

Changing the Doppler variable yields $d\phi = (1/2\pi)\omega_0 d\beta$ and

$$\begin{aligned} V_{W-N} &= \left(\frac{\omega_0}{2\pi}\right)\int_{-\infty}^{\infty}\int_0^{\infty} \Phi_{fg}(1, 0, \beta, \tau) d\beta d\tau \\ &= \left(\frac{\omega_0}{2\pi}\right) V_W = \left(\frac{\omega_0}{2\pi}\right) E_f c_g \\ &= \left(\frac{\omega_0}{2\pi}\right) E_f \int_{\omega_0 - \pi B}^{\omega_0 + \pi B} \frac{|G(\omega)|^2}{\omega} d\omega. \end{aligned} \tag{4.17}$$

This does not carry time dimension and is consistent with the narrowband volume measure. The bandpass spectrum is $G(\omega) = \tilde{G}(\omega_b)$ where $\omega_b = \omega - \omega_0$ is the baseband frequency and $\tilde{G}$ represents the envelope spectrum with bandwidth W Hz. Writing the volume in terms of these quantities produces

$$
\begin{aligned}
V_{W-N} &= \frac{\omega_0 E_f}{2\pi} \int_{-\pi W}^{\pi W} \frac{|\tilde{G}(\omega_b)|^2}{\omega_b + \omega_0} d\omega_b \\
&= \frac{E_f}{2\pi} \int_{-\pi W}^{\pi W} \frac{|\tilde{G}(\omega_b)|^2}{1 + \frac{\omega_b}{\omega_0}} d\omega_b.
\end{aligned}
$$

For $2\pi W/\omega_0 << 1$, the function becomes increasingly narrowband and

$$
\lim_{2\pi W/\omega_0 \to 0} V_{W-N} = \frac{E_f}{2\pi} \int_{-\pi W}^{\pi W} |\tilde{G}(\omega_b)|^2 d\omega_b = E_f E_g = V_N.
$$

This is consistent with the limiting relationship (4.2) between the admissibility constant and the energy.

The ratio V_{W-N}/V_N is an indicator of wideband volume expansion as a function of fractional bandwidth. As an example it has been computed for the simple case of a rectangular spectrum with centroid ω_0 [20]. Substituting (4.2) with $c_x = c_g$ into (4.17) and dividing by V_N (4.15) yields

$$
\frac{V_{W-N}}{V_N} = \left(\frac{1}{2W_f}\right) \ln\left(\frac{1+W_f}{1-W_f}\right)
$$

where $W_f = W/2f_0 \leq 1$ is the fractional bandwidth and $f_0 = \omega_0/2\pi$. The volume expansion is small for quite large fractional bandwidths approaching only ten percent at about fifty percent bandwidth as illustrated in Figure 4.6. The volume begins to expand rapidly as the lower band edge approaches zero reflecting the rapid increase of the admissibility constant.

4.3.2 Auto AF Properties at the Origin

AAF properties at the origin are important because they determine the ultimate resolution and estimation properties of the match filtered waveform. Auto ambiguity functions are maximized at the origin and the region in the neighborhood of this point has historically been called the ambiguity *mainlobe* region. This roughly corresponds to the area in which the AF monotonically decreases with distance from the origin. The region outside of the main lobe is not surprisingly called the *sidelobe* region and while it may contain subsidiary maxima, none will exceed that at the origin. AF properties in the mainlobe region are of

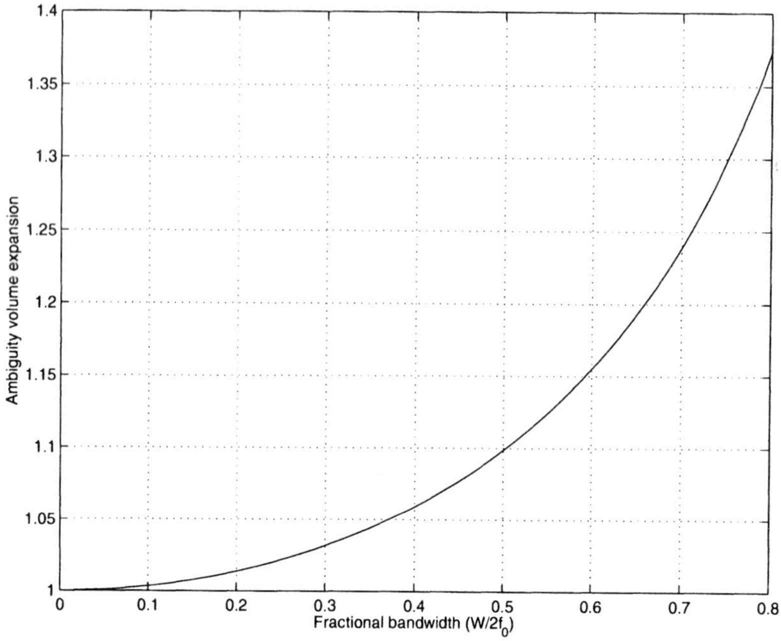

Figure 4.6: Wide band AF volume expansion

particular interest because the shape and "peakedness" of the mainlobe determines the accuracy and resolution with which parameter estimates may be made. While the origin may be defined at any point in the parameter plane, it is convenient to define it at the $\mathbf{p}_0$ point and examine $\Phi_{ff}(\mathbf{p}_0, \mathbf{p})$ for incremental values of $\mathbf{p}$ using a 2-d Taylor expansion about $\mathbf{p}_0$.

$f(t)$ is normalized for unit energy such that $\Phi_{ff}(\mathbf{p}_0, \mathbf{p}_0) = E_f = 1$ and the wide and narrowband slowly fluctuating AF expansions taken to second order are

$$\begin{aligned}
&\Phi_{ff}(1, 0, \beta, \tau) \approx \\
&\quad 1 + \frac{1}{2}\left(\tau^2 \frac{\partial^2 \Phi_{ff}}{\partial \tau^2} + 2(\beta - 1)\tau \frac{\partial^2 \Phi_{ff}}{\partial \beta \partial \tau} + (\beta - 1)^2 \frac{\partial^2 \Phi_{ff}}{\partial \beta^2}\right) \quad (\text{ wideband}) \\
&\Phi_{ff}(0, 0, \phi, \tau) \approx \\
&\quad 1 + \frac{1}{2}\left(\tau^2 \frac{\partial^2 \Phi_{ff}}{\partial \tau^2} + 2\phi\tau \frac{\partial^2 \Phi_{ff}}{\partial \phi \partial \tau} + \phi^2 \frac{\partial^2 \Phi_{ff}}{\partial \phi^2}\right) \quad (\text{ narrowband}). \qquad (4.18)
\end{aligned}$$

The first order derivatives do not appear as they are zero at the origin.

The generic AF derivatives for a single parameter p (e.g. τ or β) are

$$\begin{aligned}\frac{\partial \Phi_{ff}}{\partial p} &= \frac{\partial |\chi_{ff}|^2}{\partial p} = 2\Re\left(\chi_{ff}\frac{\partial \chi_{ff}^*}{\partial p}\right)\\ \frac{\partial^2 \Phi_{ff}}{\partial p^2} &= \frac{\partial^2 |\chi_{ff}|^2}{\partial p^2} = 2\Re\left(\chi_{ff}\frac{\partial^2 \chi_{ff}^*}{\partial p^2}\right) + 2\left|\frac{\partial \chi_{ff}}{\partial p}\right|^2\\ &= 2\Re\left(\frac{\partial^2 \chi_{ff}^*}{\partial p^2}\right) + 2\left|\frac{\partial \chi_{ff}}{\partial p}\right|^2 \qquad (4.19)\end{aligned}$$

when $p = p_0$ and $\chi_{ff}(p_0, p_0) = 1$.

The calculation of the expansion derivatives in terms of $f(\mathbf{p}, t)$ is a rather tedious undertaking especially for the wideband case.[2] The expansion and its associated derivatives for the wideband and narrowband cases are found in [18] and [13] respectively. The derivation of the wideband coefficient for the delay parameter τ is shown in detail here in order to illustrate the procedure. The other wideband coefficients and the narrowband expansion are presented without elaboration with details left to the references.

The Wideband Expansion Derivation for τ

The coefficient $\partial^2\Phi_{ff}/\partial\tau^2$ evaluated from (4.19) requires the first and second partial derivatives of χ_{ff} with respect to τ. Using (4.4) with $\hat{\mathbf{p}} = \mathbf{p}$ and $g = f$, and letting $u = \beta(t - \tau)$, the first derivative is

$$\begin{aligned}\frac{\partial \chi_{ff}}{\partial \tau} &= \frac{\partial}{\partial \tau}\left(\sqrt{\beta}\int_{\Omega_t} f(t) f^*(u)dt\right)\\ &= \sqrt{\beta}\int_{\Omega_t} f(t)\frac{\partial f^*(u)}{\partial u}\cdot\frac{\partial u}{\partial \tau}\,dt\end{aligned}$$

[2]The wideband AF with dilation parameter β rather than the frequency equivalent ϕ (4.16) will be used for consistency with common usage and the wideband literature. If it is desired to use the alternate frequency definition in the expansions, it is straightforward to change the variables. For example:

$$\begin{aligned}\frac{\partial \Phi_{ff}}{\partial \phi} &= \frac{\partial \Phi_{ff}}{\partial \beta}\cdot\frac{\partial \beta}{\partial \phi} = \frac{2\pi}{\omega_0}\frac{\partial \Phi_{ff}}{\partial \beta}\\ \frac{\partial^2 \Phi_{ff}}{\partial \phi^2} &= \frac{\partial}{\partial \phi}\left(\frac{2\pi}{\omega_0}\frac{\partial \Phi_{ff}}{\partial \beta}\right) = \frac{2\pi}{\omega_0}\left(\frac{\partial}{\partial \beta}\left(\frac{\partial \Phi_{ff}}{\partial \beta}\right)\cdot\frac{\partial \beta}{\partial \phi}\right)\\ &= \left(\frac{2\pi}{\omega_0}\right)^2\frac{\partial^2 \Phi_{ff}}{\partial \beta^2}.\end{aligned}$$

$$= -\beta\sqrt{\beta}\int_{\Omega_t} f(t)\frac{\partial f^*(u)}{\partial u}\,dt. \tag{4.20}$$

Let $\dot{f}(u)$ denote the derivative of f with respect to its argument then, at the origin ($\beta = 1$, $\tau = 0$), $u = t$ and

$$\frac{\partial \chi_{ff}}{\partial \tau} = -\int_{\Omega_t} f(t)\dot{f}^*(t)dt. \tag{4.21}$$

The second derivative is derived using (4.20) where

$$\begin{aligned}
\frac{\partial^2 \chi_{ff}}{\partial \tau^2} &= -\beta\sqrt{\beta}\int_{\Omega_t} f(t)\left(\frac{\partial \dot{f}^*(u)}{\partial \tau}\right)dt \\
&= -\beta\sqrt{\beta}\int_{\Omega_t} f(t)\left(\frac{\partial \dot{f}^*(u)}{\partial u}\cdot\frac{\partial u}{\partial \tau}\right)dt \\
&= \beta^{5/2}\int_{\Omega_t} f(t)\ddot{f}^*(u)dt \\
&= \int_{\Omega_t} f(t)\ddot{f}^*(t)dt
\end{aligned} \tag{4.22}$$

at the origin. Substituting (4.21,4.22) into (4.19) provides the delay expansion coefficient

$$\left.\frac{\partial^2 \Phi_{ff}(\mathbf{p}_0,\mathbf{p})}{\partial \tau^2}\right|_{\mathbf{p}=\mathbf{p}_0} = 2\Re\left(\int_{\Omega_t} f(t)\ddot{f}^*(t)dt\right) + 2\left|\int_{\Omega_t} f(t)\dot{f}^*(t)dt\right|^2. \tag{4.23}$$

An alternate form is obtained by integrating the first term by parts and making the assumption that $f(t)\dot{f}(t) = 0$ for $t = t_{max}, t_{min}$ at the limits of the time range Ω_t. This results in

$$\left.\frac{\partial^2 \Phi_{ff}(\mathbf{p}_0,\mathbf{p})}{\partial \tau^2}\right|_{\mathbf{p}=\mathbf{p}_0} = -2\left(\int_{\Omega_t}\left|\dot{f}(t)\right|^2 dt - \left|\int_{\Omega_t} f(t)\dot{f}^*(t)dt\right|^2\right) \tag{4.24}$$

Since the first term is a real non-negative quantity and equivalent to the first term of (4.23), the $\Re$ operator is redundant in that relationship.

These equations are expressed in the frequency domain using Parseval's rule and the Fourier relationships

$$\begin{aligned}
\mathcal{F}\,f(t) &= F(\omega) \\
\mathcal{F}\,\dot{f}(t) &= j\omega F(\omega).
\end{aligned}$$

This provides the important result

$$\left.\frac{\partial^2 \Phi_{ff}(\mathbf{p}_0, \mathbf{p})}{\partial \tau^2}\right|_{\mathbf{p}=\mathbf{p}_0} = -2\left(\frac{1}{2\pi}\int_{-\infty}^{\infty} \omega^2 |F(\omega)|^2 d\omega - \left|\frac{1}{2\pi}\int_{-\infty}^{\infty} \omega F(\omega) d\omega\right|^2\right).$$

The first term is the mean square radian frequency and the second is the frequency centroid. The second derivative is therefore proportional to the mean square radian bandwidth W_ω^2

$$\left.\frac{\partial^2 \Phi_{ff}(\mathbf{p}_0, \mathbf{p})}{\partial \tau^2}\right|_{\mathbf{p}=\mathbf{p}_0} = -\frac{1}{\pi}\int_{-\infty}^{\infty} (\omega - \omega_0)^2 |F(\omega)|^2 d\omega = -2W_\omega^2. \quad (4.25)$$

It was shown in Section 3.5 that estimation variance is bounded by the inverse of the Fisher matrix that has as its elements the second derivatives of the ambiguity function. Delay measurement accuracy is proportional to the mean square signal bandwidth but this is not the whole story because accuracy is also affected by cross products appearing in the matrix as indicated by Example 3.6. For these, the remaining partial derivatives in (4.18) are required. These have been derived by Altes [18] and are:

$$\left.\frac{\partial^2 \Phi_{ff}(\mathbf{p}_0, \mathbf{p})}{\partial \beta^2}\right|_{\mathbf{p}=\mathbf{p}_0} = -2\left(\int_{\Omega_t} t^2 |\dot{f}(t)|^2 dt - \left|\int_{\Omega_t} t f(t) \dot{f}^*(t) dt\right|^2\right)$$

$$\left.\frac{\partial^2 \Phi_{ff}(\mathbf{p}_0, \mathbf{p})}{\partial \beta \partial \tau}\right|_{\mathbf{p}=\mathbf{p}_0} = -2\left(\int_{\Omega_t} t |\dot{f}(t)|^2 dt - \Re\left(\int_{\Omega_t} \dot{f}(t) f^*(t) dt \int_{\Omega_t} t f(t) \dot{f}^*(t) dt\right)\right). \quad (4.26)$$

The AF expansion (4.18) at $\mathbf{p} = \mathbf{p}_0$ is expressed in terms of the delay, Doppler, and coupling coefficients λ^2, η^2, and γ [18, 21] as

$$\Phi_{ff}(1, 0, \beta, \tau) \approx 1 - \left(\lambda^2 \tau^2 + 2\gamma\tau(\beta - 1) + \eta^2(\beta - 1)^2\right). \quad (4.27)$$

The wideband fluctuating model expressions for the coefficients are

$$
\begin{aligned}
\lambda^2 &= -.5\frac{\partial^2 \Phi_{ff}}{\partial \tau^2} \\
&= \int_{\Omega_t} \left|\dot{f}(t)\right|^2 dt - \left|\int_{\Omega_t} f(t)\dot{f}^*(t)dt\right|^2 \\
&= \frac{1}{2\pi}\int_{-\infty}^{\infty} (\omega - \omega_0)^2 |F(\omega)|^2 d\omega = W_\omega^2 \\
\gamma &= -.5\frac{\partial^2 \Phi_{ff}}{\partial \beta \partial \tau} \\
&= \int_{\Omega_t} t|\dot{f}(t)|^2 dt - \Re\left(\int_{\Omega_t} \dot{f}(t)f^*(t)dt \int_{\Omega_t} tf(t)\dot{f}^*(t)dt\right) \\
&= \left(\frac{1}{2\pi}\int_{-\infty}^{\infty} \omega|F(\omega)|^2 d\omega \times \Im\left[\frac{1}{2\pi}\int_{-\infty}^{\infty} \omega\dot{F}(\omega)F^*(\omega)d\omega\right]\right) \\
&\quad -\Im\left[\frac{1}{2\pi}\int_{-\infty}^{\infty} \omega^2\dot{F}(\omega)F^*(\omega)d\omega\right] \\
\eta^2 &= -.5\frac{\partial^2 \Phi_{ff}}{\partial \beta^2} = \int_{\Omega_t} t^2|\dot{f}(t)|^2 dt - \left|\int_{\Omega_t} tf(t)\dot{f}^*(t)dt\right|^2 \\
&= \frac{1}{2\pi}\int_{-\infty}^{\infty} \omega^2|\dot{F}(\omega)|^2 d\omega - \left|\frac{1}{2\pi}\int_{-\infty}^{\infty} \omega\dot{F}(\omega)F^*(\omega)d\omega\right|^2 .
\end{aligned} \tag{4.28}
$$

Wideband AF expansion coefficients for the random model

The magnitudes of the coefficients λ^2 and η^2 are measures of receiver sensitivity to delay and Doppler (dilation) mismatch while γ indicates the amount of cross coupling that exists between delay and Doppler errors. This is zero for waveforms that are symmetric about the time origin.

The Deterministic Model

The wideband AAF expansion for the deterministic echo model is useful as an illustration of the role of the carrier in the determination of estimation accuracy. Recall from Chapter 3 that the deterministic signal AAF is

$$
\begin{aligned}
\Phi_{ff}(\beta,\tau) &= \Re\sqrt{\beta}\int_{\Omega_t} f(t)f^*(\beta(t-\tau))dt \\
&= \frac{1}{2}\left(\chi_{ff}(\beta,\tau) + \chi_{ff}^*(\beta,\tau)\right)
\end{aligned} \tag{4.29}
$$

and it is only necessary to evaluate the derivatives of χ_{ff}. The τ derivative is (4.22) and integrating by parts with $dv = \ddot{f}^* dt$, $u = f$ produces

$$\frac{\partial^2 \chi_{ff}}{\partial \tau^2} = \int_{\Omega_t} f(t)\ddot{f}^*(t)dt = f(t)\dot{f}^*(t)\Big|_{t_1}^{t_2} - \int_{\Omega_t} |\dot{f}(t)|^2 dt.$$

It is assumed that the waveform function is zero at its limits and by Parseval's rule,

$$\frac{\partial^2 \chi_{ff}}{\partial \tau^2} = -\left(\frac{1}{2\pi}\right) \int_{-\infty}^{\infty} \omega^2 |F(\omega)|^2 d\omega.$$

This is real so that it is also equivalent to the second partial derivative of (4.29). By recalling from (4.25) that the mean square radian bandwidth is

$$W_\omega^2 = \left(\frac{1}{2\pi}\right) \int_{-\infty}^{\infty} (\omega - \omega_0)^2 |F(\omega)|^2 d\omega$$

and that the carrier or radian frequency centroid is

$$\omega_0 = \left(\frac{1}{2\pi}\right) \int_{-\infty}^{\infty} \omega |F(\omega|^2 d\omega,$$

it is clear that

$$\frac{\partial^2 \chi_{ff}}{\partial \tau^2} = \frac{\partial^2 \Phi}{\partial \tau^2} = -(W_\omega^2 + \omega_0^2)$$

and

$$\lambda^2 = \frac{1}{2}(W_\omega^2 + \omega_0^2) \quad \text{(deterministic model)}.$$

This reflects the fact that the deterministic complex echo AF like that of the real signal model (4.9) has an oscillating $\cos \omega_0$ term (grating lobes) on the τ axis. Consequently the main AF lobe as determined by the second order expansion has a τ width determined by the period of the carrier not the mean square bandwidth. This grating lobe structure illustrates the possible pitfall in reliance solely upon the origin expansion and the associated CR bound as an accurate measure of the ability to estimate delay for real bandpass signals or under the deterministic model assumption. In most cases the bandwidth to carrier ratio of sonar waveforms is small and the grating structure appears as multiple closely spaced peaks within the broader envelope mainlobe. These multiple peaks are comparable in magnitude and make it difficult to locate the global maximum especially in the presence of interference. As a practical matter, the τ resolution and ability to estimate delay is governed by the overall mainlobe support which is determined by the mean square envelope bandwidth.

The wideband Doppler expansion for the deterministic AAF requires the second partial of (4.29) and hence χ_{ff} with respect to β.

$$\chi_{ff} = \sqrt{\beta} \int_{\Omega_t} f(t) f^*(\beta(t-\tau))dt$$

and its derivative with respect to β is

$$\begin{aligned} \frac{\partial \chi_{ff}}{\partial \beta} &= \sqrt{\beta} \int_{\Omega_t} f(t) \dot{f}^*(t-\tau)dt + \frac{1}{2\beta}\chi_{ff} \\ &= \int_{\Omega_t} t f(t) \dot{f}^*(t)dt + \frac{1}{2} \end{aligned}$$

at the origin defined by $\tau = 0$ and $\beta = 1$. Differentiating again and evaluating at the origin produces

$$\frac{\partial^2 \chi_{ff}}{\partial \beta^2} = \int_{\Omega_t} t^2 f \ddot{f}^* dt + \int_{\Omega_t} t f \dot{f}^* dt - \frac{1}{4} \tag{4.30}$$

where the time argument of $f(t)$ is dropped for convenience. Integrating the first term by parts and assuming that $f(t) = 0$ at the limits yields

$$\int_{\Omega_t} t^2 f \ddot{f}^* dt = -\int_{\Omega_t} t^2 |\dot{f}|^2 dt - 2\int_{\Omega_t} t f \dot{f}^* dt \tag{4.31}$$

and combining (4.30, 4.31) produces

$$\frac{\partial^2 \chi_{ff}}{\partial \beta^2} = -\int_{\Omega_t} t^2 |\dot{f}|^2 dt - \int_{\Omega_t} t f \dot{f}^* dt - \frac{1}{4}.$$

Integrating the second integral by parts and invoking the unit energy assumption yields

$$\begin{aligned} \int_{\Omega_t} t f \dot{f}^* d &= -\int_{\Omega_t} t \dot{f} f^* dt - 1 \\ &\Rightarrow 2\Re \int_{\Omega_t} t f \dot{f}^* dt = -1. \end{aligned}$$

The wideband Doppler expansion derivative is then

$$\frac{\partial^2 \Phi_{ff}}{\partial \beta^2} = \frac{1}{2}\left(\frac{\partial^2 \chi_{ff}}{\partial \beta^2} + \frac{\partial^2 \chi_{ff}^*}{\partial \beta^2}\right)$$

$$= \frac{1}{2}\left(-2\int_{\Omega_t} t^2|\dot{f}|^2 dt - \int_{\Omega_t} tf\dot{f}^* dt - \int_{\Omega_t} t\dot{f}f^* dt - \frac{1}{2}\right)$$
$$= \frac{1}{2}\left(-2\int_{\Omega_t} t^2|\dot{f}|^2 dt - 2\Re\int_{\Omega_t} t\dot{f}f^* dt - \frac{1}{2}\right)$$
$$= -\int_{\Omega_t} t^2|\dot{f}|^2 dt + \frac{1}{4}$$

and the wideband Doppler coefficient for the deterministic model is

$$\eta^2 = \frac{1}{2}\left(\int_{\Omega_t} t^2|\dot{f}|^2 dt - \frac{1}{4}\right).$$

The cross coupling coefficient γ is determined by a similar tedious calculation and to summarize, the wideband coefficients for the deterministic model are:

$$\begin{aligned} \lambda^2 &= \frac{1}{2}\left(W_\omega^2 + \omega_0^2\right) \\ \eta^2 &= \frac{1}{2}\left(\int_{\Omega_t} t^2|\dot{f}|^2 dt - \frac{1}{4}\right) \\ \gamma &= \int_{\Omega_t} t|\dot{f}|^2 dt \end{aligned} \tag{4.32}$$

Wide band AF expansion coefficients for the deterministic model

The cross coupling coefficient is zero for symmetric waveforms as was the case for the slowly fluctuating model (4.28). The Doppler coefficient contains the constant -1/4 but is positive which is shown by using the Schwartz inequality

$$\int_{\Omega_t} |f|^2 dt \int_{\Omega_t} t^2|\dot{f}|^2 dt \geq \left|\int_{\Omega_t} t\dot{f}f^* dt\right|^2$$

where the leftmost integral is unity for unit energy waveforms. Integrating by parts by $\int udv = uv - \int vdu$ with $u = tf^*$, $v = f$, $dv = \dot{f}dt$, and $du = t\dot{f}^* + f^*$ yields

$$\int_{\Omega_t} t\dot{f}f^* dt = t|f|^2\Big|_{t_1}^{t_2} - \int_{\Omega_t} (t\dot{f}^* + f^*)f dt.$$

The first term on the rhs is zero if it is assumed that the waveform is zero at its limits (or that it is symmetric). Therefore

$$\Re\int_{\Omega_t} tf^*\dot{f}dt = -\frac{1}{2} \Rightarrow \left|\int_{\Omega_t} tf^*\dot{f}dt\right|^2 \geq \frac{1}{4}$$

and

$$\int_{\Omega_t} t^2 |\dot{f}|^2 dt \geq \frac{1}{4} - \frac{1}{4} \geq 0.$$

The Narrowband Expansion for the Fluctuating Model

The narrowband expansion is obtained by using (4.5) with $g = f$, (4.18), and the generic derivative relationship (4.19). The result for the delay τ is the same as for the wideband case and is given by (4.24, 4.25). The Doppler and cross coupling terms are derived in [13] and are

$$\begin{aligned} \frac{\partial^2 \Phi_{ff}}{\partial \phi \partial \tau} &= 2\Im\left(2\pi \int_{-\infty}^{\infty} t f(t) \dot{f}^*(t) dt\right) \\ \frac{\partial^2 \Phi_{ff}}{\partial \phi^2} &= -2(2\pi)^2 \int_{-\infty}^{\infty} (t - t_0)^2 |f(t)|^2 dt \\ &= -8\pi^2 T_{rms}^2. \end{aligned} \tag{4.33}$$

T_{rms} is the rms pulse duration and t_0 is the time centroid of the function that is set to zero without loss of generality. As was done for the wideband model (4.28), narrowband expansion coefficients are defined as

$$\boxed{\begin{aligned} \lambda^2 &= -.5\frac{\partial^2 \Phi_{ff}}{\partial \tau^2} = \left(\frac{1}{2\pi}\right)\int_{-\infty}^{\infty} (\omega - \omega_0)^2 |F(\omega)|^2 d\omega = W_\omega^2 \\ \gamma &= -.5\frac{\partial^2 \Phi_{ff}}{\partial \phi \partial \tau} = -2\pi\Im \int_{-\infty}^{\infty} t f(t) \dot{f}^*(t) dt \\ \eta^2 &= -.5\frac{\partial^2 \Phi_{ff}}{\partial \phi^2} = (2\pi)^2 \int_{-\infty}^{\infty} (t - t_0)^2 |f(t)|^2 dt = 4\pi^2 T_{rms}^2 \qquad (4.34) \end{aligned}}$$

Narrowband AF expansion coefficients for the fluctuating model

and the resulting AF expansion is

$$\Phi_{ff}(0, 0, \phi, \tau) \approx 1 - \left(\lambda^2 \tau^2 + 2\gamma\tau\phi + \eta^2 \phi^2\right). \tag{4.35}$$

The delay resolution of both narrow and wideband receivers is proportional to the mean square bandwidth and Doppler resolution is proportional to the mean square pulse duration for the narrowband model. From the narrowband point of view, simultaneous high range and Doppler resolution requires a waveform with a large time-bandwidth (TW) product. Overall bandwidth also plays a role in determining Doppler resolution for the wideband model but in addition, the time distribution of the frequency modulation $\omega(t)$ within the pulse is an important factor. This is explored in the next section by considering the waveform as a complex analytic function with both amplitude and phase or frequency modulation.

4.4 THE BANDPASS WAVEFORM

4.4.1 Ambiguity properties at the origin

The analytic bandpass waveform model is commonly used and an understanding of the relationship between its defining parameters and ambiguity properties is important for the determination of detection and estimation performance. The general form of the waveform function is

$$\begin{aligned} f(t) &= a(t)\exp(j\Psi(t)) \\ &= a(t)\exp j(\psi(t)+\omega_0 t) \\ &= \tilde{f}(t)\exp j\omega_0 t, \ -T/2 \le t \le T/2 \end{aligned} \tag{4.36}$$

where $a(t)$ and $\psi(t)$ are the real amplitude and phase modulation functions respectively, ω_0 is the radian carrier frequency, and T is the waveform duration or *pulse width.* Taken together the modulation functions form the complex baseband envelope $\tilde{f}(t)$. Waveforms are often alternatively described in terms of the instantaneous frequency

$$\begin{aligned} \Omega(t) &= \dot{\Psi}(t) \\ &= \dot{\psi}(t) + \omega_0 \\ &= \omega(t) + \omega_0 \end{aligned} \tag{4.37}$$

where $\omega(t)$ represents the time varying *frequency modulation.*

Window functions

The real function $a(t)$ determines the waveform energy and the amplitude distribution of the envelope. It is most often employed as a window or shading function to smoothly taper the ambiguity sidelobes much as such functions are used to shape spectra in Fourier analysis.

Often, the same functions are used and descriptions of many common window functions are found in [22–25]. Rectangular shading is used when amplitude modulation of the transmitter power amplifiers is not possible or when maximization of the received energy under a peak power constraint is of paramount importance. This occurs when the interference is white noise and maximum SIR for detection is desired. The rectangular window has first order discontinuities at the pulse endpoints that produce ($\text{sinc}(x) = \sin x/x$) ambiguity sidelobes along the Doppler axis. The narrowband AF of the rectangularly weighted *continuous wave* (CW) waveform $f(t) = 1/\sqrt{T} \exp j\omega_0 t \; t \in [0,T]$ is derived in [13] and obtained from

$$\Phi_{ff}(\phi,\tau) = \left| \frac{1}{T} \int_{\tau}^{T} \exp -j(2\pi\phi t)dt \right|^2 .$$

Changing variables with $u = t - .5(\tau + T)$ and new limits $L = \pm(T-\tau)/2$ yields

$$\begin{aligned}\Phi_{ff}(\phi,\tau) &= \left| \frac{1}{T} \int_{-L}^{L} \exp -j[2\pi\phi(u + .5(\tau + T))]du \right|^2 \\ &= \left| \frac{1}{T} \int_{-L}^{L} \exp -j2\pi\phi u du \right|^2 \\ &= \left[\left(1 - \frac{|\tau|}{T}\right) \text{sinc}\left(\pi\phi T \left(1 - \frac{|\tau|}{T}\right)\right)\right]^2 . \qquad (4.38)\end{aligned}$$

$|\tau|$ is used because the AF is delay symmetric. The AF tapers parabolically to zero at $\tau = \pm T$ on the delay axis $\phi = 0$ and has a $\sin x/x$ characteristic on the Doppler axis. The pulse overlap decreases with τ and a slice of the AF parallel to the Doppler axis is equivalent to the squared Fourier transform of the foreshortened pulse. The $\sin x/x$ ridge spacing therefore increases as $\tau \to T$ and becomes infinite at the $\tau = T$.

A well defined ambiguity mainlobe without adjacent sidelobe peaks is often desired hence a tapered window that is a compromise between energy maximization and sidelobe level is used. Amplitude windows are symmetric, unimodal (single peak), non-negative, real functions and most but not all taper gradually to zero at the end points so that $a(-T/2) = a(T/2) = 0$. The symmetric taper limits the order and amplitude of the discontinuities and hence the rate at which the Doppler sidelobes decay with frequency. Windows are best applied to both the transmitted and matched filter replica functions although the transmitter power amplifiers must be designed to accommodate amplitude modulation. The square root of a window function may be used rather than the function itself in order to avoid over windowing when implementing a matched filter. CW ambiguity functions for the rectangular weight-

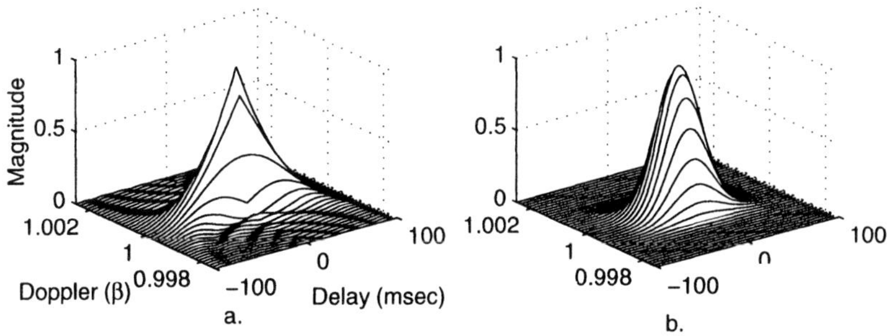

Figure 4.7: CW ambiguity functions: a. rectangular weighting, b. Hamming weighting

ing (4.38) and tapered *Hamming* weighting are shown in Figure 4.7. The windows are defined by the unit energy amplitude functions on $-T/2$, $T/2$ with

$$
\begin{aligned}
a(t) &= \frac{1}{\sqrt{T}} \quad \text{(rectangular weighting)} \\
&= \frac{1.59}{\sqrt{T}}\left(.54 + .46 \cos 2\pi \frac{t}{T}\right) \quad \text{(Hamming weighting)}
\end{aligned}
$$

respectively. The Hamming window [22,23] is a severely tapered function that requires about 2.5 times greater peak power than a rectangular window with the same energy. It reduces sidelobe effects as shown but because it has less rms time duration its AF mainlobe is broadened in the Doppler direction in comparison to that of the rectangular window of equal absolute duration.

Analytic waveform expansion coefficients

The wideband ambiguity expansion coefficients (4.28) for the complex analytic waveform defined by (4.36, 4.37) are determined by evaluating the partial derivatives (4.24, 4.26) yielding [21, 26–28]

$$\begin{aligned}
\lambda^2 &= \int_{-T/2}^{T/2} (\dot{a}^2(t) + a^2(t)\Omega^2(t))dt \\
&\quad - \left[\int_{-T/2}^{T/2} a^2(t)\Omega(t)dt\right]^2 = W_\omega^2 \\
\gamma &= \int_{-T/2}^{T/2} t(\dot{a}^2(t) + a^2(t)\Omega^2(t))dt \\
&\quad - \left[\int_{-T/2}^{T/2} a^2(t)\Omega(t)dt\right]\left[\int_{-T/2}^{T/2} ta^2(t)\Omega(t)dt\right] \\
\eta^2 &= \int_{-T/2}^{T/2} t^2(\dot{a}^2(t) + a^2(t)\Omega^2(t))dt \\
&\quad - \left[\int_{-T/2}^{T/2} ta^2(t)\Omega(t)dt\right]^2 - \frac{1}{4}\int_{-T/2}^{T/2} a^2(t)dt. \qquad (4.39)
\end{aligned}$$

Wideband fluctuating model coefficients for the bandpass waveform

Waveform functions are energy normalized and time limited with

$$\int_{-T/2}^{T/2} a^2(t)dt = 1$$

so that the last term of the third equation is (1/4).

The Doppler and coupling coefficients η^2, γ are derived from (4.33) for the narrowband case and the delay coefficient λ^2 is the same as that of the wideband case. These are

$$\begin{aligned}
\lambda^2 &= W_\omega^2 \\
\eta^2 &= (2\pi)^2 \int_{-T/2}^{T/2} t^2 a^2(t)dt = (2\pi)^2 T_{rms}^2 \\
\gamma &= 2\pi \int_{-T/2}^{T/2} ta^2(t)\Omega(t)dt \qquad (4.40)
\end{aligned}$$

Narrowband fluctuating model coefficients for the bandpass waveform

and the expansion is given by (4.35). λ^2 and η^2 establish the second order ambiguity properties along the delay and Doppler axis and γ represents

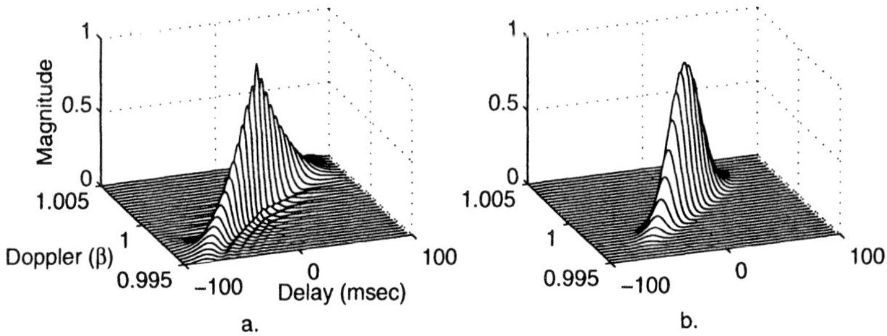

Figure 4.8: LFM ambiguity functions: a. rectangular weighting, b. Hamming weighting

the amount of cross coupling between delay and Doppler estimates. Ambiguity functions for which $\gamma = 0$ are called *uncoupled* ambiguity functions. Time delay resolution depends upon the mean square bandwidth for both models and for the narrowband model, Doppler resolvency as determined by η^2 is proportional only to the mean square time distribution of the window function $a(t)$. The phase or frequency distribution plays no direct role but has an indirect effect if $\gamma \neq 0$. Examination of the narrow and wideband relationships (4.39, 4.40) reveals that $\gamma = 0$ if $a(t)$ and $\Omega(t)$ have <u>even or odd</u> symmetry about the time origin for the wideband case and $\Omega(t)$ is <u>even</u> for the narrowband case. However, $\omega(t)$ must be <u>even</u> for both cases because the carrier ω_0 is a constant (hence even). These are important design considerations when it is desired to minimize parameter estimation error. The evenly windowed CW waveform AF's shown in Figure 4.7 are examples of uncoupled AF's. The linear frequency modulated (LFM) waveform for which $\Omega(t) = \omega_0 + 2\pi Wt/T$ and $\gamma = \pi^2 TW/3$ is an example of a highly coupled function. Examples of rectangular and Hamming windowed LFM AF's are shown in Figure 4.8. Note that the rectangularly weighted LFM AF also exhibits the sinc(x) sidelobe pattern.

4.4.2 Estimation Accuracy and the Ambiguity Error Ellipse

Estimation error is bounded by the inverse of the Fisher matrix which is related to the second gradient of the ambiguity function by (3.118) for the fluctuating scatterer in WGN. Invoking the definitions (4.28, 4.34) yields

$$\begin{aligned} J_{11} &= 2S_n\lambda^2 \\ J_{12} = J_{21} &= 2S_n\gamma \\ J_{22} &= 2S_n\eta^2 \end{aligned}$$

where SNR $= E_s/N_0$ and

$$S_n = \frac{\text{SNR}^2}{1+\text{SNR}}.$$

The estimation variances for delay (σ_τ^2) and dilation (σ_β^2) are thus bounded from below by

$$\boxed{\begin{aligned} \sigma_\tau^2 &\geq \left(\frac{1}{2S_n}\right)\left(\frac{\eta^2}{\eta^2\lambda^2-\gamma^2}\right) \\ \sigma_\beta^2 &\geq \left(\frac{1}{2S_n}\right)\left(\frac{\lambda^2}{\lambda^2\eta^2-\gamma^2}\right). \end{aligned} \quad (4.41)}$$

Lower bounds for the delay and dilation estimate variances

The AF has a global maximum at its origin that requires

$$\left(\frac{\partial^2\Phi_{ff}}{\partial\tau\partial\beta}\right)^2 - \frac{\partial^2\Phi_{ff}}{\partial\tau^2}\frac{\partial^2\Phi_{ff}}{\partial\beta^2} < 0$$

for functions of two variables [29] and

$$\lambda^2\eta^2 > \gamma^2.$$

The minimum variance bounds are

$$\begin{aligned} \sigma_\tau^2\,(min) &= \frac{1}{2S_n\lambda^2} \\ \sigma_\beta^2\,(min) &= \frac{1}{2S_n\eta^2} \end{aligned} \qquad (4.42)$$

and occur when $\gamma = 0$ and the AF is uncoupled.

The effect of parameter coupling is illustrated by Figure 4.9 which represents the elliptical constant level contour of a wideband AF near

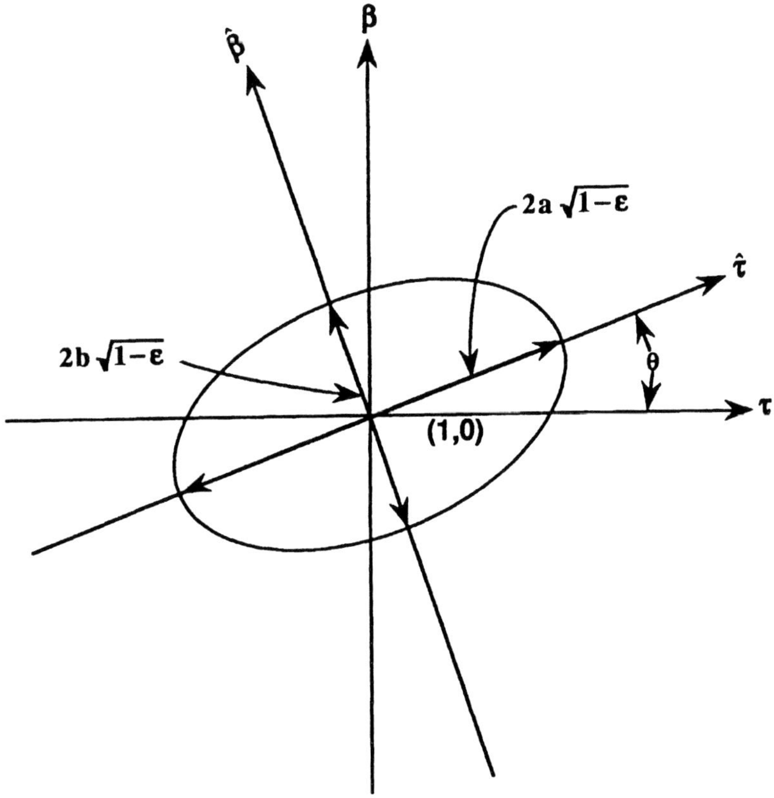

Figure 4.9: Ambiguity error ellipse

the origin at $\beta = 1, \tau = 0$ [18]. It corresponds to the locus of β, τ for constant $\Phi_{ff} < 1$ defined by the expansion (4.27)

$$\lambda^2\tau^2 + 2\gamma\tau(\beta - 1) + \eta^2(\beta - 1)^2 = 1 - \Phi_{ff} = \epsilon.$$

This has been found to be reasonably accurate for $\epsilon < .5$ [30]. It represents a coupled ellipse because of the cross term in the expansion and is equivalent to an axis rotation of the uncoupled ellipse

$$\frac{\hat{\tau}^2}{a^2} + \frac{(\hat{\beta} - 1)^2}{b^2} = \epsilon$$

where

$$\begin{bmatrix} \hat{\tau} \\ (\hat{\beta} - 1) \end{bmatrix} = \begin{bmatrix} \cos\theta & \sin\theta \\ -\sin\theta & \cos\theta \end{bmatrix} \begin{bmatrix} \tau \\ (\beta - 1) \end{bmatrix}$$

and θ is the angle of rotation. It is computed from

$$\theta = \frac{1}{2}\tan^{-1}\left(\frac{2\gamma}{\lambda^2 - \eta^2}\right)$$

and is zero for $\gamma = 0$. The uncoupled parameters a, b are likewise functions of the coupled parameters and

$$a^2 = \frac{2}{\lambda^2 + \eta^2 + \sqrt{4\gamma^2 + (\lambda^2 - \eta^2)^2}} = \frac{1}{\lambda^2},$$

$$b^2 = \frac{2}{\lambda^2 + \eta^2 - \sqrt{4\gamma^2 + (\lambda^2 - \eta^2)^2}} = \frac{1}{\eta^2}, \quad \text{when } \gamma = 0.$$

The equivalent narrowband ambiguity ellipse expressed in terms of the parameters τ, ϕ is given in references [13, 14].

Echos are random by virtue of the reflection and scattering processes and the presence of various forms of interference. Consequently, parameter estimates made from the received data are random and when they are made by seeking the peak response of a matched filter using a waveform with a coupled AF, they are correlated in proportion to the amount of ambiguity coupling. For example, it was shown in Chapter 3 that the CR bound becomes an equality and the covariance matrix of the parameter estimates is equal to the inverse of the Fisher matrix when an estimator is efficient. In that case the covariance between the estimated parameters $(\hat{\tau}, \hat{\beta})$ is equal to the off-diagonal element of the inverse Fisher matrix $\mathbf{J}^{-1}$ and

$$\mathcal{E}\{(\hat{\beta} - 1)\hat{\tau}\} = -\frac{\gamma}{2S_n(\lambda^2\eta^2 - \gamma^2)}.$$

Consequently and because the individual parameter error increases with the degree of coupling, echo-location waveforms used for delay and Doppler estimation should have uncoupled ambiguity functions unless other factors are involved. This requires that both the transmitted pulse amplitude $a(t)$ and the frequency modulation $\omega(t)$ be even for both models.

4.4.3 Wideband Doppler Resolvency

Range estimation accuracy is proportional to the mean square bandwidth for both models and Doppler resolution is proportional to the mean square time duration for the narrowband model. Wideband dilation resolution and velocity estimation accuracy also depend upon the time distribution of the amplitude $a(t)$ and frequency $\Omega(t)$ modulation [30]. In order to examine this it is necessary to consider the wideband Doppler sensitivity coefficient η^2 in detail.

Following the development of Altes and Titlebaum [21] and assuming that the waveform is energy normalized, η^2 (4.39) is separated into two parts $\eta^2 = \kappa_1(a) + \kappa_2(a, \Omega)$. The first of which

$$\kappa_1(a) = \int_{-T/2}^{T/2} t^2 \dot{a}^2(t)dt - \frac{1}{4}$$

depends only upon the window function. The second term is

$$\kappa_2(a, \Omega) = \int_{-T/2}^{T/2} t^2 a^2(t)\Omega^2(t)dt - \left[\int_{-T/2}^{T/2} ta^2(t)\Omega(t)dt\right]^2 \tag{4.43}$$

and depends upon the window as well as the frequency modulation. Both κ_1 and κ_2 are non-negative as may be shown by using the Schwarz inequality

$$\begin{aligned}\left(\int_{-T/2}^{T/2} ta(t)\dot{a}(t)dt\right)^2 &\leq \int_{-T/2}^{T/2} a^2(t)dt \int_{-T/2}^{T/2} t^2\dot{a}^2(t)dt \\ &= \int_{-T/2}^{T/2} t^2\dot{a}^2(t)dt. \end{aligned} \tag{4.44}$$

Integrating by parts by letting $u = a(t)t$, $v = a(t)$ and using the formula

$$\int udv = uv - \int vdu$$

yields

$$\int_{-T/2}^{T/2} ta(t)\dot{a}(t)dt = a^2(t)t\Big|_{T/2}^{T/2} - \int_{-T/2}^{T/2} a^2(t)dt - \int_{-T/2}^{T/2} ta(t)\dot{a}(t)dt.$$

Re-arranging and invoking the unit energy condition produces

$$\int_{-T/2}^{T/2} ta(t)\dot{a}(t)dt = .5a^2(t)t\Big|_{-T/2}^{T/2} - \frac{1}{2}$$

and if it is assumed that the amplitude function is either symmetric or zero at the pulse edges,

$$a^2(t)t\Big|_{-T/2}^{T/2} = 0$$

and

$$\left(\int_{-T/2}^{T/2} ta(t)\dot{a}(t)dt\right)^2 = \frac{1}{4}.$$

Using (4.44) and assuming that a(t) has unit energy on (-T/2,T/2) yields

$$\left(\int_{-T/2}^{T/2} ta(t)\dot{a}(t)dt\right)^2 = \frac{1}{4} \leq \int_{-T/2}^{T/2} t^2\dot{a}^2(t)dt$$

and $\kappa_1(a) \geq 0$. Equation (4.44) becomes an equality and $\kappa_1(a) = 0$ if $t\dot{a}(t) = c_1 a(t)$ the solution of which is

$$a(t) = c_2 t^{c_1}$$

where c_1 and c_2 are constants. Likewise, applying the Schwarz inequality to the last term in (4.43) produces

$$\left[\int_{-T/2}^{T/2} ta^2(t)\Omega(t)dt\right]^2 \leq \int_{-T/2}^{T/2} a^2(t)dt \int_{-T/2}^{T/2} t^2a^2(t)\Omega^2(t)dt$$

and this results in $\kappa_2(a,\Omega) \geq 0$ for unit energy $a(t)$. If $ta(t)\Omega(t) = ka(t)$ or $\Omega(t) = k/t$ with k a constant, $\kappa_2(a,\Omega) = 0$ regardless of the amplitude function. This is called hyperbolic frequency or linear period modulation (HFM, LPM) [31] and the resulting complex waveform

$$f(t) = c_2 t^{c_1} \exp jk \ln t \tag{4.45}$$

is an optimally *Doppler tolerant* waveform with an AF that is sharply delineated on the τ axis but uniformly spread in β. It is not practically realizable because its duration and bandwidth are unconstrained and hence absolute Doppler insensitivity is not possible. Altes and Titlebaum [21] derived a constrained energy, rms, and absolute time duration approximation of a(t) that is highly Doppler tolerant and was found to closely resemble the waveform used by bats. Naturally evolved acoustic echo-location used by bats and porpoises is called *bionic sonar* and in most cases is characterized by a high degree of Doppler tolerance because accurate distance estimation over a wide Doppler range is apparently important for food gathering activities [21, 27, 32]. Maximally Doppler tolerant waveforms with arbitrary amplitude windows and constrained bandwidth envelopes $\tilde{f}(t)$ have been designed [28] and although LPM waveforms are Doppler tolerant, it is still possible to measure Doppler with them [33]. Doppler tolerant waveforms are discussed in more detail in Chapter 5.

The relationship between Doppler sensitivity and the distribution of the frequency modulation is illustrated by writing (4.43) as

$$\kappa_2(a,\Omega) = \int_{-T/2}^{T/2} T_w^2(t)dt - \left[\int_{-T/2}^{T/2} a(t)T_w(t)dt\right]^2 \tag{4.46}$$

where $T_w(t) = a(t)t\Omega(t)$ is the amplitude weighted time-frequency product and κ_2 is recognized as its centralized mean square value or variance. It is evident that high sensitivity to Doppler induced time base dilation is associated with large variance of the t-f product over the pulse interval.

For most sonar applications $a(t)$ is specified as an even symmetric, energy normalized, and slowly varying real function such that $|\dot{a}(t)| << \Omega(t)$, $\kappa_1(a) << \kappa_2(a,\Omega)$ and its time variation contributes negligibly to Doppler sensitivity so that $\eta^2 \approx \kappa_2$. If $a(t)T_w(t)$ is odd,

$$\eta^2 \approx \kappa_2(a,\Omega) = \int_{-T/2}^{T/2} T_w^2(t)dt$$

and the greatest sensitivity is associated with uncoupled waveforms with even $a(t)$, $\omega(t)$ and odd $T_w(t)$. The wideband AF expansion for zero delay is

$$\begin{aligned}\Phi_{ff}(1,0,\beta,0) &= 1 - \eta^2(\beta-1)^2 \\ &\approx 1 - \left(\int_{-T/2}^{T/2} a^2(t)t^2(\omega_0 + \omega(t))^2 dt\right)(\beta-1)^2\end{aligned}$$

and it is apparent that large values of the frequency modulation $\omega(t)$ occurring at the endpoints of the pulse when $|t| = t_{max} = T/2$ will produce greater Doppler sensitivity than the same modulation occurring closer to the pulse midpoint $t = 0$. This is illustrated by the phase plots shown in Figure 4.10. Neglecting any amplitude window effects, the positive half of an odd (cubic) phase function $\Psi(t)$ of total duration T and slope (instantaneous frequency) $\Omega(t)$ is plotted according to the wide and narrowband Doppler models of Figures 4.10a,b respectively. The wideband model represents Doppler as a compression or dilation of the time base T by the factor $T(\beta-1)/\beta$ such that the total phase $2\Psi_m$ is preserved. The incremental frequency must increase in order that the same number of phase cycles exist in the shorter pulse preserving the information coded in the modulation. The narrowband model does not permit a time base dilation, only a frequency or phase slope shift. This represents the addition of a linear term $2\pi \Delta \phi t$ to the phase function so that more (or less) phase $\Psi'_m - \Psi_m$ is required at the endpoints in order that the pulse duration is maintained. The addition or subtraction of phase represents a loss or gain of information solely due to Doppler which is not possible and represents a major inconsistency of the narrowband model. The

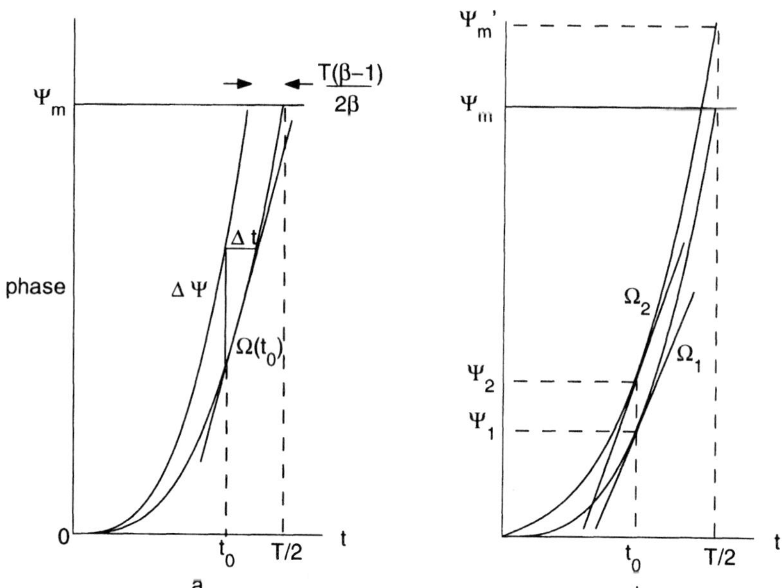

Figure 4.10: Wide and narrowband phase model examples: a. wideband phase time function, b. narrowband phase time function

stationary phase principle [13] states that the value of the ambiguity function is inversely proportional to the mean square frequency difference between a signal and its Doppler shifted or dilated replica. In other words the ambiguity loss is proportional to the integrated time variation of the phase (frequency difference) between them. If it is assumed that the function is odd and uncoupled, the incremental phase error at time

t_0 is proportional to the instantaneous frequency and

$$\triangle\Psi(t_0) = \left.\frac{\partial\Psi}{\partial t}\right|_{t_0} \triangle t = \Omega(t_0) \triangle t$$

as indicated in Figure 4.10a. $\Delta t = (t - t_0)$ is the time base difference between corresponding points on the signal and its dilated replica that have the same phase $\Psi(t_0)$. The change in the time base due to the dilation $\beta \approx 2 \triangle v/c$ caused by a velocity increment Δv is

$$\begin{aligned}\Delta t &= t_0 \left(\frac{1-\beta}{\beta}\right) \approx t_0(1-\beta) \\ &= \frac{2t_0 \triangle v}{c}.\end{aligned}$$

The instantaneous frequency Ω is the phase derivative so that the sensitivity of the phase difference to velocity is

$$\left.\frac{\Delta\Psi}{\Delta v}\right|_{t_0} = \frac{2\Omega(t_0)t_0}{c} \tag{4.47}$$

and the incremental phase error is proportional to the time weighted instantaneous frequency. By stationary phase, the ambiguity sensitivity to a change in Doppler velocity is proportional to the time integral of the square of (4.47) which is equivalent within a constant to the first term of (4.46). If T_w is not odd and the second term of (4.46) is not zero, the ambiguity peak will occur at a delay shift that minimizes the integrated mean square frequency difference. Consequently the centroid of the dilated replica will be time shifted relative to the undilated signal at the AF peak reflecting the coupling between τ and β [34]. In either case, increased slope of the phase (higher frequency) near the pulse extrema will cause greater dilation sensitivity than if it occurs near the origin.

By contrast, the narrowband model of Figure 4.10b models Doppler as the phase slope (frequency) change $2\pi \triangle \phi$ so that at time t_0,

$$\Omega_2(t_0) = \Omega_1(t_0) + 2\pi \triangle \phi$$

and

$$\begin{aligned}\Psi_2(t_0) &= \int_0^{t_0} (\Omega_1(t) + 2\pi \triangle \phi)dt \\ &= \Psi_1(t_0) + 2\pi \triangle \phi t.\end{aligned}$$

Doppler frequency is related to velocity by $\triangle\phi = 2 \triangle vf_0/c$ and consequently

$$\left.\frac{\triangle\Psi}{\triangle v}\right|_{t_0} = \frac{2\omega_0 t}{c}.$$

The instantaneous sensitivity of the phase to velocity is proportional to the time weighted constant frequency f_0 and not the frequency modulation. The duration is unchanged but the end point phase $\Psi'_m > \Psi_m$ implying an increase of the total phase of the waveform which is impossible. This illustrates the inconsistency of the narrowband model.

The delay sensitivity coefficient λ^2 is equivalent to the radian bandwidth and depends only upon the weighted integral of the modulation and not its distribution within the pulse. It is thus possible to specify a particular delay sensitivity and adjust the wideband Doppler sensitivity by placing the maximum frequency deviation either toward the endpoints or the middle of the pulse [26,30]. If $\omega(t) = 0$, the waveform is a weighted tone $a(t)\exp j\omega_0 t$ and $\kappa_2 = \omega_0^2 T_{rms}^2$. Doppler frequency is related to the dilation by (4.16) and the wide and narrowband coefficients are related by

$$\left(\frac{1}{f_0^2}\right)\frac{\partial^2\Phi_{ff}}{\partial\beta^2} = \frac{\partial^2\Phi_{ff}}{\partial\phi^2}.$$

Consequently, when there is no phase/frequency modulation, the waveform is by definition narrowband and

$$\left(\frac{1}{f_0^2}\right)\kappa_2\,(\text{wideband}) = 4\pi^2 T_{rms}^2 = \kappa_2\,(\text{narrowband}).$$

The reliance of the Doppler sensitivity upon the modulation distribution is a degree of design freedom to be considered when specifying waveforms to be used with wideband (dilation) processing. When $\gamma = 0$, delay and Doppler sensitivity and the associated estimation accuracy are determined solely by the bandwidth and waveform duration for the narrowband model. However, when it is desired to increase delay resolution by using a modulated waveform and yet maintain a degree of Doppler resolution, the time-bandwidth product TW must be increased. The incremental range resolution is proportional to c/2W and the distance a reflector moves during the time it is illuminated is vT. Unless the ratio

$$\left|\frac{vT}{c/2W}\right| = |\beta - 1|\,TW << 1,$$

the narrowband receiver response to the reflector will diminished and spread across several resolution cells reducing detector performance. In such cases the narrowband receiver is inadequate and wideband processing becomes necessary. Then, the modulation distribution must be considered as part of the waveform specification.

4.4.4 Narrowband Model Error

The issue of wideband Doppler sensitivity is important for the determination of the conditions under which it is possible to use the simpler narrowband model for the receiver structure. The performance of a matched filter detector operating in WGN against a slowly fluctuating point scatterer depends upon the maximum value of the detection index (3.62)

$$D_{\mathbf{p}\hat{\mathbf{p}}} = \frac{E_s \Phi_{ff}(\mathbf{p}, \hat{\mathbf{p}})}{N_0}$$

that occurs at the ambiguity origin $\mathbf{p} = \hat{\mathbf{p}}$. If the signal replica is exactly matched to the Doppler shifted and delayed waveform, $\Phi_{ff} = 1$ and all of the signal energy E_s is recovered by the matched filter. The wideband AF for the complex analytic waveform (4.36) is

$$\Phi_{ff}(\beta, 0, \hat{\beta}, \hat{\tau}) = \left| \sqrt{\beta\hat{\beta}} \int_{\Omega_t} \tilde{f}(\beta(t-\tau))\tilde{f}^*(\hat{\beta}(t-\hat{\tau})) \exp j2\pi\triangle\phi t dt \right|^2$$

where

$$\triangle\phi = \phi - \hat{\phi} = (\beta - \hat{\beta}) f_0$$

is the narrowband Doppler error. Clearly, except for the normalizing factor $\sqrt{\beta\hat{\beta}}$ and the appearance of the dilation parameters in the envelope functions, this is the same as the narrowband AF.

When the waveform is uncoupled and the signal is characterized by dilation β with zero delay ($\tau = 0$), the AF peak will occur along the dilation axis defined by $\hat{\tau} = 0$. When it is coupled, the peak position will be biased at some $\hat{\tau} \neq 0$ [34]. The hybrid AF of a processor that synthesizes a narrowband replica at $\hat{\tau} = 0$ but ignores the envelope dilation is

$$\Phi_h(\beta, \triangle\phi) =$$
$$\beta \left| \int_{\Omega_t} \tilde{f}(\beta t)\tilde{f}^*(t) \exp +2\pi\triangle\phi t dt \right|^2 .$$

This represents neither a wide nor narrowband model but is proportional to the expected signal output of a mismatched narrowband processor hypothesizing the Doppler shift $\Delta\phi$ when the input signal envelope is time dilated by the factor β. This always occurs when a narrowband processor is used as signals scattered from moving objects are dilated. The peak value Φ_h will be less than unity for all $\beta \neq 1$ regardless of the Doppler hypothesis $\Delta\phi$ and the peak loss due to the envelope mismatch will be reflected in a reduction in output SIR for a matched filter. The hybrid AF is

$$\Phi_h(\beta, \Delta\phi) = \beta \left| \int_{\Omega_t} a(\beta t)a(t) \exp j(\psi(\beta t) - \psi(t) + 2\pi\Delta\phi t)tdt \right|^2 \quad (4.48)$$

for the complex analytic signal model and the mismatch loss is evaluated at that value of $\Delta\phi$ maximizing this quantity. The effect of a constant or at most slowly varying window is negligible and the complex exponential with phase $\theta(t) = \psi(\beta t) - \psi(t) + 2\pi\Delta\phi t$ is the dominant factor. The value of $\Delta\phi$ that maximizes Φ_h is found by again invoking stationary phase wherein the extremal corresponds to that value of $\Delta\phi$ that minimizes the mean square frequency deviation over the interval $\Omega_t = t_{max} - t_{min}$. The frequency is the time derivative of $\theta(t)$ and the maximizing Doppler is found from (4.37) and

$$\frac{\partial}{\partial\Delta\phi} \int_{\Omega_t} (\beta\omega(\beta t) - \omega(t) + 2\pi\Delta\phi)^2 dt = 0.$$

Taking the derivative and integrating the result in time yields

$$\psi(\beta t) - \psi(t) + 2\pi\Delta\phi t \Big|_{t_{min}}^{t_{max}} = 0$$

where t_{max} and t_{min} depend upon β because the envelope window $a(t)$ is defined only on the interval $-T/2, T/2$ and is zero elsewhere. Consequently $a(\beta t) = 0$ for $|\beta t| \geq T/2$. If $\beta \geq 1$, the product $a(\beta t)a(t) = 0$ for $|t| \geq T/2\beta$ and if $\beta \leq 1$ the product is zero for $|t| \geq T/2$.

For $\beta \geq 1$, the limits are $t_{max} = -t_{min} = T/2\beta$ and the solution is

$$\Delta\phi_{opt} = \frac{\psi(T/2\beta) - \psi(-T/2\beta) - (\psi(T/2) - \psi(-T/2))}{2\pi T/\beta} \qquad \beta \geq 1.$$

For $\beta < 1$, the limits are $t_{max} = -t_{min} = T/2$ and the solution is

$$\Delta\phi_{opt} = \frac{\psi(T/2) - \psi(-T/2) - (\psi(\beta T/2) - \psi(-\beta T/2))}{2\pi T} \qquad \beta < 1.$$

For both situations, the maximizing Doppler is the frequency average represented by the pulse endpoint phase difference between the dilated and un-dilated envelope divided by the duration T. If the phase function is even (odd frequency) $\Delta\phi_{opt} = 0$. If the phase function is odd, the frequency will be even as required for uncoupled ambiguity functions and

$$\begin{aligned}\Delta\phi_{opt} &= \frac{\psi(T/2\beta) - \psi(T/2)}{\pi T/\beta} \qquad \beta \geq 1 \\ &= \frac{\psi(T/2) - \psi(\beta T/2)}{\pi T} \qquad \beta < 1.\end{aligned}$$

The first relationship may be written as

$$\Delta\phi_{opt} = \frac{\psi\left(\frac{T}{2} + \left(\frac{1}{\beta} - 1\right)\frac{T}{2}\right) - \psi\left(\frac{T}{2}\right)}{2\pi\frac{T}{2}\left(\frac{1}{\beta} - 1\right)\left(\frac{1}{1-\beta}\right)}$$

which is recognized as $(1-\beta)$ times the incremental pulse endpoint phase divided by the time increment or in other words the frequency

$$\Delta\phi_{opt} = \left(\frac{1-\beta}{2\pi}\right)\omega(T/2).$$

This is easily shown to also be the case for $\beta < 1$.

The peak value of the mismatched filter is the value of $\Phi_h(\beta, \Delta\phi)$ (4.48) at $\Delta\phi = \Delta\phi_{opt}$. The residual phase at the optimum Doppler shift is

$$\theta(t) = \psi(\beta t) - \psi(t) + 2\pi\Delta\phi_{opt}t$$

and may be written as

$$\theta(t) = (\psi(\beta t) + \tilde{\omega}_0\beta t) - (\psi(t) + \tilde{\omega}_0 t)$$

where $\tilde{\omega}_0$ is the residual carrier frequency term

$$\tilde{\omega}_0 = \frac{2\pi\Delta\phi_{opt}}{\beta - 1} = -\omega(T/2).$$

The peak value of the mismatched filter is obtained from the expansion

$$\Phi_h(\beta, \Delta\phi_{opt}) \approx 1 - \tilde{\eta}^2(\beta - 1)^2$$

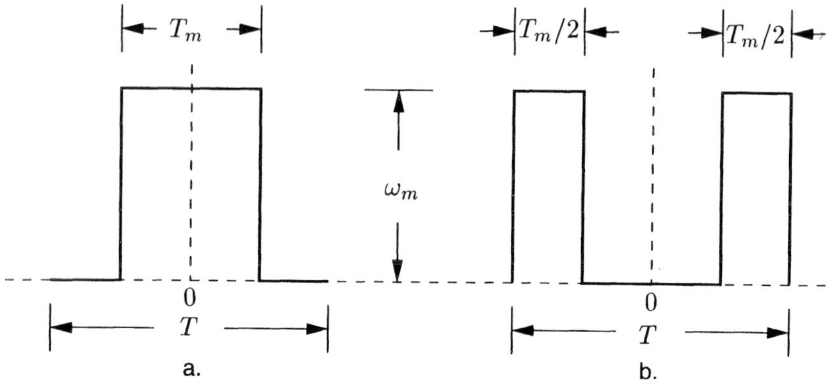

Figure 4.11: FSK modulation function examples: a. single pulse, b. double pulse

where $\tilde{\eta}$ is the envelope dilation sensitivity coefficient obtained from (4.39) with

$$\begin{aligned}\Omega(t) &= \omega(t) - \omega(T/2) \quad (\omega(t) \text{ even}) \\ &= \omega(t) \quad (\omega(t) \text{ odd}).\end{aligned}$$

The narrowband mismatch is proportional to $\tilde{\eta}$ for a fixed value of β. Waveforms with envelopes characterized by small values of this coefficient are relatively less sensitive and may be processed by a narrowband receiver over a larger Doppler range for a specified loss.

Example 4.1 Time-frequency Distribution

The effect of the time-frequency distribution of the modulation is illustrated by considering the frequency modulation functions shown in Figure 4.11. For simplicity assume that the window is the constant energy normalized function $a(t) = 1/\sqrt{T}$ defined over the full pulse length T for both modulations. The interval T_m represents the the total support for the modulation for both cases, only the distribution is different. The bandpass waveform associated with these functions is called a *Frequency Shift Keyed* (FSK) waveform where the frequency, nominally at some carrier ω_0, shifts or hops to one or more different discrete frequencies over the pulse duration. In this example, the envelope frequency merely shifts by the amount ω_m. Neglecting the effect of the window, the mean square bandwidth W_ω^2 and hence the range delay coefficient λ^2 (4.39) is

$$\lambda^2 = \frac{\omega_m^2 T_m}{T}\left(1 - \frac{T_m}{T}\right) \quad T_m \leq T$$

and is the same for both functions. $a(t)$ is even for both functions, the endpoint frequencies are assumed to be zero so that $\omega(T/2) = 0$, T_w is odd, and the Doppler coefficient (4.46) is

$$\tilde{\eta}^2 \approx \int_{-T/2}^{T/2} T_w^2(t)dt.$$

The squared time-frequency product of the modulation (4.43,4.46) is

$$T_w^2 = a(t)^2 t^2 \omega(t) = \frac{t^2 \omega_m^2}{T}$$

for the time support indicated in the figure and zero elsewhere. Carrying out the integrations for both the single and double pulses (a),(b) results in

$$\tilde{\eta}_a^2 = \frac{\omega_m^2 T_m^3}{12T}$$
$$\tilde{\eta}_b^2 = \tilde{\eta}_a^2 + \frac{\omega_m^2 T_m^2}{4}\left(\frac{T}{T_m} - 1\right).$$

Let the fractional modulation interval be $k = T_m/T \leq 1$ and denote the design bandwidth as $W = \omega_m/2\pi$. The sensitivity coefficients are then related by

$$\tilde{\eta}_b^2 = \tilde{\eta}_a^2 + \left(\frac{(2\pi)^2 T^2 W^2}{4}\right) k(1-k)$$

and

$$\tilde{\eta}_a^2 = \frac{(2\pi)^2 T^2 W^2 k^3}{12}.$$

$\tilde{\eta}_b^2 - \tilde{\eta}_a^2$ is maximized for $k = 1/2$.

The Doppler coefficients for the modulation functions of the example are

$$\tilde{\eta}_a^2 = D_{sa}(TW)^2, \quad D_{sa} = \frac{(2\pi)^2 k^3}{12}$$
$$\tilde{\eta}_b^2 = D_{sb}(TW)^2, \quad D_{sb} = \frac{(2\pi)^2(1 + (k-1)^3)}{12}$$

where D_{sa}, D_{sb} are dilation sensitivity factors that are particular to the time-frequency distributions. The dilation sensitivity of an uncoupled, bounded, and piecewise continuous frequency modulation function is expressible as $\tilde{\eta}^2 = D_s T^2 W^2$ [30] where T is the waveform duration and W represents the design bandwidth or the maximum range of the frequency about the carrier. The Doppler induced dilation or velocity range that can be tolerated by a narrowband receiver for a specified ambiguity peak loss $\Phi_h(max) = \Phi_h(\beta, \Delta\phi_{opt}) \leq 1$ (4.48) is

$$\boxed{\frac{2|v|}{c} \approx |\beta - 1| \leq \frac{1}{TW}\sqrt{\frac{1 - \Phi_h(max)}{D_s}} \qquad (4.49)}$$

Narrowband receiver velocity error tolerance

which provides a more precise specification than the usual admonition that $|\beta - 1|TW << 1$. For example, if it were determined that a ten percent loss is tolerable i.e. $\Phi_h(max) = .9$ at the peak corresponding to $\Delta\phi_{opt}$, the maximum Doppler range over which a narrowband matched filter receiver could be used would be

$$|\Delta\phi|_{max} = |\beta - 1|f_0 \leq \frac{f_0}{TW}\sqrt{\frac{.1}{D_s}}.$$

Sensitivity factors for various forms of FM and FSK modulation functions are derived in [30].

4.4.5 Quadratic Distortion (QD)

Attenuation and distortion of the delay-Doppler AF also occurs when the sonar geometry is such that the velocity vectors are not constant and collinear during the interrogation interval. Higher order time mapping terms are then required to generate an exact replica matched to the echo signal. The AF becomes a multidimensional surface and the parameter hypothesis includes acceleration and perhaps higher order terms in addition to delay and dilation (Doppler). Hypothesis testing for higher order kinematic parameters is rarely done in practice because the platform-scatterer kinematics do not often lead to sufficient AF distortion to warrant the additional computational complexity [35].

The interrogation of objects that have a significant cross range velocity component v_x that produces the signal (2.21) as shown in Figure 2.9 is a situation in which quadratic signal distortion occurs. This most commonly occurs for high resolution *synthetic aperture* sonars or radars that utilize a side-looking array with a velocity component normal to the los [36]. The cross range velocity component generates a quadratic time mapping term that is compensated for by introducing an acceleration component in the matched filter replica, a process called *focusing* [37,38]. QD also occurs when the signal is subject to frequency dependent *group delay* effects that arise most commonly with low frequency long range acoustic propagation in the ocean. [39]. In the former case the quadratic term represents the acceleration $-v_x^2/r$ and the received signal component (2.21) is modeled as

$$f_r(t) \approx f(\beta(t-\tau) - \alpha(t-\tau)^2)$$

with $\alpha = v_x^2/rc$. The WBAF computed without regard to the quadratic term is the cross ambiguity function

$$\Phi_{ff}(\mathbf{p}, \hat{\mathbf{p}}, \alpha) = \beta\hat{\beta}\left|\int_{-\infty}^{\infty} f(\beta(t-\tau) - \alpha(t-\tau)^2) f^*(\hat{\beta}(t-\hat{\tau}))dt\right|^2 \tag{4.50}$$

and assuming that the correct delay hypothesis is made so that $\tau = \hat{\tau}$,

$$\Phi_{ff}(\beta, \hat{\beta}, \alpha) = \beta\hat{\beta}\left|\int_{-\infty}^{\infty} f(\beta t - \alpha t^2) f^*(\hat{\beta}t)dt\right|^2.$$

The distortion effect is easily demonstrated for a narrowband waveform defined on $t \in [0, T]$ with a constant amplitude envelope that is unaffected by Doppler dilation or QD. The constant amplitude CW waveform $f(t) = (1/T) \exp j\omega_0 t$ is a simple example with

$$\Phi_{ff}(\Delta\beta, \alpha) = \frac{1}{T^2}\left|\int_0^T \exp j\omega_0(\Delta\beta t - \alpha t^2)dt\right|^2 \tag{4.51}$$

where $\Delta\beta = \beta - \hat{\beta}$. When there is no QD, $\alpha = 0$ and the AF is maximized for the correct Doppler hypothesis $\hat{\beta} = \beta$ resulting in an unbiased estimate of v_{los}. When $\alpha \neq 0$ the expected receiver response is maximized for that value of $\Delta\beta$ that minimizes the mean square frequency deviation over the pulse duration. The radian frequency $\omega_0(\Delta\beta + 2\alpha t)$ is the time derivative of the phase and the minimizing value of $\Delta\beta$ is obtained by solving

$$\frac{\partial}{\partial \Delta\beta}\left(\frac{1}{T}\int_0^T [\omega_0(\Delta\beta - 2\alpha t)]^2 dt\right) = 0$$

with the result $\Delta\beta = \alpha T$. This is the dilation bias introduced by the QD and the equivalent narrowband Doppler bias is $\Delta\phi = f_0 \alpha T$ where $f_0 = \omega_0/2\pi$. The Doppler deviation from that at the AF peak for any hypothesis $\hat{\beta}$ or $\hat{\phi}$ is

$$\delta\phi = f_0(\beta - \alpha T) - f_0\hat{\beta} = \Delta\phi - f_0\alpha T \tag{4.52}$$

and is the deviation from the unbiased Doppler corresponding to v_{los} when there is no QD

The QD introduces an attenuation of the AF mainlobe because the replica is mismatched to the echo signal and the maximum of the AF (4.50) is less than unity. (4.51) represents the power spectrum of an LFM waveform envelope with bandwidth $2\alpha T f_0$ while the undistorted CW waveform has a spectral width of approximately $1/T$. Ambiguity volume is conserved for narrowband waveforms and the peak response is inversely proportional to the amount of QD induced frequency spreading. In other words there is a constant amount of signal energy to be allocated among a greater number of resolution cells. Consequently the AF attenuation induced by the QD for large values of α is proportional to the ratio of the undistorted to distorted bandwidths and

$$(max)\Phi_{ff}(\Delta\phi, \alpha) \propto \frac{1/T}{2\alpha T f_0}. \tag{4.53}$$

The threshold level of cross range induced acceleration that begins to induce AF attenuation at range r for an uncompensated matched filter is

$$2\alpha T^2 f_0 > 1 \Rightarrow \alpha = \frac{v_x^2}{rc} > \frac{1}{2 f_0 T^2} \tag{4.54}$$

and the AF attenuation is given by (4.53) for

$$v_x >> \sqrt{\frac{rc}{2 f_0 T^2}}.$$

An exact expression for the attenuation is derived for intermediate conditions by writing the phase term in (4.51) as

$$\omega_0(\Delta\beta t + \alpha t^2) = -\alpha\omega_0\left(t^2 - \frac{\Delta\beta}{\alpha}t\right).$$

Completing the square yields the equivalent phase term

$$-\alpha\omega_0\left(t - \frac{\Delta\beta}{2\alpha}\right)^2 + \left(\frac{\omega_0(\Delta\beta)^2}{4\alpha}\right)$$

and substituting this in (4.51) produces

$$\Phi_{ff}(\Delta\beta, \alpha) = \frac{1}{T^2}\left|\int_0^T \exp\left[-j\alpha\omega_0\left(t - \frac{\Delta\beta}{2\alpha}\right)^2\right]dt\right|^2.$$

Let

$$z = \sqrt{\frac{2\alpha\omega_0}{\pi}}\left(t - \frac{\Delta\beta}{2\alpha}\right)$$

and change variables $t \to z$ in the AF so that

$$\Phi_{ff}(\Delta\beta, \alpha) = \frac{\pi}{2\alpha\omega_0 T^2}\left|\int_{z_1}^{z_2} \exp -j\frac{\pi z^2}{2}dz\right|^2$$

with

$$z_1 = -\Delta\beta\sqrt{\frac{f_0}{\alpha}} = -\frac{\Delta\phi}{\sqrt{\alpha f_0}}$$

$$z_2 = -\Delta\beta\sqrt{\frac{f_0}{\alpha}} + 2T\sqrt{\alpha f_0} = z_1 + 2T\sqrt{\alpha f_0}$$

where $f_0 = \omega_0/2\pi$. Define the dimensionless parameters

$$p_1 = T\sqrt{\alpha f_0} = \frac{v_x T}{\sqrt{r\lambda_0}}, \quad p_2 = (\delta\phi)T$$

where λ_0 is the wavelength associated with the carrier f_0, p_1 is a measure of the severity of QD, and p_2 is a measure of the deviation of the Doppler hypothesis from that maximizing the AF. $\delta\phi$ is defined by (4.52) hence

$$p_2 = 0 \Rightarrow \hat{\phi} = \phi - \alpha f_0 T$$

corresponding to the Doppler at the biased AF peak. $\hat{\phi} = \phi$ is the correct (unbiased) v_{los} Doppler in the absence of QD.

The integration limits are expressed in terms of these parameters as

$$z_1 = -\left(p_1 + \frac{p_2}{p_1}\right), \quad z_2 = \left(p_1 - \frac{p_2}{p_1}\right) \tag{4.55}$$

and the AF is written in terms of sine and cosine integrals

$$\Phi_{ff}(p_1, p_2) = \frac{1}{4p_1^2}\left[\left(\int_{z_1}^{z_2} \cos\frac{\pi z^2}{2}dz\right)^2 + \left(\int_{z_1}^{z_2} \sin\frac{\pi z^2}{2}dz\right)^2\right].$$

These are decomposed into separate Fresnel integrals with lower bounds of zero

$$\mathcal{C}(x) = -\mathcal{C}(-x) = \int_0^x \cos\frac{\pi}{2}t^2 dt$$
$$\mathcal{S}(x) = -\mathcal{S}(-x) = \int_0^x \sin\frac{\pi}{2}t^2 dt$$

and

$$\Phi_{ff}(p_1, p_2) = \frac{1}{4p_1^2}\left[(\mathcal{C}(z_2) + \mathcal{C}(-z_1))^2 + (\mathcal{S}(z_2) + \mathcal{S}(-z_1))^2\right].$$

Finally, substituting the limits (4.55) yields

$$\begin{aligned}\Phi_{ff}(p_1, p_2) &= \frac{1}{4p_1^2}\left(\mathcal{C}(p_1 + p_2/p_1) + \mathcal{C}(p_1 - p_2/p_1)\right)^2 \\ &+ \frac{1}{4p_1^2}\left(\mathcal{S}(p_1 + p_2/p_1) + \mathcal{S}(p_1 - p_2/p_1)\right)^2. \end{aligned} \quad (4.56)$$

The AF maximum occurs at the bias compensated Doppler hypothesis where $p_2 = 0$ and

$$\Phi_{ff}(p_1) = \frac{1}{p_1^2}\left[\mathcal{C}^2(p_1) + \mathcal{S}^2(p_1)\right] \leq 1. \quad (4.57)$$

As p_1 becomes large, the Fresnel integrals approach 1/2 and

$$\Phi_{ff}(max) \rightarrow \frac{1}{2p_1^2} = \frac{1}{2T^2\alpha f_0}$$

which is the same as the asymptotic result (4.53).

The attenuation of the AF at it maximum corresponding to the bias compensated Doppler $\hat{\phi}$ (4.57) is plotted as a function of the QD parameter p_1 in Figure 4.12a. The onset of QD (4.54) corresponds to $p_1^2 = .5$ at which point the AF peak is attenuated by about 10%. The energy removed from the peak must appear in the sidelobes because of narrowband volume conservation hence the sidelobes increase as the peak is attenuated according to (4.56) as shown in Figure 4.12b.

4.5 THE STATIONARY PHASE PRINCIPLE

For all but the simplest cases, a computer is necessary for the construction and plotting of ambiguity functions especially when it is necessary

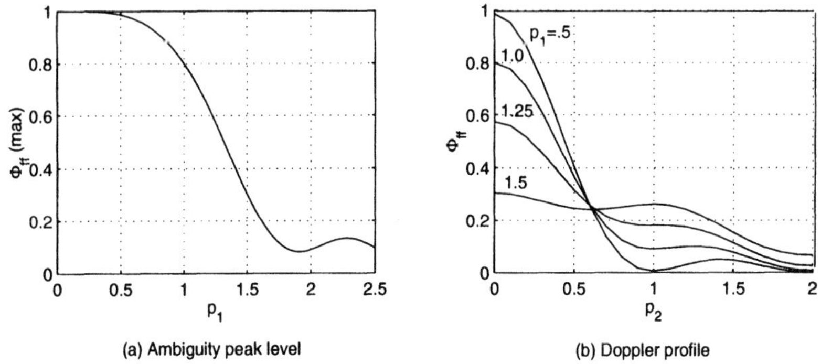

Figure 4.12: Ambiguity quadratic distortion effects plotted from (4.56,4.57): a. ambiguity peak level, b. Doppler profile

to see the fine details in the sidelobe regions as shown in Figures 4.7, 4.8. On the other hand it is often desired to merely estimate the gross features of the auto-ambiguity mainlobe or in the case of cross AF's, the magnitude and distribution of ambiguity in the delay-Doppler plane. Narrowband auto and cross AF features for waveform functions with simple time-frequency modulation profiles may be ascertained by a graphical technique based upon the stationary phase principle. The technique is also applicable to waveforms with more complex modulation profiles that can be decomposed into multiple simple forms.

The un-normalized narrowband cross-AF for the waveform functions

$$f(t) = a(t) \exp j(\psi_f(t) + \omega_0 t) \quad t \in [-T_f/2, T_f/2]$$
$$g(t) = b(t) \exp j(\psi_g(t) + \omega_0 t) \quad t \in [-T_g/2, T_g/2]$$

is $\Phi_{fg}(0, 0, \phi, \tau)$ which for simplicity is written as

$$\Phi_{fg}(\phi, \tau) = \left| \int_{\Omega_t} a(t) b(t - \tau) \exp j\theta(t)\, dt \right|^2 . \qquad (4.58)$$

$\theta(t)$ is the phase term

$$\theta(t) = \psi_f(t) - \psi_g(t - \tau) - 2\pi\phi t$$

with corresponding time varying frequency

$$\omega(t) = \omega_f(t) - \omega_g(t - \tau) - 2\pi\phi.$$

The waveform functions are not normalized so that

$$E_f = \int_{T_f/2}^{T_f/2} a^2(t)dt, \quad E_g = \int_{T_g/2}^{T_g/2} b^2(t)dt.$$

The amplitude functions $a(t)$, $b(t)$ are real and the greatest contributions to the AF integration will occur in intervals or times $(\hat{t})$ when the complex (phase) term is constant or equivalently when $\omega(\hat{t}) = 0$. $\hat{t}$ is called a stationary point and represents the time or times at which the delayed and Doppler shifted frequency function of $g(t)$ matches that of $f(t)$. This is illustrated by the following example.

Example 4.2 LFM Waveforms

Let $f(t)$ and $g(t)$, be LFM waveforms with $T_f = T_g = T$, and amplitude functions $a(t)$ and $b(t)$. They each have the same frequency modulation

$$\omega_f(t) = \omega_g(t) = \frac{2\pi W t}{T}$$

and the stationary phase point occurs at $\hat{t}$ when

$$\omega(\hat{t}) = \frac{2\pi W \hat{t}}{T} - \frac{2\pi W(\hat{t} - \tau)}{T} - 2\pi\phi = 0.$$

Attempting to solve this reveals that $\hat{t}$ is indeterminate because there is a non-zero interval over which $\omega = 0$. This occurs for

$$\phi = \frac{W\tau}{T}$$

as shown in Figure 4.13 and represents an overlap of the frequency functions of the two waveforms such that the overall phase function $\theta(t)$ is constant and does not diminish the integral. The CAF is then

$$\Phi_{fg}(\phi, \tau) = \left| \int_{\tau - T/2}^{T/2} a(t)b(t - \tau)\, dt \right|^2$$

in the overlap area that corresponds to its mainlobe region. In regions of the ϕ, τ plane where no overlap occurs (sidelobes), the CAF will be diminished by integration of the time varying phase term and hence

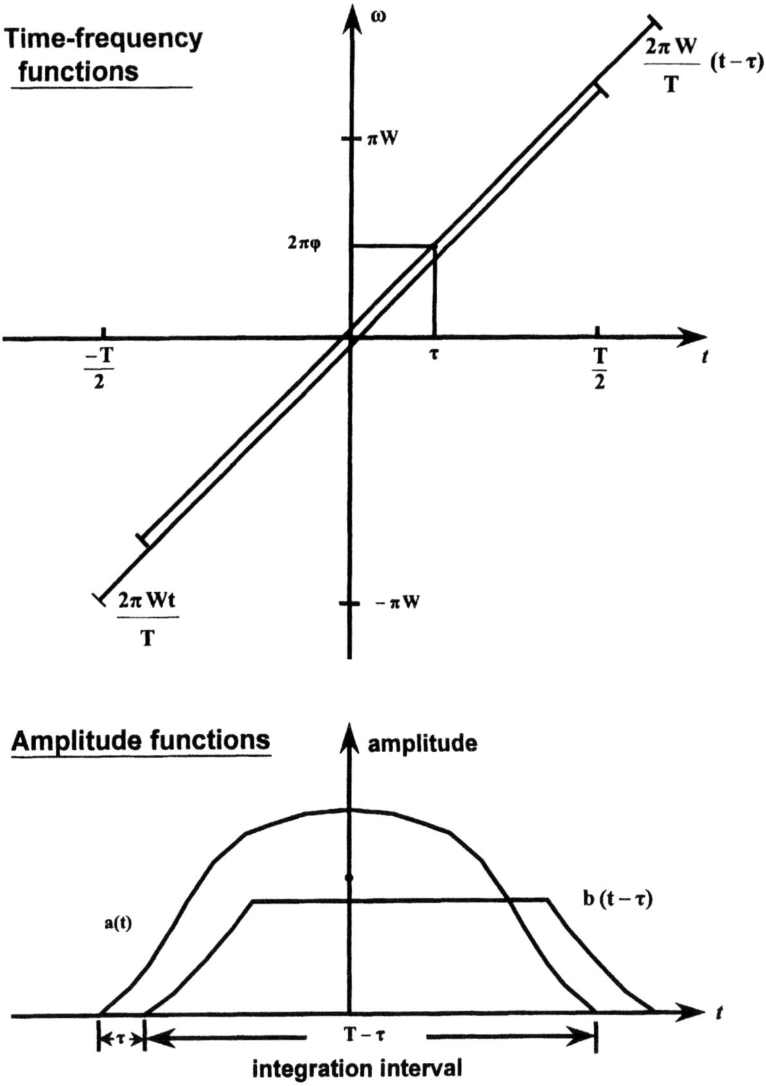

Figure 4.13: Graphical estimation of the LFM cross-ambiguity function with different amplitude windows

significantly smaller. When the amplitude functions are constant corresponding to rectangular weighting

$$a(t) = \sqrt{\frac{E_f}{T}}, \quad b(t) = \sqrt{\frac{E_g}{T}}$$

and the CAF is

$$\Phi_{fg}(\phi, \tau) = \frac{E_f E_g}{T^2}(T - \tau)^2$$

in the overlap region. It is a normalized AAF and is unity at the origin when $E_f = E_g = 1$. Its mainlobe profile is then the same as that of the computer derived AAF of Figure 4.8a. Ambiguity functions obtained graphically this way are approximations that are valid only when the time-frequency functions cross or overlay each other. The AF in regions where they do not cross cannot be easily computed in closed form and it is approximated by assuming that it is zero there.

When the the frequency functions do not overlay each other for some non-zero interval but cross each other at one or more stationary points $\hat{t}$, the ambiguity estimate can be obtained by invoking the *stationary phase principle* at each point. This was utilized to determine the Doppler maximizing the wideband ambiguity response in the previous section. Here it will be used to determine the AF for pre-specified waveforms whose frequency functions $\omega_f(t)$, $\omega_g(t)$ intersect in the time-frequency plane. Stationary phase as a method of mathematical approximation is discussed in [40, 41] and its application to ambiguity estimation is found in [13, 42, 43]. The following derivation follows that of Cook and Bernfeld [13] and it is assumed that there is only one stationary point for any particular combination of ϕ and τ.

The first step is to determine the stationary point corresponding to $\omega(\hat{t}) = 0$ and expand the phase about $\hat{t}$ in the Taylor series

$$\theta(t) \approx \theta(\hat{t}) + \frac{1}{2}\omega'(\hat{t})(t - \hat{t})^2.$$

This is substituted into (4.58) and it is assumed that the amplitude functions are slowly varying so that in the neighborhood $(\hat{t} - \delta \leq t \leq \hat{t} + \delta)$, $a(t) \approx a(\hat{t})$ and $b(t) \approx b(\hat{t})$. The AF is then approximated by

$$\Phi_{fg}(\phi, \tau) \approx \left| a(\hat{t}) b(\hat{t} - \tau) \int_{\hat{t}-\delta}^{\hat{t}+\delta} \exp j \left[\frac{1}{2} \left(\omega'(\hat{t}) \right) \left(t - \hat{t} \right)^2 \right] dt \right|^2 .$$

The integral is put into the form of a Fresnel integral by defining the argument

$$\frac{1}{2}\omega'(t-\hat{t})^2 = \frac{1}{2}\pi u^2$$

and

$$(t-\hat{t}) = u\sqrt{\frac{\pi}{\omega'(\hat{t})}}.$$

The variable of integration is changed from t to u by

$$dt = \sqrt{\frac{\pi}{\omega'(\hat{t})}}\, du$$

and by defining new limits $\pm u_l$ corresponding to $t = \hat{t} \pm \delta$. This yields

$$(\hat{t}\pm\delta) - \hat{t} = (\pm u_l)\sqrt{\frac{\pi}{\omega'(\hat{t})}}$$

and

$$\pm u_l = \pm\delta\sqrt{\frac{\omega'(\hat{t})}{\pi}}.$$

Substituting for t in the integral produces

$$\Phi_{fg}(\phi,\tau) \approx \left| a(\hat{t})b(\hat{t}-\tau)\sqrt{\frac{\pi}{\omega'(\hat{t})}} \int_{-u_l}^{u_l} \exp j\frac{\pi y^2}{2} du \right|^2$$

and since the integrand is even, the integration may be confined to $u > 0$ and

$$\Phi_{fg}(\phi,\tau) \approx \left(\frac{4\pi a^2(\hat{t})b^2(\hat{t}-\tau)}{|\omega'(\hat{t})|}\right) \left|\int_0^{u_l} \exp j\frac{\pi u^2}{2} du\right|^2.$$

u_l is proportional to $\sqrt{\omega'(\hat{t})}$ and if the aggregate rate of change of frequency is large such that

$$\omega'(\hat{t}) = \left|\omega_f'(\hat{t}) - \omega_g'(\hat{t}-\tau)\right| >> \pi,$$

the integral is

$$\int_0^{u_l} exp\, j\frac{\pi u^2}{2} du \approx \sqrt{\frac{1}{2}} \exp j\frac{\pi}{4}$$

and

$$\Phi_{fg}(\phi,\tau) \approx \left(\frac{2\pi a^2(\hat{t})b^2(\hat{t}-\tau)}{|\omega_f'(\hat{t}) - \omega_g'(\hat{t}-\tau)|}\right)$$

where $\hat{t}$, ϕ, τ are related at the stationary points by

$$\omega_f(\hat{t}) - \omega_g(\hat{t} - \tau) - 2\pi\phi = 0.$$

When the waveforms have constant amplitude as in Example 4.2, the AF at the stationary points is

$$\Phi_{fg}(\phi, \tau) \approx \frac{E_f E_g}{T_f T_g \left| \nu_f'(\hat{t}) - \nu_g'(\hat{t} - \tau) \right|} \tag{4.59}$$

where $\nu_f(t) = \omega_f(t)/2\pi$ and $\nu_g(t) = \omega_g(t)/2\pi$. The aggregate rate of frequency change will be large when the slopes of the frequency functions have the greatest difference which means that the AF approximation is most accurate when the AF is smallest.

Mehta and Titlebaum [42, 43] have related this approximation to a simple geometric construction for waveforms with linear frequency functions and at most, slowly varying amplitudes. They show that (4.59) is equivalent to the area of a parallelogram formed by plotting the intersecting frequency functions and their respective squared amplitudes as pairs of parallel lines as shown by the hatched area in Figure 4.14. The area of such a figure is equal to the product of the vertical distances between the parallel lines divided by the difference between the slopes. The slope of an LFM frequency function is the bandwidth divided by the pulse duration so this is

$$\text{Area} = \Phi_{ff}(\phi, \tau) = \frac{E_f E_g}{T_f T_g \left(\frac{W_f}{T_f} + \frac{W_g}{T_g} \right)} = \frac{E_f E_g}{W_f T_g + W_g T_f} \tag{4.60}$$

and is the same as would be obtained from (4.59) for LFM's with slopes of opposite sign. The reader is referred to the cited references for discussions extending this idea to linearly varying amplitude functions and situations involving multiple intersections in the t-f plane.

Ambiguity functions of waveforms with more complicated frequency functions may often be approximated by multiple application of the stationary phase relationship or the overlay concept of Example 4.2. This is most easily done for waveforms with simple CW or LFM segments such as the VFM frequency function illustrated in the following example.

Example 4.3 The VFM Waveform

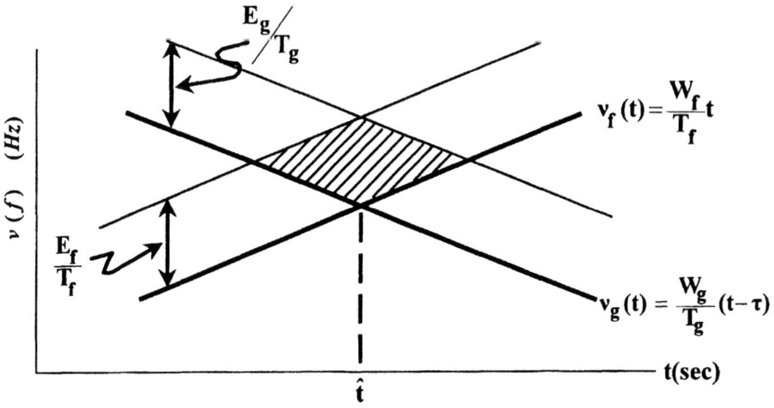

Figure 4.14: Geometric AF approximation for waveform segments with different time-frequency slopes

A VFM waveform sometimes called a *v-chirp* waveform,[3] is a combination of two LFM's of equal but opposite slope as shown in the time-frequency plot of Figure 4.15. For this example, the waveform has an even frequency function, its total duration and bandwidth are T (sec.) and W (Hz.) respectively, and it has unit energy with constant amplitude $\sqrt{T}$. The AF is estimated by applying the frequency overlay and stationary phase approximations for the delay and Doppler conditions under which the LFM segments either overlay or cross each other. An easy way to visualize the overlap for any narrowband AF is to copy the plot onto a transparency and overlay it on the original. The VFM frequency functions overlap when the individual LFM segments overlay or cross each other as illustrated in parts (c) and (d) of the figure. The waveform is symmetric thus the AF will have 4-quadrant symmetry and it is only necessary to estimate the AF for one overlay segment and one oblique crossing segment. The LFM overlay segments are computed as in Example 4.2 except that each LFM has duration $T/2$. The integration for the left half of the pulse as shown in the figure is

$$\Phi_{ff}(\phi,\, \tau) = \left| \int_{-\frac{T}{2}+\tau}^{0} \left(\frac{1}{T}\right) dt \right|^2$$

[3]LFM and VFM waveforms are often called chirp signals because of the bird-like chirp sound they make when translated to audio frequencies.

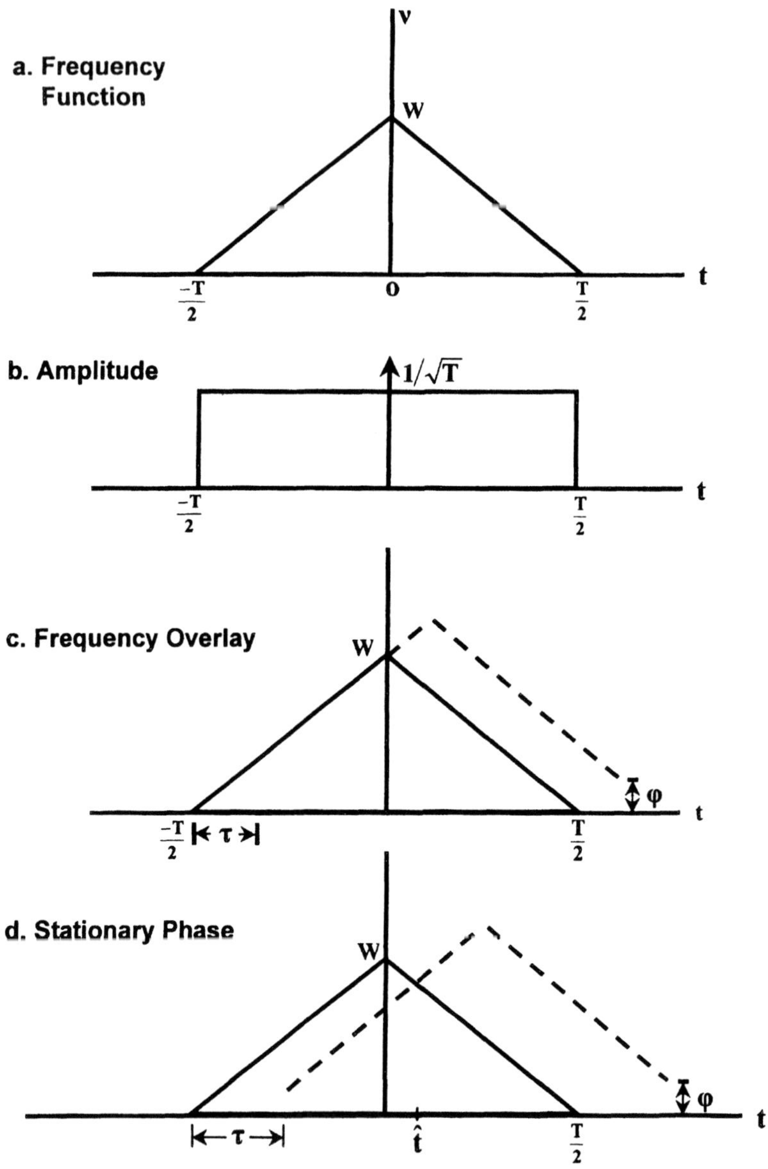

Figure 4.15: Example of the time-frequency shift of a VFM waveform: a. frequency function, b. amplitude function, c. frequency overlay of the same segments, d. frequency overlay of opposite segments

$$= \frac{\left(\frac{T}{2} - \tau\right)^2}{T^2}$$

for ϕ and τ related by

$$\phi = \frac{2W\tau}{T}.$$

By symmetry, the same result is obtained when the right half overlays itself and therefore the overlay segments contribute a pair of ridges defined by

$$\Phi_{ff}(\phi, \tau) = \left(\frac{1}{2} - \frac{|\tau|}{T}\right)^2$$

that cross when $\tau = 0$ at the origin with a peak value of $1/4$.

The AF in the areas where the left and right LFM segments cross is estimated by stationary phase from (4.60) where each segment carries half of the energy so that $E_f = E_g = 1/2$, $T_f = T_g = T/2$, and

$$\Phi_{ff}(\phi, \tau) = \frac{1}{4TW}.$$

This constant level region is called a *pedestal* and occupies the bow tie shaped area shown in Figure 4.16. The complete AF estimate consists of the ridges indicated by the solid lines in the figure, the pedestal bounded by the ridges and the dashed lines, and a central spike of unit height at the origin. The spike occurs because the unit energy AF must have unit height at the origin yet the peak ridge height is only $1/4$. The estimated AF regions of the ϕ, τ plane outside of the hatched area are estimated to have zero ambiguity volume. The actual computer generated AF of this waveform is shown in Figure 4.17.

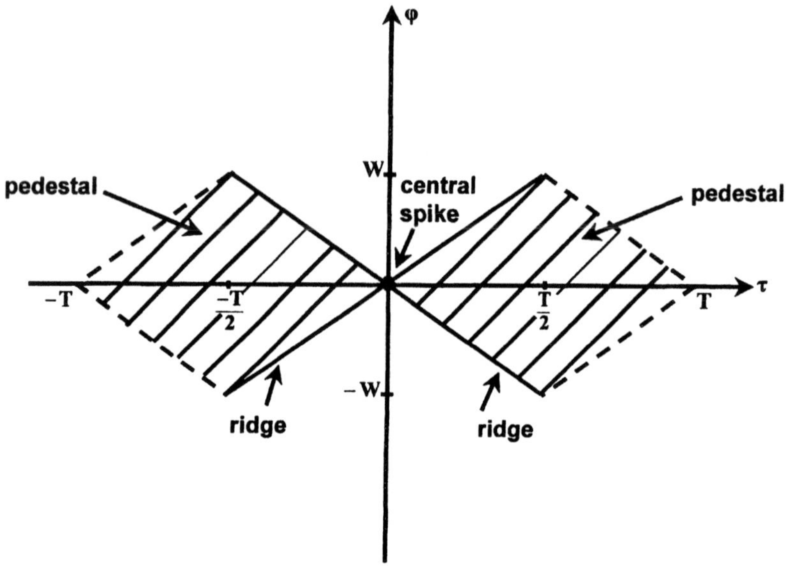

Figure 4.16: VFM ambiguity function graphical estimate

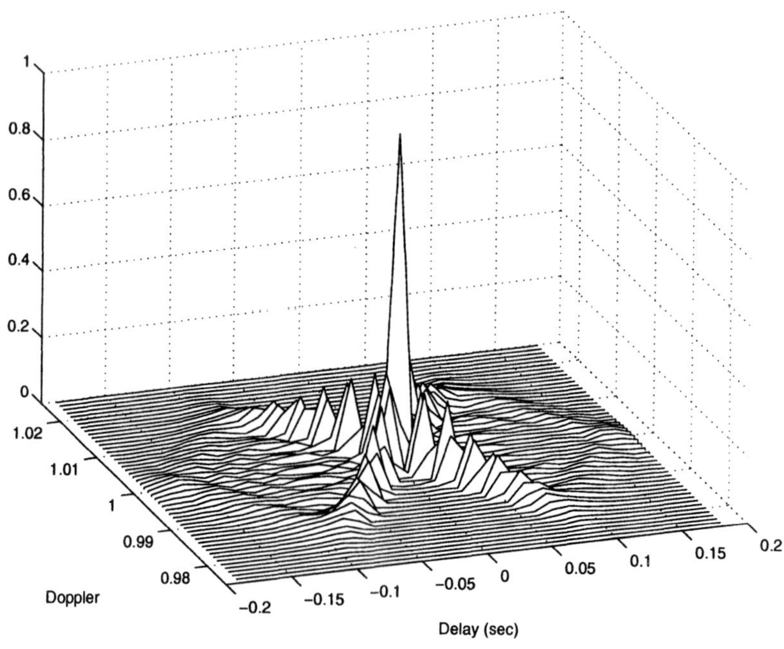

Figure 4.17: Computed VFM ambiguity function

4.6 SUMMARY

The ambiguity function (AF) arises as a consequence of the maximum likelihood point scatter detector (matched filter) formulation for Gaussian processes. This involves a replica correlation process wherein parameterized replicas of the waveform are correlated with the received signal and the magnitude square of the result plotted as a function of delay and Doppler. The resulting surface represents the AF when the signal emanates from a single point reflector and its shape and distribution determines the accuracy with which delay and Doppler are estimated.

Echo-location in the presence of platform and scatter motion results in a dilation of the signal time base that that must be accurately modeled in the replicas. Linear (Doppler) dilation can be approximated by a fixed (undilated) complex envelope replica and a frequency shift model for narrowband waveforms over a small Doppler range. A full dilation (wideband) model must be used when an appreciable range of scatter velocity in relation to 1/TW is anticipated. The advantage of the narrowband model is that the receiver Doppler hypotheses can be rapidly implemented using an FFT applied to sequential time series segments. The wideband receiver requires the generation of time dilated envelope replicas that correspond to the desired Doppler hypothesis increments. This is computationally expensive but the implementation can be simplified by using fixed dilated replicas in Doppler bands and the narrowband implementation within each band. The Doppler range of each band is determined by the tolerable ambiguity loss (4.49) and the dilation sensitivity of the waveform. The dilation sensitivity was shown to be a function of the time distribution of the frequency modulation within a pulse and that it is possible to reduce the sensitivity by concentrating the rapidly varying modulation components close to the pulse centroid.

A side by side comparison of the general properties of both narrow and wideband representations obtained from the wavelet and classical ambiguity literature were discussed. In particular it was shown that the important AF volume property differs significantly between the two models. The narrowband AF volume is constant and unity at the origin for normalized waveforms leading to what is known as the "waterbed" effect in that it can only be redistributed in the parameter plane. The wideband AF volume integrated in the delay-Doppler frequency plane has volume that increases with fractional bandwidth. The volume increment is not significant for many applications amounting to only about a .3% increase for 10% fractional bandwidth.

The shape and distribution of the AF at the origin determines the accuracy with which parameter estimates can be made. The accuracy is determined by the CR bound and the associated Fisher matrix that are functions of the shape as expressed by the second order AF expansion at the origin. This has implications for waveform design as it was shown that even waveform functions minimize ambiguity delay-Doppler coupling and hence estimation error.

The narrowband ambiguity expansions that were discussed appear in several radar related references such as: Cook and Bernfeld [13], Van Trees [14], and Rihaczek [15]. The wideband relationships may be found in the work of Kelley and Wishner [17], Altes [18, 19], and Sibul and Titlebaum [20].

Ambiguity mismatch occurs whenever the time base of the echo signal is not accurately represented in the replication process. The most common source of mismatch is the quadratic distortion (QD) caused by a second order acceleration term (2.21) usually induced by a cross range velocity component as illustrated by Figure 2.9. Uncompensated QD redistributes the ambiguity volume by reducing the ambiguity peak at the origin and spreading the sidelobes as shown in Figure 4.12. The effect is to compromise both the detection and estimation capability of the receiver.

The stationary phase concept is introduced at the end of the chapter to provide a simple model for the approximation of ambiguity shapes. This is particularly useful for the prediction and visualization of the gross features of multicomponent waveforms that are discussed in the next chapter.

Bibliography

[1] L. Cohen. Time-Frequency Distibutions; a Review . *Proceedings of the IEEE*, 77(7):941–981, July 1989.

[2] A.H. Nuttall. Wigner Distribution Function: Relation to Short-Term Spectral Estimation, Smoothing, and Performance in Noise . Technical report, Naval Underwater Syst. Center, Newport RI, Feb. 1988.

[3] G.F.B. Bartels. *Time-Frequency Signal Processing Algorithms: Analysis and Synthesis using Wigner Distributions* . PhD Thesis in Electrical Engineering, Rice Univ., Houston, TX, Dec. 1983.

[4] R.K. Young. *Wavelet Theory and its Applications* . Kluwer, Norwell MA, 1993.

[5] L.G. Weiss. Wavelets and Wideband Correlation Processing . *IEEE Sig. Proc. Magazine*, Jan. 1994.

[6] D.M. Drumheller. Theory and Application of the Wavelet Transform to Signal Processing . Technical report, Naval Research Lab., July 1991. NRL TR 9316.

[7] F. Hlawatsch and G.F. Boudreaux-Bartels. Linear and Quadratic Time-Frequency Signal Representations . *IEEE Signal Processing Magazine*, 9:21–67, April 1992.

[8] L.G. Weiss. *Wideband Inverse Scattering and Wideband Deconvolution of Acoustic Signals Using Wavelet Transforms* . PhD Thesis in Acoustics, Penna. State Univ., May 1993.

[9] S.G. Mallat. A Theory for Multi-resolution Signal Decomposition: The Wavelet Representation . *IEEE Trans. on Pattern Anal. and Machine Intel.*, 11(7):674–693, July 1989.

[10] I. Daubechies. The Wavelet Transform, Time-Frequency Localization and Signal Analysis . *IEEE Trans. Info. Theory*, 36(5):961–1005, Sept. 1990.

[11] H.I. Choi and W.J. Williams. Improved Time-Frequency Representation of Multi-component Signals Using Expoenential Kernels . *IEEE Trans. Acoust.,Speech, and Signal Proc.*, 37(6):862–871, 1989.

[12] P.M. Woodward. *Probability and Information Theory with Applications to Radar* . Pergamon Press, London, 1953.

[13] C.E. Cook and M. Bernfeld. *Radar Signals* . Academic Press, New York NY, 1967.

[14] H.L. Van Trees. *Detection, Estimation, and Modulation Theory-Part III* . Wiley, 1971.

[15] A.W. Rihaczek. *Principles of High Resolution Radar* . McGraw-Hill, New York NY, 1969.

[16] J.L. Stewart and E.C. Westerfield. A theory of active sonar detection. *Proceedings of the IRE*, 47:872–881, May 1959.

[17] E.J. Kelly and R.P. Wishner. Matched Filter Theory for High Velocity Accelerating Targets . *IEEE Trans. on Military Electronics*, MIL-9(1):56–69, 1965.

[18] R.A. Altes. *Methods of Wideband Signal Design for Radar and Sonar Systems* . PhD Thesis in Electrical Engineering, University of Rochester, 1970. ASTIA doc. no. AD732494.

[19] Altes R.A. Some Invariance Properties of the Wideband Ambiguity Function . *Journ. of the Acoust. Soc. Amer.*, 53(4):1154–1160, 1973.

[20] L.H. Sibul and E.L. Titlebaum. Volume Properties for the Wideband Ambiguity Function . *IEEE Trans. on Aerosp. and Elect. Systs.*, 17(1):83–86, Jan. 1981.

[21] R.A. Altes and E.L. Titlebaum. Bat Signals as Optimally Doppler Tolerant Waveforms. *Journ. of the Acoust. Soc. Amer.*, 48((4)2):1014–1020, 1970.

[22] D.C. Rife and G.A. Vincent. Use of Discrete Fourier Transform in the Measurement of Frequencies and Levels of Tones . *Bell Sytem Tech. Journ.*, 49(2):197–228, Feb. 1970.

[23] F.J. Harris. On the Use of Windows for Harmonic Analysis with the Discrete Fourier Transform . *Proceedings of the IEEE*, 66(1):51–83, Jan. 1978.

[24] N.C. Geckinli and D.Yavuz. Some Novel Windows and a Concise Tutorial Comparison of Window Families . *IEEE Trans. Acoust.,Speech, and Signal Proc.*, 26(6):501–507, Dec. 1978.

[25] H. Rohling and J. Schurmann. Discrete Time Window Functions with Arbitrarily Low Sidelobe Levels . *Signal Processing*, 5(2):127–139, Mar. 1983.

[26] Altes R.A. Optimum waveforms for Sonar Velocity Discrimination . *Proceedings of the IEEE*, 59:1615, Nov. 1971.

[27] P. Flandrin, P. Cros and G. Mange. Sensitivity of Doppler Tolerance to the Structure of Bat-like Signals . *Acoustica*, 62:40–47, 1986.

[28] D.W. Ricker. Constrained Bandwidth Waveforms with Minimal Dilation Sensitivity . *IEEE Trans. on Aerosp. and Elect. Systs.*, 29(3):666–679, July 1993.

[29] W. Kaplan. *Advanced Calculus* . Addison-Wesley, 1959.

[30] D.W. Ricker. The Doppler Sensitivity of Large TW Phase Modulated Waveforms . *IEEE Trans. Sig. Proc.*, 40(10):2406–2413, Oct. 1992.

[31] Zhen biao Lin. Wideband Ambiguity Functions of Broadband Signals . *Journ. of the Acoust. Soc. Amer.*, 83(6):2108–2116, June 1988.

[32] R.A. Altes and W.D. Reese. Doppler Tolerant Classification of Distributed Targets–A Bionic Sonar . *IEEE Trans. on Aerosp. and Elect. Systs.*, 11(5):708–715, Sept. 1975.

[33] R.A. Altes and D.P. Skinner. Sonar Velocity Resolution With a Linear-Period Modulated Pulse . *Journ. of the Acoust. Soc. Amer.*, 61(4):1019–1030, April 1977.

[34] M. Mamode. Estimation of the Arrival Time of a Sonar Echo and Choice of a Time Reference . *Annals of Telecommunications*, 37:3939–398, 1982.

[35] S.A. Kramer. Doppler and Acceleration Tolerances of High-Gain Wideband Linear FM Correlation Receivers . *Proceedings of the IEEE*, 55(5):627–636, May 1967.

[36] M.P. Hayes and P.T. Gough. Broad-band Synthetic Aperture Sonar. *IEEE Journ. of Oceanic Engineering*, 17:80–94, Jan. 1992.

[37] W.M. Brown and C.J. Palormo. Effects of Phase Errors on Resolution . *IEEE Trans. on Military Electronics*, MIL-9:4–9, Jan. 1965.

[38] R.O. Harger. *Synthetic Aperture Radar Systems* . Academic Press, 1970.

[39] S.C. Clay and H. Medwin. *Acoustical Oceanography* . Wiley Interscience, 1977.

[40] A. Erdelyi. *Asymptotic Expansions* . Dover Publications, New York NY, 1956.

[41] A. Papoulis. *The Fourier Integral and its Applications* . McGraw-Hill, New York NY, 1987.

[42] E.L. Titlebaum and S.K. Mehta. Asymptotic and Geometric Procedures for Estimating Correlation and Ambiguity Functions . In *Proc. Intnl. Symp. on Information Theory*, San Diego CA, Jan. 1990.

[43] S.J. Mehta. Signal Design Issues for the Wigner Distribution and a Twin Processor for the Measurement of Target and Channel Structures . Technical report, Univ. Rochester Dept. of Elect. Eng., Nov. 1991. TR EE-91-13.

Chapter 5

WAVEFORMS

5.1 INTRODUCTION

Waveform specification is an integral part of the overall echo-location system design process because the transmitted waveform represents the "voice" of the system and its properties largely determine the amount and quality of information obtained from an active interrogation. The waveform ambiguity properties that are determined by the amplitude and phase/frequency modulation functions define both the detection performance and the parameter estimation characteristics of systems based upon coherent matched or mismatched filter processing.

The concept of waveform design originated in the early fifties [1] and was motivated by a need to maintain radar range resolution while increasing the pulse energy and hence the SIR and detection range. Because of peak power hardware limitations in radar and cavitation in sonar, energy could only be increased by increasing the pulse duration. This was accompanied by a loss in range resolution for CW waveforms because, in the absence of modulation the resolution as measured by the delay width of the ambiguity function is inversely proportional to the pulse duration as shown in Figure 4.7. The earliest attempts at what came to be known as "pulse compression" a term coined by R.H. Dicke [2] utilized a linear frequency modulation (LFM) waveform to achieve a matched filter "compression ratio" proportional to the inverse of the time-bandwidth product. Radars based upon this principle were developed by the Sperry Gyroscope Company and research was carried out simultaneously at General Electric and Bell and Lincoln Laboratories.

While LFM can be used to advantage for pulse compression in the range dimension, its frequency modulation function is an odd function of time and hence its AF is highly coupled as discussed in Chapter 4. This results in ambiguous range and Doppler estimates and precludes the use of LFM when accurate range and velocity measurements are required. In 1953, P.M. Woodward published his seminal book on ambiguity theory [3] and it was soon realized that more complex modulation functions that were "noise like" could be used for accurate simultaneous range and velocity measurements. The requirement for unambiguous range-Doppler measurements arose from the development of missile tracking radars and the first such waveforms were used in the AN/FPS-17 coded pulse radar developed by Lincoln Labs in 1954 [1]. These were biphase coded pulses similar to today's pseudo-random noise (PRN) sequences discussed later in this chapter.

While much of the early waveform theory was radar related and largely concerned with detection and estimation under white noise conditions, the issue of detection in *clutter* that is, backscattered interference from the ground, water surface, rain, etc. soon arose. Unwanted backscatter of sonar signals is called *reverberation* and includes backscatter from inhomogeneities in the water column as well as continuous backscatter from the surface and bottom.

Stewart and Westerfield [4] were among the first to address waveform design from a sonar point of view by analyzing the reverberation detection performance of short and long pulse CW, LFM, and pseudo random waveforms. They also addressed the fundamental issue of Doppler induced time dilation that arises because of the relatively low acoustic propagation speed in air and water. This issue was of less concern for radar designers because the condition $2v/c << 1/TW$ was almost always met at the time. More recently, time dilation has become more important as radar is used to track high speed objects such as satellites. In 1960, much of the then current results pertaining to waveforms and matched filters for both radar and sonar were summarized in a special issue of the *IRE Transactions* [5].

The description of waveform time base distortion was extended to higher order (acceleration) [6] and the concepts of the time mapping and the trajectory diagram as discussed in Chapter 2 were introduced by several authors during the 60's and 70's as a means of waveform design and analysis [7–9]. During this period there was considerable interest in *bionic* sonar or the use of echo-location by bats and porpoises and it was realized that in some cases, in order to maximize the efficiency of their hunt for prey, these animals had evolved sonar waveforms that were insensitive to Doppler or *Doppler tolerant* [9, 10]. These are characterized by ambiguity functions that are broad in the Doppler direction yet sharply peaked in delay and are applicable where Doppler estimation is of unnecessary or of secondary interest. The HFM waveform mentioned in Chapter 4 is an example.

The overall waveform ambiguity distribution largely determines detection performance in correlated interference while the AF mainlobe resolution properties determine the parameter estimation performance of matched receivers. Indeed assuming that the optimal processor for either WGN or CGN is used, the waveform is the only item under the control of the designer [11]. Some of the waveform design issues are: the placement of ambiguity volume in the delay-Doppler plane for the maximization of SIR when the interference is nonuniform, design for Doppler tolerance or sensitivity, controlling resolution in delay, Doppler or both, imaging multiple scatterers, wideband (dilation) design, and cross talk minimization in a multiuser environment. These all relate specifically to matched and mismatched filter processing for which ambiguity functions are defined but other receiver structures exist for which the AF properties of the complete waveform are of secondary importance. Diversity receivers that coherently process multiple waveform components and then incoherently integrate the filtered components are an example [12, 13]. These will be discussed later when the issue of random fading is addressed.

The waveforms to be discussed are mathematically defined, energy normalized, and electronically generated time functions composed of one or more complex analytic bandpass subfunctions defined on a finite time interval (pulse width) of duration T sec. They are assumed to be band limited by virtue of design or the physical limitations of the transmitter even though it is mathematically impossible for a function to be simultaneously time and band limited. These distinctions are made to exclude sonar and seismic waveforms generated by impulsive sources such as explosives. While the set of all possible waveform functions is unbounded, those commonly employed for radar/sonar echo-location fall into one of the following four categories [7, 14–17]:

Amplitude weighted tones

These are the complex simusoids

$$f(t) = a(t) \exp j\omega_0 t$$

where $a(t)$ is usually a real, unimodal, and symmetric weighting function defined on the interval $[0, T]$. It is sometimes called a *window function* because it is most often associated with the tapering of Fourier transforms. [18–20] Uniform weighting often called "boxcar" weighting, is the simplest window and maximizes the signal energy when the peak power is limited. It is defined as

$$a(t) = \frac{1}{\sqrt{T}} \left(\mathcal{U}(t) - \mathcal{U}(t - T) \right)$$

where $\mathcal{U}(t)$ is the unit step function. Window functions are also used to smooth the sidelobe characteristics of tone and other ambiguity functions. Examples are shown in Figures 4.7,4.8 and in [7, 17]. More complex amplitude modulation is possible but not commonly used because it reduces the signal energy for a given peak power level.

Continuous phase modulation

These waveforms have the general complex envelope form (4.37)

$$\begin{aligned} f(t) &= \tilde{f}(t) \exp j\omega_0 t \\ \tilde{f}(t) &= a(t) \exp j\psi(t) \end{aligned}$$

where $a(t)$ is the energy normalized window and $\psi(t)$ is a real phase modulation function with continuous derivatives in the pulse interval. The LFM waveform with the AF shown in Figure 4.8 is an example of one of these.

Concatenated discontinuous modulation

These consist of concatenated discrete subpulses $f_i(t)$ such that

$$\begin{aligned} f(t) &= f_1(t) \quad 0 \leq t < T_1 \\ &= f_2(t - T_1) \quad T_1 \leq t < T_2 \\ &\vdots \\ f(t) &= f_N(t - T_{N-1}) \quad T_{N-1} \leq t \leq T_N. \end{aligned} \tag{5.1}$$

The individual subpulse intervals $T_{i+1} - T_i$ may or may not be equal and the subpulse functions may be zero padded, that is portions of the subpulse interval are zero and the waveform becomes a train of distinct time separated pulses.

The subpulses can have different modulation functions such as the VFM discussed in the previous chapter however *frequency shift keyed*

(FSK) or *frequency hop* (HOP) codes consisting of concatenated sequences of equal duration CW segments exp $j\omega_i t$ with different frequencies are most common. The time-frequency allocations for these include discrete versions of continuous waveforms such as the LFM or VFM. Other examples include sparse CW sequences designed to produce an ambiguity function with a localized high resolution mainlobe and a sidelobe region with subsidiary peaks of uniform height. Examples of the latter are the Costas codes [21] and the various types of congruence (Titlebaum) codes [22, 23]. The subpulses of low order HOP codes can be individually windowed if the code elements are to be individually filtered as for a diversity processor.

Discrete phase coded sequences that maintain a constant carrier frequency but switch the phase between subpulses in discrete steps according to a prescribed algorithm [14, 24–27] are members of this class. Pseudo random noise PRN sequences are common examples and like the high resolution HOP codes are characterized by what are often called thumb tack ambiguity functions. These approach the ideal of flat sidelobes with a highly resolvent needle-like mainlobe.

Simultaneous Transmission

Simultaneous waveforms unlike concatenated subpulse waveforms consist of the additive combination of one or more functions each of duration T. When these are tones the waveform is called a *simultone* or *frequency comb*. The addition of multi-frequency components produces a strongly amplitude modulated waveform and the average to peak power ratio decreases approximately as $1/N$. If N becomes large and the frequencies are distributed in a band, the waveforms become noise-like resembling a band limited Gaussian process. The ambiguity function of equispaced tones is the "bed of nails" discussed by Rihaczec [7].

5.2 AMBIGUITY RESOLUTION

The simultaneous interrogation of multiple point scatterers that are displaced from one another in range and velocity generates multiple replications of the waveform at different delays and Dopplers. Processing such a composite echo with a matched filter receiver and plotting the response in the 2-d domain of the parameters creates a delay-Doppler snapshot image of the scatter distribution. If the scatterers are slowly fluctuating the snapshot represents a random realization. The waveform joint delay-Doppler ambiguity characteristics are important for 2-d imaging but Doppler is sometimes of secondary interest and in some cases considered a nuisance parameter. Doppler tolerant waveforms that are minimally affected by scatter velocity are then sought because only a 1-d range delay (radar A-scan) image is required and delay resolution is of paramount importance.

The ambiguity delay-Doppler (phase plane) resolution of a matched filtered waveform is an AF mainlobe property that reflects the ability to estimate delay and Doppler and depends upon mean square duration, bandwidth, and waveform symmetry as discussed in Chapter 4. Ambiguity properties away from the mainlobe (sidelobe region) become important when it is desired to discriminate or "image" multiple scatterers distributed in the plane. The image is obtained by plotting the MF response as a function of $\hat{\tau}$ and $\hat{\beta}$ or $\hat{\phi}$ as these parameters are tested at uniform intervals. Ideally the resulting image is an unambiguous map of the scatter locations. For this to occur, the AF mainlobe resolution must be sufficient to distinguish individual scatterers and the sidelobes should not contribute artifacts that generate spurious responses in the plane. Waveform design in this context is an attempt to balance these requirements and achieve specified ambiguity main and sidelobe properties when constrained to a specified time duration and bandwidth.

Mathematically the situation is illustrated by considering the echo signal generated when multiple slowly fluctuating point scatterers (3.49) each described by a random spreading function $\mathcal{S}_{spi}$ are illuminated by the same interrogation. The signal time function is

$$\mathsf{s}(t) = \sum_i \sqrt{E_0}\mathcal{S}_{spi} f(\beta_i(t - \tau_i))$$

and assuming that a matched filter with parameter hypothesis $\hat{\mathbf{p}}$ is used to process the data, its signal only response is

$$\begin{aligned} |\mathsf{s}_1|^2 &= \left| \int_{\Omega_t} \mathsf{s}(t) f^*(\hat{\beta}(t - \hat{\tau}))dt \right|^2 \\ &= \left| \sum_i \mathcal{S}_{spi} \chi_{ff}(\mathbf{p}_i, \hat{\mathbf{p}}) \right|^2 \\ &= \sum_i \sum_j \mathcal{S}_{spi} \mathcal{S}^*_{spj} \chi_{ff}(\mathbf{p}_i, \hat{\mathbf{p}}) \chi^*_{ff}(\mathbf{p}_j, \hat{\mathbf{p}}) \\ &= \sum_i E_0 |\mathcal{S}_{spi}|^2 \Phi_{ff}(\mathbf{p}_i, \hat{\mathbf{p}}) + \text{cross terms.} \end{aligned}$$

The expected signal only response depends upon the relationship between the random $\mathcal{S}_{spi}$. The expected cross term contributions drop out if the scatterers are uncorrelated leaving only energy weighted realizations of the ambiguity function centered at the scatter parameter locations $\mathbf{p}_i$. If they are correlated with correlation coefficient ρ_{ij}, the expected cross terms are the real quantities

$$\chi_{ij} = 2\sqrt{E_i E_j}\,\Re[\rho_{ij}\chi_{ff}(\mathbf{p}_i, \hat{\mathbf{p}})\chi^*_{ff}(\mathbf{p}_j, \hat{\mathbf{p}})]$$

and only become negligible if the products of the uncertainty functions at all parameter locations are negligible. The delay-Doppler plane response generated for a single realization as $\hat{\mathbf{p}}$ is varied will reflect all of the contributing terms and the cross terms will contribute to the extent that the uncertainty functions overlap.

Delay-Doppler imaging is illustrated by the interrogation of the collection of three Doppler and delay spread point scatterers shown in Figure 5.1 by the four different waveforms whose ambiguity functions are shown in Figure 5.2. The waveforms are a CW, LFM, VFM and HOP code constructed from a series of frequency shift keyed tone pulses. The waveforms exhibit considerable diversity in their ambiguity functions that is reflected in their ability to resolve and image the three scatterers as shown in Figure 5.3. Clearly it is possible to recognize that three objects generated echos for all of the waveforms in the noise free environment. Only the VFM and HOP code were able to localize the scatterers reasonably well although both have severe sidelobe interaction problems as evidence by the significant number of lower level peaks and ridges in the illustrations. Clearly the ideal AF would be a 2-d delta function with vanishing sidelobes that would resolve individual scatterers no matter how close they were to each other in the phase plane. This is not possible because it suggests unbounded resolution that in turn would require unbounded time-bandwidth as discussed in Chapter 4. The question is; what criteria can be applied in the design of waveforms such that the ideal of an uncoupled mainlobe with maximally flat sidelobes is approached within a bounded time-bandwidth constraint?

5.2.1 Woodward's Time Resolution Constant

The problem of waveform optimization was perhaps first addressed by Woodward [3] where he sought a waveform specification that would maximize AF mainlobe delay resolution yet maintain minimal AF sidelobe area on the delay axis given a design bandwidth constraint. In other

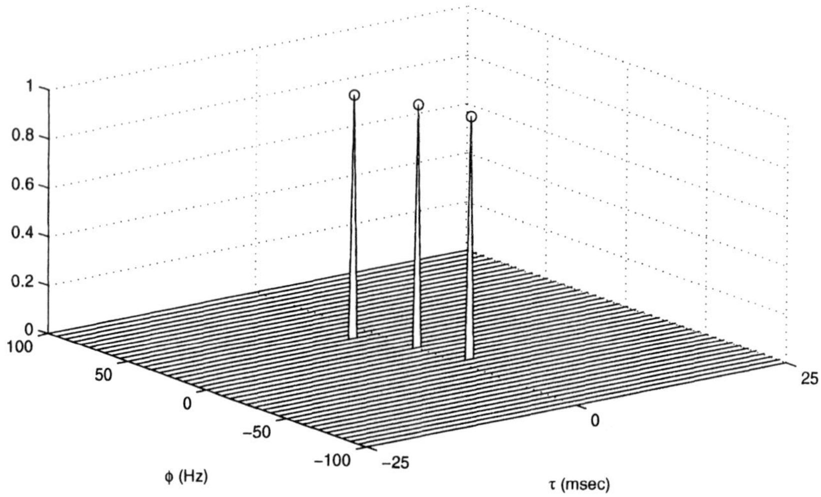

Figure 5.1: Delay-Doppler distribution of three point scatterers

words, how should waveform spectral energy be allocated to maximize the delay resolution at the origin yet minimize the delay sidelobes? He defined delay resolution by the *time resolution constant* [3, 14]

$$T_r = \frac{\int_{-\infty}^{\infty} \Phi_{ff}(0,\tau)d\tau}{\Phi_{ff}(0,0)}$$

where $\Phi_{ff}(\phi, \tau)$ is the narrowband ambiguity function of a waveform $f(t)$ specified with duration T and design bandwidth W. ϕ and τ are the Doppler frequency and time delay repectively. W is the design bandwidth of the frequency components specified by the underlying modulation not the realized bandwidth that is theoretically infinite. T_r represents the normalized area under the zero Doppler (τ) axis slice of the AF. The AF magnitude at the origin is unity for normalized waveforms thus minimization of T_r maximizes mainlobe resolution with minimum delay sidelobes. The frequency domain representation of the AF is

$$\Phi_{ff}(\phi, \tau) = \left| \frac{1}{2\pi} \int_{-\infty}^{\infty} F(\omega)F^*(\omega - 2\pi\phi) \exp j\omega\tau d\omega \right|^2$$

and

$$\Phi_{ff}(0,0) = \left| \frac{1}{2\pi} \int_{-\infty}^{\infty} |F(\omega)|^2 d\omega \right|^2 = E_f^2$$

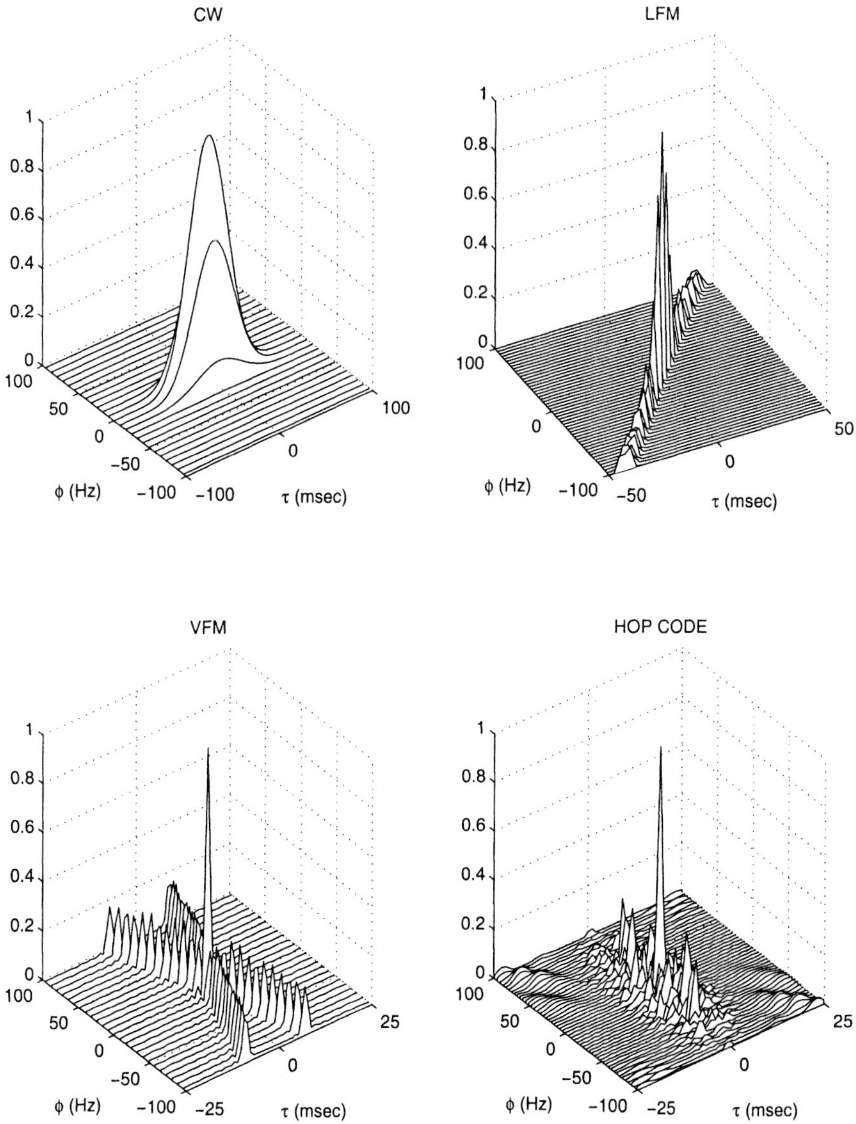

Figure 5.2: Ambiguity function examples

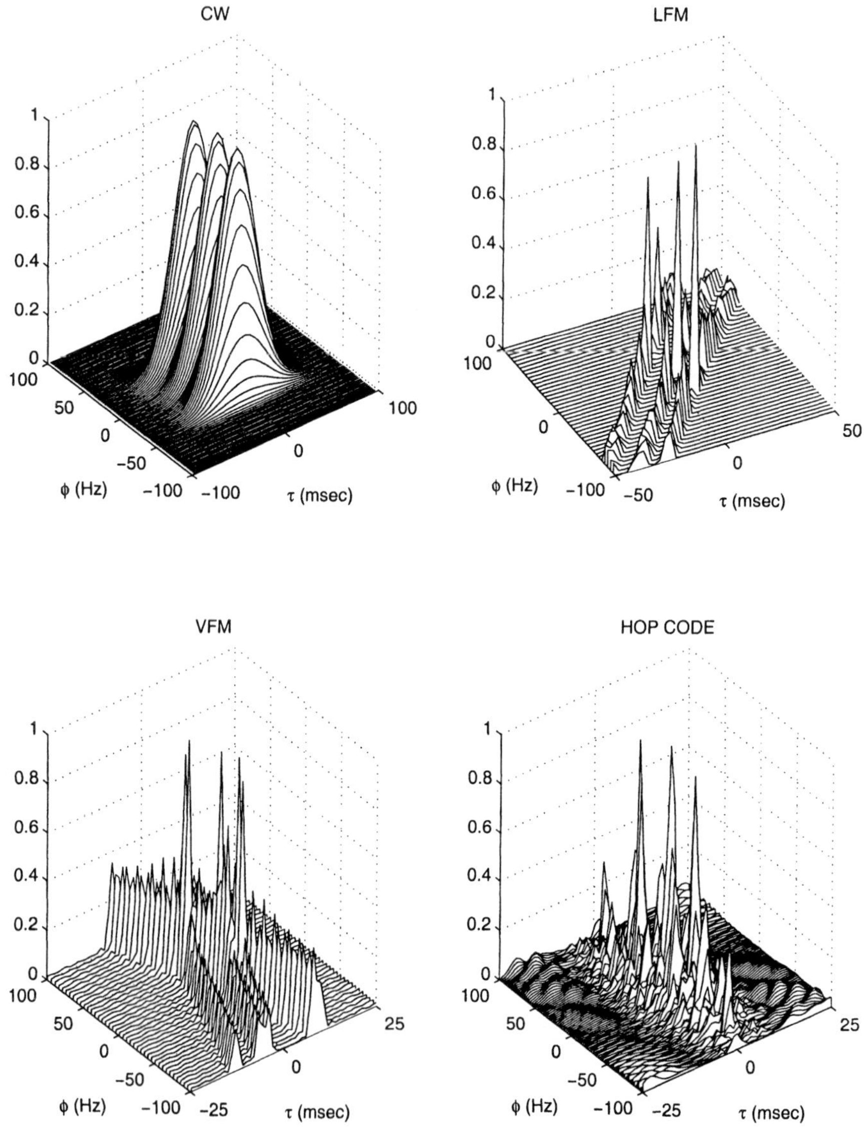

Figure 5.3: Delay-Doppler images of the Figure 5.1 scatterer distribution

which is the square of waveform energy that is unity for normalized waveforms.

Denoting the spectral energy density as $E_F(\omega) = |F(\omega|^2$ and integrating the NBAF along the τ axis yields

$$\int_{-\infty}^{\infty} \Phi_{ff}(0,\tau)d\tau = \int_{-\infty}^{\infty} \left| \frac{1}{2\pi} \int_{-\infty}^{\infty} E_F(\omega) \exp j\omega\tau \right|^2 d\tau$$
$$= \left(\frac{1}{2\pi}\right)^2 \int_{-\infty}^{\infty} \int_{-\infty}^{\infty} E_F(\omega) E_F(\nu) \int_{-\infty}^{\infty} \exp j\tau(\omega - \nu) d\tau d\omega d\nu$$

but the inner integral is the delta function and

$$\int_{-\infty}^{\infty} \Phi_{ff}(0,\tau)d\tau = \frac{1}{2\pi} \int_{-\infty}^{\infty} E_F^2(\omega)d\omega.$$

The total waveform energy is

$$E_f = \frac{1}{2\pi} \int_{-\infty}^{\infty} E_F(\omega)d\omega$$

and the constrained minimization of T_r is that E_F minimizing the goal function

$$\mathcal{J} = \frac{1}{2\pi} \int_{-\infty}^{\infty} (E_F^2 - \lambda E_F)d\omega + \lambda E_F$$

with Lagrange multiplier λ. The solution is obtained from the Euler equation [28]

$$\frac{\partial(E_F^2 - \lambda E_F)}{\partial E_F} = 0$$

and is $E_F(\omega) = \lambda/2$ indicating that the energy spectral density E_F is constant over the occupied bandwidth. A similar calculation [3] defining a frequency (Doppler) resolution constant reveals that the optimal solution occurs for a constant magnitude $|f(t)|^2$ in the time domain. The conclusion is that for a bounded time-bandwidth region WT, the highest resolution along the delay and Doppler axes occurs for waveforms with uniform magnitude over T and spectral density over W. The problem is that the time and frequency resolution constants only provide for optimum resolution along the axis. Serious sidelobe problems can occur in other regions of the (ϕ, τ) phase plane. The linear FM waveform is a good example of a waveform with a nearly uniform spectrum and magnitude that has excellent resolution in delay and Doppler but is highly coupled with resulting poor resolution properties along the ridge $\phi = W\tau/T$.

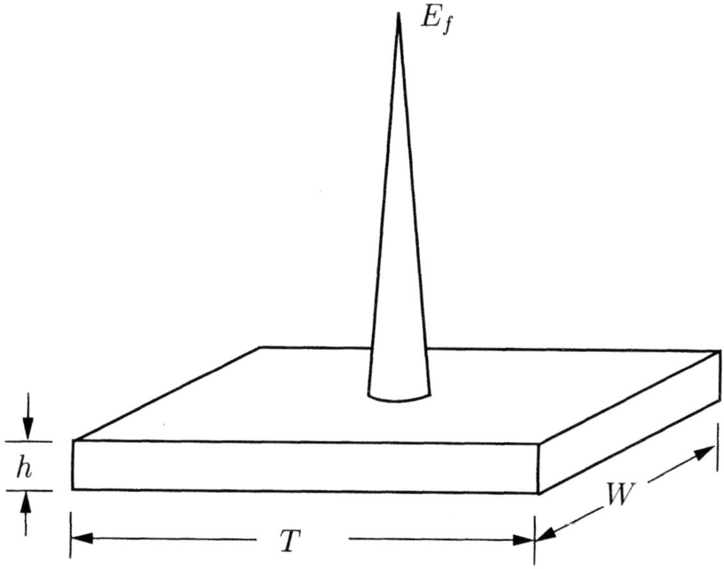

Figure 5.4: Idealized thumbtack ambiguity function

The extension of this idea to two dimensions in order to define the characteristics of an ideal imaging waveform is suggested by the graphical approach to ambiguity estimation discussed in Chapter 4. It is shown there that the magnitude of the AF at a particular delay and Doppler is proportion to the degree of overlap of the shifted waveform frequency functions at that point. Consequently, if it were possible to design a waveform such that its basebanded time-frequency representation is uniform in the bounded region $-T/2$, $T/2$ and $-W/2$, $W/2$, its AF would be expected to resemble a thumbtack with a resolvent central "spike" at the origin surrounded by a uniform sidelobe (pedestal) region as illustrated in Figure 5.4. The total volume and mainlobe height would be equal to the waveform energy E_f and the mainlobe would have resolution or area of approximately E_f/TW. The pedestal area is approximately the TW support of the waveform thus its height (h) is found from $(E_f/WT) + (hWT) = E_f$ or

$$h = \left(\frac{TW - 1}{(TW)^2}\right) E_f.$$

5.2.2 The Wigner-Ville Time-Frequency Criterion

This simple heuristic argument can be put on a firmer mathematical footing by generalizing Woodward's constant for resolution in 2-d. Minimization of the area under Φ along the τ and ϕ axis provides criteria for the waveform spectral density and envelope but the more general problem of AF sidelobe minimization over the whole delay-Doppler plane can be similarly approached by seeking criteria for minimizing the area under the AF for <u>any</u> slice that passes through the origin. If this area can be made constant as a function of the "angle" at which the slice is made, it would serve as a necessary condition for uniform sidelobes throughout the plane.

To start, consider the symmetric form of the narrowband uncertainty function (NBUF)

$$\chi(\tau,\phi) = \int_{-\infty}^{\infty} f(t+\tau/2) f^*(t-\tau/2) \exp -j2\pi\phi t dt$$

and the associated AF

$$\Phi(\tau,\phi) = |\chi(\tau,\phi)|^2.$$

Define dimensionless parameters $\tau_n = \tau/T$, $t_n = t/T$, $\phi_n = \phi/W$ where T and W are the duration and design bandwidth of the waveform, then

$$\Phi(\tau,\phi) = \left| T\int_{-\infty}^{\infty} f(T(t_n+\tau_n/2)) f^*(T(t_n-\tau_n/2)) \exp -j2\pi WT(\phi_n t_n) dt_n \right|^2.$$

It is now possible to define the area of the AF slice at angle θ with respect to the τ_n axis indicated by the hatched area in Figure 5.5. The angle is geometrically consistent because it is defined with respect to a dimensionless axis. Rotation of the coordinants by θ as indicated in Figure 5.6 transforms the delay and Doppler by

$$\begin{aligned} \tau_n &= \tau_n' \cos(\theta) - \phi_n' \sin(\theta) \\ \phi_n &= \tau_n' \sin(\theta) - \phi_n' \cos(\theta) \end{aligned}$$

and the AF is expressed as

$$\Phi(\tau,\phi) = \Phi[T(\tau_n' \cos(\theta) - \phi_n' \sin(\theta)), W(\tau_n' \sin(\theta) + \phi_n' \cos(\theta))].$$

The slice at θ is along the τ_n' axis of the rotated system where $\phi_n' = 0$

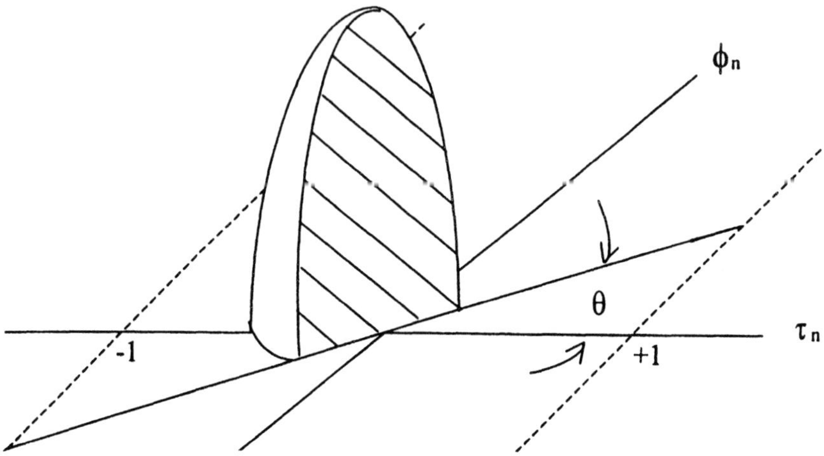

Figure 5.5: Ambiguity delay-Doppler slice

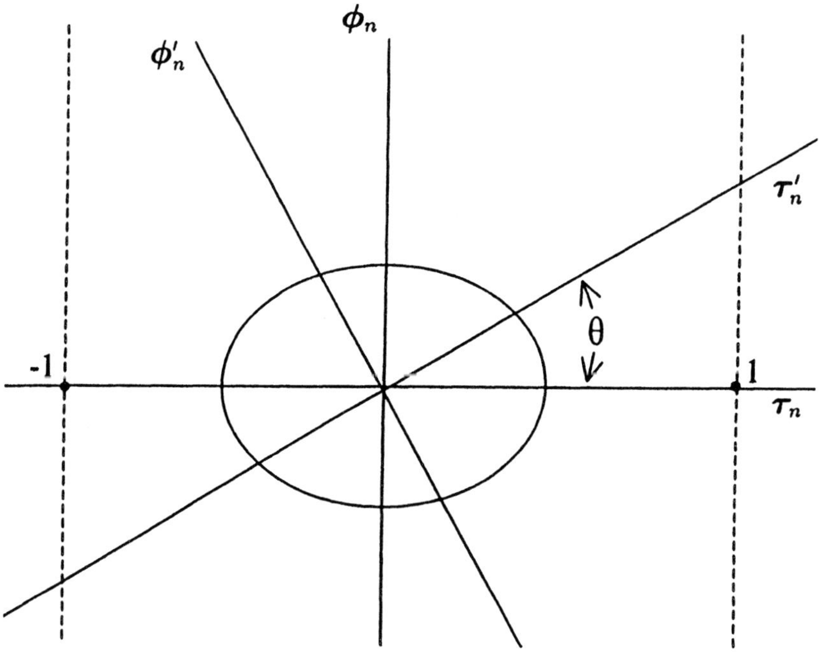

Figure 5.6: Ambiguity function coordinant rotation

and

$$\begin{aligned}\tau_n &= \tau_n' \cos(\theta) \\ \phi_n &= \tau_n' \sin(\theta).\end{aligned} \tag{5.2}$$

The area to be minimized is

$$A_a = T \int_{-\infty}^{\infty} \Phi(T\tau_n' \cos(\theta), W\tau_n' \sin(\theta)) d\tau_n' \tag{5.3}$$

and subject to the origin constraint $\Phi(0,0) = E_f^2$.

Minimization of the AF area along the τ or ϕ axis results in Woodward's conditions for waveform power spectrum and envelope. Minimization along an arbitrary line through the origin results in a specification for the waveform joint time(t)-frequency(ξ) characteristic or Wigner-Ville distribution (WVD) [29–31]

$$V_w(t,\xi) = \int_{-\infty}^{\infty} f(t+\tau/2) f^*(t-\tau/2) \exp -j2\pi\xi\tau d\tau.$$

The WVD is a real function and related to the NBUF by the 2-d Fourier transform

$$\chi(\tau,\phi) = \int_{-\infty}^{\infty}\int_{-\infty}^{\infty} V_w(t,\xi) \exp j2\pi(\xi\tau - \phi t) dt d\xi. \tag{5.4}$$

It has the marginals

$$\begin{aligned}\int_{-\infty}^{\infty} V_w(t,\xi) dt &= |F(\xi)|^2 \\ \int_{-\infty}^{\infty} V_w(t,\xi) d\xi &= |f(t)|^2.\end{aligned}$$

and

$$\int_{-\infty}^{\infty}\int_{-\infty}^{\infty} V_w(t,\xi) dt d\xi = \int_{-\infty}^{\infty}\int_{-\infty}^{\infty} |\chi(\tau,\phi)| d\tau d\phi = E_f. \tag{5.5}$$

The AF corresponding to (5.4) becomes

$$\Phi(\tau,\phi) = \\ \left| WT \int_{-\infty}^{\infty}\int_{-\infty}^{\infty} V_w(Tt_n, W\xi_n) \exp j2\pi WT(\xi_n\tau_n - \phi_n t_n) dt_n d\xi_n \right|^2$$

by defining the normalized frequency $\xi_n = \xi/W$. The area of the AF slice is expressed in terms of the WVD by using (5.2) and substituting for the AF in (5.3)

$$\begin{aligned} A_a &= T\int_{-\infty}^{\infty}\Bigg|WT\int_{-\infty}^{\infty}\int_{-\infty}^{\infty} V_w(Tt_n, W\xi_n) \\ &\qquad \cdot \exp j2\pi WT(\xi_n\tau_n' \cos(\theta) - t_n\tau_n' \sin(\theta))dt_n d\xi_n\Bigg|^2 d\tau_n' \\ &= (WT)^2\int_{-\infty}^{\infty}\cdots\int_{-\infty}^{\infty} V_w(Tt_n, W\xi_n)V_w(Tu_n, W\zeta_n) \\ &\quad \cdot\Bigg[\int_{-\infty}^{\infty} \exp j2\pi WT\tau_n'\Big([\xi_n\cos(\theta) - t_n\sin(\theta)] \\ &\quad - [\zeta_n\cos(\theta) - u_n\sin(\theta)]\Big)d\tau_n'\Bigg] dt_n\, du_n\, d\xi_n\, d\zeta_n. \end{aligned} \tag{5.6}$$

The integration with respect to τ_n' is equivalent to $\delta([\xi_n\cos(\theta) - t_n\sin(\theta)] - [\zeta_n\cos(\theta) - u_n\sin(\theta)])$ and consequently the normalized time and frequency are related by

$$\xi_n\cos(\theta) - t_n\sin(\theta) = \zeta_n\cos(\theta) - u_n\sin(\theta) = r$$

which defines the line

$$\xi_n = \frac{r + t_n\sin(\theta)}{\cos(\theta)}$$

in the WVD time-frequency plane as shown in Figure 5.7. Letting $r = \xi_0\cos(\theta)$ where ξ_0 is the ξ_n axis intercept, the integration is along the contour $\xi_n = \xi_0 + t_n\tan\theta$ and (5.6) becomes

$$A_a(\theta) = W^2\int_{-\infty}^{\infty}\left(T\int_{-\infty}^{\infty} V_w(Tt_n, W(\xi_0 + t_n\tan\theta))dt_n\right)^2 d\xi_0.$$

The bracketed integration represents the area of the WVD along the contour that is projected on the t_n axis and denoting this by $A_{wp}(\xi_0, \theta)$, the ambiguity area is

$$\begin{aligned} A_a(\theta) &= W^2\int_{-\infty}^{\infty} A_{wp}^2(\xi_0, \theta)d\xi_0 \\ A_{wp}(\xi_0, \theta) &= T\int_{-\infty}^{\infty} V_w(Tt_n, W(\xi_0 + t_n\tan\theta))dt_n. \end{aligned} \tag{5.7}$$

The object is the determination of the WVD that minimizes $A_a(\theta)$ and clearly a constraint is required to avoid the trivial solution $A_{wp} = 0$.

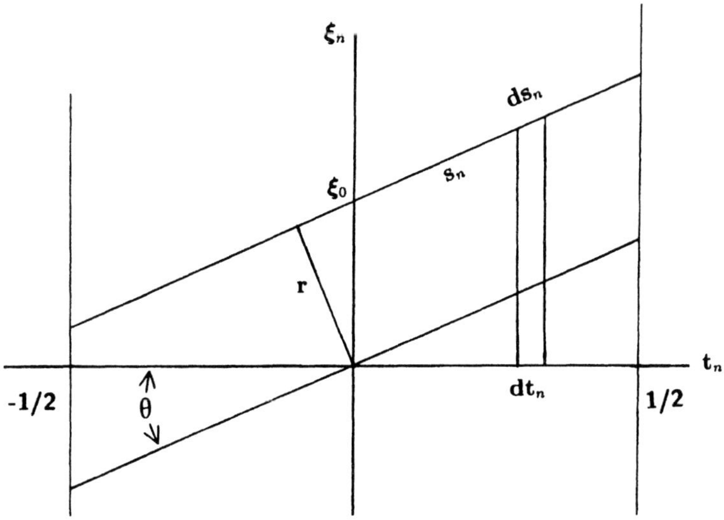

Figure 5.7: WVD time-frequency plane

This is established from the volume property of the WVD

$$\begin{aligned} V_w &= TW \int_{-\infty}^{\infty} A_{wp}(\xi_0, \theta) d\xi_0 \\ &= WT \int_{-\infty}^{\infty} \int_{-\infty}^{\infty} V_w(Tt_n, W(\xi_0 + t_n \tan\theta)) dt_n d\xi_0 \end{aligned}$$

by changing variables back to the original time and frequency $t = Tt_n$, $\xi = W(\xi_0 + t - n\tan\theta)$. The Jacobian of the transformation is

$$\begin{vmatrix} \frac{\partial t_n}{\partial t} & \frac{\partial t_n}{\partial \xi} \\ \frac{\partial \xi_0}{\partial t} & \frac{\partial \xi_0}{\partial f} \end{vmatrix} = \frac{1}{TW}$$

and from (5.5)

$$V_w = \int_{-\infty}^{\infty} \int_{-\infty}^{\infty} V_w(t, \xi) dt d\xi = W \int_{-\infty}^{\infty} A_{wp}(\xi_0, \theta) d\xi_0 = E_f. \quad (5.8)$$

The minimization of A_a is then equivalent to the minimization of the goal function

$$\mathcal{J} = W^2 \int_{-\infty}^{\infty} A_{wp}(\xi_0, \theta)^2 d\xi_0 + \lambda \left(E_f - W \int_{-\infty}^{\infty} A_{wp}(\xi_0, \theta) d\xi_0 \right)$$

$$= \int_{-\infty}^{\infty} \left(W^2 A_{wp}(\xi_0, \theta)^2 - \lambda W A_{wp}(\xi_0, \theta) \right) d\xi + \lambda E_f$$

subject to a constant energy constraint with Lagrange multiplier λ. The Euler equation solution [28]

$$A_{wp}(\xi_0, \theta) = \frac{\lambda}{2W} = \text{constant in } \xi_o$$

provides the necessary condition and is interpreted to mean that the time axis projection of the area under the WVD surface and along the contour $\xi_n = \xi_0 + t_n \tan\theta$ must be constant as a function of the intercept ξ_0. This is consistent with the the uniform spectral density result for the time-resolution constant established along the delay axis for $\theta = 0$. Assuming $A_{wp}(\xi_0, \theta)$ is constant with ξ_0,

$$A_{wp}(\xi_0, 0) = A_{wp}(0, 0)$$

and

$$T \int_{-\infty}^{\infty} V_w(Tt_n, W\xi_0))dt_n = \int_{-\infty}^{\infty} V_w(t, 0))dt = |F(\xi)|^2 \text{ constant}$$

which is equivalent to the previous result.

The necessary condition applies to the projected area and is a function of θ but ideally a condition is sought that minimizes the area under the AF at all angles. If this is can be done such that A_a is minimized and constant as a function of θ, the AF will have uniform sidelobes over its phase plane support. Integration with respect to the dimensionless distance s_n along the contour $\xi_n = t_n \tan\theta = s_n \cos\theta$ shown in Figure 5.7 is the area $A_w(\theta)$ under the WVD. WV distributions have the same time-frequency support as their underlying waveforms [29–31] thus for $t \in [-T/2, T/2]$, $t_n \in [-1/2, 1/2]$, changing variables $t_n \to s_n$ in A_{wp} produces

$$\begin{aligned} A_{wp} &= T\cos\theta \int_{-1/2\cos\theta}^{1/2\cos\theta} V_w(Ts_n\cos\theta, W(\xi_0 + s_n\sin\theta))ds_n \\ &= T\cos\theta \int_{-1/2\cos\theta}^{1/2\cos\theta} V_w(Ts_n\cos\theta, Ws_n\sin\theta)ds_n \\ &= A_w(\theta)\cos\theta \end{aligned}$$

where the $\xi_0 = 0$ because of the minimum AF area condition. A sufficient condition for the WVD area to be independent of θ is that the WVD be a constant v_0 such that

$$\begin{aligned} A_{wp} &= T\cos\theta \int_{-1/2\cos\theta}^{1/2\cos\theta} V_w(Ts_n\cos\theta, Ws_n\sin\theta)ds_n \\ &= T\cos\theta\frac{v_0}{\cos\theta} = v_0 T. \end{aligned}$$

Substituting this in the volume constraint (5.8) and assuming WVD bandwidth support on $\pm W/2$ such that $\xi_0 \in [-1/2, 1/2]$ yields $v_0 = E_f/WT$. In turn, using this in (5.7) and assuming the same frequency limits yields $A_a = E_f^2$ independent of θ, the desired result. The generation of minimal uniform ambiguity sidelobes and retaining the central "spike" is the 2-d analog of the 1-d Fourier pair of the delta function and a uniform spectrum. The formal minimization assures that the sidelobe area is minimized for a particular θ if the projected area of the WVD is constant in frequency. If the WVD itself is constant over its support which is the 2-d analog of a uniform spectrum or envelope, the AF sidelobes are uniform and minimal throughout the phase plane. This result suggests that the ideal thumbtack AF might be attainable by waveforms that have a nearly uniform distribution in time-frequency. Such waveforms would have both a uniform time domain envelope and uniform instantaneous frequency spectrum within their support.

5.3 SPARSE SIGNAL SEQUENCES

The approximation of a uniform t-f distribution can be made by generating time and bandwidth limited random time series that would on average have a uniform distribution but for any given realization might have poor ambiguity properties. Moreover, the generation of replicas for matched filtering would not be the simple operation of generating time dilated and delayed versions of a well defined mathematical function. A reasonable approach that retains design control over the transmitted signal and its ambiguity properties is to seek deterministic mathematically defined functions or collections of functions that have nearly uniform t-f distributions. One way to accomplish this is to define the signal as a composite of identical subfunctions that are non-overlapping and sparse in the t-f plane. The subfunction AAF's are identical and form distributed equal amplitude sidelobes of the composite.

Composite signals are collections of bounded duration and bandwidth subfunctions that are transmitted as a group either sequentially as concatenated subpulses or simultaneously in different frequency bands. The subfunctions can be the identical or concatenations of different waveform functions such as multiple LFM's or LFM and CW combinations. A benefit of this is that the collection can be processed as a whole or each subfunction can be processed independently of the others by filtering and time segmentation even though the collection as a whole comprises a single interrogation. This permits multiple signals with different characteristics to be processed simultaneously on a single interrogation.

Composite signals are sparse [23, 32–34] if:

1. The time-bandwidth support of the subfunctions occupy disjoint non-overlapping regions Ω_i in time-frequency (phase) space. Formally, $\Omega_i \cap \Omega_j = \mathcal{N}$ where $\mathcal{N}$ denotes the null space.

2. Any narrow or wideband Doppler and delay shift of the phase space with respect to itself within the Doppler bounds imposed by the physics can produce at most one overlap of different subfunction support regions.

A subfunction support region is modeled as a rectangle in phase space roughly of dimension $T \times W$ where T and W are the subfunction duration and bandwidth. An example of a three subfunction array is shown in Figure 5.8 and can be shown to be sparse by overlaying it with a transparency of itself. Theoretically, the first condition can only be met by subfunctions that are time separated as all finite duration signals must have infinite bandwidth. As a practical matter there can be time overlap if the band separation exceeds the sum of the individual signal *design* bandwidths defined as the highest and lowest frequencies specified by the waveform modulation plus a factor for band edge rolloff.

A composite signal is a collection of subfunctions denoted by

$$s_i(\mathbf{p}_i, t) = a_i f_i(\mathbf{p}_i, t)$$

where each has duration T_i and $f_i(\mathbf{p}, t)$ is an energy normalized complex waveform. The amplitudes are assumed to be constant over the durations T_i so that each subfunction has energy $E_i = a_i^2 T_i$. $\mathbf{p}_i$ represents a parameter vector that localizes each in phase space and

$$\begin{aligned} s_i(\mathbf{p}_i, t) &= \sqrt{\beta_i} s_i(\beta_i(t - \tau_i)) \\ &= \sqrt{\beta_i} a_i f_i(\beta_i(t - \tau_i)), \quad \mathbf{p}_i = [\beta_i, \tau_i] \end{aligned}$$

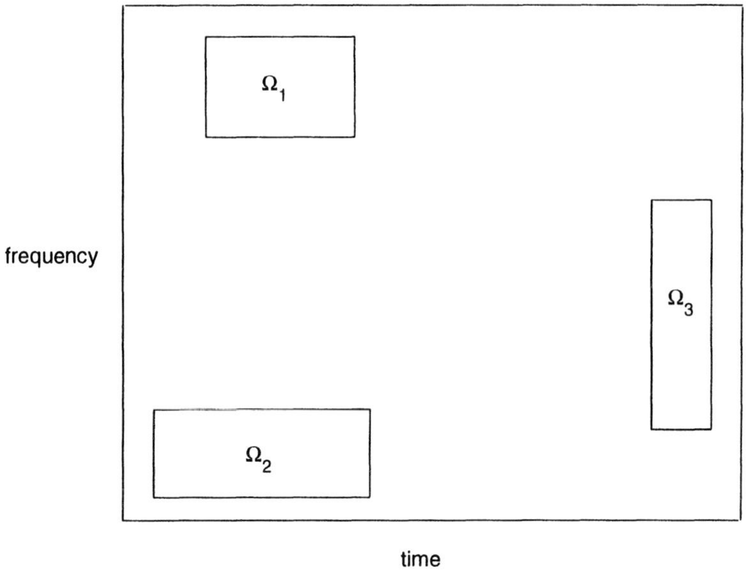

Figure 5.8: Example of the delay-Doppler support regions of a sparse signal

is the wideband (wavelet) model with β_i and τ_i as the dilation and delay parameters. The corresponding narrowband model is

$$s_i(\mathbf{p}_i, t) = a_i f_i(t - \tau_i) \exp j2\pi\phi_i t, \quad \mathbf{p}_i = [\phi_i, \tau_i]$$

where ϕ_i represents Doppler frequency. The underlying waveform functions are assumed to be energy normalized bandpass functions comprised of an envelope $\tilde{f}(t)$ and carrier term $\exp j\omega_0 t$ so that the wideband signal representation is

$$s_i(\mathbf{p}_i, t) = \sqrt{\beta_i} a_i \tilde{f}_i(\beta_i(t - \tau_i)) \exp j\omega_0 t \exp j(\beta_i - 1)\omega_0 t.$$

The narrowband representation is the same except that β_i appears only in the carrier term and $2\pi\phi = (\beta_i - 1)\omega_0$. Either representation locates the subfunction in a region of the phase plane centered at the delay τ_i and Doppler ϕ_i or more generally β_i.

The cross uncertainty function (CUF) of two subfunctions is

$$\chi_{ij}(\mathbf{p}_i, \mathbf{p}_j) = \int_{-\infty}^{\infty} s_i(\mathbf{p}_i, t) s_j^*(\mathbf{p}_j, t) dt \tag{5.9}$$

and the corresponding signal cross ambiguity function (CAF) is

$$\Phi_{ij}(\mathbf{p}_i, \mathbf{p}_j) = |\chi_{ij}(\mathbf{p}_i, \mathbf{p}_j)|^2.$$

These are auto uncertainty/ambiguity functions (AUF,AAF) when the underlying waveform functions are identical. The AAF's will have peak values

$$\Phi_{ii} = a_i^2 a_j^2 T_i^2 = E_i E_j$$

when the parameters are brought into coincidence by shifting their phase plane support such that $\mathbf{p}_j \to \mathbf{p}_i$. The CAFs do not necessarily have a unique peak if the signals are not identical but will have volume that is distributed over a support region that covers an area equal to the sum of the individual subfunction support areas $(T_i + T_j) \times (W_i + W_j)$. For example, the "mesa" function shown in Figure 5.9 is the CAF between 500Hz bandwidth-100 msec duration up and downchirp LFMs and is of nearly uniform height $E_i E_j / (T_i W_j + T_j W_i)$. Each subfunction pair of a composite signal contributes an ambiguity function at a point in the plane where their parameters coincide and each is distinct if the total array of signals is sparse. The ambiguity functions have volume equal to the product of the subfunction energies $E_i E_j$ and those corresponding to identical subfunctions will generate AAF's with peaks with magnitude equal to their volume E_i^2 at the parameter coincidence points.

5.3.1 Wideband Ambiguity Consistency

The wideband Doppler dilation representation is appropriate if it is desired that the subfunction ambiguity properties be consistent at different center frequencies. [34,35]. The issue is important if for example, a parent signal is designed for use at one frequency and it is desired to translate it to another band but maintain its AF characteristics in $[\beta, \tau]$. Let $s_i(t)$ be a parent and the wideband CUF between it and another parent signal $s_j(t)$ corresponding to (5.9) is

$$\chi_{ij}(\mathbf{p}_0, \hat{\mathbf{p}}) = \sqrt{\hat{\beta}} \int_{-\infty}^{\infty} s_i(t) s_j^*(\hat{\beta}(t - \hat{\tau}))dt \tag{5.10}$$

where $\mathbf{p}_0$ represents the pair $\beta = 1, \tau = 0$ and the hypotheses pair $\hat{\mathbf{p}} = [\hat{\beta}, \hat{\tau}]$ are the dilation and delay relative to the phase space position (origin) of the first parent. Both signals are assumed to have the same duration T and bandwidth W for simplicity. Let the offspring $x_i(t), x_j(t)$

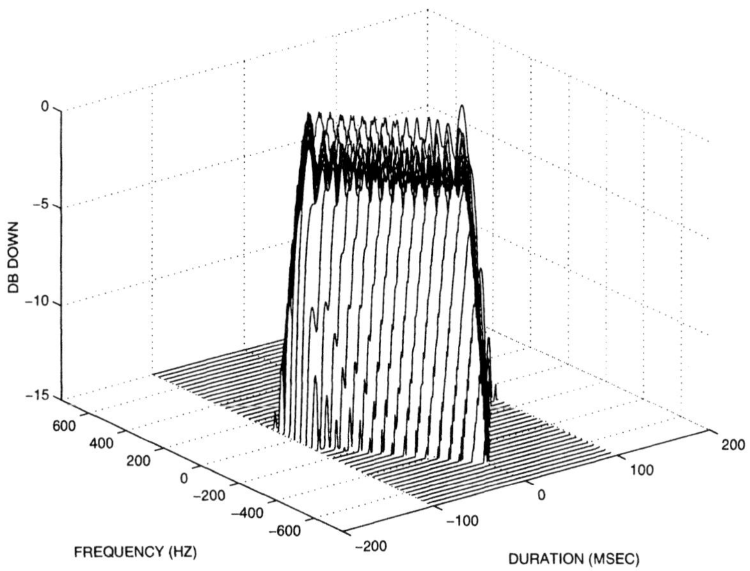

Figure 5.9: LFM mesa cross ambiguity function

be Doppler scaled versions of their parents with duration T/β and bandwidth βW

$$\begin{aligned} x_i(t) &= \sqrt{\beta} s_i(\beta t) \\ x_j(t) &= \sqrt{\beta} s_j(\beta t). \end{aligned}$$

The CUF of the offspring is

$$\begin{aligned} \tilde{\chi}_{ij}(\mathbf{p}_0, \hat{\mathbf{p}}) &= \sqrt{\hat{\beta}} \int_{-\infty}^{\infty} x_i(t) x_j^*(\hat{\beta}(t - \hat{\tau}))dt \\ &= \beta \sqrt{\hat{\beta}} \int_{-\infty}^{\infty} s_i(\beta t) s_j^*(\beta \hat{\beta}(t - \hat{\tau}))dt \end{aligned}$$

changing variables with $u = \beta t$ yields

$$\tilde{\chi}_{ij}(\mathbf{p}_0, \hat{\mathbf{p}}) = \sqrt{\hat{\beta}} \int_{-\infty}^{\infty} s_i(u) s_j^*(\hat{\beta}(u - \beta\hat{\tau}))du \quad u \in [0, T/\beta]$$

which is the same as (5.10) except that the delay hypothesis is scaled according to the duration of the offspring and the ambiguity function will be foreshortened by $1/\beta$ on the τ axis. The profile in β will be unchanged. This will not be the case if the offspring signals are merely shifted in frequency.

5.3.2 Global Ambiguity Properties

Define the composite signal as the sum

$$s(t) = \sum_{i=1}^{N} s_i(\mathbf{p}_i, t)$$

and a shift of the phase space is denoted by writing

$$s(\delta, t) = \sum_{i=1}^{N} s_i(\delta \mathbf{p}_i, t)$$

where δ is a shift operator for the signal time base Doppler and delay parameters. If the narrowband model is used, the shift is an additive increment of the Doppler frequency and delay

$$\delta = [\delta_\phi, \delta_\tau]$$

and

$$\begin{aligned} \phi_i &\to \phi_i + \delta_\phi \\ \tau_i &\to \tau_i + \delta_\tau . \end{aligned}$$

The wideband shift for delay is the same but is multiplicative for scale and

$$\beta_i \to \delta_\beta \beta_i .$$

For example, to bring $s_j(\mathbf{p}_j, t)$ into parameter coincidence with $s_i(\mathbf{p}_i, t)$ requires that $\delta_\beta = \beta_i / \beta_j$ and $\delta_\tau = \tau_j - \tau_i$. Then

$$\begin{aligned} s_j(\delta \mathbf{p}_j, t) &= \sqrt{\delta_\beta \beta_j} a_j f_j(\delta_\beta \beta_j (t - (\tau_j - \delta_\tau)) \\ &= s_j(\mathbf{p}_i, t). \end{aligned}$$

Denote the component signals $s_i(\mathbf{p}_i, t)$ as simply s_i for convenience. The uncertainty function (UF) of the collective signal is then

$$\begin{aligned} \chi_{ss} &= \int_{-\infty}^{\infty} \sum_{i=1}^{N} s_i \sum_{j=1}^{N} s_j^* dt \\ &= \sum_{i=1}^{N} \sum_{j=1}^{N} \int_{-\infty}^{\infty} s_i s_j^* dt \\ &= \sum_{i=1}^{N} \sum_{j=1}^{N} \chi_{ij} . \end{aligned}$$

The shift of the collective signal with respect to itself defines the global signal ambiguity properties. The UF becomes

$$\chi_{ss}(\delta) = \sum_{i=1}^{N}\sum_{j=1}^{N} \chi_{ij}(\delta) \tag{5.11}$$

where all of the individual UF's are shifted by the same 2-d operator δ

$$\chi_{ij} = \int_{-\infty}^{\infty} s_i(\mathbf{p}_i, t)s_j^*(\delta\mathbf{p}_j, t)dt \quad \forall\,(i,j).$$

The χ_{ij} represent the sidelobes of the global UF each with support over the range of δ that causes overlap of the support regions Ω_i and Ω_j. This is called a support coincidence in order to discriminate it from a parameter coincidence which is an exact match of the parameters. At most only one coincidence of either type will occur for any δ because the subfunction collection is assumed to be sparse. If two components s_i and s_j share the same waveform function $f(t)$ but are at different phase space positions, a parameter coincidence will generate a sidelobe of χ_{ss} with height E_iE_j that is the AUF of f. The χ_{ii} are already in coincidence when $\delta = 0$ and small perturbations of δ less than the inverse of global TW product define the mainlobe of the collective signal AF.

The AF corresponding to (5.11) is

$$\begin{aligned}\Phi_{ss} &= \left|\sum_{i=1}^{N}\sum_{j=1}^{N}\chi_{ij}\right|^2 \\ &= \left|\sum_{i=1}^{N}\chi_{ii}\right|^2 + \left|\sum_{i=1}^{N}\sum_{j=1}^{N}\chi_{ij}\right|^2_{i\neq j} \\ &\quad +2\Re\left[\left(\sum_{i=1}^{N}\chi_{ii}\right)\left(\sum_{i=1}^{N}\sum_{j=1}^{N}\chi_{ij}\right)^*_{i\neq j}\right]\end{aligned} \tag{5.12}$$

and all of the AUF and CUF cross products are

$$\begin{aligned}\chi_{ii}(\delta)\chi_{ij}^*(\delta)|_{i\neq j} &= 0 \\ \chi_{ij}(\delta)\chi_{kl}^*(\delta)|_{\substack{i\neq k\\ j\neq l}} &= 0.\end{aligned}$$

because the array is sparse. Consequently the third term of (5.12) is zero as well as the cross products in the second term so that

$$\Phi_{ss}(\delta) = \left|\sum_{i=1}^{N} \chi_{ii}(\delta)\right|^2 + 2\sum_{i=1}^{N}\sum_{j=i+1}^{N} |\chi_{ij}(\delta)|^2 .$$

At the origin ($\delta = 0$), only the auto terms $\chi_{ii} = E_i$ contribute, and

$$\Phi_{ss}(0) = \left|\sum_{i=1}^{N} \chi_{ii}\right|^2 = \left(\sum_{i=1}^{N} E_i\right)^2 = E_s^2$$

which is the square of the collective signal energy. The total volume is the aggregate of the volume in the neighborhood of the origin plus that contributed by all of the sidelobes. Only the subfunction AAF's contribute at the origin and the mainlobe volume

$$V_{ml} = \sum_{i=1}^{N} E_i^2$$

is the sum of the individual AAF volumes which are the squares of the composite energies E_i^2. Sidelobe volume is the contribution of all of the cross AF's generated by shifts that bring subfunction support regions into coincidence. The volume of each is E_iE_j and there are two contributions for each pair so that the sidelobe volume is

$$V_{sl} = 2\sum_{i=1}^{N}\sum_{j=i+1}^{N} E_iE_j$$

and the total volume is the same as the peak value at the origin

$$\begin{aligned} V_s &= V_{ml} + V_{sl} \\ &= \sum_{i=1}^{N} E_i^2 + 2\sum_{i=1}^{N}\sum_{j=i+1}^{N} E_iE_j \\ &= \left(\sum_{i=1}^{N} E_i\right)^2 = E_s^2. \end{aligned}$$

When all of the signal waveforms and amplitudes are identical with amplitude a but are sparse in the phase plane, $s_i(\mathbf{p}_i, t) = s(\mathbf{p}_i, t)$ and all coincidences generate identical AAF's that comprise the sidelobes of the collective signal. The effective duration[1] of the collective signal is T and the subfunctions each have duration T/N. The height of the collective AF mainlobe is the square of the total energy $h_{ml} = (a^2T)^2$. Its sidelobe heights and volumes are equivalent to the individual subfunction energies $h_{sl} = V_{sl} = (a^2T/N)^2$ so that the ratio of sidelobe to mainlobe height is $(1/N)^2$. The mainlobe volume of the collective signal is the sum of the subfunction volumes $V_{ml} = NV_{sl}$ and its total sidelobe volume is equal to the number of sidelobes times the individual volumes. Each subfunction pair contributes 2 sidelobes and the number of sidelobes generated by N subfunctions is $2N/2(N-2) = N(N-1)$. The total sidelobe volume is

$$V_{sl(total)} = \frac{N(N-1)}{N^2}(a^2T)^2$$

and the ratio of mainlobe to sidelobe volume is $1/(N-1)$ [23].

Example 5.1 Repeated Subfunction

The waveform of Figure 5.10a is composed of the time concatenation of two identical subfunctions of duration T_a separated by a different one of duration T_b so that the total duration is $T = (2T_a + T_b)$. The amplitude is constant and it is of interest to determine the relative height of the sidelobes caused by the AAF's of the identical segments assuming that the CAF between the different segments is negligible in relation to the total energy. The ratio of sidelobe to mainlobe height is the ratio of the squared energies and

$$\begin{aligned}\frac{h_{sl}}{h_{ml}} &= \frac{a^4T_a^2}{a^4(2T_a + T_b)^2} \\ &= (\frac{T_a}{T})^2.\end{aligned}$$

The range delay AF profile will exhibit a unit height mainlobe at the origin and two sidelobes corresponding to delay shifts of $\tau = \pm(T_a + T_b)$ that bring the identical functions into coincidence as indicated in the Figure 5.10c. No significant sidelobes will occur for overlap of the different functions occurring at $\tau = \pm(T_a + T_b)/2$ as in Figure 5.10b.

[1]This is the total non-zero time support of the collection.

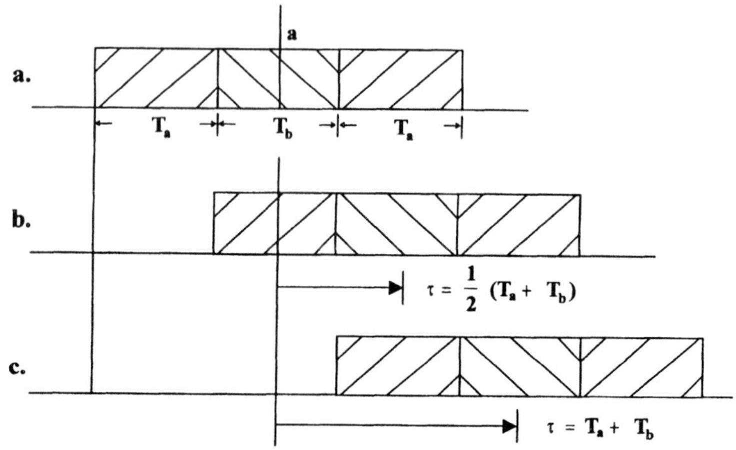

Figure 5.10: Delay shift of a three component concatenated waveform

This is demonstrated by synthesizing two 5 kHz bandwidth pseudo random noise (PRN) waveforms that have a thumbtack ambiguity function and are concatenated as in Figure 5.10a. The total duration is 300 msec with $T_a = .87T_b$, so that T_a=95.3 msec, T_b=109.5 msec, and the sidelobes occur at

$$\tau = \pm(T_a + T_b) = \pm 205 \text{ msec.}$$

The relative sidelobe height does not depend upon the total duration so that arbitrarily setting $T_b = 1$, $T_a = .87$ produces

$$\frac{h_{sl}}{h_{ml}} = \left(\frac{.87}{2(.87)+1}\right)^2 = .1011(-9.95\text{dB})$$

as shown in the AAF τ axis plot of Figure 5.11

5.4 FREQUENCY HOP CODES

Waveforms comprised of N concatenated CW subpulses each of duration T/N that are sparsely distributed in scale/frequency are often used for echo-location. (Non-uniform subpulse durations are possible but rarely

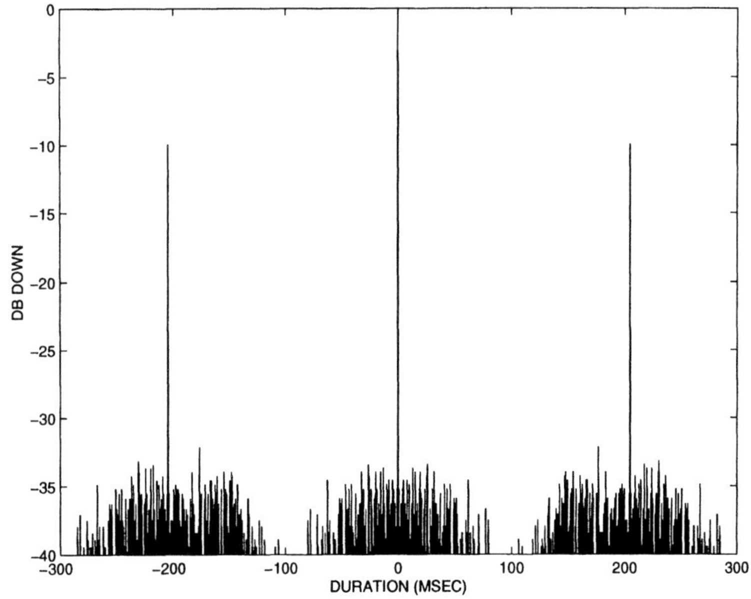

Figure 5.11: Delay axis slice of a three component PRN ambiguity function

used.) They are called frequency shift keyed (FSK) sequences, frequency hop codes, or simply hop codes and the individual CW components are sometimes called "chips" that may be individually and identically windowed. They have the form (5.1)

$$f(t) = \sum_{i=0}^{N-1} f_i(t - iT/N) \tag{5.13}$$

with identically weighted CW subpulses

$$f_i(t) = \begin{cases} a(t) \exp j\omega_i(t + \theta_i) & \text{for} \quad 0 \le t < T/N, \\ 0 & \text{otherwise.} \end{cases} \tag{5.14}$$

The $\theta_i = \omega_{i-1}T/N$ are included to assure phase continuity at the subpulse transitions. The $a(t)$ are real window functions that are normalized by

$$\int_0^{T/N} a^2(t)dt = 1/N$$

so that the overall waveform has unit energy

$$\int_0^T |f(t)|^2 dt = \int_0^T \left| \sum_{i=0}^{N-1} f_i(t - iT/N) \right|^2 dt = 1$$

and $a(t) = \sqrt{1/T}$ for rectangular weighting. The time-frequency support of the individual chips is roughly $T/N \times N/T$ centered at times $(2i + 1)T/2N$, $i = 0, \cdots, N-1$ and at specified frequencies distributed over a design bandwidth W. W is the difference between the maximum and minimum frequencies so that the chip frequencies occur at intervals of $W/(N-1)$.

Hop codes can be configured in a multitude of ways and can be used to generate discrete versions of continuous waveforms such as the linear FM. They are considered to be sparse if the minimum frequency spacing exceeds the chip bandwidth N/T. Perhaps the greatest sonar application of hop codes has been the generation of waveforms whose ambiguity functions approach the ideal "thumbtack" [15] shown in Figure 5.4. This requires that the discrete AF sidelobes be of equal height so that a normalized waveform with N identical chips should have $N(N-1)$ distinct sidelobes of height $1/N^2$. Moreover, the WVD criterion suggests that the time-frequency distribution of the waveform should be as uniform as possible. Constructing such a waveform from a sequence of identical frequency shifted chips requires that the distribution be sparse and only one overlap between chips occurs for any time-frequency shift.

5.4.1 Costas Codes

The design of hop code waveforms with these ideal ambiguity properties was considered by H.P. Costas during the 60's but his work did not reach the general literature until the 80's. Costas expressed the ideal chip placement in terms of a square matrix in which the rows and columns represent discrete equispaced frequencies and time positions. His design criterion [21] specifies that N ones representing CW chips are placed in an $N \times N$ matrix of zeros (null matrix) such that each row and column contains only a single one. The placement is such that at most only one pair of ones will coincide for all possible non-zero x-y permutations (shifts) of the matrix relative to itself but all N ones coincide at the origin x=y=0. Codes with this property are called *Costas codes* and represent specific permutations of the row (frequency) indices but comprise only a small subset of all $N!$ possible row permutations especially for large

N. They are also called "full" codes because every row and column has exactly one entry.

Coincidences of code elements have been dubbed as "hits" by Titlebaum [36] and 2-d plots of the number of hits vs the index shifts are called auto hit arrays [32, 37]. Auto hit arrays for sparse chip distributions closely approximate the auto ambiguity distribution. An example of a Costas matrix or code time-frequency pattern and its accompanying hit array is shown in Figure 5.12. The code matrix can be easily shown to have the single coincidence property by overlaying it with a transparency of itself and each coincidence corresponds to an entry in the hit array matrix at the appropriate shift relative to its origin at (11,11). There are always N^2 total auto hits or coincidences of which N occur at the origin and $N(N-1)$ are apportioned in the sidelobes. The grey squares in the hit array figure represent single hits of coincident CW chips and correspond to sparse code ambiguity sidelobes of height $1/N^2$. The black square at the origin corresponds to the mainlobe of unity height. Cross hit arrays are generated by plotting coincidences between different code sequences and are closely related to cross ambiguity functions. They are important when it is desired to design waveforms that minimize mutual interference when used simultaneously.

At first glance it would seem that the design of such permutation matrices could simply be attacked by pencil and paper methods. However one soon finds that codes of order much higher than the example shown are exceedingly difficult to construct by hand because only a small subset of the $N!$ possible permutations have the desired property. This remained a serious limitation until it was discovered that the codes could be constructed by using the algebra of finite or Galois fields [38, 39]. A Galois field is defined as the set of integers

$$GF(q) = [0, 1, 2, \cdots, q-1]$$

where q is a prime number. Addition and multiplication on the set are defined as the ordinary function mod(q) such that for elements a, b of the set, $(a + b) \triangleq (a + b)\text{mod } q$, and $(ab) \triangleq (ab)\text{mod } q$. The nonzero elements

$$GF'(q) = GF(q) - [0] = [1, 2, \cdots, q-1]$$

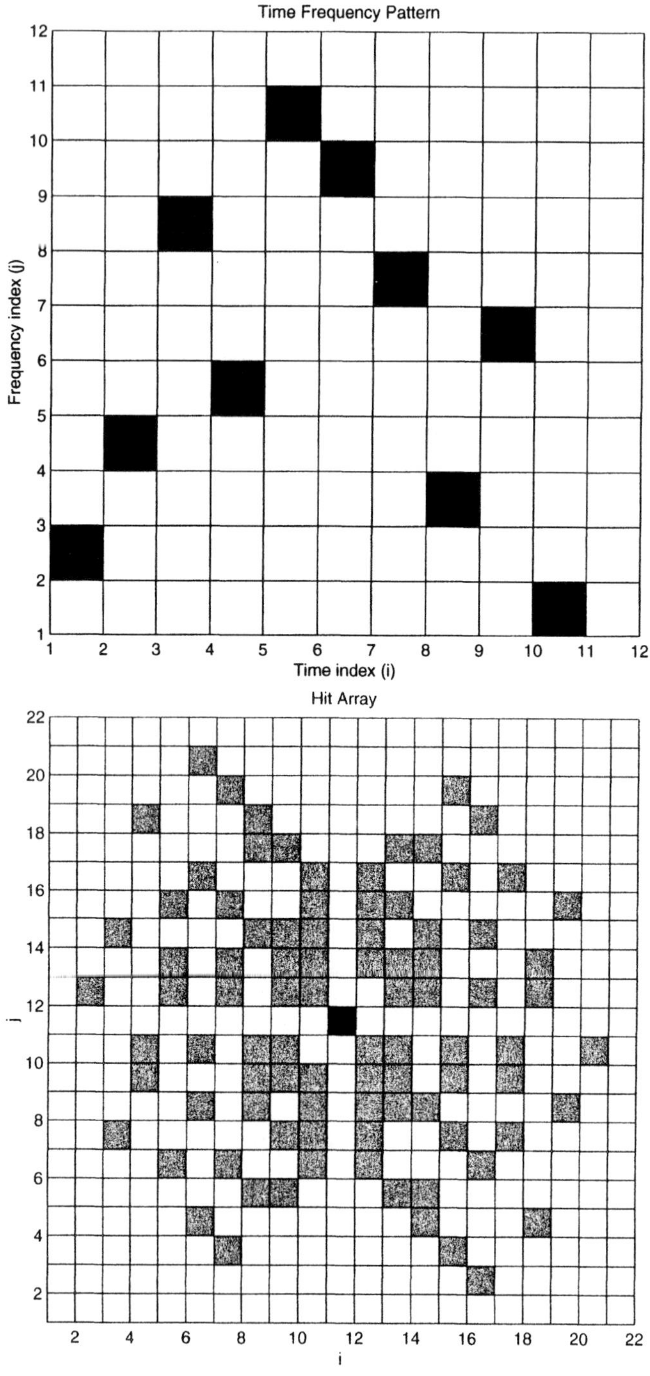

Figure 5.12: Costas code pattern and hit array example

form a group under multiplication which means that the product mod q of any two of them belongs to the group and cannot be zero. A primitive element of the group is an element e_p of the group such that a permutation of all members of the group is generated by the placement operator

$$j(i) = e_p^i \text{ (mod q)}, \; i = 1, \cdots, q-1 \tag{5.15}$$

which is called the Welch construction [36, 38] and generates a Costas array where j and i represent the frequency and time indices respectively. In other words, all of the integers $[1, \cdots, q-1]$ are uniquely determined by raising e_p to integer powers $i = 1, \, q-1$ and each primitive element results in a different permutation of the group and hence a different Costas array. The number of possible arrays for any prime q corresponds to the number of its primitive elements. The primitive elements for the first 25 primes are provided in the accompanying table. For example the pattern of Figure 5.12 is generated with $q = 11$, e_p=2, and i=1,10. The code sequence is

$$\begin{aligned} j(i) &= 2^i \text{ (mod q)}, \; i = 1, \, 10 \\ &= 2, 4, 8, 5, 10, 9, 7, 3, 6, 1 \end{aligned}$$

and was used to generate the hop code ambiguity function shown in Figure 5.2. The total duration and bandwidth were 200 msec and 500 Hz respectively so that the waveform is composed of 10-20 msec chips with incremental frequency spacings of 56 Hz.

Primitive Elements for Galois Fields GF(q)

GF(3)	GF(13)
2	2 6 7 11
GF(5)	**GF(17)**
2 3	3 5 6 7 10 11 12 14
GF(7)	**GF(19)**
3 5	2 3 10 13 14 15
GF(11)	**GF(23)**
2 6 7 8	5 7 10 11 14 15 17 19 20 21

GF(29)
2 3 8 10 11 14 15 18 19 21 26 27

GF(31)
3 11 12 13 17 21 22 24

GF(37)
2 5 13 15 17 18 19 20 22 24 32 35

GF(41)
6 7 11 12 13 15 17 19 22 24 26 28
29 30 34 35

GF(43)
3 5 12 18 19 20 26 28 29 30 33 34

GF(47)
5 10 11 13 15 19 20 22 23 26 29 30
31 33 35 38 39 40 41 43 44 45

GF(53)
2 3 5 8 12 14 18 19 20 21 22 26
27 31 32 33 34 35 39 41 45 48 50 51

GF(59)
2 6 8 10 11 13 14 18 23 24 30 31
32 33 34 37 38 39 40 42 43 44 47 50
52 54 55 56

GF(61)
2 6 7 10 17 18 26 30 31 35 43 44
51 54 55 59

GF(67)
2 7 11 12 13 18 20 28 31 32 34 41
44 46 48 50 51 57 61 63

GF(71)
7 11 13 21 22 28 31 33 35 42 44 47
52 53 55 56 59 61 62 63 65 67 68 69

GF(73)
5 11 13 14 15 20 26 28 29 31 33 34
39 40 42 44 45 47 53 58 59 60 62 68

GF(79)
3 6 7 28 29 30 34 35 37 39 43 47
48 53 54 59 60 63 66 68 70 74 75 77

GF(83)
2 5 6 8 13 14 15 18 19 20 22 24
32 34 35 39 42 43 45 46 47 50 52 53
54 55 56 57 58 60 62 66 67 71 72 73
74 76 79 80

GF(89)
3 6 7 13 14 15 19 23 24 26 27 28
29 30 31 33 35 38 41 43 46 48 51 54
56 58 59 60 61 62 63 65 66 70 74 75
76 82 83 86

GF(97)
5 7 10 13 14 15 17 21 23 26 29 37
38 39 40 41 56 57 58 59 60 68 71 74
76 80 82 83 84 87 90 92

GF(101)
2 3 7 8 11 12 15 18 26 27 28 29
34 35 38 40 42 46 48 50 51 53 55 59
61 63 66 67 72 73 74 75 83 86 89 90
93 94 98 99

In order to prove that the Welch algorithm generates a Costas array it is only necessary to show that the array is full (no skipped frequency indices) and any shift in the time and/or frequency indices (i, j) produces only one hit. Raising a primitive element to all integer powers $1, \cdots, q-1$ produces a permutation of all members of the group thus the code must be full. The code shifted by k time and m frequency increments is

$$l(i) = e_p^{i+k} + m \text{ (mod q)}, \; k, m \neq 0$$

and suppose that multiple hits occur such that $l(i) = j(i)$ at times $i = (i_1, i_2, \cdots, i_n)$, $0 \leq n \leq q-1$. At i_1 a hit implies that

$$\begin{aligned} l(i_1) &= e_p^{i_1+k} + m = j(i_1) = e_p^{i_1} \\ \Rightarrow e_p^{i_1} &= -\frac{m}{e_p^k - 1}. \end{aligned}$$

If there are multiple hits,

$$e_p^{i_1} = e_p^{i_2} = e_p^{i_3} \text{ etc.} = -\frac{m}{e_p^k - 1}$$

contradicting the fact that the elements of the field are unique powers of e_p. Consequently $i_1 = i_2 = i_3, \cdots$, and a shifted code can generate only one hit.

A Costas code can be rotated and shifted cyclicly to generate other patterns that are also Costas [40] and codes generated by different primitive elements often yield patterns that are inverted or cyclic shifts of each other. For example, the pattern resulting from the Welch construction with $e_p = 6, q = 11$ is a left-right reversal and shift of Figure 5.12. Costas codes are also generated by the Golomb construction [38,39] where $i, j = 1, \cdots, (q-2)$

$$j(i) = \log_{e_{p1}}(1 - e_{p2}^i)$$

and e_{p1}, e_{p2} are primitive elements $GF(q)$. This is equivalent to placing a one in the $(q-2) \times (q-2)$ matrix at positions (i, j) corresponding to

$$e_{p2}^i + e_{p1}^j = 1$$

using modulo q arithmetic. The special case of $e_{p1} = e_{p2}$ is called the Lempel construction.

5.4.2 Congruence Codes

The Welch, Golomb, and Lempel constructions are Costas codes based upon finite field representations of exponential functions and their single hit auto ambiguity property is analogous to the single crossing property of shifted continuous exponential functions. The single hit and full code properties are important considerations for the design of "thumbtack" waveforms where maximum flatness of the sidelobes and best utilization of the time-frequency plane is desired. These properties may not be of primary importance for applications where minimal cross-hit properties are desired and it is necessary to reduce the mutual interference of multiple sonars. For these application other constructions have been developed each of which represents a tradeoff between auto and cross ambiguity properties. Most are what are known as congruence or Titlebaum codes for E. L. Titlebaum who with his students and collaborators developed most of them. They are based upon other finite analogs of continuous waveforms with single crossing characteristics such as straight lines, quadratic, cubic, and hyperbolic functions. Titlebaum [36] has provided descriptions of the congruence codes that are specified by a placement operator $j(i)$ as for the Costas codes.

- Linear congruences (LC): These are straight lines defined on the finite field by

  $$j(i,a,b) = [ai + b]\mathrm{mod}(q), \ i, b \in GF(q), a \in GF'(q). \quad (5.16)$$

  They are full codes that have at most one cross hit between codes but can have as many as $q - 1$ auto hits.

- Quadratic congruences (QC): These are parabolas defined on the finite field by

  $$j(i,a,b,c) = [.5(ai^2 + bi) + c]\mathrm{mod}(q), \ i, b, c \in GF(q), \\ a \in GF'(q). \quad (5.17)$$

  They are not full codes but have at most one auto hit and two cross hits. They all have $(q - 1)/2$ open frequency slots (missing frequencies) and $(q - 1)/2$ double frequencies. This means that they have $q - 1$ single hits on the τ axis. They are also generated by the recursion of a linear congruence

  $$j(i + 1, a, b) = [j(i, a, b) + ai + .5(a + b)]\mathrm{mod}(q).$$

- Cubic congruences (CC): Cubic congruences are defined by the cubic polynomial

  $$j(i,a,b,c) = [a(i + b)^3 + c]\mathrm{mod}(q), \ i, b, c \in GF(q), \ a \in GF'(q)$$

Only codes for which the defining prime number is of the form q=3k+2 (q=5,11,17,23, etc.) are full codes. The cubics have at most two auto hits and three cross hits.

- Hyperbolic congruences (HC): The placement operator for these is the discrete analog of the hyperbola

$$j(i, a) = [aI(i)]\text{mod}(q), \; i, a \in GF'(q) \tag{5.18}$$

where $I(i)$ is the inverse of the integer i defined by

$$[i \times I(i)]\text{mod}(q) = 1.$$

For GF$'$(11)=[1,2,3,4,5,6,7,8,9,10] the inverses are

$$I(i) = [1, 6, 4, 3, 9, 2, 8, 7, 5, 10], \; i = 1, \cdots, 10$$

where for example, when i=6, I(6)=2 because $6 \times 2 = 12\text{mod}(11) = 1$. The code corresponding to a=3 is

$$j(i, 3) = 3I(i)\text{mod}(11) = [3, 7, 1, 9, 5, 6, 2, 10, 4, 8].$$

These are full codes with a maximum of two auto and two cross hits.

Examples of LC,QC, and HC patterns and their accompanying hit arrays for GF(11) are shown in Figures 5.13, 5.14, 5.15. The time axis indices are increased by 1 for consistency. The LC is generated from (5.16) with $a = b = 3$, the QC is generated from (5.17) with $a = b = c = 1$, and the HC is generated from (5.18) with $a = 3$. The LC pattern is obviously the worst from an auto hit standpoint with a maximum of 8 hits in the vicinity of the origin with no hits on the delay and Doppler axis. Its linearity is clear from the figure. The QC pattern is clearly parabolic but has 5 skipped frequencies and 10 hits on the delay axis.

In a single user echo-location context, the Costas codes are the best choice because all are full codes with a single auto hit. Aside from the choice of a specific construction the waveform design issue is one of the choice of code order for a specified TW product. The ideal thumbtack AF has unit magnitude at the origin and uniformly bounded sidelobes. As a rule of thumb, Costas codes with $N = q - 1$ chips are considered to be sparse if the individual chip frequency support (N/T) is smaller than the interchip frequency spacing $W/(N-1)$ or roughly when $N^2 << TW$. The sidelobes then approach the distinct ambiguity functions of the individual chips distributed in delay-Doppler according to the hit array. Each of these is the AF of a CW pulse of duration T/N with maximum height $1/N^2$ given by (4.38).

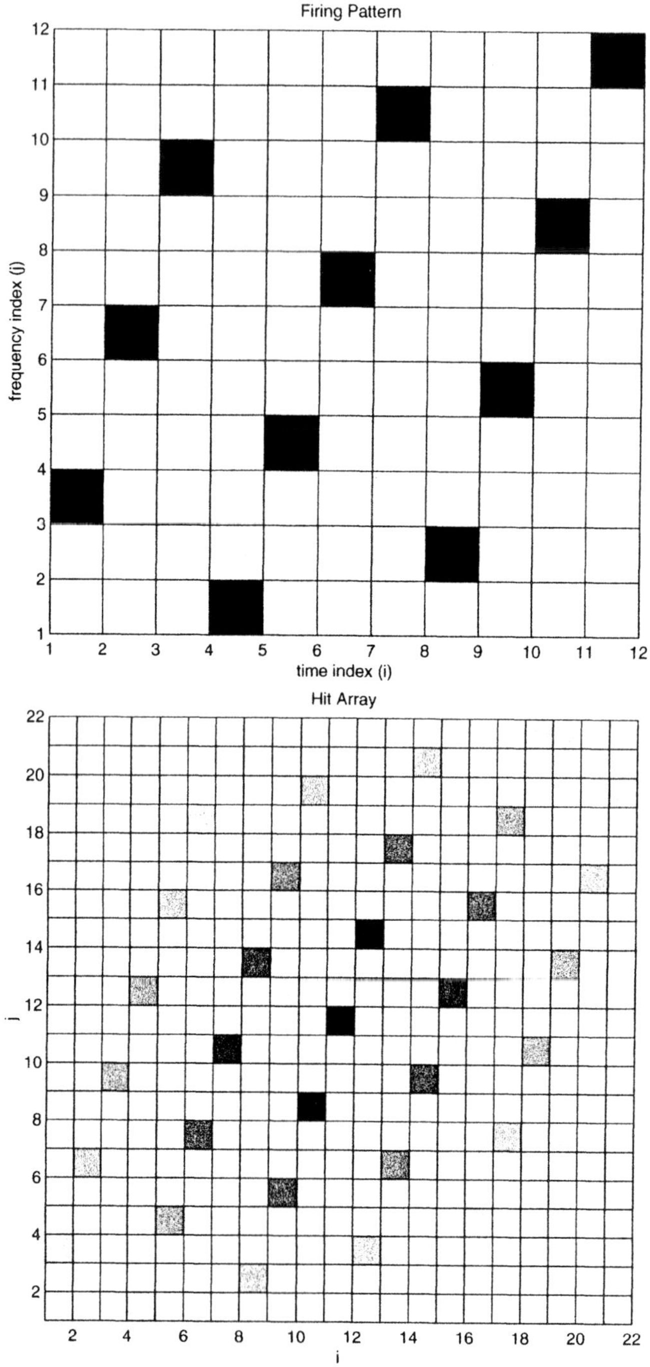

Figure 5.13: Linear congruence pattern and hit array for GF(11), $a = b = 3$

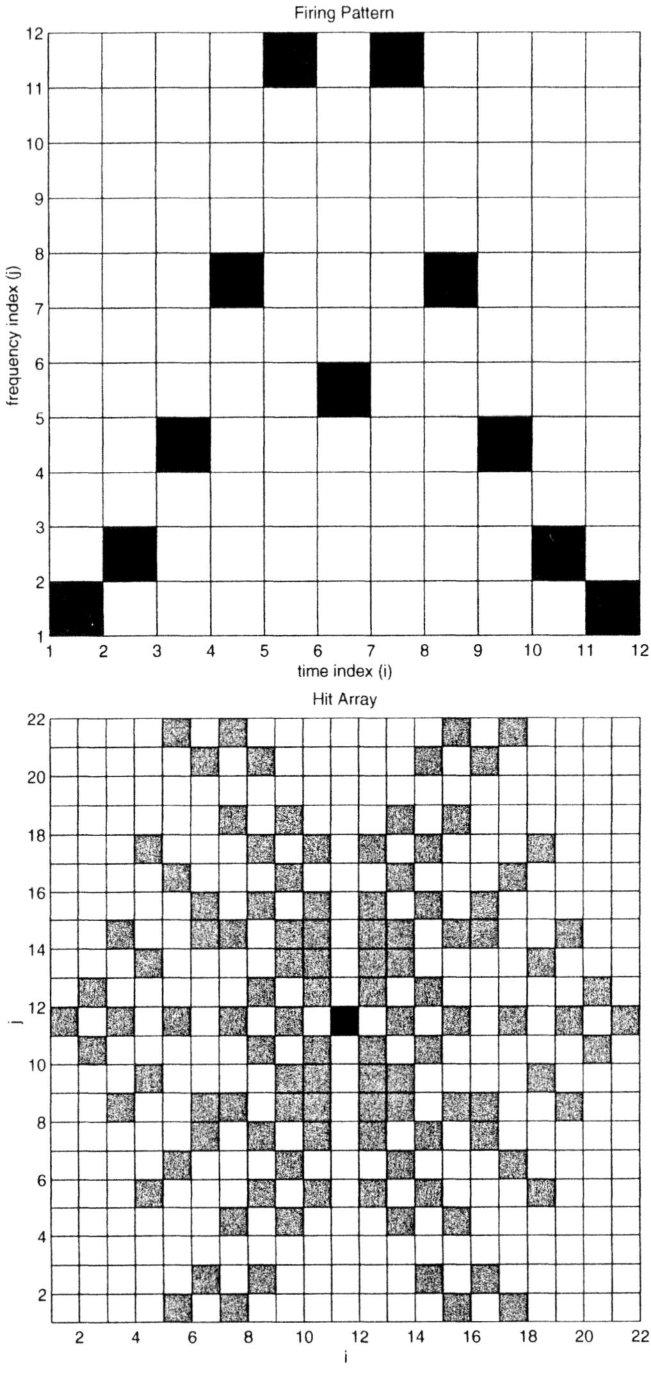

Figure 5.14: Quadratic congruence pattern and hit array for GF(11), $a = b = c = 1$

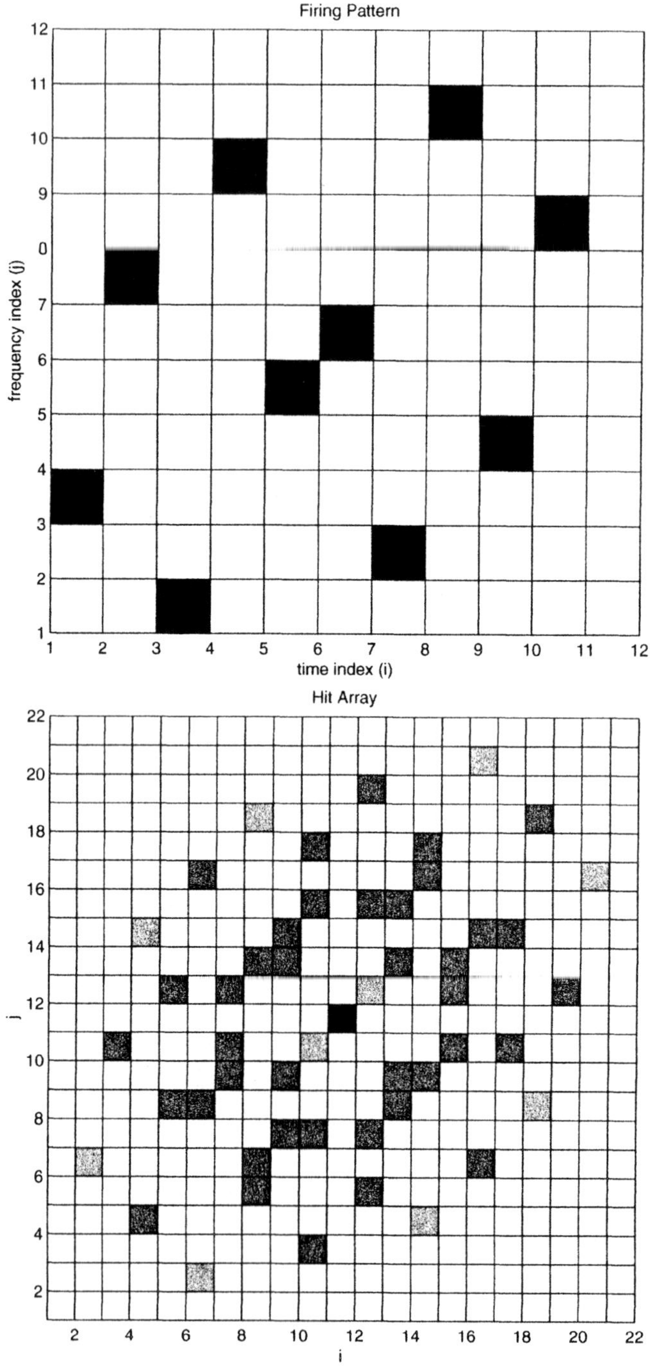

Figure 5.15: Hyperbolic congruence pattern and hit array for GF(11), $a = 3, b = 0$

Each of these "miniature" AF's has volume equivalent to its height, there are $N(N-1)$ of them, and their aggregate volume (sidelobe volume) is $(N-1)/N$. The AF of the N-chip waveform has unity volume and therefore the mainlobe has volume $1/N$. The mainlobe height is unity and as N increases, its effective delay-Doppler area decreases improving its resolution properties while reducing the sidelobes as $1/N^2$. Clearly there is a tradeoff because N cannot be increased without limit for fixed TW. At some point the waveform ceases to be sparse and the sidelobes will begin to overlap and increase.

5.4.3 AF Sidelobe Bounds for Hop Codes

An upper bound for the ambiguity sidelobes of a HOP code of order $N = q-1$ has been derived by Titlebaum [22,36] by evaluating the cross correlation between a code sequence and a shifted version of itself. The waveform of each CW chip denoted by $f_j(t)$ is

$$f_j(t) = \frac{1}{\sqrt{T}} \exp\, j(\omega_j t + \theta)$$

on an interval of duration T/N. The magnitude of the cross-correlation of two such chips delay shifted by τ with respect to each other is

$$|c_{jk}| = \left| \frac{1}{T} \int_{\tau}^{T/N} \exp\, j \,\Delta\, \omega_{jk} t dt \right|$$

and is the magnitude of their uncertainty function $|\chi(\Delta\omega_{jk}, \tau)|$ with $\Delta\omega_{jk} = \omega_j - \omega_k$. This is the square root of the AF (4.38) with T replaced by T/N and ϕ replaced by $\Delta\omega_{jk}/2\pi$. The magnitude of the cross-correlation is maximum for $\tau = 0$ and is otherwise bounded by

$$|c_{jk}| \leq \begin{cases} 1/N & \text{if } j = k, \\ \frac{2}{|\Delta\omega_{jk}|T} & \text{if } j \neq k. \end{cases}$$

$\Delta\omega_{jk} \approx 2\pi W|j-k|/N$ for full codes with equispaced frequency chips.

Two hop code pulse trains with delay τ and Doppler $\phi = 0$ where each chip interacts with two other chips are shown in Figure 5.16. The overall correlation is bounded from above by the sum of all of the subpulse bounds c_{jk} and the worst case and (greatest) upper bound occurs when the code frequency difference indices are

$$m = j - k = [\pm 1, \pm 2, \cdots, \pm N/2].$$

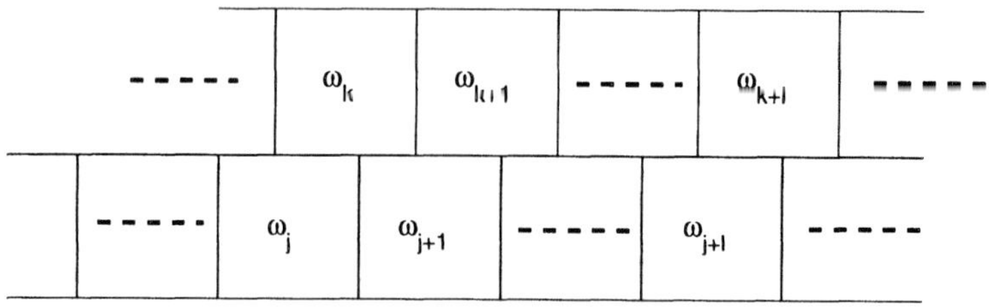

Figure 5.16: Costas pulse train correlation

The overall bound for shifts along the τ axis is then

$$\begin{aligned}
|C(\tau \neq 0, \phi = 0)| &\leq 2 \sum_{m=-N/2}^{N/2} \left(\frac{N}{\pi WT|m|} \right), \quad m \neq 0 \\
&= \frac{4N}{\pi WT} \sum_{m=1}^{N/2} \frac{1}{m}, \quad m > 0 \\
&\leq \frac{4N}{\pi WT}(1 + \ln(N/2)) \qquad (5.19)
\end{aligned}$$

since

$$\sum_{m=1}^{N/2} (1/m) \leq (1 + \ln(N/2)).$$

The bound is conservative because chips at the ends of the sequence will not overlap for $\tau \neq 0$ and contribute to the sidelobe level.

When there is both a delay and Doppler shift, there can be a hit with $|j-k| = 0$ that contributes an additional $1/N$ so that the overall bound is

$$|C((\tau, \phi) \neq 0)| \leq \frac{1}{N} + \frac{4N}{\pi WT}(1 + \ln(N/2)). \qquad (5.20)$$

When the shift occurs on the Doppler axis, all of the frequency differences are constant and depend only upon the number of vertical (Doppler) shifts $m = 1, N - 1$. The bound is then

$$|C(\tau = 0, \phi = Wm/N)| \leq \frac{N^2}{m\pi WT}. \tag{5.21}$$

The ambiguity function is $\Phi = |C|^2$ and its bound is the square of (5.19, 5.20, 5.21) depending upon the parameters. The bound corresponding to (5.20) is plotted as a function of code order and TW product in Figure 5.17. For a fixed TW product, the ambiguity bound decreases as the

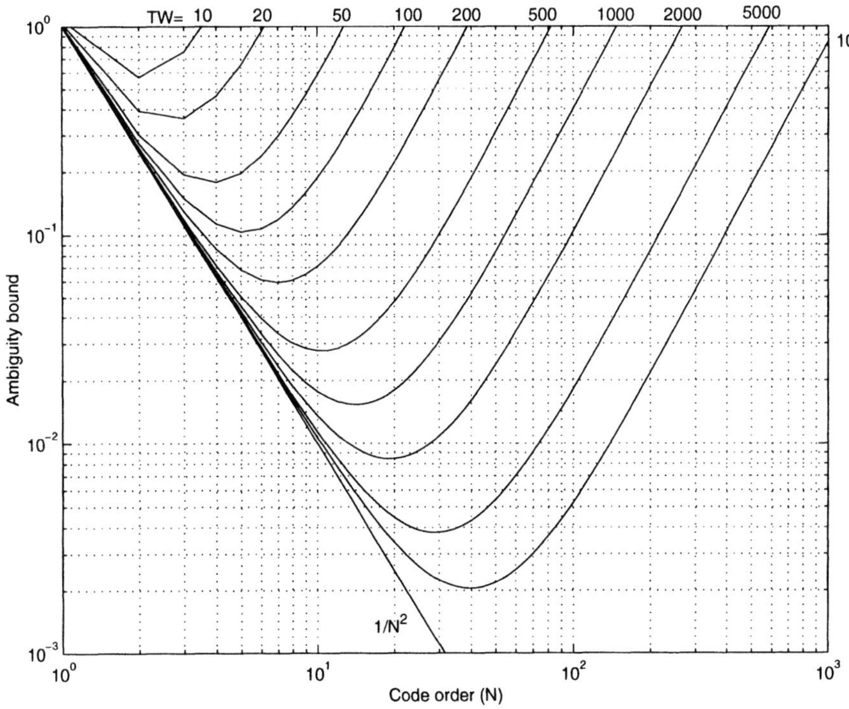

Figure 5.17: Upper bound for the sidelobes of a Costas code ambiguity function

the sparse code sidelobe level $1/N^2$ but begins to deviate from this as the waveform becomes less sparse and the pulse frequency spacing approaches the individual subpulse bandwidth. The bound has a minimum value at the code order N_{opt} where the interaction between subpulses overcomes the downward trend. It is clear from the figure that N_{opt} is significantly less than the "rule of thumb" code order $N^2 = TW$ that defines a sparse waveform. A relationship between N_{opt} and the TW

product is found by differentiating (5.20) with respect to N and setting the result to zero yielding

$$TW = \frac{4N_{opt}^2}{\pi}(2 + \ln(N_{opt}/2)).$$

Substituting this back into (5.20) results in the least upper uncertainty bound

$$|C((\tau, \phi) \neq 0)| \leq \frac{1}{N_{opt}}\left(1 + \frac{1 + \ln(N_{opt}/2)}{2 + \ln(N_{opt}/2)}\right)$$

that approaches $2/N_{opt}$ for large N_{opt}. Squaring this reveals that the ambiguity sidelobe maximum of a code with N_{opt} components is bounded for small and large values of N_{opt} respectively by

$$\frac{2.25}{N_{opt}^2} \leq \Phi_{max}(\tau, \phi \neq 0) \leq \frac{4}{N_{opt}^2}.$$

The bound always exceeds $1/N^2$ reflecting the effects of pulse interactions even for relatively sparse codes. As an example let TW=2000. $N_{opt} \approx 19$ which is considerably less than $\sqrt{2000} = 45$. $\Phi_{max} \approx .0085$ from Figure 5.17 which is bracketed by $2.25/19^2 = .0062$ and $4/19^2 = .011$.

The waveform spectrum reflects the effect of increasing the density of a code by maintaining its bandwidth but crowding more pulses into it. This is illustrated by the spectra of a sequence of hop code waveforms generated by the Welch-Costas code (5.15) with fixed $T = .2$ sec and W=500 Hz shown in Figures 5.18 a-d. The codes are denoted by GF:(7)3, (11)2, (23)5, and (41)6 where the first number is the field order (q) and the second is the primitive element e_p. These have orders $N = q - 1$=6,10,22, and 40 respectively and represent codes of increasing density because the TW=100 is fixed. $N_{opt} \approx 6$ for a code with TW=100 and the spectrum of Figure 5.18a is representative of a sparse code with distinct and separable subpulse components with a spacing of 100 Hz. The rule of thumb condition $N^2 = TW$ for sparseness generates the spectrum of Figure 5.18 b with N=10. At this point and beyond the code can no longer be regarded as sparse because the individual subpulses overlap spectrally and lose their distinct identities. Increasing the density produces spectra that begin to appear like those of band limited noise as illustrated by Figures 5.18 c,d.

5.4.4 Hop Code AF Main Lobe Characteristics

The resolution of a complex bandpass waveform for the purpose of maximum likelihood parameter estimation is determined by its expansion

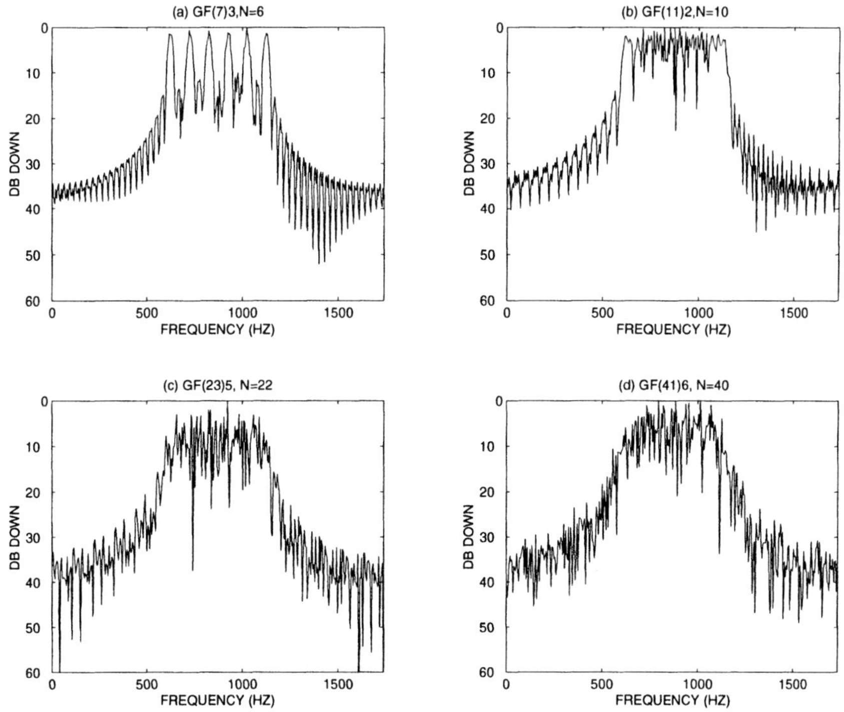

Figure 5.18: Welch-Costas spectral characteristics as a function of code density

properties at the AF mainlobe. Delay and Doppler resolution defined by the support of the central peak are functions of the coefficients λ^2, γ, η^2 given by (4.39, 4.40, 4.42) that relate to the mean square waveform bandwidth, coupling, and duration for the narrowband model and waveform modulation for the wideband model. Ambiguity characteristics in the neighborhood of the central peak are a consideration when it is desired to unambiguously discriminate closely spaced scatterers. Hop codes approach the ideal thumbtack but have fine structure near the mainlobe that consists of localized subsidiary peaks the number and amplitude of which are functions of the code order and TW product [21]. If of sufficient magnitude, these generate artifacts in the delay-Doppler map that can be mistaken for actual scatterers.

The mainlobe neighborhood of a Costas code is the region surrounding the central peak in which the delay and Doppler shift are confined to the support of an individual chip. In other words it is the region $\tau < 2T/N$ and $\phi < 2N/T$ that roughly defines the individual CW chip ambiguity function where the chips overlap themselves but not each other. The narrowband auto-uncertainty function (4.5) is

$$\chi_{ff}(\phi,\tau) = \int_{-\infty}^{\infty} f(t)f^*(t-\tau)\exp -j2\pi\phi t dt \tag{5.22}$$

and substituting (5.13 yields

$$\begin{aligned}\chi_{ff}(\phi,\tau) = &\int_{-\infty}^{\infty}\left(\sum_{i=0}^{N-1} f_i(t-iT/N)\right)\left(\sum_{k=0}^{N-1} f_k^*(t-kT/N-\tau)\right)\\ &\cdot\exp -j2\pi\phi t dt\\ = &\sum_{i=0}^{N-1}\int_{-\infty}^{\infty} f_i(t-iT/N)f_i^*(t-iT/N-\tau)\exp -j2\pi\phi t dt\\ &+\sum_{i=0}^{N-1}\sum_{k\neq i}^{N-1}\int_{-\infty}^{\infty} f_i(t-iT/N)f_k^*(t-kT/N-\tau)\exp -j2\pi\phi t dt.\end{aligned}$$

The subpulses are disjoint and when the subpulse waveforms (5.14) are substituted, the integration over the total duration T reduces to integration over the subpulse duration T/N in time segments $iT/N \leq t < (i+1)T/N$ and

$$\begin{aligned}&\chi_{ff}(\phi,\tau) =\\ &\sum_{i=0}^{N-1}\int_{iT/N+\tau}^{(i+1)T/N} a(t-iT/N)a(t-iT/N-\tau)\exp j(\omega_i\tau - 2\pi\phi t)dt\\ &+\sum_{i=0}^{N-1}\sum_{k\neq i}^{N-1}\int_{-\infty}^{\infty} a(t-iT/N)a(t-kT/N-\tau)\exp j((\omega_i-\omega_k-2\pi\phi)t\\ &+\omega_k\tau+\theta_i-\theta_k)dt.\end{aligned}$$

The first term is the aggregate contribution of the subpulse auto uncertainty functions in the mainlobe region when the subpulses overlap themselves. The second represents the interpulse contribution that occurs when the τ and ϕ are sufficient to shift one pulse into another. If the code order $N \leq N_{opt}$ as defined by Figure 5.17, the code is sparse and negligible interpulse overlap occurs in the neighborhood of the mainlobe. Consequently the second term is dropped and the UF approximated by the first term. The time integrations are reduced to a single integration on $0 \leq t < T/N$ by changing variables by $u = t - iT/N$ for each term in the sum. The result is

$$\chi_{ff}(\phi,\tau) \approx \sum_{i=0}^{N-1} \int_{\tau}^{T/N} a(u)a(u-\tau) \exp j(\omega_i \tau - 2\pi(u + iT/N)\phi)du$$

$$= \chi_{aa}(\phi,\tau) \sum_{i=0}^{N-1} \exp j(\omega_i \tau - 2\pi i \phi T/N) \tag{5.23}$$

where

$$\chi_{aa}(\phi,\tau) = \int_{\tau}^{T/N} a(u)a(u-\tau) \exp -j2\pi\phi u du$$

is the auto-uncertainty function of the window function.

The AF on the Doppler axis is obtained from (5.22) as

$$\Phi_{ff}(\phi,0) = |\chi_{ff}(\phi,0)|^2 = \left| \int_0^{\mathrm{T}} |f(t)|^2 \exp -j2\pi\phi dt \right|^2$$

and if the pulse transitions are phase coherent and the weighting is rectangular

$$\Phi_{ff}(\phi,0) = \operatorname{sinc}^2(\pi\phi T) \tag{5.24}$$

which is identical to the AF Doppler profile of the rectangularly weighted CW of duration T (4.38).

The AF on the τ axis is obtained from (5.23) with $\phi = 0$ and

$$\Phi_{ff}(0,\tau) = \Phi_{aa}(0,\tau) \left| \sum_{i=0}^{N-1} \exp j\omega_i \tau \right|^2$$

since $a(t)$ is assumed to be a real window function. Costas codes have a uniform frequency allocation and it is possible to reorder the summation and write it as

$$\sum_{i=0}^{N-1} \exp j\omega_i \tau = \sum_{n=0}^{N-1} \exp jn\delta\omega\tau$$

where $\delta\omega$ is the incremental radian frequency $2\pi W/N$. The AF is written as

$$\Phi_{ff}(0,\tau) = \Phi_{aa}(0,\tau) N^2 \left| \frac{\exp jN\delta\omega\tau - 1}{\exp j\delta\omega\tau - 1} \right|^2$$

by using the algebraic relationship

$$\sum_{i=0}^{N-1} x^N = \frac{x^N - 1}{x - 1}.$$

Note that by Euler's formula

$$\frac{\sin Nx/2}{\sin x/2} = \exp - \left(\frac{(N-1)x}{2}\right)\left(\frac{\exp jNx - 1}{\exp jx - 1}\right)$$

so that

$$\left|\frac{\exp jNx - 1}{\exp jx - 1}\right|^2 = \left(\frac{\sin Nx/2}{\sin x/2}\right)^2.$$

Substituting $\delta\omega\tau$ for x yields the AF

$$\Phi_{ff}(0,\tau) = \Phi_{aa}(0,\tau)N^2\left(\frac{\sin N\delta\omega\tau/2}{\sin \delta\omega\tau/2}\right)^2.$$

$a(t) = 1/\sqrt{T}$ for a rectangular unit energy window and

$$\Phi_{aa}(0,\tau) = \left(\frac{T/N - |\tau|}{T}\right)^2.$$

The AF on the τ axis is then

$$\Phi_{ff}(0,\tau) = \left(\frac{N(T/N - |\tau|)}{T}\right)^2\left(\frac{\sin N\delta\omega\tau/2}{\sin \delta\omega\tau/2}\right)^2 \qquad (5.25)$$

and is unity for $\tau = 0$. This and the Doppler axis relationship (5.24) indicate that Costas codes and hop codes in general have considerable mainlobe structure. The mainlobe support region is defined by the individual chips and as the code order increases it narrows in delay but expands in Doppler for fixed T and W.

This is illustrated by the AF plots of Figures 5.19a-d that are of the same fixed TW=100 Welch-Costas waveforms used to generate the spectra of Figure 5.18a-d. The optimum order for minimum interpulse sidelobes is about 6 according to Figure 5.17 and should produce maximum sidelobes of about -10 dB from the peak. Clearly the sidelobes in the vicinity of the mainlobe exceed this for the low (close to optimum) order GF(7) code shown in Figure 5.19a. This occurs because the bound represents sidelobes generated from interpulse correlations and not those generated in the mainlobe neighborhood by (5.24, 5.25). As the code

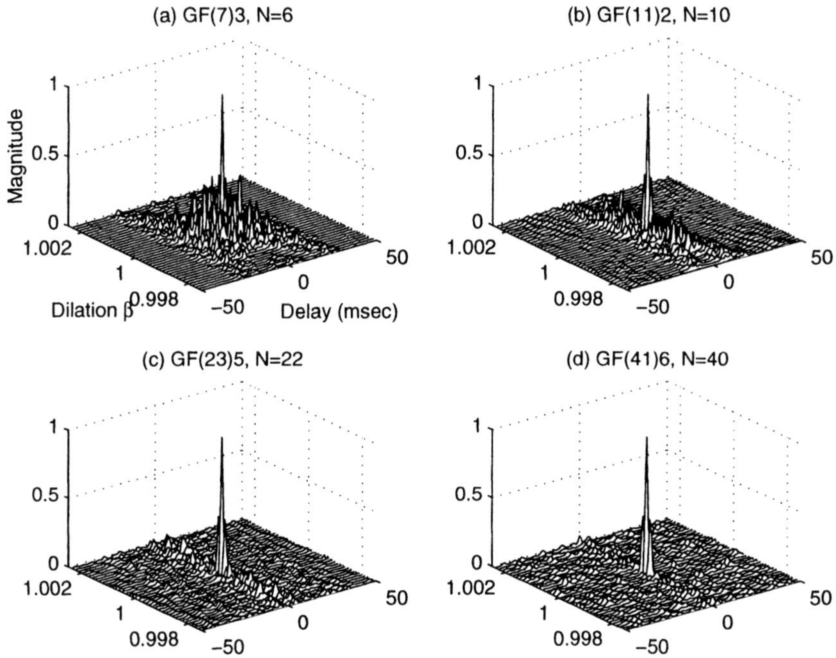

Figure 5.19: Welch-Costas AF mainlobe characteristics as a function of code density (TW=100)

order increases the mainlobe support region broadens in Doppler and contracts in delay and the AF becomes more like the proverbial thumbtack as shown in Figures 5.19b-d. This occurs because an increase in code density tends to spread the waveform more uniformly in frequency as shown in Figure 5.18 and the AF approaches the ideal thumbtack as predicted by the Wigner-Ville criterion. The downside of this is that by increasing the density beyond the optimum, the occurrence of interpulse sidelobes greater than the bound away from the mainlobe becomes more likely. These are called ambiguity "pop ups", are usually isolated, and may not pose any difficulty depending upon the application. Often dense hop code waveforms with good resolvent mainlobe characteristics have pop ups that can be tolerated because they are in "don't care" regions of the delay Doppler plane. However, a search for them as a function of code order, construction, and TW product can be a formidable task.

5.4.5 Logarithmic Frequency Allocation of Hop Codes

The frequency allocation corresponding to the various code placement operators $j(i)$ must be logarithmic if the resulting waveforms are to be dilation consistent as discussed in Section 5.3.1. This is illustrated by defining the linear allocation as a function of the placement operator $j(i)$

$$\begin{aligned}
\omega_i &= \omega_0 + 2\pi W J(i) \quad \text{(linear allocation)} \\
J(i) &= \left[\frac{j(i) - j_{min}}{j_{max} - j_{min}} - \frac{1}{2}\right], \; i = 1, \cdots, N \\
\omega_0 &= \frac{\omega_{max} + \omega_{min}}{2} \\
2\pi W &= \omega_{max} - \omega_{min}.
\end{aligned} \tag{5.26}$$

j_{min}, j_{max} are the minimum and maximum values of $j(i)$ and their difference is thc total number of available frequency bins not all of which are necessarily filled as illustrated by the QC code of Figure 5.14.

Consider two frequency pairs with the same difference

$$\omega_i - \omega_j = \omega_k - \omega_l = 2\pi W \triangle J$$

and

$$\begin{aligned}
\omega_i &= \omega_j + 2\pi W \triangle J \\
\omega_k &= \omega_l + 2\pi W \triangle J.
\end{aligned}$$

Let the code be dilated such that $\beta\omega_j = \omega_i$, and $\beta\omega_l = \omega_k$. Then,

$$\begin{aligned}
\beta &= \frac{\omega_i}{\omega_j} = \frac{\omega_k}{\omega_l} \\
&\Rightarrow 1 + \frac{2\pi W \triangle J}{\omega_j} = 1 + \frac{2\pi W \triangle J}{\omega_l} \Rightarrow \omega_j = \omega_l.
\end{aligned}$$

The linear allocation is inconsistent with the Doppler dilation model because frequency element pairs cannot have the same difference and ratio unless they are the same.

A dilation consistent logarithmic allocation assigns the frequencies according to

$$\log \omega_i = \log \omega_c + J(i) \log R_b \quad \text{(logarithmic allocation).} \tag{5.27}$$

where $\omega_c = \sqrt{\omega_{max}\omega_{min}}$ is the geometric radian frequency centroid [34]. $R_b = \omega_{max}/\omega_{min}$ is the band ratio that is analogous to bandwidth W

and related to it and the arithmetic centroid ω_0 by

$$R_b = \frac{\omega_0 + \pi W}{\omega_0 - \pi W}.$$

Codes generated by this algorithm are dilation consistent because the frequencies are allotted as ratios with respect to the geometric centroid ω_c rather than additive increments with respect to the arithmetic centroid ω_0. Table 5.1 illustrates the difference between the linear and logarithmic allocations for the Welch code of Figure 5.12 with $j(i) = 2^i (\text{mod } 11)$. The arithmetic frequency centroid and bandwidth are specified as 5000 and 500 (Hz) respectively and

i	j(i)	J(i)	f_{lin}(Hz)	f_{log}(Hz)
1	2	-.389	4806	4803
2	4	-.167	4917	4911
3	8	.278	5139	5135
4	5	-.056	4972	4966
5	10	.5	5250	5250
6	9	.389	5194	5192
7	7	.167	5084	5078
8	3	-.278	4861	4857
9	6	.056	5028	5022
10	1	-.5	4750	4750

Table 5.1: Welch GF(11)2 linear and logarithmic frequency allocations

$$J(i) = \frac{j(i) - 5.5}{9}.$$

The geometric frequency centroid is 4993.7 (Hz), $R_b = 1.1053$, log $R_b =$ .04348, and

$$f_i(lin) = 5000 + 500J(i)$$
$$f_i(log) = 4993.710^{.04348J(i)}.$$

There is not a great deal of absolute difference between the allocations however taking the ratios of corresponding code pairs reveals that the logarithmic allocation preserves the frequency ratios between chips. The ratio of the lowest and highest pairs is 4803/4750=5250/5192=1.0112 for the log allocation and the corresponding linear allocation ratios are 4806/4750=1.0118 and 5250/5194=1.0108. As a rule the linear allocation can be used whenever the narrowband model is appropriate and the expected Doppler velocity is $v_{los} << c/2TW$.

5.5 PRN WAVEFORMS

Pseudo random noise (PRN) waveforms have the ambiguity attributes of a random noise process and are generated by the same deterministic algorithms that are used to generate pseudo random binary sequences for spread spectrum communications. These systems spread the frequency content of relatively narrowband information over a wide bandwidth so that it appears as random noise to all but the intended receiver. It is widely used for multiple user cellular phone, mobile, and wireless systems where it is desired to accommodate multiple channels with minimal interference and a maximum of privacy. Code-division multiple access (CDMA) systems accommodate multiple users within the same bandwidth by assigning each user a unique PRN code that has very low cross correlation with all other codes. Spread spectrum offers advantages such as resistance to fading, high spectral efficiency, and graceful degradation of performance in the presence of interference [41].

Waveforms derived from PRN sequences may be utilized for echo location and like the hop codes are used for the generation of high resolution thumb tack ambiguity functions [24]. They are similar to the frequency hop codes but rather consist of M individual CW subpulses or chips all with the same frequency. The carrier frequency f_0 is modulated by introducing phase discontinuities between chips according to a deterministic binary sequence. The ambiguity delay resolution is proportional to the inverse bandwidth which is approximately the inverse of the chip duration $W = M/T$, the narrowband Doppler resolution is proportional to the inverse of the total duration T, and the TW product which is the overall resolution measure is equal to the number of chips M.

Different schemes may be used to generate sequences for spread spectrum/CDMA [26, 27] but the most common family of algorithms used for the generation of PRN sonar waveforms are called by various names including biphase shift keyed (BPSK) codes, shift register codes, and maximal length sequence codes. The complex waveform generated by such a binary sequence is

$$
\begin{aligned}
f(t) &= \sum_{m=1}^{M} f_m(t) \\
f_m(t) &= \frac{1}{\sqrt{T}} \exp j \left[J_m \omega_0 (t - t_{m-1}) + \theta_{m-1} \right] \quad t \in [t_{m-1}, t_m] \\
\theta_m &= \theta_{m-1} + J_m \omega_0 (t_m - t_{m-1})
\end{aligned}
\tag{5.28}
$$

where $J_m = 2I_m - 1$ and I_m, $m = 1, M$ is a deterministic binary (0,1) sequence that closely resembles a random binary process with switch points corresponding to the chip intervals. The recursion in θ assures that the phase is continuous at the transition points although the frequency can reverse. Note that $t_m - t_{m-1} = T/M = 1/W$ so that

$$\theta_m = \theta_{m-1} + J_m \left(\frac{2\pi f_0}{W} \right).$$

The number of cycles in a chip is equal to the ratio $f_0/W = f_0 T/M$ and if it is an integer, the chip transitions occur at the zero phase points and the θ recursion is unnecessary. Wideband dilation of the waveform for the generation of Doppler hypotheses generates the replicas

$$f_m(\hat{\beta} t) = \sqrt{\frac{\hat{\beta}}{T}} \exp j \left[J_m \omega_0 (\beta t - t_{m-1}) + \theta_{m-1} \right], \quad t \in \left[\frac{t_{m-1}}{\hat{\beta}}, \frac{t_m}{\hat{\beta}} \right].$$

5.5.1 Shift Register Sequences

The most common binary code generator is an N stage shift register with linear feedback as shown in Figure 5.20. All of the terms are binary (0,1) and arithmetic is modulo two (exclusive OR) where 0+0=1+1=0, 1+0=0+1=1 and $0 \times 0 = 0 \times 1 = 1 \times 0 = 0$, $1 \times 1 = 1$. Subtraction is equivalent to addition because -1+1=0=1+1. The set of numbers 0 and 1 together with modulo two arithmetic is called a binary or Galois field of two elements and is denoted by GF(2) [42]. The shift register is defined by the linear feedback algorithm

$$a_i = c_1 a_{i-1} + c_2 a_{i-2} + \cdots + c_N a_{i-N} = \sum_{k=1}^{N} c_k a_{i-k} \tag{5.29}$$

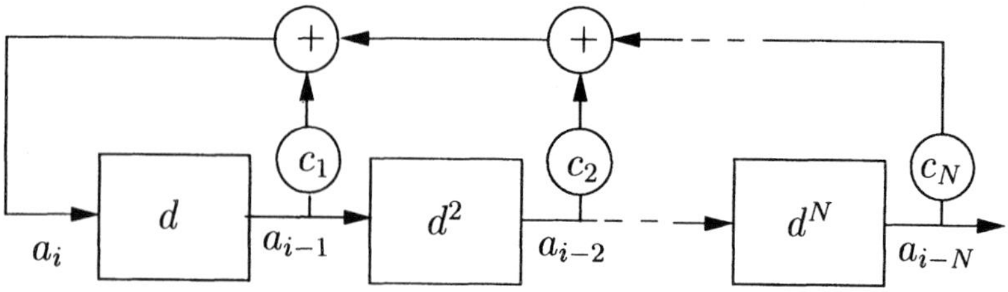

Figure 5.20: Shift register for the generation of maximal length PRN sequences

where the c_k are the elements of a *connection vector* and are binary variables that act as switches that tap the register at particular stages and determine the number of feedback elements. The chips are defined by clock times t_m occurring at intervals T/M when the shift register shifts its contents one step to the right. Consequently $a_{i-k}(t_m) = a_{i-k+1}(t_{m-1})$ and at any instant t_m the input to the first register is $a_i(t_m)$, the input to the second is $a_i(t_{m-1})$, and so on. The delay operator d shown in the figure defines the operation

$$a_{i-k}(t_{m-1}) = da_{i-k}(t_m)$$

and consequently at any time, the first register input is a_i, the second is da_i, the third is d^2a_i etc. The algorithm (5.29) is expressed in operator notation as

$$a_i = \sum_{k=1}^{N} c_k d^k a_i.$$

Setting this to zero defines the *characteristic polynomial* [26]

$$p(d) = 1 - \sum_{k=1}^{N} c_k d^k$$

that depends solely on the connection variables. It is related to the *generating function* defined by the infinite sequence

$$g(d) = a_0 + a_1 d + a_2 d^2 + \cdots = \sum_{i=0}^{\infty} a_i d^i.$$

This is reduced to a finite recursion by using (5.29) where

$$g(d) = \sum_{i=0}^{\infty} \sum_{k=1}^{N} c_k a_{i-k} d^i$$

$$- \sum_{k=1}^{N} c_k d^k \left[\sum_{i=0}^{\infty} a_{i-k} d^{i-k} \right]$$

$$= \sum_{k=1}^{N} c_k d^k [a_{-k} d^{-k} + a_{1-k} d^{1-k} + \cdots g(d)].$$

Solving for $g(d)$ yields the ratio of polynomials

$$g(d) = \frac{\sum_{k=1}^{N} c_k d^k [a_{-k} d^{-k} + a_{1-k} d^{1-k} + \cdots a_{-1} d^{-1}]}{1 - \sum_{k=1}^{N} c_k d^k}$$

$$= \frac{p_0(d)}{p(d)}. \tag{5.30}$$

The numerator can be written as

$$p_0(d) = \sum_{k=1}^{N} c_k \left[a_{-k} + a_{1-k} d + a_{2-k} d^2 + \cdots + a_{-1} d^{k-1} \right]$$

$$= c_1 a_{-1} + c_2 (a_{-2} + a_{-1} d) + c_3 (a_{-3} + a_{-2} d + a_{-1} d^2) +$$

$$\vdots$$

$$+ c_N (a_{-N} + a_{1-N} d + a_{2-N} d^2 + \cdots + a_{-1} d^{N-1})$$

and depends upon the initial condition (IC) $(a_{-1}, a_{-2}, \cdots, a_{-N})$ that is the content of the register at the initiation of the sequence when $i = 0$ and

$$a_0 = c_1 a_{-1} + c_2 a_{-2} + \cdots + c_N a_{-N} = \sum_{k=1}^{N} c_k a_{-k}.$$

c_N must be unity because the register is defined as having N stages. If $a_{-N} = 1$ and $a_{-1} = a_{-2} = \cdots a_{-N+1} = 0$,

$$p_0(d) = 1, \quad g(d) = \frac{1}{p(d)}.$$

Each stage of the shift register can have value of either zero or unity consequently there are 2^N possible states for the register as a whole. If all of the states are zero the register will remain at zero thus there are $M = 2^N - 1$ possible non-zero states. There are at most M clock cycles that result in distinct states of the register and each state is an IC for all successive states. When any one state repeats after M cycles all successive states will repeat in order and M is the maximum possible period. Sequences with this period are called *maximal length shift register* (MLSR) sequences. A necessary but not sufficient condition for a register to be maximal length is that its characteristic polynomial be irreducible (non factorable) [26]. A table of irreducible polynomials expressed as shift register connection vectors or taps is provided by Dixon [27] for $N \leq 89$ stages. While such a large number of stages is clearly overkill for sonar waveform design, large code lengths are often required for spread spectrum communication in multi-user environments.

Each register state corresponds to a particular code sequence (5.28) and the number of stages is related to the waveform TW product by

$$N = \frac{\log(1+TW)}{\log 2}.$$

A PRN code with the same TW product as the Costas code examples of Figure 5.19 with T=.2 s and W=500 Hz requires a sequence of $M = TW = 100$ chips of 2 msec duration which is obtainable from a register of N=7 stages. $2^7 - 1 = 127$ and 27 elements of the shift register sequence are superfluous and any sequence of 100 consecutive (0,1) elements from the output (or any other stage) of the register may be chosen as the generating sequence I_m for the waveform (5.28). Connection vectors for $N \leq 12$ stages obtained from [27] are provided in Table 5.2. Many more are available in the cited reference but these should suffice for sonar waveform applications with $TW \leq 4095$.

Runs and Repeated Segments

Once the number of stages and a connection vector is chosen, the generating sequence I_m is determined by the initial condition (IC) and obtained by clocking the register M times and recording the output bit. A complete PRN shift register sequence contains $M_1 = 2^{N-1}$ ones and $M_0 = 2^{N-1} - 1$ zeros and if the IC is $a_1 = a_2 = \cdots = a_{N-1} = 0$ and $a_{-N} = 1$, the state of the register is 0, 0, 0, $\cdots$, 0, 1 and a string or run of $N-1$ zeros will occur. Likewise, a string of $N_r = N-1$ ones occurs if the IC is all ones except for $a_{-N} = 0$. A run of $N_r = N$ zeros or ones occurs if the IC is all zero or one but the all zero case is degenerate because it is the stationary null state. The IC must always have no more than $N-1$ zeros in order to avoid this. The fraction of all runs in a string with $N_r = N-1$ is therefore $2/2^N = 1/2^{N-1}$ and is also equal to the fraction of runs of order N even though the null string is degenerate. There are 2^{N_r} possible binary sequences in a string of $N_r < N$ bits however only two result in runs of all zeros or ones.The total number of

Stages (N)	Code length (M)	$c_k = 1$ for $k = [\ \ ]$
2	3	[1,2]
3	7	[1,3]
4	15	[1,4]
5	31	[1,5][2,3,4,5][1,2,4,5]
6	63	[1,6][1,2,5,6][2,3,5,6]
7	127	[1,7][1,2,3,7][2,3,4,7][2,4,6,7][1,3,6,7]
8	255	[2,3,4,8][3,5,6,8][2,5,6,8][1,3,5,8]
9	511	[4,9][3,4,6,9][4,5,8,9][1,4,8,9][2,3,5,9]
10	1023	[3,10][2,3,8,10][1,3,4,10][1,5,8,10]
11	2047	[1,11][2,5,8,11][2,3,7,11][2,3,10,11]
12	4095	[1,4,6,12][2,3,9,12][1,2,5,10,11,12]

Table 5.2: Maximal Length Shift Register Connection Vectors

possibilities for a run of N_r is 2×2^{N_r} thus the fraction of runs of order N_r is $1/2^{N_r}$. The conclusion is that of the total number of runs of all orders in a string of N binary elements, the fraction of order N_r is $1/2^{N_r}$ for $N_r < N$ and is $1/2^{N-1}$ for $N_r = N$. This rule is exact if the fractions give an integral number of runs [25, 26].

It is always possible to use a register length longer than the minimum required by the TW product and chose a shorter code segment from the string of M binary digits. The disadvantage is that runs represent strings of chips with continuous phase and hence are CW segments with duration equal to the run length. These contribute coherent sidelobes to the ambiguity function that may produce subsidiary peaks away from the mainlobe. The maximum run length is minimized by using the smallest possible number of stages. Conversely, it is necessary to use at least the minimum number of stages otherwise the code will repeat generating sidelobes caused by correlation of the repeated segments. This is illustrated by the repeated subfunction example of Figures 5.10, 5.11 for which a PRN waveform of $TW = M = 1500$ chips requiring at least $N = 11$ stages was synthesized by a 10 stage register. The code repeated after 1023 chips and the first 1500-1023=477 bits of the string were appended to the end. Two identical code segments of 477 bits were separated by another string of 546 bits producing the sidelobes shown in the figure.

Generating a PRN Code Sequence

The generating function as its name implies can be used to generate the output sequence corresponding to a particular connection vector and set of initial conditions. The sequence is generated by expanding (5.30) using long division and binary arithmetic and recording the string of (0,1) coefficients of the expansion. As an example consider a three stage register with connections $c_1 = c_3 = 1$, $c_2 = 0$ and initial conditions $a_{-1} = a_{-2} = 0$, $a_{-3} = 1$. The recursion is given by

$$a_i = a_{i-1} + a_{i-3}$$

and $p(d) = 1 - d - d^3$. The expanded generating function is

$$\begin{aligned} g(d) &= \frac{1}{1 - d - d^3} \\ &= 1 + d + d^2 + d^4 + d^7 + d^8 + d^9 + d^{11} + d^{14} + d^{15} + \cdots \end{aligned}$$

and the coefficient string is $(1, 1, 1, 0, 1, 0, 0, 1, 1, 1, 0, 1, 0, 0, 1, 1, \cdots)$ with period $M = 7$. The corresponding shift register sequence is given in Table 5.3 where the same periodic string occurs. A string of N bits can be chosen from any point in the repeated sequence to form a code. Choosing the first seven from the last column of the table yields the string (0,0,1,1,1,0,1) and there is one run of three, one run of two, and two runs of one for a total of 4 four runs. These numbers are consistent with the predicted run lengths where the fraction is 1/4 for $N_r = 3$ and $N_r = 2$, and 1/2 for $N_r = 1$. The run lengths and total number of runs vary depending upon the segment chosen as illustrated by choosing the last 7 elements of the first column which is the string (1,0,1,0,0,1,1). This has no runs of three, two runs of two, and three runs of one = five total. In this case the fractions do not yield integral run predictions.

Autocorrelation and Power Spectrum

The autocorrelation of a PRN shift register sequence approximates that of a random binary process that is triangular over the interval $|\tau| \leq T/M$ and zero otherwise [26, 43]. The PRN sequence differs because it is deterministic with period T and has a fixed number of M_0 zeros and M_1 ones in a period of $M = 2^N - 1$ chips. Consequently the autocorrelation

Clock cycle (i)	a_{i-1}	a_{i-2}	a_{i-3}
0 (IC)	0	0	1
1	1	0	0
2	1	1	0
3	1	1	1
4	0	1	1
5	1	0	1
6	0	1	0
7	0	0	1
8	1	0	0
9	1	1	0
10	1	1	1
11	0	1	1
12	1	0	0
13	0	1	0
14	0	0	1
15	1	0	0
16	1	1	0

Table 5.3: Shift register sequence for $a_i = a_{i-1} + a_{i-3}$

for shifts greater than the chip duration has a negative bias [42]

$$\begin{aligned} R_p(\tau) &= \frac{M_0 - M_1}{M} \\ &= \frac{2^{M-1} - 1 - 2^{M-1}}{M} = -\frac{1}{M}, \quad \frac{T}{M} \leq |\tau| < T\left(\frac{M-1}{M}\right) \end{aligned}$$

and approaches the autocorrelation of the random process as the code order $M \to \infty$. The overall autocorrelation is

$$R_p(\tau) = \begin{cases} 1 - \frac{(M+1)|\tau|}{T} & \text{if } |\tau| \leq \frac{T}{M}, \\ -\frac{1}{M} & \text{if } |\tau| > \frac{T}{M} \end{cases}$$

as shown in Figure 5.21. The corresponding power spectral density neglecting the bias is the Fourier transform of the triangular autocorrelation function

$$S_{ff}(\omega) = \left(\frac{2M}{\omega T}\right)^2 \sin^2\left(\frac{\omega T}{2M}\right) = \text{sinc}^2\left(\frac{\omega T}{2M}\right).$$

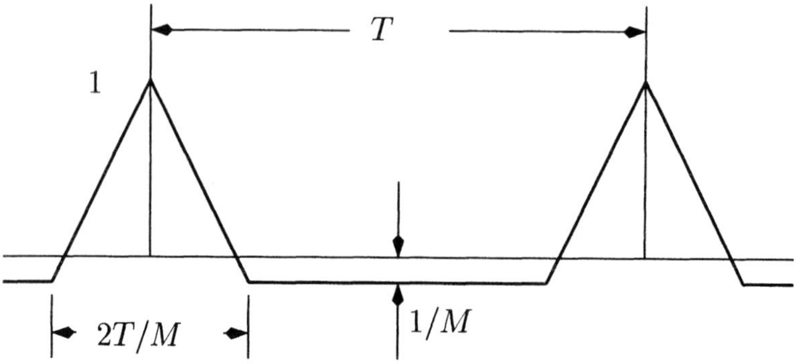

Figure 5.21: Autocorrelation of a deterministic PRN sequence

As an example, the power spectrum of a PRN waveform with T=.2 s, W= 500 Hz, and M=100 is shown in Figure 5.22. The waveform was generated using a seven stage register and the [1,7] connection vector given in Table 5.2. Its duration and bandwidth were chosen to match those of the Costas example of Figure 5.19. Its ambiguity function plotted in dB is shown in Figure 5.23. The AF of the corresponding Welch-Costas waveform of Figure 5.19d plotted in dB is shown in Figure 5.24 for comparison. This waveform is considered to be a dense hop code because $N = 40 >> 10 = \sqrt{TW}$ and its sidelobes are more uniform than the other more sparse codes illustrated in Figure 5.19. Both ambiguity functions are approximations to the thumbtack AF of Figure 5.4 and their predicted average sidelobe height is -10logTW=-20dB for $TW = 100$. Sidelobe pop ups or local peaks above the average always occur and the floor of both ambiguity figures is set to -18 dB to illustrate the sidelobe behavior in relation to the predicted average. Both waveforms have sidelobe pop ups that exceed the -20 dB average level but the PRN sidelobes are more uniform with a lower peak (pop up) value. None of the PRN pop ups exceed -16 dB but some of those of the Costas code exceed -15 dB.

It is possible to construct a Welch-Costas waveform of order $N =$ 100, chip duration (2 msec), and design bandwidth (500 Hz) as that of the PRN with $M = N$ chips by choosing GF(101)2 from the primitive element table. The resulting hop code waveform spectrum and ambiguity functions are shown in Figures 5.25, 5.26. It is evident from a comparison of Figures 5.18d and 5.25 that once a hop code is dense with $N^2 >$ TW, a further increase in density by increasing its order within the

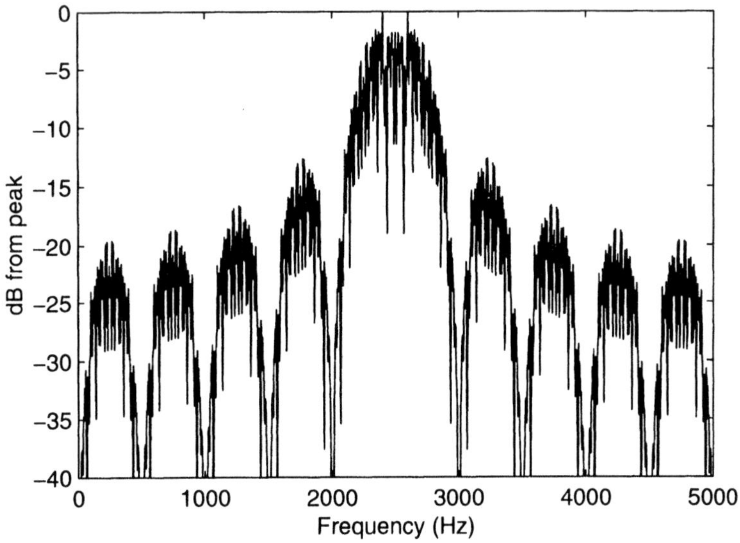

Figure 5.22: PRN [1,7] power spectrum (T=.2 sec, W=500 Hz)

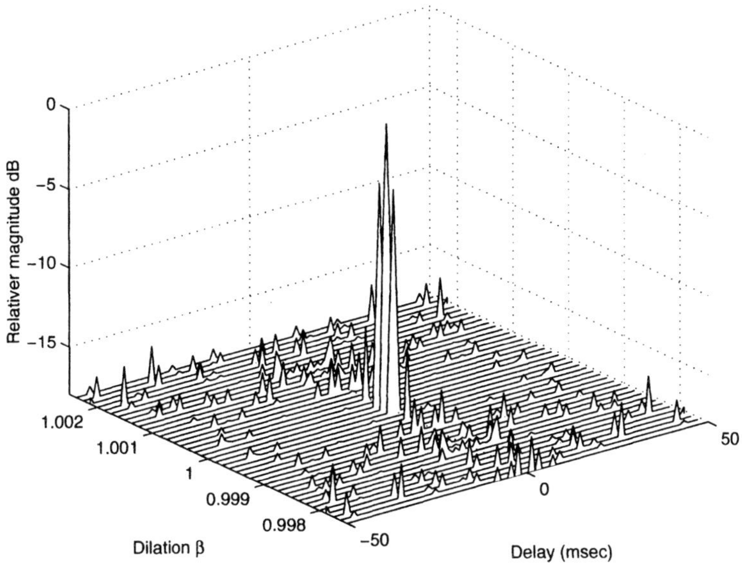

Figure 5.23: dB plot of the PRN [1,7] ambiguity function (T=.2 sec, W=500 Hz, M=100)

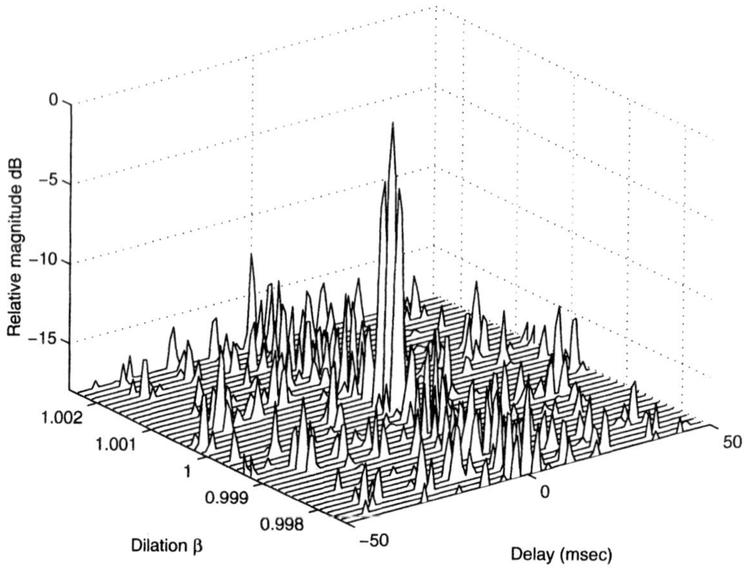

Figure 5.24: dB plot of the Welch-Costas GF(41)6 ambiguity function (T=.2 sec, W=500 Hz)

same duration-bandwidth constraint does little to alter its spectrum. Likewise, the ambiguity functions are grossly unaffected as is evident by a comparison of Figures 5.23 and 5.26. In both cases the maximum sidelobe pop up level is only slightly greater than -15 dB. Both PRN and dense hop codes possess Wigner-Ville (WVD) or Choi-Williams (CWD) distributions that are reasonably uniform in time-frequency, an attribute that was shown in Section 5.2.2 to be associated with a jointly resolvent ambiguity mainlobe. The CWD's with $\sigma = 1$ (4.11) corresponding to the ambiguity functions of Figures 5.23, 5.24 are shown in Figures 5.27, 5.28. These codes are dense in time-frequency and the CWD provides a somewhat more accurate and smoother representation than the WVD because it suppresses cross terms in the distribution. The CWD of the PRN appears to be somewhat more uniform in comparison to that of the FSK waveform which is consistent with the slightly higher sidelobe levels of the FSK ambiguity function. PRN waveforms are generally acknowledged to have slightly better sidelobe behavior with lower pop ups than comparable high density hop codes [24]. Sparse low order hop codes have isolated discrete sidelobes of approximately equal height and

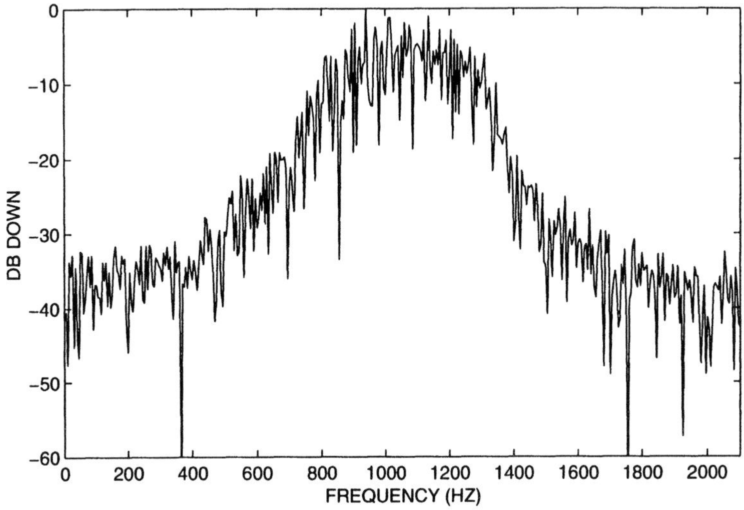

Figure 5.25: Welch-Costas GF(101)2 waveform spectrum (T=.2 sec, W=500 Hz)

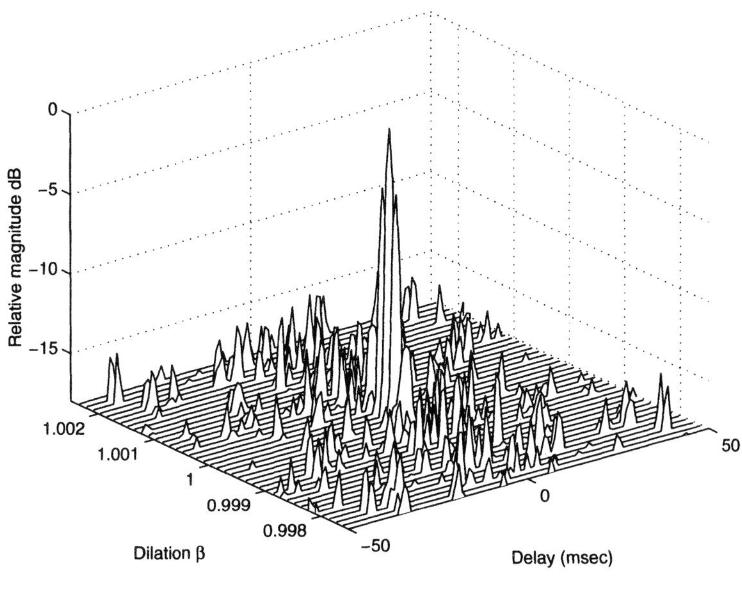

Figure 5.26: dB plot of the Welch-Costas GF(101)2 ambiguity function (T=.2 sec, W=500 Hz)

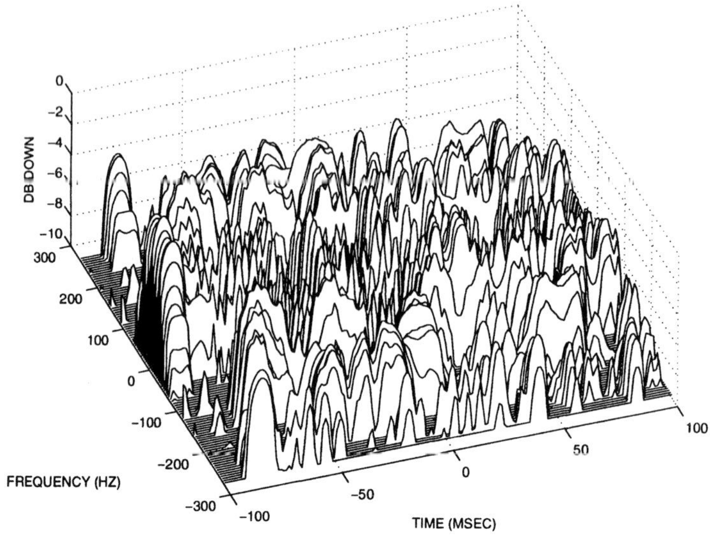

Figure 5.27: PRN [1,7] (T=.2 sec, W=500 Hz, M=100) Choi-Williams distribution

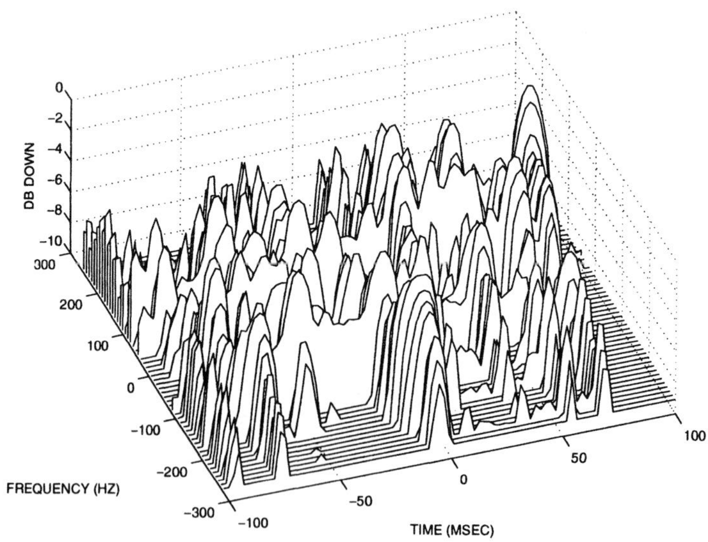

Figure 5.28: Welch-Costas GF(41)6 (T=.2 sec, W=500 Hz) Choi-Williams distribution

considerable ambiguity volume in the neighborhood of the mainlobe as shown in Figure 5.19.

5.6 DOPPLER TOLERANCE

High resolution jointly resolvent waveforms with "thumbtack " ambiguity functions while suitable for delay-Doppler imaging, are computationally expensive for detection applications if it is necessary to search a large region of the phase space. Without prior information regarding the parameter location of the scattering process a large number of receiver hypotheses may be required. It is often more efficient to perform detection using a singularly resolvent waveform that has high resolution in delay or Doppler but not in both. Such a waveform is said to be "tolerant" of one of the parameters and is characterized by a ridge-like ambiguity function that is confined to either the Doppler or delay axis. It is not possible for a waveform to be simultaneously delay and Doppler tolerant in the sense that its ambiguity function is uniform with unit magnitude over a non zero delay-Doppler region because of the ambiguity volume conservation property. Cross-ambiguity functions can be made very uniform over an arbitrarily large area as illustrated by the LFM mesa of Figure 5.9 but their height is reduced.

The simple CW waveform can be made either delay or Doppler tolerant by merely increasing or decreasing its pulse duration. Its narrowband ambiguity function (4.38) is $(1-|\tau|/T)^2$ and $\mathrm{sinc}(\pi\phi T)$ on the delay and Doppler axis for rectangular weighting and it becomes ridge like on the Doppler axis for small values of T and vice versa. The drawback is that the transmitted signal energy is inversely proportional to its Doppler tolerance. A short duration, range resolvent, but Doppler tolerant CW signal will have relatively little energy for detection in noise limited situations. Detection must precede Doppler measurement and it is desirable to maintain signal energy in the presence of noise hence the short low energy tone is at a disadvantage.

In the early days of radar this problem led to the pulse compression concept and the utilization of the LFM waveform to obtain range resolution without energy loss. The LFM recovers signal energy but its ambiguity function is the highly coupled ridge shown in Figure 4.8. It is not Doppler tolerant because a slice of the AF on the Doppler axis has a width that is proportional to the inverse of its duration. However the ridge will intersect the delay axis and a Doppler shifted echo can generate a significant response on the axis if the Doppler shift ϕ is small with respect to the waveform bandwidth. The delay at which the zero Doppler response occurs is shifted from the actual reflector delay by $\tau = \phi T/W$. When the Doppler shift $\phi = (\beta - 1)\omega_0/2\pi$ is small with respect to the bandwidth, the delay uncertainty is correspondingly small and can be tolerated in many cases. In this sense the LFM waveform may be considered as Doppler tolerant because a measurable response can be obtained by a zero Doppler matched filter for echos with a considerable Doppler range.

5.6.1 HFM/LPM Waveforms

There are situations in which Doppler induced delay error can not be tolerated an example of which is the the search for food by echo locating mammals such as bats and porpoises. Bats have evolved a Doppler tolerant (bionic) sonar that is particularly adapted to the accurate echo ranging of rapidly moving insects. Tests with live bats revealed that they utilize a hyperbolic frequency modulation (HFM) or equivalently, linear period modulation (LPM) waveform of the form (4.45) [2]

$$f(t) = a(t) \exp jk \ln t.$$

It has a nominal frequency centroid of about 70 kHz, bandwidth of 60 kHz [10,44,45] and would be considered to be a very wideband waveform by medium and high frequency sonar standards. These waveforms have been mathematically proven to be optimally Doppler tolerant by Altes and Titlebaum [10].

The dilated and delayed HFM waveform function with a slowly varying window $a(t)$ is

$$\begin{aligned} f(\beta(t-\tau)) &= \sqrt{\beta} a(\beta(t-\tau)) \exp j[k \ln(\beta(t-\tau))] \\ &\approx [a(t-\tau) \exp jk \ln(t-\tau)] \exp jk \ln \beta \\ &= (\exp jk \ln \beta) f(t-\tau). \end{aligned}$$

The dilation effect enters as a phase shift and the square law detected ambiguity function is

$$\Phi_{ff}(\tau) = |\chi_{ff}(\beta,\tau)|^2 = \left| \int_{-\infty}^{\infty} f(t) f^*(t-\tau) dt \right|^2$$

[2]The terms HFM and LPM are equivalent and used interchangeably

and only a function of τ and hence is Doppler tolerant. Extraction of velocity information from an LPM echo is done by finding the peak of Φ_{ff} in delay τ and then evaluating the phase ($\exp jk \ln \beta$) [46].

The waveform function is defined in terms of pulse duration and bandwidth by defining the band ratio $r_w = \pi W/\omega_0 \leq 1$ of a hyperbolic frequency function with centroid ω_0. The minimum and maximum radian frequencies are ω_{min}, ω_{max} respectively and the design bandwidth is $W = |\omega_{max} - \omega_{min}|/2\pi$. The HFM/LPM radian frequency function is [7, 47]

$$\omega(t) = \frac{\omega_0(1 - r_w^2)}{1 + r_w\left(1 - \frac{2t}{T}\right)} \quad t \in [0, T] \tag{5.31}$$

and defines an *upsweep* HFM where ω_{min} occurs at $t = 0$. The corresponding upsweep HFM waveform function with the real window $a(t)$ and complex phase function $\psi(t)$ is[3]

$$\begin{aligned} f(t) &= a(t) \exp j\psi(t) \\ \psi(t) &= \frac{\omega_0(r_w^2 - 1)T}{2r_w} \ln\left[1 + r_w\left(1 - \frac{2t}{T}\right)\right]. \end{aligned} \tag{5.32}$$

LPM/HFM waveforms have the unique property that a dilated segment of the frequency function corresponds to a time shifted segment of the undilated function. The time translation is determined by equating the dilated and undilated frequency functions with an added delay increment τ_s. The dilated phase is

$$\psi(\beta t) = \frac{\omega_0(r_w^2 - 1)T}{2r_w} \ln\left[1 + r_w\left(1 - \frac{2\beta t}{T}\right)\right]$$

and the dilated frequency is the time derivative

$$\omega(\beta t) = \dot{\psi}(\beta t) = \frac{\omega_0(1 - r_w^2)\beta}{1 + r_w\left(1 - \frac{2\beta t}{T}\right)}, \quad t \in [0, T/\beta].$$

τ_s is obtained by equating a shifted version of this to (5.31) where

$$\frac{\omega_0(1 - r_w^2)\beta}{1 + r_w\left(1 - \frac{2\beta(t-\tau_s)}{T}\right)} = \frac{\omega_0(1 - r_w^2)}{1 + r_w\left(1 - \frac{2t}{T}\right)}. \tag{5.33}$$

[3]The discussion pertains to the upsweep case but comparable relationships for a *downsweep* HFM with ω_{max} occurring at $t = 0$ are obtained by changing the sign of r_w.

The resulting coincidence time shift for the upsweep HFM is

$$\tau_s = \frac{(\beta - 1)(1 + r_w)T}{2r_w\beta} \approx \frac{v(1 + r_w)T}{r_w c}. \quad (5.34)$$

$v > 0$ is associated with closing range, $\beta > 1$, and positive Doppler frequency shift.

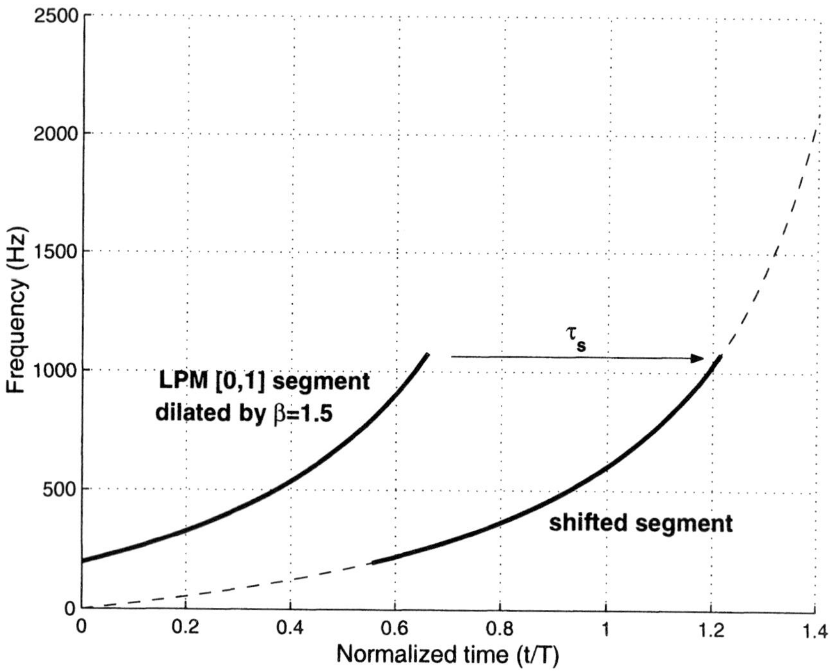

Figure 5.29: Illustration of dilated and delay shifted LPM frequency segment matching for $\beta = 1.5$ and a waveform with r_w=.43

τ_s is positive (greater delay) for an upsweep HFM echo reflected from a closing scatterer as illustrated by Figure 5.29. The plot represents the frequency function for an upsweep HFM waveform with unspecified duration T, W=600Hz, f_0=700Hz and dilated by β=1.5. The ordinant of the figure is scaled relative to the minimum frequency which is 400 Hz and represents $(\omega(t/T) - \omega_{min})/2\pi$. The undilated HFM frequency function is the $t/T \in [0, 1]$ segment of the unbounded hyperbola $\omega(t)$ defined on $(t/T) \in [-\infty, (1 + r_w)/(2r_w)]$ and indicated by the dashed curve in the figure. The dilated segment is indicated in bold and exhibits the shortening of duration and bandwidth expansion consistent with $\beta =$

1.5. This corresponds to an echo reflected from an object traveling at c/4 or a speed of 375 m/sec in water and 75 m/sec in air. It is perhaps outside the realm of most sonar applications but easily attained in air by many aircraft and some race cars.

The dilated frequency function segment $\omega(\beta t)$ is identical to the undilated segment $\omega(t)$ for $t \in [\tau_s, \tau_s + (T/\beta)]$ and is indicated in the figure by the coincidence of the two segments. The quantity $(1 + r_w)/2r_w \geq 1$ because $r_w \leq 1$ and consequently $\tau_s \geq T(\beta - 1)/\beta$ or equivalently, $\tau_s + (T/\beta) \geq T$. The trailing edge of the dilated and shifted segment exceeds the trailing edge of the undilated segment when they are shifted into coincidence and the waveforms overlap only for $t \in [\tau_s, T]$. The HFM/LPM ambiguity ridge is defined in delay-Doppler by (5.34). Since the frequency segments match in the overlap region, the AF amplitude on the ridge is proportional to the square of the time overlap by the principle of stationary phase and

$$\Phi_{ff}(\beta, \tau_s) \propto \left(1 - \frac{\tau_s}{T}\right)^2 .$$

This is proven directly from the ambiguity function by evaluating it at the τ_s corresponding to a particular β. Let the waveform be uniformly weighted by

$$a(t) = \frac{1}{\sqrt{T}}, \quad t \in [0, T]$$

for simplicity and define

$$c_1 = \frac{\omega_0 (r_w^2 - 1) T}{2 r_w}$$

for the constant in the phase function (5.32). Note that the phase function can be written

$$\psi(t) = c_1 \ln \left[\beta + r_w \beta \left(1 - \frac{2t}{T}\right)\right]$$

and that

$$c_1 \ln \left[\beta + r_w \beta \left(1 - \frac{2t}{T}\right)\right] = c_1 \ln \left[1 + r_w \left(1 - \frac{2\beta(t - \tau_s)}{T}\right)\right]$$

by the coincidence shift relationship (5.33). The AF is

$$\Phi_{ff}(\beta, \tau_s) = \left| \frac{\sqrt{\beta}}{T} \int_{\tau_s}^{T} \exp j[\psi(t) - \psi(\beta(t - \tau_s))] dt \right|^2$$

$$= \left| \frac{\sqrt{\beta}}{T} \int_{\tau_s}^{T} \exp j c_1 \left\{ -\ln\beta + \ln\left[\beta + r_w \beta \left(1 - \frac{2t}{T}\right)\right] \right. \right.$$
$$\left. \left. - \ln\left[1 + r_w\left(1 - \frac{2\beta(t-\tau-_s)}{T}\right)\right] \right\} dt \right|^2.$$

The logarithmic time functions within the exponential are equal and cancel leaving

$$\Phi_{ff}(\beta, \tau_s) = \left| \frac{\sqrt{\beta}}{T} \int_{\tau_s}^{T} \exp j(-c_1 \ln\beta) dt \right|^2 = \beta\left(1 - \frac{\tau_s}{T}\right)^2 \tag{5.35}$$

which is consistent with the stationary phase result.

The delay shift τ_s=.555 obtained from (5.34) corresponds to $r_w =$ 6/14 or.4285 for the example in the figure and

$$\Phi_{ff}(1.5, .555) = 1.5(1 - (.555/1))^2 = .3.$$

This is demonstrated by the ambiguity function illustrated in Figure 5.30 that exhibits the narrow ridge of width $1/W$ along the dilation-delay contour determined by (5.34) and amplitude determined by (5.35).

The shift τ_s decreases inversely with the band ratio r_w for a fixed Doppler velocity and consequently there is more overlap with the shifted segment as r_w increases. There is then less attenuation of the ambiguity function, it becomes less sensitive to Doppler, and approaches a constant amplitude "razor blade" aligned along the Doppler axis.

The slope of the undilated HFM frequency function is

$$\omega'(t) = \frac{\frac{2\omega_0 r_w}{T}\left(1 - r_w^2\right)}{\left(1 - r_w\left(\frac{2t}{T} - 1\right)\right)^2}$$
$$= \frac{2\omega_0 r_w(1 - r_w^2)}{T} \quad \text{at } t = T/2$$

and

$$\lim_{r_w \to 0} \omega'(T/2) = \frac{2\pi W}{T}.$$

The frequency function approaches that of the linear FM with constant slope for small values of the band ratio $r_w = \pi W/\omega_0$. Likewise the LFM and LPM ambiguity functions are virtually identical under narrowband conditions but diverge in form as the bandwidth to carrier ratio and the range of β becomes large. This occurs because the LFM time-frequency

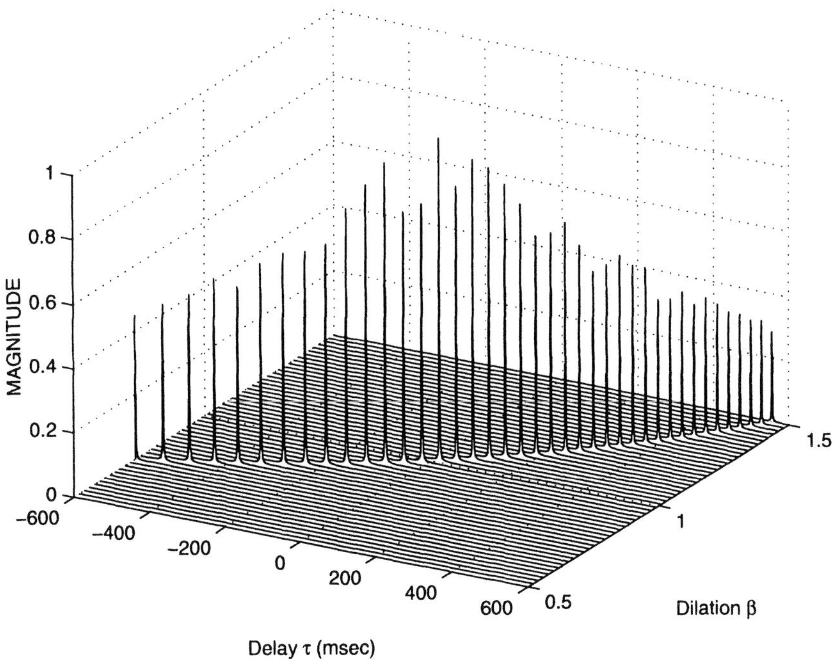

Figure 5.30: Ambiguity function of the LPM/HFM waveform with T=1 s, W=600 (Hz), and r_w=.43 illustrated in Figure 5.29

slope increases with β and the dilated LFM frequency function is coincident with the undilated function at only one point. The dilated phase and frequency functions for an LFM waveform defined on $[0, T]$ are

$$\psi(\beta t) = \frac{\pi W \beta^2 t^2}{T} + \omega_0 \beta t$$
$$\omega(\beta t) = \dot{\psi}(\beta t) = \frac{2\pi W \beta t}{T/\beta} + \beta\omega_0, \quad t \in [0, T/\beta].$$

The time translation that brings the dilated and undilated LFM frequency functions into coincidence is determined by solving $\omega(\beta(t-\tau_s)) = \omega(t)$ for τ_s yielding

$$\tau_s = \frac{(\beta^2 - 1)t}{\beta^2} + \frac{T}{2r_w}\left(\frac{\beta - 1}{\beta^2}\right).$$

τ_s is not constant as it is for the LPM (5.34) but a function of time and the dilation of an LFM segment is not equivalent to a uniform time translation of a different segment. No matter how much the LFM band ratio is increased, the dilated and undilated waveforms will have different time-frequency slope and cannot be brought into coincidence everywhere by a fixed time shift as is the case with the LPM.

By the stationary phase principle, the wideband auto-ambiguity function of a dilated and shifted LFM will experience attenuation in addition to that caused by partial time overlap because the frequencies are not coincident in the overlap region. Its magnitude is approximated by (4.59) where the duration and bandwidth of the dilated waveform are T/β and βW and the AF within the ridge is

$$\begin{aligned}\Phi_{ff}(\beta,\tau) &\approx \left(\frac{T^2}{\beta}\left|\frac{W}{T}-\frac{\beta^2 W}{T}\right|\right)^{-1} \\ &= \frac{\beta}{TW|1-\beta^2|} = \frac{1}{TW}\left(\frac{1-(v/c)^2}{4v/c}\right), \ (\beta,\, v) \neq 0. \quad (5.36)\end{aligned}$$

The ambiguity function of an LFM with the same parameters as the LPM example of Figures 5.30, 5.29 is plotted in Figure 5.31.

The figure is plotted in dB in order to illustrate the severe attenuation caused by the dilated LFM frequency slope mismatch. The AF with $TW = 600$ computed from (5.36) at ($\beta = 1.5$, $\tau_s = .555$) is down by 27 dB in relation to the origin and is considerably spread in delay as is evident from the figure. The LPM at the same point is attenuated by only about 5 dB and suffers no delay spreading because the shifted segments match exactly.

The extreme Doppler range condition $\beta \in [.5,\, 1.5]$ illustrates the difference between finite duration LFM and LPM/HFM waveforms. The example of Figures 5.29, 5.30, 5.31 demonstrates that while neither is Doppler tolerant in the sense that the ambiguity ridge aligns exactly along the Doppler axis, the LPM is much more tolerant of Doppler because it is not attenuated by the frequency slope mismatch of the LFM. A comparison between LFM and LPM ambiguity functions for more reasonable sonar conditions is shown in Figure 5.32. The expected Doppler range is $\beta = \pm.02$ corresponding to about ± 15 m/s in water and the delay resolution is 1.5 meters which translates to a bandwidth of 500 Hz. The illustrations represent LPM and LFM waveforms operating at 1 and 10 kHz with durations of .2 sec and 2 sec respectively. The range-velocity error of the LFM is determined by the slope of the ambiguity ridge

$$\frac{\Delta r}{\Delta v} = \frac{\Delta \tau}{\Delta \beta} = \frac{T}{2r_w} \quad \text{(LFM)}.$$

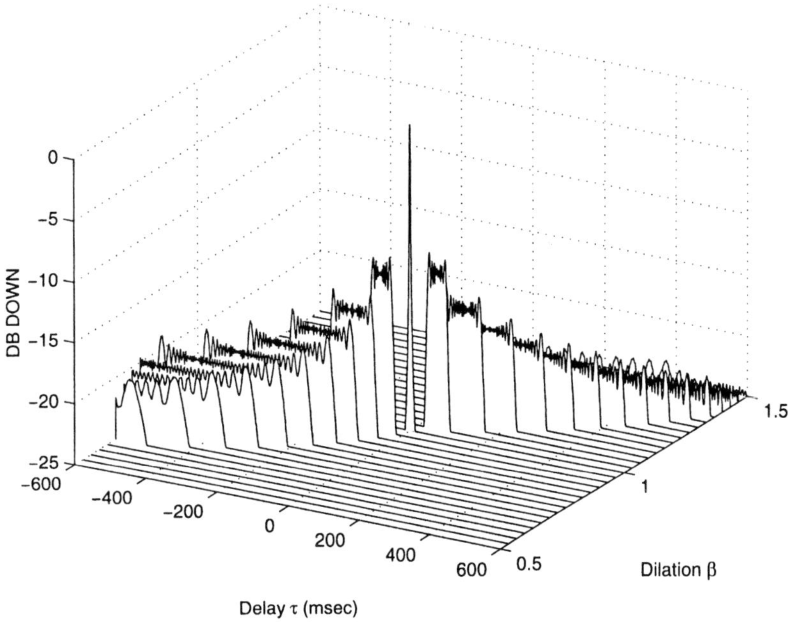

Figure 5.31: Ambiguity function of an LFM waveform with T=1 s, W=600 (Hz), and r_w=.43 that is comparable to the LPM of Figure 5.30

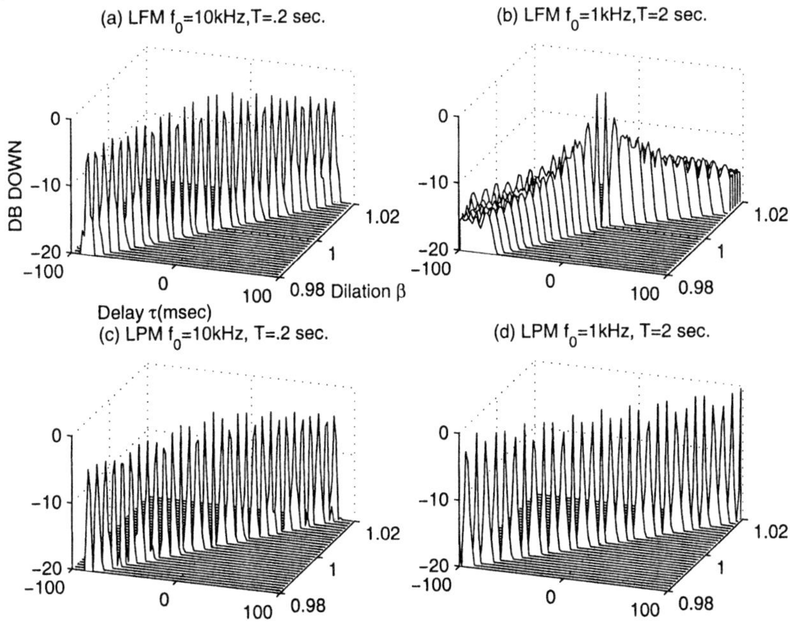

Figure 5.32: Comparison of LFM and LPM/HFM ambiguity functions with r_w = .025, .25, $\triangle v = \pm$ 15 m/sec., and W=500 Hz

This is 4 (m/(m/sec)) for both the low and high frequency LFM examples and is interpreted to mean that a matched filter formed from a zero Doppler replica will respond to an object with a range error of 4 meters for every meter/sec velocity.

The ridge of the LPM/HFM

$$\frac{\Delta r}{\Delta v} \approx \frac{f_0 T(1+r_w)}{W} \quad \text{(LPM)}$$

is obtained from the nonlinear relationship (5.34) and approximates that of the LFM for small r_w as indicated by subplots (a,c). Subplots (a,c) and (b,d) represent relatively narrowband and wideband conditions with $r_w = .025, .25$ respectively. Under the narrowband condition that is more typical of high frequency sonar, the ambiguity functions are nearly identical. The LFM becomes severely attenuated with Doppler under the more wideband condition. The stationary phase LFM approximation can be used as a design tool for the determination of the useful velocity range for the LFM by specifying the tolerable $\Phi_{ff} \leq 1$ and solving (5.36) for the velocity.

5.6.2 Doppler Resolvent Dilation Tolerant Waveforms

There is another side to the issue of Doppler tolerance that is concerned with the design of waveforms for matched filtering that resolve Doppler yet permit more efficient narrowband implementation of multiple matched filter hypotheses in both dimensions. Such waveforms are characterized by a low value of the dilation sensitivity coefficient D_s (4.49). A receiver that implements the wideband (dilation) model is inherently computationally expensive because a dilated replica time function is required for each Doppler hypothesis. The narrowband receiver for the slowly fluctuating point scatterer is simpler because it assumes the signal model (2.29) and forms the statistic (3.58

$$|\mathsf{r}_1|^2 = \left| \int_{\Omega_t} \mathsf{r}(t) f^*(t-\tau) \exp j2\pi\phi t dt \right|^2$$

that is implemented by advancing the delayed waveform replica through the data time series $\mathsf{r}(t)$ and using the fast Fourier transform (FFT) to test Doppler at each delay τ. This approach is sometimes called *dechirping* which referred to the demodulation of an LFM (chirp) waveform in the early days of radar and is carried out in two stages. First the data is "dechirped" by multiplying by a single conjugate replica $f^*(t-\tau)$ and then passed to a spectrum analyzer for the Doppler search. The assumption is made that to some measurable degree $f(t) \approx \sqrt{\beta} f(\beta t)$ and that a certain amount of performance degradation is acceptable as a tradeoff to making an efficient narrowband Doppler search about a nominal dilation β. The full Doppler hypothesis range is searched with a few dilated replicas $\sqrt{\beta_k} f(\beta_k(t-\tau))$ about which FFT processing is implemented.

LPM/HFM and to some extent LFM waveforms with ridge-like ambiguity functions are Doppler tolerant because a significant response occurs on a relatively short delay interval over a wide range of Dopplers. On the other hand, it may be desired to estimate both delay and Doppler with precision as is done with FSK or PRN sequences yet use a waveform with a low value of D_s that permits narrowband receiver implementation over a range of Dopplers with minimal performance degradation. Such a waveform resolves Doppler but is called *dilation tolerant* to distinguish it from Doppler tolerant waveforms such as HFM that do not.

$f(t) = a(t) \exp \Psi(t)$ is the bandpass waveform function with slowly varying even amplitude window $a(t)$ and odd phase $\Psi(t)$ and consists of envelope and carrier phase terms $\Psi(t) = \psi(t) + \omega_0 t$. The complex envelope is $\tilde{f}(t) = a(t) \exp j\psi(t)$ and dilation but not Doppler tolerance suggests that it be uncoupled with an even frequency function $\omega(t) = \dot{\psi}(t)$, specified TW product, a low dilation coefficient D_s, and specified tolerance range (4.49). The dilated waveform is

$$f(\beta t) = \tilde{f}(\beta t) \exp j\omega_0 \beta t = a(\beta t) \exp j\psi(\beta t) \exp j\omega_0 t \exp j2\pi\phi t$$

and dilation tolerance implies that for some range of β, $\tilde{f}(\beta t) \approx \tilde{f}(t)$ and

$$\begin{aligned} f(\beta t) &\approx [a(t) \exp j\psi(t) \exp j\omega_0 t] \exp j2\pi\phi t \\ &= f(t) \exp j2\pi\phi t \end{aligned}$$

with $\beta \approx 1 + 2v/c$ and $\phi = (\beta - 1) f_0$. Doppler frequency resolution of the bandpass function is then determined by the rms pulse duration (4.40).

A symmetric version of the LPM/HFM waveform with the form $f(t) = a(t) \exp j\Psi(t)$ with

$$\begin{aligned} a(t) &= \alpha_1 |t|^{-1/2} \exp[-\alpha_2 (\ln|\alpha_3 t|)^2] \\ \Psi(t) &= \alpha_4 \ln|t| + \alpha_5 \qquad -\infty < t < \infty \end{aligned} \tag{5.37}$$

has been proposed [46] where α_{1-5} are real constants. This version of the LPM has the advantage that its real and imaginary parts are a

Hilbert transform pair and phase analysis can be used to extract Doppler information. It is also an even function so that its ambiguity function is uncoupled and is hence a minimum variance estimator. It has a frequency pole at the origin but that portion of the waveform is attenuated by the window function.

An alternate approach that does not require attenuation at the origin seeks a dilation tolerant even bandpass frequency function by the constrained minimization of the Doppler sensitivity coefficient (4.43) [48].

$$\eta^2 \approx \kappa_2(a, \nu(t)) = \int_{-T/2}^{T/2} t^2 a^2(t)\nu^2(t)dt - \left[\int_{-T/2}^{T/2} ta^2(t)\nu(t)dt\right]^2$$

where $\nu(t)$ is the radian frequency function to be determined and the second (squared integral) term is zero for even $\nu(t)$. The desired waveform is a minimum variance estimator, has a specified delay resolution, and utilizes an even but unspecified window function that is not necessarily zero at the origin. The mean square delay resolution $\sigma_\tau^2 \propto 1/W_\omega^2$ (4.42) is specified by imposing a constraint on the mean square radian bandwidth

$$W_\omega^2 = \int_{-T/2}^{T/2} a^2(t)\nu^2(t)dt - \left[\int_{-T/2}^{T/2} a^2(t)\nu(t)dt\right]^2 \tag{5.38}$$

that is obtained from (4.39) assuming that the window contribution is negligible. The optimum $\nu(t)$ minimizes the goal function

$$\begin{aligned}\mathcal{J}(\nu(t)) = &\int_{-T/2}^{T/2} t^2 a^2(t)\nu^2(t)dt \\ &+ \mu\left\{\int_{-T/2}^{T/2} a^2(t)\nu^2(t)dt - \left[\int_{-T/2}^{T/2} a^2(t)\nu(t)dt\right]^2 - W_\omega^2\right\}\end{aligned}$$

with Lagrange multiplier μ under the assumption that the frequency and window functions are even. The squared integral within the constraint is the amplitude weighted frequency and is denoted as L^2 so that the goal function is compactly written as

$$\begin{aligned}\mathcal{J}(\nu(t)) &= \int_{-T/2}^{T/2} a^2(t)[t^2\nu^2(t) + \mu\nu^2(t) - \mu L\nu(t)]dt - \mu W_\omega^2 \\ &= \int_{-T/2}^{T/2} \mathcal{J}_1(\nu(t))dt - \mu W_\omega^2.\end{aligned}$$

The optimum $\nu(t)$ is found as the solution to the Euler equation [28]

$$\frac{\partial \mathcal{J}_1(\nu)}{\partial \nu} = a^2(t)\left[2t^2\nu(t) + 2\mu\nu(t) - \mu L - \mu\nu(t)\frac{\partial L}{\partial \nu}\right] = 0$$

where

$$\frac{\partial L}{\partial \nu} = \int_{-T/2}^{T/2} a^2(t)dt = 1$$

for energy normalized waveforms. Solving for $\nu(t)$ yields

$$\nu(t) = \frac{\mu L}{2t^2 + \mu} > 0, \quad t \in [-T/2, T/2] \tag{5.39}$$

which is a bandpass function that does not necessarily correspond to a global minimum of η^2 because the frequency at the end points is not specified. The global minimum is determined by subtracting a constant frequency offset to form the baseband (envelope) frequency function

$$\omega(t) = \nu(t) - \bar{\nu}$$

and minimizing the envelope sensitivity coefficient

$$\tilde{\eta}^2 = \int_{-T/2}^{T/2} a^2(t)t^2(\nu(t) - \bar{\nu})^2 dt$$

with respect to $\bar{\nu}$. The result is

$$\begin{aligned}\bar{\nu} &= \frac{\int_{-T/2}^{T/2} a^2(t)t^2\nu(t)dt}{\int_{-T/2}^{T/2} a^2(t)t^2 dt} \\ &= 3\int_0^1 u^2\nu'(u)du \end{aligned} \tag{5.40}$$

for the uniform window $a(t) = 1/T$ and normalized time $u = 2t/T$. The frequency function (5.39) is expressed as

$$\nu'(u) = \nu\left(\frac{\mu T}{2}\right) = \frac{2\pi W\zeta^2(1+\zeta^2)}{u^2 + \zeta^2} \tag{5.41}$$

by defining the shape parameter $\zeta = \sqrt{2\mu}/T$ and the design bandwidth or maximum frequency range

$$\begin{aligned}W &= (\omega(0) - \omega(T/2))/2\pi \\ &= \frac{L}{2\pi(1+\zeta^2)}.\end{aligned}$$

The baseband frequency offset

$$\bar{\nu} = 3(2\pi W)(1+\zeta^2)\zeta^2[1-\zeta\tan^{-1}(1/\zeta)] \tag{5.42}$$

is then obtained by carrying out the integration (5.40) using (5.41). The dilation tolerant envelope modulation

$$\begin{aligned} \omega(u) &= \nu'(u) - \bar{\nu} \\ &= 2\pi W \zeta^2 (1+\zeta^2)\left[\frac{1}{u^2+\zeta^2} - 3(1-\zeta\tan^{-1}(1/\zeta))\right] \\ &\text{for } u \in [-1,1] \end{aligned} \tag{5.43}$$

is obtained by combining (5.41, 5.42). Adding a carrier frequency term ω_0 produces the bandpass waveform frequency function $\Omega(u) = \omega(u)+\omega_0$ (4.37) that largely determines the waveform Doppler resolution. The corresponding phase function is obtained from

$$\Psi(t) = \int_{-T/2}^{T} \nu(x)dx - (\bar{\nu} - \omega_0)(t - T/2)$$

and is expressed in terms of ζ, normalized time u, and the bandwidth to carrier ratio $r_w = 2\pi W/\omega_0$ as

$$\begin{aligned} \Psi(u) = &\frac{\omega_0 T r_w}{2}\left[\zeta(1+\zeta^2)\tan^{-1}(u/\zeta) - 3\zeta^2(1+\zeta^2)(1-\zeta\tan^{-1}(1/\zeta))u\right] \\ &+ \left(\frac{\omega_0 T}{2}\right)u + \psi_0 \end{aligned} \tag{5.44}$$

where

$$\psi_0 = \pi T W \zeta(1+\zeta^2)\tan^{-1}(1/\zeta) + (\bar{\nu} - \omega_0)(T/2)$$

is a constant phase term.

The passband waveform dilation resolution is $\sigma_\beta^2 \propto 1/\eta^2$ and

$$\begin{aligned} \eta^2 &= \frac{T^2}{4}\int_0^1 u^2(\omega(u)+\omega_0)^2 du \\ &= \frac{T^2\omega_0^2}{4}\int_0^1 u^2\left(\frac{\omega(u)}{\omega_0}+1\right)^2 du \\ &\approx \omega_0^2\frac{T^2}{12} = \omega_0^2 T_{rms}^2 \end{aligned}$$

for the uniform window if the design bandwidth W is small in relation to the carrier frequency and $\omega(u)/\omega_0 << 1$. The narrowband Doppler frequency ϕ (4.40) is

$$\eta_{nb}^2 = (2\pi)^2 T_{rms}^2$$

and $\eta^2 = \omega_0^2 \eta_{nb}^2 / (2\pi)^2$. This illustrates the dilation tolerant property whereby the envelope resolves delay in proportion to its mean square bandwidth yet is optimally dilation insensitive permitting narrowband implementation over a maximum Doppler range. Doppler frequency resolution is determined by the rms pulse duration when the bandwidth to carrier ratio is small and the waveform can be regarded as narrowband. Delay and Doppler estimates are unbiased because the bandpass waveform is even and hence uncoupled. The normalized envelope frequency

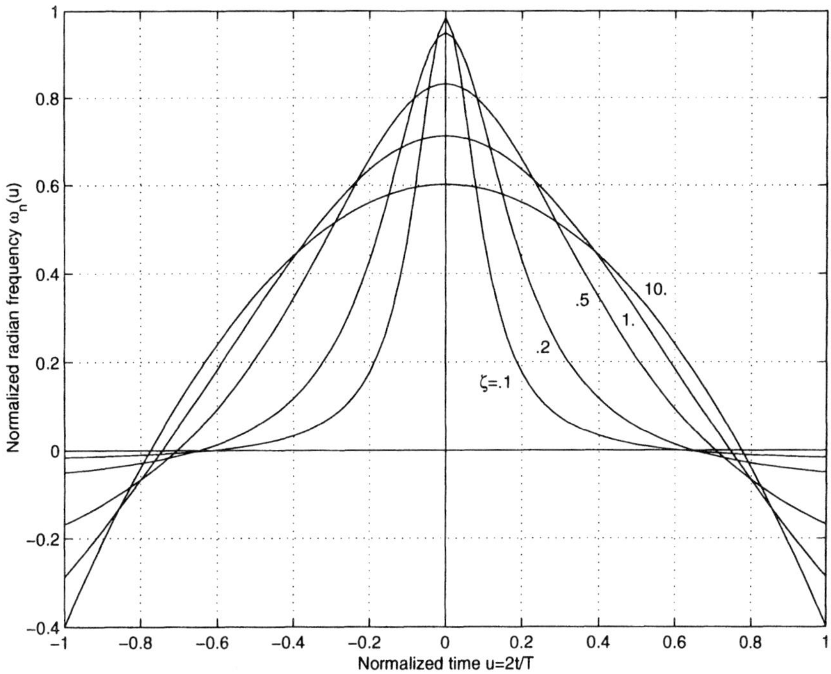

Figure 5.33: Normalized baseband modulation $\omega(u)/2\pi W$ of the even maximally dilation tolerant waveform (5.43) plotted as a function of the shape parameter ζ

function $\omega_n(u) = \omega(u)/2\pi W$ is plotted for several values of ζ in Figure 5.33. Note that while the frequency span (design bandwidth) of each function is the same, the frequency distribution varies with the shape parameter ζ that determines the mean square radian bandwidth W_ω (5.38). It in turn determines the delay resolution of the waveform. Small values of ζ are associated with a sharply "peaked" frequency profile that approaches a bounded impulse as $\zeta \to 0$. $\omega(u)$ becomes parabolic for large values of ζ because $\nu'(u) = 2\pi W(1+\zeta^2-u^2)$ which is derived by expanding (5.41) as a Taylor series about $u=0$. The frequency profiles are consistent with the observation made in Chapter 4 that dilation tolerance of symmetric waveforms is increased by concentrating the higher frequency components near the pulse mid point and vice versa. Note that reversing the sign of μ in (5.39) or ζ in (5.41) produces a frequency function with its maximum magnitude at the end points implying greater dilation sensitivity. The waveform delay resolution $\sigma_\tau^2 \propto 1/W_\omega^2$ and the mean square bandwidth (5.38) is related asymptotically to the shape parameter ζ by [48]

$$W_\omega^2 \approx \begin{cases} .089(2\pi W)^2 & \text{for } \zeta >> .1, \\ (\pi/4)\zeta(2\pi W)^2 & \text{for } \zeta << .1. \end{cases} \tag{5.45}$$

Delay resolution thus increases in proportion to ζ for $\zeta << .1$ but approaches a constant maximum for fixed W when $\zeta >> .1$ and the frequency profile becomes parabolic.

The envelope dilation sensitivity for uniform weighting is obtained from (5.41, 5.42) where

$$\begin{aligned}
\tilde{\eta}^2 &= \frac{T^2}{4}\int_0^1 u^2(\nu'(u)-\bar{\nu})^2 du \\
&= \frac{T^2}{4}\int_0^1 u^2\nu'(u)^2 du - \frac{T^2}{12}\bar{\nu}^2 \\
&= (TW)^2 D_s \\
D_s &= \pi^2\zeta^2(1+\zeta^2)^2 \\
&\cdot \left[.5\left(\zeta\tan^{-1}(1/\zeta) - \frac{\zeta^2}{1+\zeta^2}\right) - 3\zeta^2(1-\zeta\tan^{-1}(1/\zeta))^2\right]
\end{aligned} \tag{5.46}$$

D_s is the dilation sensitivity factor (4.49) that is plotted as a function of ζ in Figure 5.34. It is asymptotic to

$$D_s = \frac{(\pi\zeta)^3}{4}$$

for small values of ζ and approaches a maximum of about .22 for $\zeta >> 1$ corresponding to the limiting parabolic frequency profile. The bandpass frequency function is $\Omega(t) = \omega(t) + \omega_0$ and as ζ becomes small, the envelope frequency profile (5.33) becomes impulsive, D_s decreases (dilation tolerance increases), and the bandpass frequency function becomes dominated by the steady carrier term except for an impulsive frequency shift

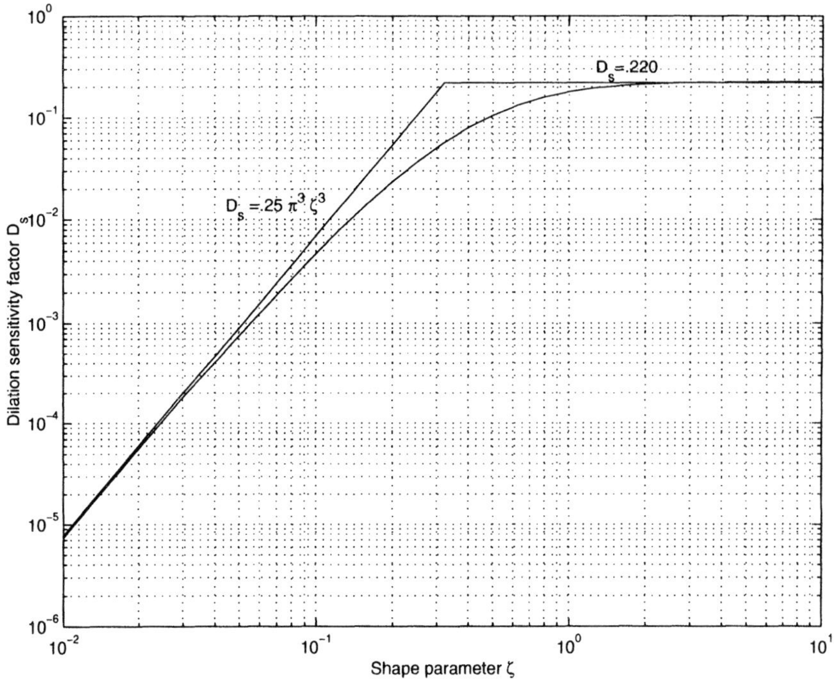

Figure 5.34: The dilation sensitivity factor D_s (5.46) as a function of the shape parameter ζ

of W (Hz) around the origin. The energy contribution of the bounded frequency impulse at the origin decreases with ζ, the waveform approaches a CW of duration T, and the delay resolution decreases with the mean square bandwidth (5.45). The phase modulation has less of an effect on the Doppler resolution as D_s is diminished and it becomes a function of only the rms duration and carrier frequency.

The velocity range over which it is possible to use the narrowband approximation is (4.49)

$$|\triangle v| = \frac{c}{2TW}\sqrt{\frac{1-\Phi_{ff}(max)}{D_s}}.$$

As an example, assume that it is desired that the ambiguity attenuation not exceed 10% for the dilation tolerant waveform with $TW = 100$

and $\zeta > 1$. The velocity range is

$$|\triangle v| \leq \frac{750}{100}\sqrt{\frac{1-.9}{.220}} = 5.05(\text{m/sec}).$$

In order to bound the attenuation when covering a greater velocity range $\pm V$ it is necessary to synthesize several wideband replicas $\sqrt{\beta_k}\tilde{f}(\beta_k t)$ using the exact phase function $\psi(\beta_k t)$. These are spaced at intervals

$$\frac{\beta_{k+1}}{\beta_k} = \frac{c + 2\triangle v}{c - 2\triangle v}$$

where the number of discrete intervals is $2V/\triangle v$. Within any interval centered at β_k, Doppler is tested using the narrowband replica $\sqrt{\beta_k}\tilde{f}(\beta_k)\exp j2\pi\phi$ with

$$\phi \in \pm\left(\frac{2\triangle v\beta_k f_0}{c}\right).$$

This can result in substantial computational saving because wideband testing over the whole range requires approximately $(1+V/c)^2 f_0 T$ dilated replicas. This implementation approach is not limited to optimally dilation tolerant waveforms but can be applied to any phase modulated waveform. For example, $D_s = .123$ for VFM waveforms and $.014 \leq D_s \leq .57$ for FSK codes [49]. The advantage of the optimal waveforms is that they minimize $\tilde{\eta}^2$ and D_s for a given mean square bandwidth.

Dilation tolerant ambiguity function examples corresponding to the envelope phase function (5.44) for ζ=.1,10 are shown in Figure 5.35. The waveform is configured as a bandpass function with a rectangular window, r_w=.05, and $TW = 100$ for comparison with the high resolution PRN and Costas codes of Figures 5.23, 5.24, 5.26. The ambiguity functions illustrate the variation in delay resolution with the shape parameter ζ. Small values of ζ are associated with a more impulse-like frequency profile and lower mean square bandwidth as indicated by (5.45) and Figure 5.33. The corresponding AF of Figure 5.35a exhibits broad delay sidelobes that approach those of a CW waveform as $\zeta \to 0$. The frequency profile becomes more parabolic for large values of ζ and the sidelobes as indicated by Figure 5.35b begin to resemble those of the VFM of Figure 5.2.

It is also possible to specify an optimal dilation tolerant odd waveform function by minimizing the goal function [48]

$$\mathcal{J}(\nu(t)) = \int_{-T/2}^{T/2} t^2 a^2(t)\nu^2(t)dt - \left[\int_{-T/2}^{T/2} a^2(t)t\nu(t)dt\right]^2$$

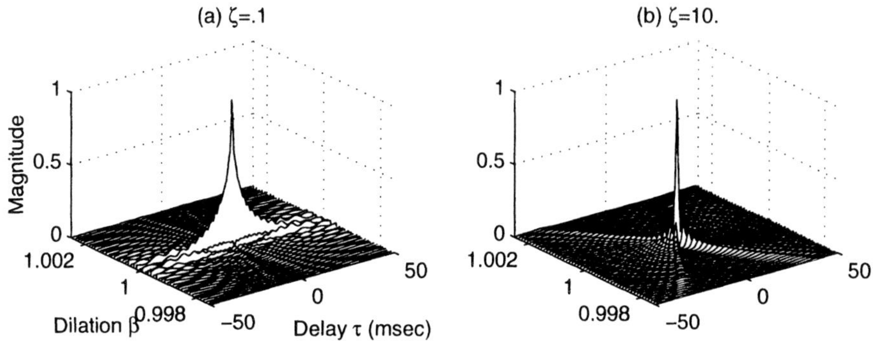

Figure 5.35: Dilation tolerant ambiguity functions for $T = .2$ sec, $W = 500$ Hz, $r_w = .05$

$$+\mu\left\{\int_{-T/2}^{T/2} a^2(t)\nu^2(t)dt - W_\omega^2\right\}$$

yielding

$$\nu(t) = \frac{Lt}{2(t^2+\mu)}, \quad t \in [-T/2, T/2].$$

This approaches the odd version of the symmetric LPM/HFM (5.37) as $\mu \to 0$ and resembles an LFM near the origin and an HFM as $t >> \mu$. It is somewhat less dilation tolerant than the even function owing in part to the fact that it is not an unbiased estimator of delay and Doppler.

5.7 SIMULTANEOUS TRANSMISSION

The transmission of multiple time overlapping waveform functions permits multiple receivers to be simultaneously synthesized. The individual waveforms occupy different frequency bands and are directly summed prior to transmission. Upon reception the signal components are separated by bandpass filtering and the filtered time sequences are individually processed. Examples of applications of this technique are the simultaneous transmission of a Doppler resolvent CW and a delay resolvent LFM without generating an overly long concatenated pulse and frequency diversity detection. Any combination of waveforms that can be individually transmitted can be directly added to form a simultaneous waveform. The most common combinations are sums of $N \geq 2$ CW waveforms of duration T/N sometimes called *simultones or frequency combs* that are used for detection in reverberant or fading environments [50]. They are members of the class of FSK waveforms but are transmitted simultaneously rather than as a concatenated pulse train. The advantage is that the Doppler resolution of the individual subpulses is proportional to $1/T$ not N/T as it is for concatenated pulses. The disadvantage is that addition introduces a strongly amplitude modulated waveform and decreased signal energy when peak power is bounded. A rectangularly weighted simultone of order N is defined as the complex sum

$$f(t) = F_0 \sum_{k=1}^{N} \exp j\omega_k t, \quad t \in (0,T)$$

where the frequency increments may be arbitrarily chosen to cover a desired band. More specifically the uniform frequency increment passband form is

$$f(t) = F_0(\exp j\omega_0 t) \sum_{k=-(N-1)/2}^{(N-1)/2} \exp jk\Delta\omega t, \quad t \in (0,T)$$

where ω_0 is the frequency centroid, $\Delta\omega = 2\pi \Delta f$ is the radian frequency increment, and N may be either even or odd depending upon whether it is desired to retain (N odd) or suppress the carrier ω_0 (N even). The envelope function is real because the tones are evenly spaced about the carrier giving rise to the amplitude modulated (AM) waveforms

$$\begin{aligned} f(t) &= 2F_0 \exp j\omega_0 t \sum_{k-1}^{N/2} \cos \frac{(2k-1)}{2} \Delta\omega t, \quad (N \text{ even}) \\ &= 2F_0 \exp j\omega_0 t \left(1 + \sum_{k=1}^{(N-1)/2} \cos k\Delta\omega t \right), \quad (N \text{ odd}). \end{aligned}$$

The spectrum is symmetric about the carrier with lines at $\omega_0 \pm (k + 1/2)\Delta\omega$ for N even and at $\omega_0 \pm k\Delta\omega$ for N odd where $k = 0, \cdots, (N-1)/2$ for both cases..

The magnitude of the time series appears as a series of N_p discrete peaks at time intervals of $\Delta t = 2\pi/\Delta\omega$. ($N_p$) is equal to one plus the number of time intervals $1/\Delta f = (N-1)/W$ in the duration T or

$$N_p = 1 + \Delta f T = 1 + \frac{TW}{(N-1)}. \tag{5.47}$$

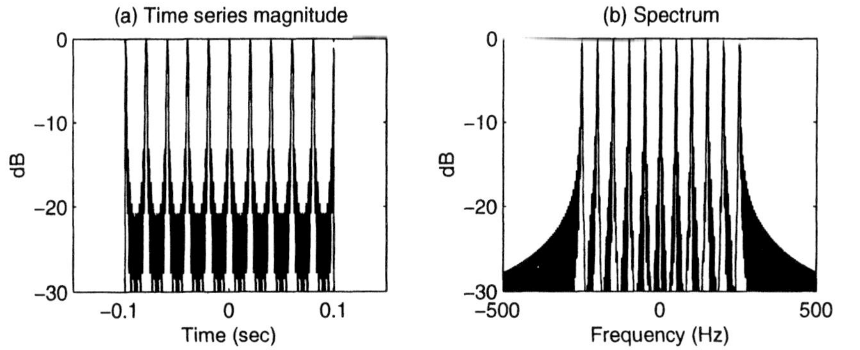

Figure 5.36: Equispaced $N_p = 11$ simultone time series and spectrum with T=.2 s, W=500 (Hz), $\triangle$ f=50 (Hz)

The time series and spectral magnitude of a simultone with $N_p = 11$, $T = .2$ sec, $W = 500$ Hz, and frequency spacing $\triangle f = 50$ Hz are illustrated in Figures 5.36a,b respectively. The spectrum of the impulse series reflects the spectra of the individual impulses which are sinc (sin x/x) functions in the frequency domain. The tonal sum approximates a series of time domain sinc functions and the corresponding spectrum approaches the rectangular. In the limit as $N \to \infty$ and $\triangle f \to 0$, the line spectrum becomes continuous as for an LFM pulse and the time series becomes a single sinc function representing the autocorrelation or equivalently the zero Doppler slice of the LFM ambiguity function.

The peak time domain magnitude is $|f_p| = NF_0$ and $P_p = |f_p|^2$ is the corresponding peak power. The energy in the waveform is

$$\begin{aligned} E_f &= P_p T_p N_p \\ &= F_0^2 N^2 T_p N_p \end{aligned}$$

where T_p is the effective duration of each time domain impulse that is approximately equal to the inverse waveform bandwidth. $T_p \approx 1/(N\triangle f)$ and from (5.47)

$$E_f = \frac{F_0^2 N(1 + \triangle f T)}{\triangle f}.$$

If it is desired to transmit a signal with energy E_f

$$F_0 = \sqrt{\tfrac{E_f \triangle f}{N(1+\triangle f T)}}$$

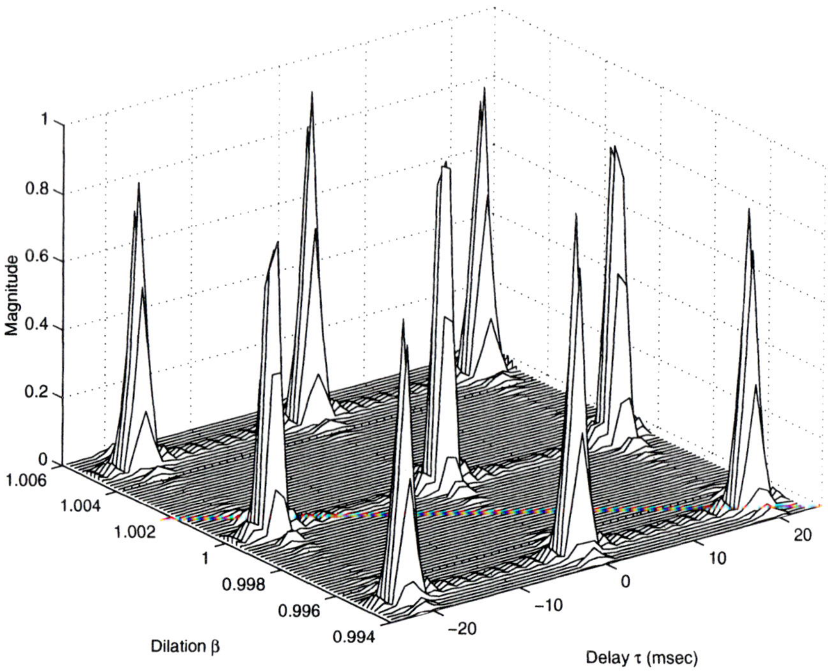

Figure 5.37: Simultone bed of nails ambiguity function corresponding to the time series and spectrum of Figure 5.36

$$\approx \sqrt{\frac{E_f}{NT}}$$

for $\triangle fT >> 1$. The peak power is then NE_f/T, the average power is E_f/T, and the ratio of peak to average power is the number of tones summed. $F_0 = 1/\sqrt{NT}$ for the corresponding unit energy waveform function.

The AF of a series of equispaced impulses in the time and frequency domains is the classic "bed of nails" discussed by Richaczek [7]. It consists of impulses in 2 dimensions with Doppler and delay spacing of $\triangle f$ or $\triangle\beta = \triangle f/f_0$ and $1/\triangle f$ respectively. The impulses are surrounded by ambiguity clear areas of unit time-frequency area as illustrated in Figure 5.37 for the waveform of Figure 5.36.

Simultaneous transmission can be achieved without introducing amplitude modulation by using phase or sub-band modulation in which the transmitted waveform has the form

$$f(t) = \exp j\left(\omega_0 + k_p \sum_{m=1}^{N} f_m(t)\right).$$

The multiple waveforms are encoded as the phase functions $f_m(t)$ that are each confined to a separate frequency band about the carrier ω_0. The received signal is demodulated by a discriminator and the individual signal components recovered by filtering [51]. The signal bandwidth depends upon the deviation constant k_p and can be made much larger than the totality of the component bandwidths in effect achieving a form of spread spectrum. The detection and estimation performance of a system utilizing sub-band modulated signals reflects the SIR threshold effect typical of FM systems. Output SIR tends to remain reasonably constant as the input SIR decreases but drops precipitously beyond a certain point.

5.8 SUMMARY

A study of waveform design for echo-location must be motivated by the application. Much of this chapter has been devoted to a discussion of some broad waveform classes and the criteria for the design of high resolution waveforms that have ambiguity functions that approach the ideal "thumbtack" for parameter estimation or imaging applications. It was shown that the best imaging waveforms have a flat WVD implying a uniform distribution of energy in time-frequency. Two waveform classes; HOP codes and PRN sequences that have this attribute and are widely used and were extensively covered. Waveform design for detection was not emphasized because it is not an issue when scattering is pointlike and the interference is WGN. The matched filter is the optimum detector for this ideal condition and its performance is determined solely by signal energy and parameter localization error.

The issue of waveform and receiver design for detection arises because echo-location especially sonar rarely involves the ideal point scatterer and WGN. Propagation in a non-homogeneous ocean environment produces ray bending and multipath effects in shallow water. Scatterers may be spatially extended but even if pointlike the echo may consist of multiple delayed multipath replications. Moreover the interference is often dominated by backscattered energy from interfering scattering mechanisms producing reverberation or clutter that is non-stationary and correlated with spectral characteristics determined by the environment and the waveform. Sonar design for this more general problem requires different detector strategies and waveform designs that attempt to mitigate the effects of spread scattering and non-stationary interference. Approaches to this problem are discussed in the following chapter.

Bibliography

[1] C.E. Cook and W.M. Siebert. The Early History of Pulse Compression Radar . *IEEE Trans. on Aerosp. and Elect. Systs.*, 24(6):825–837, Nov. 1988.

[2] R.H. Dicke. Object Detection System . US patent 2624876.

[3] P.M. Woodward. *Probability and Information Theory with Applications to Radar* . Pergamon Press, London, 1953.

[4] J.L. Stewart and E.C. Westerfield. A theory of active sonar detection. *Proceedings of the IRE*, 47:872–881, May 1959.

[5] Multiple authors. Special Issue on Matched Filtering . *IRE Trans. Info. Theory*, IT-6(3), June 1960.

[6] E.J. Kelly and R.P. Wishner. Matched Filter Theory for High Velocity Accelerating Targets . *IEEE Trans. on Military Electronics*, MIL-9(1):56–69, 1965.

[7] A.W. Rihaczek. *Principles of High Resolution Radar* . McGraw-Hill, New York NY, 1969.

[8] R.A. Altes and E.L. Titlebaum. Graphical Derivations of Radar, Sonar, and Communication Signals . *IEEE Trans. on Aerosp. and Elect. Systs.*, AES 11(1), Jan. 1975.

[9] E.L. Titlebaum. Echolocation Systems in Nature; A Theory of Optimum Insensitivity to Motion for Bats . In *Proc. 1978 Ultrasonics Symposium*, volume IEEE cat. no. 78CH1344-1SU, 1978.

[10] R.A. Altes and E.L. Titlebaum. Bat Signals as Optimally Doppler Tolerant Waveforms. *Journ. of the Acoust. Soc. Amer.*, 48((4)2):1014–1020, 1970.

[11] L.H. Sibul and E.L. Titlebaum. Signal Design for Detection of Targets in Clutter . *Proceedings of the IEEE*, 69(4):481–482, April 1981.

[12] J.V. DiFranco and W.L. Rubin. *Radar Detection* . Artech House, Dedham, MA, 1980.

[13] D.K. Barton. *Radars: Frequency Agility and Diversity* , volume 6. Artech House, Dedham, Mass., 1977.

[14] C.E. Cook and M. Bernfeld. *Radar Signals* . Academic Press, New York NY, 1967.

[15] H.L. Van Trees. *Detection, Estimation, and Modulation Theory-Part III* . Wiley, 1971.

[16] B.L. Lewis, F.F. Kretschmer and W.W. Shelton. *Aspects of Radar Signal Processing* . Artech House, Norwood MA, 1986.

[17] R.O. Nielsen. *Sonar Signal Processing.* Artech House, Boston/London, 1991.

[18] D.C. Rife and G.A. Vincent. Use of Discrete Fourier Transform in the Measurement of Frequencies and Levels of Tones . *Bell Sytem Tech. Journ.*, 49(2):197–228, Feb. 1970.

[19] F.J. Harris. On the Use of Windows for Harmonic Analysis with the Discrete Fourier Transform . *Proceedings of the IEEE*, 66(1):51–83, Jan. 1978.

[20] A.W. Nuttall. Some Windows with Very Good Sidelobe Behavior . *IEEE Trans. Acoust.,Speech, and Signal Proc.*, 29(1):84–91, Feb. 1981.

[21] J. Costas. A Study of a Class of Detection Waveforms Having Nearly Ideal Range-Doppler Ambiguity Properties . *Proceedings of the IEEE*, 72:996–1009, Aug. 1984.

[22] E.L. Titlebaum. Time-Frequency Hop Signals Part I: Coding based Upon the Theory of Linear Congruences . *IEEE Trans. on Aerosp. and Elect. Systs.*, 17(4):490–493, July 1981.

[23] E.L. Titlebaum and L.H. Sibul. Time-Frequency Hop Signals Part II: Coding Based Upon the Theory of Quadratic Congruences . *IEEE Trans. on Aerosp. and Elect. Systs.*, 17(4):494–499, July 1981.

[24] G. Jourdain and J.P. Henrioux. Use of Large TW Binary Phase Shift Keying Signals in Target Delay Doppler Measurements . *Journ. of the Acoust. Soc. Amer.*, 90(1):299–309, July 1991.

[25] MacWilliams and Sloan. Pseudo-Random Sequences and Arrays . *Proceedings of the IEEE*, 64(12):1715–1729, Dec. 1976.

[26] A.J. Viterbi. *Principles of Spread Spectrum Communication* . Addison-Wesley, 1995.

[27] R.C. Dixon. *Spread Spectrum Systems* . Wiley Interscience, 1994.

[28] F.B. Hildebrand. *Methods of Applied Mathematics* . Prentice-Hall, 1965.

[29] S.J. Mehta. Signal Design Issues for the Wigner Distribution and a Twin Processor for the Measurement of Target and Channel Structures . Technical report, Univ. Rochester Dept. of Elect. Eng., Nov. 1991. TR EE-91-13.

[30] L. Cohen. Time-Frequency Distibutions; a Review . *Proceedings of the IEEE*, 77(7):941–981, July 1989.

[31] G.F.B. Bartels. *Time-Frequency Signal Processing Algorithms: Analysis and Synthesis using Wigner Distributions* . PhD Thesis in Electrical Engineering, Rice Univ., Houston, TX, Dec. 1983.

[32] J.R. Bellegarda and E.L.Titlebaum. The Hit Array: An Analysis Formalism for Multiple Access Frequency Hop Coding . *IEEE Trans. on Aerosp. and Elect. Systs.*, 27(1):30–37, Jan. 1991.

[33] J.R. Bellegarda and E.L.Titlebaum. Time-Frequency Hop Codes Based Upon Extended Quadratic Congruences . *IEEE Trans. on Aerosp. and Elect. Systs.*, 24(6):726–742, Nov. 1988.

[34] D.W. Ricker. Logarithmic Frequency Allocation Algorithm for Wideband Discrete Frequency Pulse Trains . *IEEE Trans. on Aerosp. and Elect. Systs.*, 18(3):347–349, 1982.

[35] L.G. Weiss. Wavelets and Wideband Correlation Processing . *IEEE Sig. Proc. Magazine*, Jan. 1994.

[36] E.L. Titlebaum. Frequency and Time Hop Coded Signals for use in Radar and Sonar Systems and Multiple Access Communications

Systems . In *Proc. 27th Asilomar Conf. on Signals, Systems, and Computers*, Nov. 1993.

[37] J.R. Bellegarda et al. The Hit Array: A Synthesis Tool for Multiple Access Frequency Hop Signals . *IEEE Trans. on Aerosp. and Elect. Systs.*, 29(3):624–635, July 1993.

[38] S.W. Golomb and H. Taylor. Constructions and Properties of Costas Arrays . *Proceedings of the IEEE*, 72(9):1143–1163, Sept. 1984.

[39] S.W. Golomb. Algebraic Constructions for Costas Arrays . *Journ. of Combinatorial Theory*, A37:13–21, 1984.

[40] D.M. Drumheller and E.L. Titlebaum. Cross-Correlation Properties of Algebraically Constructed Costas Arrays . *IEEE Trans. on Aerosp. and Elect. Systs.*, 27(1):2–10, Jan. 1991.

[41] E.H. Dinan and B.Jabbari. Spreading codes for Direct Sequence CDMA and Wideband CDMA Cellular Networks. *IEEE Communications Magazine*, 36(9):48–54, Sept. 1998.

[42] D.J.Torrieri. *Principles of Secure Communication Systems* . Artech House, Norwood MA, 1985.

[43] A. Papoulis. *Probability, Random Variables, and Stochastic Processes* . McGraw-Hill, New York NY, 1965.

[44] Zhen biao Lin. Wideband Ambiguity Functions of Broadband Signals . *Journ. of the Acoust. Soc. Amer.*, 83(6):2108–2116, June 1988.

[45] P. Flandrin, P. Cros and G. Mange. Sensitivity of Doppler Tolerance to the Structure of Bat-like Signals . *Acoustica*, 62:40–47, 1986.

[46] R.A. Altes and D.P. Skinner. Sonar Velocity Resolution With a Linear-Period Modulated Pulse . *Journ. of the Acoust. Soc. Amer.*, 61(4):1019–1030, April 1977.

[47] D.M. Drumheller. Uniform Spectral Amplitude Windowing for HFM Waveforms . Technical report, Naval Research Lab., July 1994. NRL TR 94-9713.

[48] D.W. Ricker. Constrained Bandwidth Waveforms with Minimal Dilation Sensitivity . *IEEE Trans. on Aerosp. and Elect. Systs.*, 29(3):666–679, July 1993.

[49] D.W. Ricker. The Doppler Sensitivity of Large TW Phase Modulated Waveforms . *IEEE Trans. Sig. Proc.*, 40(10):2406–2413, Oct. 1992.

[50] H. Lai H. Cox. Geometric Comb Waveforms for Reverberation Suppression . In *28th Asilomar Conf. Record*, volume 2 of 2, Pacific Grove, CA, 1994.

[51] E.R. Boeker. Sub-band Modulation in Active Sonar. MS Thesis in Acoustics, Penna. State Univ., Univ. Park, PA, Dec. 2001.

Chapter 6

SPREAD SCATTERING AND PROPAGATION

6.1 INTRODUCTION

The sonar environment rarely produces the ideal conditions of point scattering in white (uncorrelated) Gaussian interference because the ocean is not a homogeneous medium. The salinity and temperature vary as a function of depth and location due to weather changes, solar heating, and fresh water influx from rivers and estuaries. These induce density variations that change the refractive index of the water and hence the propagation speed causing sound wave refraction. This is commonly called *ray bending* for sonars operating at medium to high frequency (1-100kHz) [1–3] and can cause a transmitted pulse and the resulting echo to propagate over several paths with different delays. Multiple boundary reflections from the surface and bottom are possible and the combined phenomenon is called *multipath* propagation. Figure 6.1 is an example of a summertime near surface downward refracting sound velocity profile (SVP) and raypath plot. The sun warms the surface waters generating a gradient of decreasing temperature with depth to about 100 ft depth. Thereafter temperature slowly increases with depth. Sound speed increases with water temperature causing the acoustic wavefront to refract downward near the surface but upward below creating a "duct" of converging raypaths at about 200 ft depth. Multiple reflections are also occurring at the surface and bottom. A continuum of propagation paths results from the width of the transmit and receive beams and multipath occurs where the paths intersect. The intersection of two paths

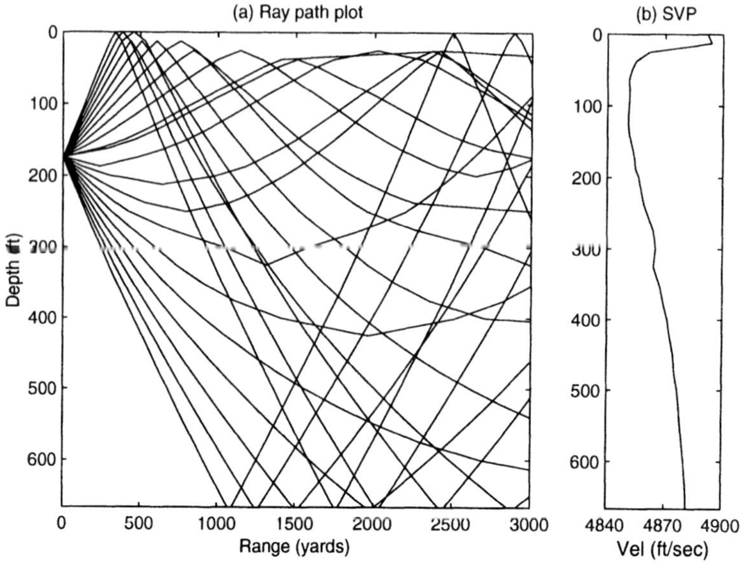

Figure 6.1: A summertime sound velocity profile and ray path plot with both surface and bottom reflections

often generates more than two multipath echoes because the both the transmitted pulse and reflected echo are affected. The pulse travels out and back along each distinct path but a component also travels out on one and back on the other and vice versa. The example exhibits both multiple direct path and boundary reflected propagation because the water is sufficiently shallow. The pulse may also be reflected from a physically extended object with multiple facets or discrete sub-reflectors called *highlights* that contribute additional delayed echoes that are often indistinguishable from multipath. The combined reflection and propagation effects produce a *delay spread* echo that may have a duration greatly exceeding that of the original transmitted waveform. In addition to delay spreading, *Doppler spreading* occurs if the transmitter/receiver is moving and/or the reflector has multiple components that are moving at different speeds such as a school of fish. If for example the source shown in Figure 6.1 is moving, multipath components will have Doppler shifts that are a function of the arrival angle in relation to the platform velocity. This is most evident in the Doppler spread of reverberation observed from a moving platform. Signal spreading in an echo-location context reflects the combined spread channel effects of propagation and extended reflection. The channel is called *doubly or singly spread* according to whether the spreading occurs in both delay and Doppler or just one parameter [4, 5]. More general models include bearing spread [6–9].

Echo spreading wherein a single point scatter can give rise to multiple delayed and Doppler shifted echoes complicates the sonar detection and parameter estimation process. The simple matched filter/replica correlator predicated upon a point scatter echo in WGN is no longer the optimum detector and different strategies are required. Moreover the concept of parameter estimation needs to be re-examined because one object can generate a multitude of delayed and Doppler shifted echoes. In such cases prior information about the scattering process is required and incorporated in the implementation of the receiver structure. This chapter explores these issues beginning with a discussion of the widely used linear models for propagation and reflection.

6.2 THE LINEAR SPREADING MODELS

Random delay and Doppler spreading is a two dimensional process and can be modeled by a time varying impulse response $\mathsf{h}(t, t-\tau)$ that is usually written as $\mathsf{h}(\tau, t)$. It represents a channel's response at time t to an impulse applied at time $t-\tau$ with the input-output (IO) signal relationship

$$\mathsf{s}(t) = \sqrt{E_0}\int_{-\infty}^{\infty} f(t-\tau)\mathsf{h}(\tau, t)d\tau \tag{6.1}$$

that reverts to the familiar

$$\mathsf{s}(t) = \sqrt{E_0}\int_{-\infty}^{\infty} f(t-\tau)\mathsf{h}(\tau)d\tau$$

for random time-invariant channels. Time invariance in an echo location context implies at most slow fluctuation in that the channel impulse response may vary only with intervals greater than the scatterer illumination interval. $\mathsf{h}(\tau)$ describes a channel that only varies between interrogations but may introduce delay spreading such that the echo is longer than the transmitted waveform. If $\mathsf{h}(\tau)$ collapses to an impulse it describes the slowly fluctuating point scatterer $\mathcal{S}_{sp}\delta(\tau-\tau_0)$ at range delay τ_0 and the echo signal becomes (3.49). $\mathsf{h}(\tau, t)$ describes a channel with an impulse response that may fluctuate within time intervals that are short with respect to the illumination interval introducing Doppler as well as delay spreading. Such channels are called rapidly fading or fluctuating channels. The IO relationship can represent any two points of an echo energy cycle such as described by Figure 2.1 for point scattering. The simple sonar equation point propagation loss and target strength parameters (L_{xr}, T_s) are replaced by individual impulse response functions for spread scattering. The overall echo cycle is represented by a concatenation of individual convolutions. For convenience in the following discussions it is assumed that the IO under discussion is the overall cycle, the input is the transmitted signal function $\sqrt{E_0} f(t)$, and the output $\mathsf{s}(t)$ is the signal appearing at the echo processor input with energy E_s. Unless otherwise noted, the separate propagation loss, scattering, and reflection spreading effects are combined in the single impulse response $\mathsf{h}(\tau, t)$.

This is permitted because a concatenation of several IO relationships is equivalent to a single relationship with an impulse response formed by the convolution of the individual impulse responses. This is demonstrated by considering a concatenated channel with impulse responses h_1, h_2 and output responses

$$\mathsf{s}_1(t) = \int_{-\infty}^{\infty} f(t-\tau)\mathsf{h}_1(\tau, t)d\tau$$
$$\mathsf{s}_2(t) = \int_{-\infty}^{\infty} \mathsf{s}_1(t-\tau')\mathsf{h}_2(\tau', t)d\tau'.$$

Substituting the expression for s_1 into that for s_2 produces

$$\mathsf{s}_2(t) = \int_{-\infty}^{\infty}\int_{-\infty}^{\infty} f(t-(\tau'+\tau))\mathsf{h}_1(\tau, t-\tau')\mathsf{h}_2(\tau', t)d\tau' d\tau.$$

Let $\tilde{\tau} = \tau' + \tau$ and change variables producing

$$\begin{aligned}\mathsf{s}_2(t) &= \int_{-\infty}^{\infty} f(t-\tilde{\tau}) \left[\int_{-\infty}^{\infty} \mathsf{h}_1(\tau, t-(\tilde{\tau}-\tau))\mathsf{h}_2(\tilde{\tau}-\tau, t)d\tau\right] d\tilde{\tau} \\ &= \int_{-\infty}^{\infty} f(t-\tilde{\tau})\mathsf{h}_3(\tilde{\tau}, t)d\tilde{\tau}\end{aligned}$$

where h_3 is the convolution of h_1 and h_2 denoted by $\mathsf{h}_3 = \mathsf{h}_1 * \mathsf{h}_2$. Note that $\mathsf{h}(\tau, t) \to \mathsf{h}(\tau)$ and

$$\mathsf{h}_3(\tilde{\tau}) = \int_{-\infty}^{\infty} \mathsf{h}_1(\tau)\mathsf{h}_2(\tilde{\tau}-\tau)d\tau$$

for the time invariant case.

$\mathsf{h}(\tau, t)$ is related by Fourier transformation to three other functions each of which provides a different perspective of the scattering process [9]. The time varying random *transfer function* and h are the Fourier pair

$$\begin{aligned} \mathsf{H}(\zeta, t) &= \int_{-\infty}^{\infty} \mathsf{h}(\tau, t) \exp -j2\pi\zeta\tau d\tau \\ \mathsf{h}(\tau, t) &= \int_{-\infty}^{\infty} \mathsf{H}(\zeta, t) \exp j2\pi\zeta\tau d\zeta \end{aligned} \tag{6.2}$$

where ζ represents the frequencies associated with the variation of h with respect to the time lag or delay τ and t represents its time dependence. The random signal *spreading function* denoted by $\mathcal{S}_s(\phi, \tau)$ and $\mathsf{h}(\tau, t)$ are the Fourier transform pair

$$\begin{aligned} \mathcal{S}_s(\phi, \tau) &= \int_{-\infty}^{\infty} \mathsf{h}(\tau, t) \exp -j2\pi\phi t dt \\ \mathsf{h}(\tau, t) &= \int_{-\infty}^{\infty} \mathcal{S}_s(\phi, \tau) \exp j2\pi\phi t d\phi \end{aligned} \tag{6.3}$$

where the frequency ϕ is associated with the time variation of the impulse response. In an echo-location context it represents the narrowband Doppler shift and/or spreading frequency(s) induced by time (t) variations (motion) of the scattering process. It is important to note the difference between it and the frequencies ζ that are associated with the delay (τ) or range variation of the process.

These two frequency variables appear together in the *bifrequency function* which is the transform of the transfer function with respect to the time variable or equivalently that of the spreading function with respect to τ

$$\begin{aligned} \mathsf{B}(\zeta, \phi) &= \int_{-\infty}^{\infty} \mathsf{H}(\zeta, t) \exp -j2\pi\phi t dt \\ &= \int_{-\infty}^{\infty} \mathcal{S}_s(\phi, \tau) \exp -j2\pi\zeta\tau d\tau. \end{aligned}$$

The relationship between the random signal spectrum

$$\mathsf{S}(\nu) = \int_{-\infty}^{\infty} \mathsf{s}(t) \exp -j2\pi\nu t dt$$

and the bifrequency function is obtained by substituting (6.3) in (6.1) and expressing the signal time function in terms of its spreading function as

$$\mathsf{s}(t) = \sqrt{E_0} \int_{-\infty}^{\infty} \int_{-\infty}^{\infty} \mathcal{S}_s(\phi, \tau) f(t - \tau) \exp j2\pi\phi t d\phi d\tau.$$

The delayed waveform time function expressed in terms of its spectrum is

$$f(t - \tau) = \int_{-\infty}^{\infty} F(\zeta) \exp j2\pi\zeta(t - \tau) d\zeta.$$

Substituting the signal and waveform time functions in the signal spectrum expression and gathering terms yields

$$\mathsf{S}(\nu) = \sqrt{E_0} \int_{-\infty}^{\infty} \int_{-\infty}^{\infty} F(\zeta) \left[\int_{-\infty}^{\infty} \mathcal{S}_s(\phi, \tau) \exp -j2\pi\zeta\tau d\tau \right] \cdot \left[\int_{-\infty}^{\infty} \exp j2\pi(\phi - (\nu - \zeta))tdt \right] d\phi d\zeta.$$

The first bracketed integral is $\mathsf{B}(\zeta, \phi)$, the second is the delta function $\delta(\phi - (\nu - \zeta))$, and

$$\begin{aligned} \mathsf{S}(\nu) &= \sqrt{E_0} \int_{-\infty}^{\infty} \int_{-\infty}^{\infty} F(\zeta) \mathsf{B}(\zeta, \phi) \delta(\phi - (\nu - \zeta)) d\phi d\zeta \\ &= \sqrt{E_0} \int_{-\infty}^{\infty} F(\zeta) \mathsf{B}(\zeta, \nu - \zeta) d\zeta. \end{aligned}$$

The signal spectrum is the frequency domain convolution of the waveform spectrum and the bifrequency function and its bandwidth is the combined waveform bandwidth and the frequency spread $\phi = \nu - \zeta$ associated with the time variation of the spreading process. If the waveform is a monotone $F(\zeta)\delta(\zeta - \zeta_0)$, the output spectrum is $\mathsf{B}(\zeta_0, \nu - \zeta_0)$, centered at ζ_0, and its frequency distribution is determined solely by the spreading process. When the backscatter is generated by a constant velocity object with no other spreading $\phi = \nu - \zeta_0$ is a constant Doppler frequency shift, and the signal is also a monotone with frequency $\nu = \zeta_0 + \phi$. If the waveform is not a monotone the output will have the same spectrum shifted by ϕ.

The spectrum is also related to the spreading and transfer functions by

$$\begin{aligned} \mathsf{S}(\nu) &= \sqrt{E_0} \int_{-\infty}^{\infty} \int_{-\infty}^{\infty} F(\zeta) \mathsf{H}(\zeta, t) \exp -j2\pi(\nu - \zeta) t dt d\zeta \\ &= \sqrt{E_0} \int_{-\infty}^{\infty} \int_{-\infty}^{\infty} F(\zeta) \mathcal{S}_s(\tau, \nu - \zeta) \exp -j2\pi\zeta\tau d\tau d\zeta \end{aligned} \tag{6.4}$$

and if the channel is time invariant

$$\begin{aligned}
\mathsf{S}(\nu) &= \sqrt{E_0}\int_{-\infty}^{\infty} F(\zeta)\mathsf{H}(\zeta)\left[\int_{-\infty}^{\infty} \exp -j2\pi(\nu-\zeta)t dt\right] d\zeta \\
&= \sqrt{E_0}\int_{-\infty}^{\infty} F(\zeta)\mathsf{H}(\zeta)\delta(\nu-\zeta)d\zeta \\
&= F(\nu)\mathsf{H}(\nu)
\end{aligned}$$

which is the frequency domain time invariant channel IO relationship.

6.2.1 Time-Frequency Correlation and Scattering Functions

The four linear model descriptions of narrowband signal scattering ($\mathsf{h}(\tau, t)$, $\mathcal{S}_s(\phi, \tau)$, $\mathsf{H}(\zeta, t)$, and $\mathsf{B}(\zeta, \phi)$) are zero mean random functions of their parameters. Their second order properties are described by their autocorrelation functions, the more important of which for echo-location applications are the *time-frequency correlation* (TFCF) and *scattering functions* (SF)

$$\begin{aligned}
R_H(\zeta, t, \zeta', t') &= \mathcal{E}\{\mathsf{H}(\zeta, t)\mathsf{H}^*(\zeta', t')\} \quad \text{(TFCF)} \\
\mathcal{R}_s(\phi, \tau, \phi', \tau') &= \mathcal{E}\{\mathcal{S}_s(\phi, \tau)\mathcal{S}_s^*(\phi', \tau')\} \quad \text{(SF)}.
\end{aligned}$$

The TFCF describes the correlation of the spreading process in time and frequency and hence the correlation between time and frequency separated echo components. The SF describes the correlation between echoes from scattering sources at different delays and Doppler shifts. These as well as the autocorrelations of the impulse response and bifrequency function are all related by Fourier transformation [9]. The transform relationship between the TFCF and the SF is obtained by combining (6.2, 6.3) yielding

$$\mathsf{H}(\zeta, t) = \int_{-\infty}^{\infty}\int_{-\infty}^{\infty} \mathcal{S}_s(\phi, \tau) \exp j2\pi(\phi t - \zeta\tau) d\phi d\tau$$

and

$$\begin{aligned}
R_H(\zeta, \zeta', t, t') = \int_{-\infty}^{\infty} \cdot\cdot \int_{-\infty}^{\infty} & \mathcal{R}_s(\phi, \tau, \phi', \tau') \exp j2\pi(\phi t - \phi' t') \\
& \cdot \exp j2\pi(\zeta'\tau' - \zeta\tau) d\phi d\phi' d\tau d\tau'.
\end{aligned}$$

If the scattering is uncorrelated in delay and Doppler,

$$\mathcal{R}_s(\phi, \tau, \phi', \tau') = \mathcal{R}_s(\phi, \tau)\delta(\tau - \tau')\delta(\phi - \phi')$$

and

$$R_H(\triangle\zeta, \triangle t) = \int_{-\infty}^{\infty}\int_{-\infty}^{\infty} \mathcal{R}_s(\phi, \tau) \exp j2\pi(\phi \triangle t - \tau \triangle \zeta) d\phi d\tau. \quad (6.5)$$

with $\triangle t = t - t'$, $\triangle\zeta = \zeta' - \zeta$. The TFCF is wide sense stationary (WSS) because it is a function of only the time and frequency differences which is a direct consequence of uncorrelated scattering (US) in delay and Doppler. The combined effect is called the WSSUS condition [5, 9] which is fairly restrictive since it means that the scattering process depends only upon the local scattering properties of the reflecting object or propagation paths. For example a fish school where each fish scatters independently of its neighbors would be WSSUS but an interconnected structure that permits acoustic energy to propagate through it and re-radiate from various points would not be. Random low frequency (<1kHz) sonar scattering and propagation channels are generally not WSSUS. The SF is the non-negative real function

$$\mathcal{R}_s(\phi, \tau) = \mathcal{E}\{|\mathcal{S}_s(\phi, \tau)|^2\}$$

that is the 2-d Fourier transform of the TFCF (6.5). It is the average delay-Doppler distribution of the scattering process under the WSSUS condition and describes the amount of spreading induced by the channel. The TFCF is generally complex and describes the correlation between time and frequency separated components of the scattered echo signal [9].

Signal fluctuation or fading and that which occurs predominantly in time and/or frequency is called *time/frequency selective fading.* Channels that induce fluctuation over short time or at close frequency intervals are said to have a high fade rate. A high time selective fade rate implies that the fluctuations of random time separated signal components are uncorrelated over short time intervals commensurate with the echo signal duration. Likewise, a high frequency selective fade rate implies uncorrelated fading between relatively closely spaced signal spectrum components. Fading is a form of signal distortion because a random realization of a signal propagated through a fading channel has time and frequency components that are selectively attenuated or amplified with respect to those in the waveform. Moreover additional spectral components may be introduced by the convolutional nature of the linear scattering process. The frequency selective fading caused by a channel consisting of multiple range distributed, fluctuating discrete reflectors (highlights) is an easily visualized example. If the channel is interrogated by an FSK waveform consisting of several monotone components with duration greater than the highlight spacing, selective phase interference and re-enforcement between the overlapping echoes causes frequency selective fluctuations that vary randomly between interrogations. The analogous phenomenon occurs for time separated components in a Doppler spread channel.

Fading and spreading are intimately related because the SF and TFCF are a Fourier pair in the WSSUS channel. A channel described by a scattering function with large delay spread significantly elongates the time duration of the transmitted pulse. Likewise, a Doppler spread channel broadens the waveform bandwidth. Because of the transform relationship, a high degree of spreading in delay τ and Doppler ϕ implies a high fade rate (short correlation interval) in frequency $\Delta\zeta$ and time Δt [4]. Let Ω_ϕ and Ω_τ denote the Doppler and delay spreading support of the SF. Fading (uncorrelated fluctuation) between signal components separated in time by Δt and frequency by $\Delta\zeta$ occurs when $\Delta t > 1/\Omega_\phi$ and $\Delta\zeta > 1/\Omega_\tau$ respectively. As a rule of thumb, the time and bandwidth support for moderate spreading are approximately those of the original waveform and $\Delta t \approx T$, $\Delta\zeta \approx W$. Time selective fading thus occurs if $\Omega_\phi > 1/T$, frequency selective fading occurs if $\Omega_\tau > 1/W$, and the combined fading condition is $\Omega_\tau\Omega_\phi > 1/TW$. Conversely, fading is negligible and the signal is undistorted if $\Omega_\tau\Omega_\phi < 1/TW < 1$. Such a channel is called *under spread*. Conversely, channels for which $\Omega_\tau\Omega_\phi > 1$ are called *over spread* [4].

The signal time domain autocorrelation is

$$\begin{aligned} R_s(t,u) &= \mathcal{E}\{\mathsf{s}(t)\mathsf{s}^*(u)\} \\ &= E_o \int_{-\infty}^{\infty} \cdot\cdot \int_{-\infty}^{\infty} \mathcal{R}_s(\phi,\tau,\phi',\tau') f(t-\tau) f^*(u-\tau') \\ &\quad \cdot \exp j2\pi(\phi t - \phi' u) d\phi d\phi' d\tau d\tau' \end{aligned}$$

and simplifies to

$$R_s(t,u) = E_0 \int_{-\infty}^{\infty}\int_{-\infty}^{\infty} \mathcal{R}_s(\phi,\tau) f(t-\tau) f^*(u-\tau) \exp j2\pi\phi(t-u) d\phi d\tau$$

when the scattering is WSSUS. The expected value of the signal energy is

$$\begin{aligned} E_s &= \int_{-\infty}^{\infty} \mathcal{E}\{|\mathsf{s}(t)|^2\} dt = \int_{-\infty}^{\infty} R_s(t,t) dt \\ &= E_0 \int_{-\infty}^{\infty}\int_{-\infty}^{\infty} \mathcal{R}_s(\phi,\tau) \left[\int_{-\infty}^{\infty} |f(t-\tau)|^2 dt\right] d\phi d\tau \end{aligned}$$

$$= E_0 \int_{-\infty}^{\infty} \int_{-\infty}^{\infty} \mathcal{R}_s(\phi, \tau) d\phi d\tau \; < \; E_0 \tag{6.6}$$

because the waveform is assumed to be energy normalized. This indicates that the totality of the received signal energy is obtained by integrating over the parameter space support of the scattering function and is not a function of the waveform in WSSUS channels. Moreover, scattering and spreading represent dissipation of transmitted energy hence $E_s < E_0$.

For example, the slowly fluctuating point scatterer SF in a benign (non-spreading) environment discussed in Chapter 3 is the constant $\mathcal{R}_{sp} = E_s/E_0$ that includes all of the losses associated with propagation, target backscatter and the other parameters of the sonar equation (2.4). It is

$$10 \log \mathcal{R}_{sp} = T_s + D_i + G_t - L_{xr} - 10 \log z - 11 - D_t - D_r$$

as obtained from (2.1, 2.4) in dB notation. Scattering functions can be defined between any two points of the interrogation cycle and the effective overall SF is the convolution of the constituent SF's just as it is for the impulse response. If the input is a wide sense stationary random process $\mathsf{f}(t)$ with power spectral density $S_f(\zeta)$, the psd of the scattered signal is obtained by the convolution

$$S_s(\nu) = \int_{-\infty}^{\infty} \int_{-\infty}^{\infty} S_f(\zeta) \mathcal{R}_s(\tau, \nu - \zeta) d\tau d\zeta \tag{6.7}$$

which is analogous to the deterministic waveform relationship (6.4).

6.2.2 The Wideband Model

The Fourier relationship (6.3) between the time varying impulse response and the spreading function is inherently narrowband because the effect of the time variation of the impulse response arising from scatter motion is interpreted as a Doppler shift or spread rather than the time dilation that actually occurs. In order to accommodate both Doppler induced time dilation and Doppler frequency shift models, the signal is expressed by the generic spreading function relationship

$$\mathsf{s}(t) = \sqrt{E_0} \int_{\Omega} \mathcal{S}_s(\mathbf{p}) f(\mathbf{p}, t) d\mathbf{p}. \tag{6.8}$$

The integrated random realizations of the spreading function at each point in the parameter domain Ω provide a realization of the signal time series. The parameter set for delay and Doppler may take either the narrow or wideband forms (2.27) where $\mathbf{p} = [\beta, \tau]$ or $[\phi, \tau]$ and the signal is [10, 11]

$$\begin{aligned}\mathsf{s}(t) &= \sqrt{E_0} \int_{-\infty}^{\infty} \int_{-\infty}^{\infty} \mathcal{S}_s(\phi, \tau) f(t-\tau) \exp j2\pi\phi t \, d\phi d\tau \quad \text{(narrowband)} \\ &= \sqrt{E_0} \int_{-\infty}^{\infty} \int_{-\infty}^{\infty} \sqrt{\beta} \mathcal{S}_s(\beta, \tau) f(\beta(t-\tau)) \, d\beta d\tau \quad \text{(wideband).} \quad (6.9)\end{aligned}$$

When the spreading function consists of M discrete scatterers (highlights) $\mathcal{S}_s(\mathbf{p}_i)\delta(\mathbf{p} - \mathbf{p}_i)$ distributed in the domain of the parameters Ω, the signal is

$$\begin{aligned}s(t) &= \sqrt{E_0} \sum_{i=1}^{M} \mathcal{S}_s(\mathbf{p}_i) f(\mathbf{p}_i, t) \\ &= \sqrt{E_0} \sum_{i=1}^{M} \sqrt{\beta_i} \mathcal{S}_s(\beta_i, \tau_i) f(\beta_i(t-\tau_i)) \quad \text{(wideband model)} \\ &= \sqrt{E_0} \sum_{i=1}^{M} \mathcal{S}_s(\phi_i, \tau_i) f(t-\tau_i) \exp j2\pi\phi_i t \quad \text{(narrowband model).}\end{aligned}$$

The $\mathcal{S}_s(\mathbf{p}_i)$ are random slowly fluctuating coefficients and if $M = 1$ the coefficient is equivalent to the slowly fluctuating coefficient $\mathcal{S}_{sp}$ of the point scatter model (3.49). The linear spread scattering model (6.8) generates an echo signal that is a continuum and/or discrete collection of waveform realizations spread over a range of Dopplers and delays hence the echo can occupy much greater delay and Doppler support than the original waveform.

The wideband spreading function does not conveniently relate to analogous bifrequency, transfer, and impulse response functions by Fourier transformation. It can be related to an impulse response by letting $z = \beta(t - \tau)$ and changing variables with respect to the integration in τ in the wideband model spreading model (6.9) [12, 13]. Then

$$\mathsf{s}(t) = \sqrt{E_0} \int_{\Omega_\beta} \int_{\Omega_\tau} \frac{1}{\sqrt{\beta}} \mathcal{S}_s \left(\beta, t - \frac{z}{\beta} \right) f(z) \, dz d\beta.$$

Let $f(z) = \delta(z - (t - \tau))$ which is an impulse applied at time $t - \tau$ and integrate with respect to z. By definition $\mathsf{s}(t)$ is the impulse response and

$$\mathsf{s}(t) = \mathsf{h}(\tau, t) = \int_{\Omega_\beta} \frac{1}{\beta} \mathcal{S}_s \left(\beta, \frac{\tau - (1-\beta)t}{\beta} \right) d\beta.$$

The spreading function model is the more intuitively satisfying echo model because it describes the scattering process as the integrated contribution of spatially and velocity distributed reflectors. The corresponding wideband signal scattering function (SF) is

$$\mathcal{R}_s(\mathbf{p}, \mathbf{p}') = \mathcal{R}_s(\beta, \tau, \beta', \tau') = \mathcal{E}\{\mathcal{S}_s(\beta, \tau)\mathcal{S}_s^*(\beta', \tau')\}$$

and reduces to $\mathcal{R}_s(\mathbf{p}) = \mathcal{E}\{|\mathcal{S}_s(\beta, \tau)|^2\}$ when the scattering is uncorrelated. The narrow and wideband scattering models are used interchangeably. The narrowband model for R_H is more convenient when describing the correlation properties of signals expressed in terms of their time-frequency distribution. The wideband spreading function model more accurately describes the dilation effects of Doppler spreading.

6.3 RECEIVER RESPONSE

It was shown in Chapter 3 that the optimum (maximum likelihood) receiver for point scattering in Gaussian noise is a correlation receiver that is a replica correlator or matched filter (3.58) when the interference is white Gaussian noise (WGN) and a mismatched filter (3.74) when the interference is correlated. In either case, detector performance is proportional to the amount of signal energy realized at the correlator output. It is maximized for point scattering in WGN by a matched filter when the hypothesized parameters $\hat{\mathbf{p}}$ are equal to the actual parameters $\mathbf{p}$ and the signal energy is weighted by unity at the ambiguity function origin. When the echo is spread in delay-Doppler it is not possible to uniformly weight the energy because the ambiguity function is unity only at its origin and the signal response is diminished in comparison to what it would be if the energy was concentrated at a single point in the delay-Doppler (phase) plane.

Point scattering is more the exception than the rule for sonar because reflectors are often extended in range and/or Doppler, propagation usually involves some degree of multipath, and spreading can vary and often cannot be predicted. Moreover, echoes scattered from stationary or at most slowly moving reflectors are usually masked by reverberation which is also a spread scattering process. Correlation receivers are not necessarily optimal under these conditions but their average performance is easily determined by a convolutional relationship between the processor ambiguity function and the scattering functions of the reflection process and interference. It is based upon the assumption that the spreading process is random, zero mean and the fluctuating model (3.49) is appropriate.

The relationship is derived by starting with the sufficient statistic (3.51) the signal component of which is

$$\mathsf{s}_1(\hat{\mathbf{p}}) = \int_{-\infty}^{\infty} \mathsf{s}(t) f^*(\hat{\mathbf{p}}, t) dt \tag{6.10}$$

with hypothesized wideband or narrowband parameters $\hat{\mathbf{p}} = [\hat{\beta}, \hat{\tau}]$, $[\hat{\phi}, \hat{\tau}]$ respectively. Substituting (6.8) for the point scatter signal $\mathsf{s}(t)$ in (3.49) produces

$$\begin{aligned}\mathsf{s}_1(\hat{\mathbf{p}}) &= \sqrt{E_0} \int_{\Omega} \mathcal{S}_s(\mathbf{p}) \int_{-\infty}^{\infty} f(\mathbf{p}, t) f^*(\hat{\mathbf{p}}, t) dt d\mathbf{p} \\ &= \sqrt{E_0} \int_{\Omega} \mathcal{S}_s(\mathbf{p}) \chi_{ff}(\mathbf{p}, \hat{\mathbf{p}}) d\mathbf{p}.\end{aligned} \tag{6.11}$$

Square law detection produces

$$|\mathsf{s}_1(\hat{\mathbf{p}})|^2 = E_0 \int_{\Omega} \int_{\Omega'} \mathcal{S}_s(\mathbf{p}) \mathcal{S}_s^*(\mathbf{p}') \chi_{ff}(\mathbf{p}, \hat{\mathbf{p}}) \chi_{ff}^*(\mathbf{p}', \hat{\mathbf{p}}) d\mathbf{p} d\mathbf{p}'$$

and the expected MF response for the parameter hypothesis $\hat{\mathbf{p}}$ is

$$\begin{aligned}\mathcal{E}\{|\mathsf{s}_1(\hat{\mathbf{p}})|^2\} = E_{fs}(\hat{\mathbf{p}}) &= E_0 \int_{\Omega} \mathcal{R}_s(\mathbf{p}) \Phi_{ff}(\mathbf{p}, \hat{\mathbf{p}}) d\mathbf{p} \\ &= E_0 \int_{\Omega_\beta} \int_{\Omega_\tau} \mathcal{R}_s(\beta, \tau) \Phi_{ff}(\hat{\beta}/\beta, \beta(\hat{\tau} - \tau)) d\beta d\tau \quad \text{(wideband)} \\ &= E_0 \int_{\Omega_\phi} \int_{\Omega_\tau} \mathcal{R}_s(\phi, \tau) \Phi_{ff}(\Delta\phi, \Delta\tau) d\phi d\tau \quad \text{(narrowband)}\end{aligned} \tag{6.12}$$

where $\Delta\phi = \hat{\phi} - \phi$, $\Delta\tau = \hat{\tau} = \tau$ and the scattering is uncorrelated. This is the multi-dimensional convolution of the scattering and waveform ambiguity functions in the parameter domain that is applicable to both narrow and wideband Doppler models [4, 5, 9–11]. It is also valid for mismatched receivers and their associated cross-ambiguity functions Φ_{fg} and is often written in the shorthand forms

$$E_{fs}(\hat{\mathbf{p}}) = E_0 \mathcal{R}_s ** \Phi_{ff} \text{ or } E_0 \mathcal{R}_s ** \Phi_{fg}.$$

Interference $\mathsf{n}(t)$ consists of noise plus reverberation that is also produced by a backscatter process and described by a scattering function

$\mathcal{R}_r$. The expected interference contribution to the sufficient statistic for a particular $\hat{\mathbf{p}}$ is

$$\mathcal{E}\{|\mathsf{n}_1(\hat{\mathbf{p}})|^2\} = E_{fi}(\hat{\mathbf{p}}) = E_{fr}(\hat{\mathbf{p}}) + N_0$$
$$E_{fr}(\hat{\mathbf{p}}) = E_0 \int_\Omega \mathcal{R}_r(\mathbf{p})\Phi_{ff}(\mathbf{p},\hat{\mathbf{p}})d\mathbf{p} = E_0\mathcal{R}_r ** \Phi_{ff} \quad (6.13)$$

assuming that the reverberation backscatter is uncorrelated in the parameter domain and accompanied by non-backscattered white noise with psd N_0. The expected MF receiver response to data $\mathsf{r}(t)$ when the scattering processes are uncorrelated is

$$\mathcal{E}\{|\mathsf{r}_1|^2_{H_1}\} = E_{fs} + E_{fi}$$
$$\mathcal{E}\{|\mathsf{r}_1|^2_{H_0}\} = E_{fi}.$$

6.3.1 Total Energy and the Detection Index

$E_{fs}(\hat{\mathbf{p}})$ represents the energy output of the filter at the hypothesis $\hat{\mathbf{p}}$ and integrating over all possible delay-Doppler hypotheses recovers all of the available signal energy in the echo. Using (6.12) this is

$$\int_{-\infty}^{\infty} E_{fs}(\hat{\mathbf{p}})d\hat{\mathbf{p}} = E_0 \int_\Omega \mathcal{R}_s(\mathbf{p}) \left[\int_{-\infty}^{\infty} \Phi_{ff}(\mathbf{p},\hat{\mathbf{p}})d\hat{\mathbf{p}}\right] d\mathbf{p}$$
$$= E_0 \int_\Omega \mathcal{R}_s(\mathbf{p})d\mathbf{p} = E_s \quad (6.14)$$

and is

$$\int_{-\infty}^{\infty}\int_{-\infty}^{\infty} E_{fs}(\hat{\phi},\hat{\tau})d\hat{\tau}d\hat{\phi}$$
$$= E_0 \int_{\Omega_\tau}\int_{\Omega_\phi} \mathcal{R}_s(\phi,\tau)\left[\int_{-\infty}^{\infty}\int_{-\infty}^{\infty} \Phi_{ff}(\Delta\phi,\Delta\tau)d\hat{\tau}d\hat{\phi}\right] d\phi d\tau$$
$$= E_0 \int_{\Omega_\phi}\int_{\Omega_\tau} \mathcal{R}_s(\phi,\tau)d\phi d\tau = E_s < E_0$$

for the narrowband representation (6.12) because of (6.6) and the unit volume property (4.15) of the AF. The same result is obtained for the wideband model if the Doppler variable is changed according to (4.16) and the fractional bandwidth is reasonably small as indicated by Figure 4.6. The total available signal energy E_s is also recovered by a matched filter when the scattering is point like with $\mathcal{R}_s(\phi,\tau) = \mathcal{R}_s\delta(\phi-\phi_0)\delta(\tau-\tau_0) = \mathcal{R}_{sp}$ and

$$\begin{aligned} E_{fs}(\hat{\phi},\hat{\tau}) &= E_0\mathcal{R}_{sp}\Phi_{ff}(\hat{\phi}-\phi_0,\hat{\tau}-\tau_0) \\ &= E_s \quad \text{when } \hat{\phi}=\phi_0,\ \hat{\tau}=\tau_0. \end{aligned}$$

A replica correlator can only realize a portion of the available energy when scattering is spread because the AF can be unity only at its origin and the expected receiver signal output is $E_{fs}(\hat{\phi},\hat{\tau}) \leq E_s < E_0$. Likewise, the expected reverberation output is $E_{fr}(\hat{\phi},\hat{\tau}) \leq E_r < E_0$ where E_r is the reverberation energy at the receiver input integrated over the SF phase plane support

$$E_r = E_0 \int_{\Omega_\phi}\int_{\Omega_\tau} \mathcal{R}_r(\phi,\tau)d\phi d\tau. \tag{6.15}$$

The AF-SF convolution is illustrated in Figure 6.2 where an ambiguity function illustrated by the cross hatched elliptical area is convolved by varying amounts with a distributed scattering function. Scattering functions and ambiguity functions of waveforms with small fractional bandwidth are non-negative real functions with bounded volume hence the highest expected MF response occurs when the SF and AF maximally overlap and share the greatest common volume. Lower expected response occurs when they partially overlap, and none occurs when they are disjoint. The point of maximum signal or interference response is that delay and Doppler hypothesis $\hat{\tau}$ and Doppler $\hat{\phi}$ that maximize the convolution equations (6.12) or (6.13). It occurs for point scattering when the ambiguity origin coincides with the delay and Doppler of the scatterer, $\hat{\mathbf{p}} = \mathbf{p}$, $\Phi_{ff} = 1$, and the total scattering energy is reflected in the receiver response. The spread scattering detection index corresponding to (3.62) is

$$\begin{aligned} D_{\hat{\mathbf{p}}} &= \frac{E_0\int_\Omega \mathcal{R}_s(\mathbf{p})\Phi_{ff}(\mathbf{p},\hat{\mathbf{p}})d\mathbf{p}}{E_0\int_\Omega \mathcal{R}_r(\mathbf{p})\Phi_{ff}(\mathbf{p},\hat{\mathbf{p}})d\mathbf{p} + N_0} \\ &= \frac{E_0\mathcal{R}_s **\Phi_{ff}}{E_0\mathcal{R}_r **\Phi_{ff} + N_0} = \frac{E_{fs}(\hat{\mathbf{p}})}{E_{fr}(\hat{\mathbf{p}}) + N_0} = \mathrm{SIR}_{out} \end{aligned} \tag{6.16}$$

and represents the expected output signal to expected interference ratio of a correlation receiver under spread scattering conditions. The interference includes non-backscattered ambient components consisting of electronic, mechanical, and in-water noise sources that in the aggregate

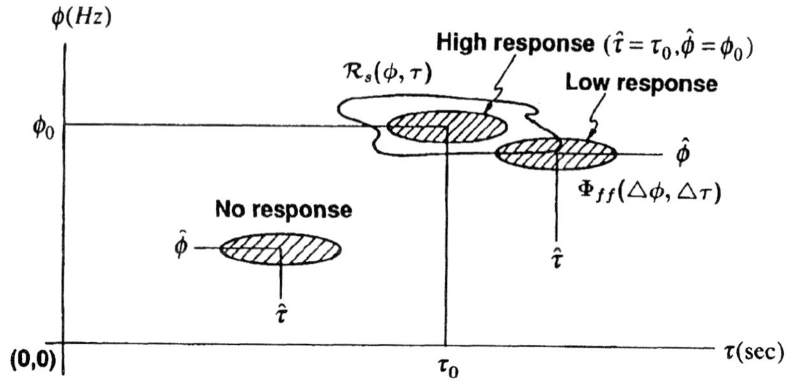

Figure 6.2: Expected matched filter response as determined by the overlap (convolution) of the scattering and ambiguity functions

are described by the covariance function (3.24). These may be spectrally uniform over the system bandwidth in which case are described by the noise power spectral density (psd) N_0 (Joules). The detection index is defined for a mismatched receiver by replacing Φ_{ff} with the cross-ambiguity function Φ_{fg}.

The detection index establishes the detection and false alarm probabilities as was the case for point scattering. The input data $\mathsf{r}(t)$ is assumed to be complex Gaussian hence the square law detected MF receiver outputs are exponentially distributed by (3.60) with $\sigma_0^2 = E_{fi}(\hat{\mathbf{p}})$ and $\sigma_1^2 = E_{fs}(\hat{\mathbf{p}}) + E_{fi}(\hat{\mathbf{p}})$. The spread scatter ROC relationship corresponding to (3.61) is

$$\wp_d = \wp_f^{\frac{1}{1+D_{\hat{\mathbf{p}}}}} \tag{6.17}$$

as plotted in Figure 3.4.

The detection index for point scattering in WGN (3.62) is denoted as $D_{\mathbf{p}\hat{\mathbf{p}}}$ because it depends upon the parameter space separation between the hypothesis $\hat{\mathbf{p}}$ and the point scatter parameters $\mathbf{p}$. MF performance in WGN is maximized when $\hat{\mathbf{p}} = \mathbf{p}$ and the hypothesized parameters match those of the scatterer at the peak of the AF. The particular waveform is immaterial from a detection point of view as $\Phi_{ff}(\mathbf{p}, \mathbf{p}) = 1$ for all ambiguity functions of normalized waveforms. This is not the case for spread signal and reverberation scattering because the detection index depends only upon the hypothesis $\hat{\mathbf{p}}$ in relation to the signal and reverberation scattering distributions and is denoted as $D_{\hat{\mathbf{p}}}$.

Maximum $D_{\hat{\mathbf{p}}}$ does not necessarily correspond to the maximum signal response when the interference SF is non-uniformly distributed. Generally, maximization of (6.16) is a difficult min-max problem requiring the simultaneous minimization of $E_{fr}(\hat{\mathbf{p}})$ and maximization of $E_{fs}(\hat{\mathbf{p}})$ as a function of $\hat{\mathbf{p}}$. It requires knowledge of the scattering function distributions, their relationship with respect to each other in the delay-Doppler plane, and the synthesis of a waveform or more generally a waveform pair that maximizes $D_{\hat{\mathbf{p}}}$. The resulting processor is not necessarily a matched filter because maximum performance may well be obtained from a cross-ambiguity function arising from a mismatched receiver [14].

The MF is optimum only for point scattering in white noise but its performance in spread scattering and correlated interference is often adequate especially if careful attention is paid to waveform design seeking ambiguity functions that maximize E_{fs} and minimizes E_{fr}. Processors that provide adequate performance over a broad set of conditions for which they are not necessarily optimized are called *robust* [15]. Optimum processors have the highest performance under the specific conditions for which they are designed but their performance often seriously degrades in relation to that of a robust processor when the conditions are not met.

6.3.2 Scattering Function Convolution

When the scattering process is concatenated as it might be when an echo is generated by a spread reflector and propagated through a spreading (multipath) channel, the expected response of a correlation receiver is the convolution of its ambiguity function with a scattering function that is itself the convolution of the constituent scattering functions. This is illustrated for the wideband model by writing (6.8) as the concatenation of two processes

$$\begin{aligned}
\mathsf{s}(t) &= \sqrt{E_0}\int_{-\infty}^{\infty}\int_{-\infty}^{\infty}\sqrt{\beta}S_{s1}(\beta,\tau)f(\beta(t-\tau))d\beta d\tau \\
\mathsf{s}'(t) &= \int_{-\infty}^{\infty}\int_{-\infty}^{\infty}\sqrt{\beta'}S_{s2}(\beta',\tau')\mathsf{s}(\beta'(t-\tau'))d\beta' d\tau' \\
&= \sqrt{E_0}\int_{-\infty}^{\infty}\cdot\cdot\int_{-\infty}^{\infty}\sqrt{\beta\beta'}S_{s1}(\beta,\tau)S_{s2}(\beta',\tau') \\
&\quad\cdot f(\beta[\beta'(t-\tau')]-\tau)d\beta d\tau d\beta' d\tau'.
\end{aligned}$$

The wideband version of the statistic (6.10) is

$$\begin{aligned}
\mathsf{s}_1(\hat{\beta},\hat{\tau}) &= \sqrt{\hat{\beta}}\int_{-\infty}^{\infty} \mathsf{s}'(t) f^*(\hat{\beta}(t-\hat{\tau}))dt \\
&= \sqrt{E_0}\int_{-\infty}^{\infty}\cdots\int_{-\infty}^{\infty}\left[\sqrt{\beta\beta'\hat{\beta}}\int_{-\infty}^{\infty} f(\beta\beta'(t-\tau'-\tau/\beta'))f^*(\hat{\beta}(t-\hat{\tau}))dt\right] \\
&\cdot \mathcal{S}_{s1}(\beta,\tau)\mathcal{S}_{s2}(\beta',\tau')d\beta\cdots d\tau'.
\end{aligned}$$

The quantity in the square brackets is an uncertainty function that is put into the standard 2-variable form (4.4)

$$\chi_{ff}(1,0,b,a) \equiv \chi_{ff}(b,a) = \sqrt{b}\int_{-\infty}^{\infty} f(t) f^*(b(t-a))dt$$

by changing variables with $u = \beta\beta'(t-\tau'-\tau/\beta')$. The statistic then becomes

$$\begin{aligned}
\mathsf{s}_1(\hat{\beta},\hat{\tau}) = \sqrt{E_0}\int_{-\infty}^{\infty}\cdots\int_{-\infty}^{\infty} \chi_{ff}\left(\frac{\hat{\beta}}{\beta\beta'}, \beta\beta'(\hat{\tau}-\tau')-\beta\tau\right) \\
\cdot\ \mathcal{S}_{s1}(\beta,\tau)\mathcal{S}_{s2}(\beta',\tau,)d\beta\cdot\cdot d\tau'.
\end{aligned}$$

Changing variables again by setting

$$\begin{aligned}
\tilde{\beta} &= \hat{\beta}/\beta\beta' \\
\tilde{\tau} &= \beta\beta'(\hat{\tau}-\tau')-\beta\tau,
\end{aligned}$$

eliminating β', τ', invoking the uncorrelated scattering assumption, and taking the expectation yields

$$\mathcal{E}\{|\mathsf{s}_1(\hat{\beta},\hat{\tau})|^2\} = E_0\int_{-\infty}^{\infty}\int_{-\infty}^{\infty}\left(\frac{1}{\tilde{\beta}^2}\right)\mathcal{R}_{s3}\left(\frac{\hat{\beta}}{\tilde{\beta}}, \hat{\tau}-\frac{\tilde{\beta}\tilde{\tau}}{\hat{\beta}}\right)\Phi_{ff}(\tilde{\beta},\tilde{\tau})d\tilde{\tau}d\tilde{\beta}$$

where the combined scattering function is

$$\begin{aligned}
\mathcal{R}_{s3}\left(\frac{\hat{\beta}}{\tilde{\beta}}, \hat{\tau}-\frac{\tilde{\beta}}{\hat{\beta}}\right) &= \int_{-\infty}^{\infty}\int_{-\infty}^{\infty}\frac{1}{\beta^2}\mathcal{R}_{s1}(\beta,\tau)\mathcal{R}_{s2} \\
&\cdot \left(\frac{\hat{\beta}}{\beta\tilde{\beta}}, \hat{\tau}-\frac{\tilde{\beta}}{\hat{\beta}}(\tilde{\tau}+\beta\tau)\right)d\beta d\tau. \qquad (6.18)
\end{aligned}$$

This result is consistent with

$$\mathsf{s}_1(\hat{\beta},\hat{\tau}) = \sqrt{E_0}\int_{-\infty}^{\infty}\int_{-\infty}^{\infty}\mathcal{S}_s(\beta,\tau)\chi_{ff}(\hat{\beta}/\beta, \beta(\hat{\tau}-\tau))d\beta d\tau$$

which is obtained from the wideband version of (6.11) with $\mathbf{p} = [\beta, \tau]$ and $\hat{\mathbf{p}} = [\hat{\beta}, \hat{\tau}]$ for one spreading function by letting $\tilde{\beta} = \hat{\beta}/\beta$ and $\tilde{\tau} = \beta(\hat{\tau} - \tau)$ and change variables to replace β, τ. The resulting expectation of the magnitude square assuming uncorrelated scattering is

$$\mathcal{E}\{|s_1(\hat{\beta}, \hat{\tau})|^2\} = \int_{-\infty}^{\infty}\int_{-\infty}^{\infty}\left(\frac{1}{\tilde{\beta}^2}\right)\mathcal{R}_s\left(\frac{\hat{\beta}}{\tilde{\beta}}, \hat{\tau} - \frac{\tilde{\beta}\tilde{\tau}}{\hat{\beta}}\right)\Phi_{ff}(\tilde{\beta}, \tilde{\tau})d\tilde{\beta}d\tilde{\tau}.$$

(6.18) is called the affine or wideband convolution [10, 11] of the constituent scattering functions $\mathcal{R}_{s1}$, $\mathcal{R}_{s2}$. A similar relationship for the more familiar narrowband convolution is derived in [16].

6.4 PROCESSING GAIN ESTIMATION

Matched filter spread scattering performance as determined by the detection and false alarm probabilities (6.17) depends upon the output SIR or detection index $D_{\hat{\mathbf{p}}}$. $D_{\hat{\mathbf{p}}}$ is related to the input SIR by the processing gain

$$G_p = \frac{\text{SIR}_{out}}{\text{SIR}_{in}} \tag{6.19}$$

given by (3.63) for point scattering in WGN and easily estimated from the SF-SF convolution relationship under spreading conditions. G_p relates sonar equation related quantities that determine SIR_{in} (2.7) such as target strength, propagation loss, source level, directivity index etc. to the ambiguity and scattering functions that determine SIR_{out} and the ROCs (6.17). It is a useful measure for the specification of waveform parameters and the assessment of correlation receiver performance in a spreading environment. When ambiguity and scattering functions are approximated by simple geometric shapes, it is possible to derive closed form G_p estimates because the signal and interference convolutions are reduced to calculations of overlapping areas. This is illustrated for matched filters using CW, LFM, low order FSK, and thumbtack waveforms interrogating point and line targets in a reverberant environment. These waveforms are often employed for echo-location applications and their ambiguity functions can be approximated by simple constant amplitude models of the mainlobe region [4, 17].

The input SIR is the power ratio

$$\text{SIR}_{in} = \frac{P_s}{P_n + P_r} \tag{6.20}$$

regardless of the specific waveform function. $P_s = E_s/T$ is the signal power and $P_n = N_0 W_s$ is the input noise power. W_s is the system or noise bandwidth that is assumed to exceed the maximum frequency support W of the waveform plus an allowance for the expected Doppler excursion. It is equivalent to the sampling frequency for sampled analytic signals. P_r is the reverberation power at the range at which the MF performance is to be estimated and may consist of volume, boundary or contributions from both mechanisms as indicated in (2.7).

6.4.1 Scattering models: the signal component

Sonar echoes are not often significantly Doppler spread but are more likely to exhibit delay spreading because of multipath or scatter extent. Their scattering functions tend to lie along the τ axis and are somewhere between the extremes of point and uniform line scattering. Point scattering is the simplest model and corresponds to a specular reflection in a homogeneous environment with no boundary interactions. Its SF is the delta function $\mathcal{R}_{sp}\delta(\tau - \tau_0)\delta(\phi - \phi_0)$ at the origin and the received echo power and energy are related to the transmitted energy and waveform duration T by

$$\begin{aligned} E_s &= E_0\mathcal{R}_{sp} \\ P_s &= E_s/T = P_0\mathcal{R}_{sp} \quad \text{(point scattering).} \end{aligned} \tag{6.21}$$

The continuous line scatterer is the other extreme and is modeled by a SF that is confined to the τ axis and

$$E_s = E_0 \int_{-\infty}^{\infty} \mathcal{R}_s(\tau)d\tau.$$

$\mathcal{R}_s(\tau)$ represents the incremental backscatter strength per unit time delay at the range delay (τ) and has units of 1/sec. or Hz. Objects that are point-like or doubly spread in delay-Doppler are characterized by SF's that represent scattering per unit time-frequency and hence are dimensionless.

Uniform line scattering on an interval τ_t represents a scatterer with range extent $\tau_t c/2$ about a nominal delay $\tau_0 = 2r_0/c$. It is represented by the constant SF

$$\mathcal{R}_s(\tau) = \mathcal{R}_{sl} \quad \tau \in [\tau_0 - \tau_t/2,\ \tau_0 + \tau_t/2]$$

and the corresponding received signal power for short ($\tau_t \leq T$), and long ($\tau_t > T$) line scattering is

$$P_s = \begin{cases} P_0 \mathcal{R}_{sl} \tau_t, & \text{short line scatt. } \tau_t \leq T \\ P_0 \mathcal{R}_{sl} T, & \text{long line scatt. } \tau_t > T. \end{cases} \tag{6.22}$$

Matched filters are optimum for point scattering in WGN but are mismatched for spread scattering and their performance can be significantly degraded. Most echos emanate from scattering processes that are neither truly specular nor uniformly range spread but lie somewhere in between. Point and uniform line scattering represent the extreme conditions of delay spreading and serve to provide performance bounds for matched filter processors.

6.4.2 Scattering models: the reverberation component

Reverberation scattering functions are determined by transmit and receive beam configuration, sonar velocity, boundary and/or scatter motion (waves and bubbles), and backscatter strength [1, 18–20]. Typically they are generated by computer models that produce reverberation time series for simulation studies and incorporate both environmental and sonar parameters. Examples are the Sonar Simulation Toolkit (SST) model developed at the University of Washington [21] and REVMOD [22]. A typical reverberation scattering function as realized for a moving, forward looking sonar platform is shown in Figure 6.3. It is characterized by a delay axis ridge with Doppler width proportional to the beam mainlobe width, orientation, platform speed, and the inherent scatter velocity distribution. Beam sidelobes contribute low Doppler reverberation scattering from moving platforms transmitting in the direction of travel. At short range and in shallow water the Doppler profile can reflect strong low Doppler surface or bottom scattering arriving nearly perpendicular to the boresight direction hence at lower Doppler. This is called the *reverberation hook* and is indicated in the figure by the expansion of the reverberation spectrum toward low Doppler at short range. The Doppler profile falls off abruptly at the maximum Doppler corresponding to scattering on the array boresight. Volume and boundary reverberation typically decrease with range by approximately $1/r^2$ and $1/r^3$ respectively as indicated by (2.5, 2.6). The variation is often regarded as negligible over time intervals commensurate with the scatterer and pulse duration [4]. The Doppler profile of $\mathcal{R}_r(\phi, \tau)$ is often modeled

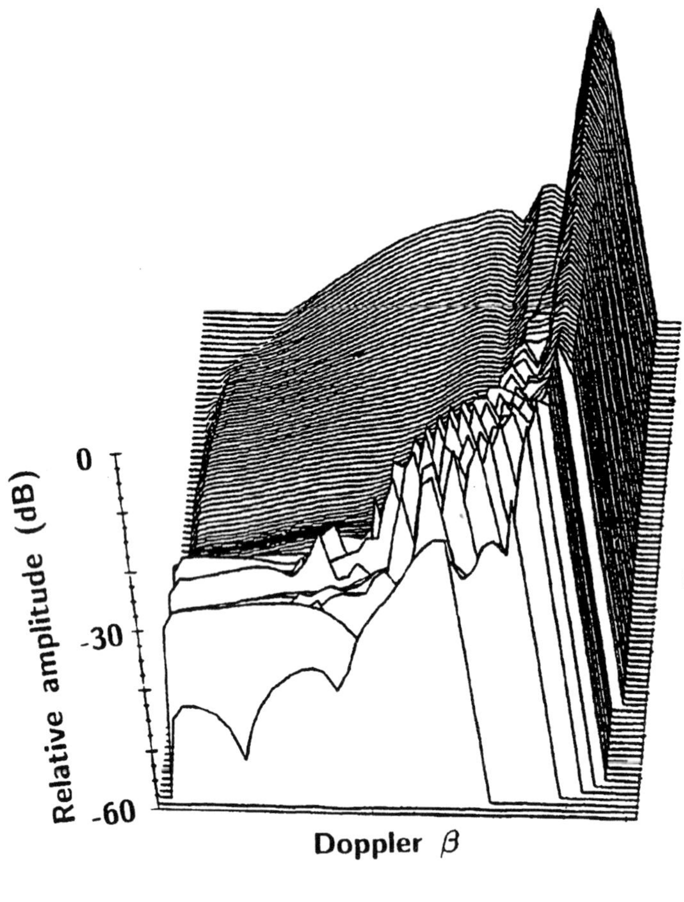

Figure 6.3: Example of the reverberation scattering function corresponding to a moving forward looking sonar platform

as a Gaussian function

$$\mathcal{R}_r(\phi, \tau_0) = A(\tau_0)\exp(-(\phi - \phi_0)^2/2\sigma_\phi^2) \tag{6.23}$$

that is constant with delay over a small increment about the nominal range delay τ_0. It is not the observed reverberation spectrum determined by the convolutions (6.4, 6.7) but rather the inherent frequency spreading induced by the scattering process that is independent of the signal. The frequency variance is

$$\sigma_\phi^2 = \sigma_{\phi v}^2 + \sigma_{\phi e}^2$$

where $\sigma_{\phi e}$ is the environmental component determined by the inherent motion of the medium such as surface and internal waves and $\sigma_{\phi v}^2$ is determined by platform motion, velocity, beam tilt (squint) angle, and beamwidth. Sea surface wave motion is the dominant contributor to the environmental component and is modeled as [23, 24]

$$\sigma_{\phi e}^2 = k_1(f_0)k_2(\theta_g, \theta_a)\sigma_h(v_w).$$

k_1 is a frequency dependent term, θ_g is the surface grazing angle ($\theta_g = 90°$ is normal to the surface), θ_w is the wave motion angle, and $\sigma_h(v_w)$ is a function of the wind speed v_w. The model is claimed to be accurate and supported by data obtained for frequencies $15 \leq f_0 \leq 60$ kHz and grazing angles $0° \leq \theta_g \leq 30°$.

The motion induced Doppler component is a function of platform speed v_p, operating wavelength λ_0, beam width θ_b, and direction. It is estimated from

$$\begin{aligned}\sigma_{\phi v}^2 &= \frac{4v_p^2}{\theta_b\lambda_0^2}\int_0^{\theta_b/2}\cos^2\theta d\theta - \frac{4v_p^2}{\lambda_0^2} \\ &= \frac{2v_p^2}{\lambda_0^2}(1 - \text{sinc }\theta_b)\end{aligned} \tag{6.24}$$

for a forward looking uniform beam aligned with the platform velocity. The relative contributions under nominal high frequency conditions are illustrated by assuming a sonar operating at 50 kHz, with v_p=5 m/sec, $\theta_b = 10°$, a wind speed of 10 m/sec. The rms motion component computed from (6.24) is $\sigma_{\phi v}$=17 Hz and the corresponding environmental component is $\sigma_{\phi e}$=21 Hz [24, 25]. The motion and environmental contributions are comparable for these parameters but the motion component will increase significantly for wider beam widths and for beams tilted away from the velocity vector. The Doppler range for a beam squinted (tilted) to an angle θ_0 is

$$\Delta\phi = \phi_{max} - \phi_{min} \approx \left(\frac{2v_p \sin\theta_0}{\lambda_0}\right)\theta_b$$

and is 41 Hz for $\theta_0 = 45°$ and the parameters of the illustration. For the purpose of processing gain estimation the reverberation SF is modeled as a rectangular delay axis ridge with indefinite extent, uniform magnitude $\mathcal{R}_r$, and Doppler spread

$$B_r = 2.35\sigma_\phi$$

equivalent to the 3 dB width of the Gaussian scattering function profile (6.23) [17]. It represents backscatter strength per unit time-frequency and hence has no dimension.

The input reverberation power is

$$\begin{aligned} P_r &= E_r/T \\ E_r &= E_0 \int_{-T/2}^{T/2} \int_{-B_r/2}^{B_r/2} \mathcal{R}_r d\phi d\tau \\ &= E_0 \mathcal{R}_r B_r T \end{aligned} \tag{6.25}$$

and obtained from (6.15) assuming a waveform with duration T. SIR_{in} (6.20) is computed from (6.21, 6.22, 6.25) for the spread target and reverberation scattering models and

$$\begin{aligned} P_i &= P_r + P_n = P_0\mathcal{R}_r B_r T + N_0 W_s \\ &= P_n(1 + r_n) = N_0 W_s(1 + r_n) \quad \text{(total interference)} \end{aligned} \tag{6.26}$$

where

$$r_n = P_r/P_n = \frac{P_0 B_r \mathcal{R}_r T}{N_0 W_s} = \frac{E_0 B_r \mathcal{R}_r}{N_0 W_s}$$

is the reverberation to noise power ratio. The input SIR relationships are summarized for convenience as

$$\begin{aligned} \text{SIR}_{in} &= \frac{P_0\mathcal{R}_{sp}}{N_0 W_s(1 + r_n)} \quad \text{(point scattering)} \\ &= \frac{P_0\mathcal{R}_{sl}\tau_t}{N_0 W_s(1 + r_n)} \quad \text{(line scattering, } \tau_t \le T) \\ &= \frac{P_0\mathcal{R}_{sl}T}{N_0 W_s(1 + r_n)} \quad \text{(line scattering, } \tau_t > T). \end{aligned} \tag{6.27}$$

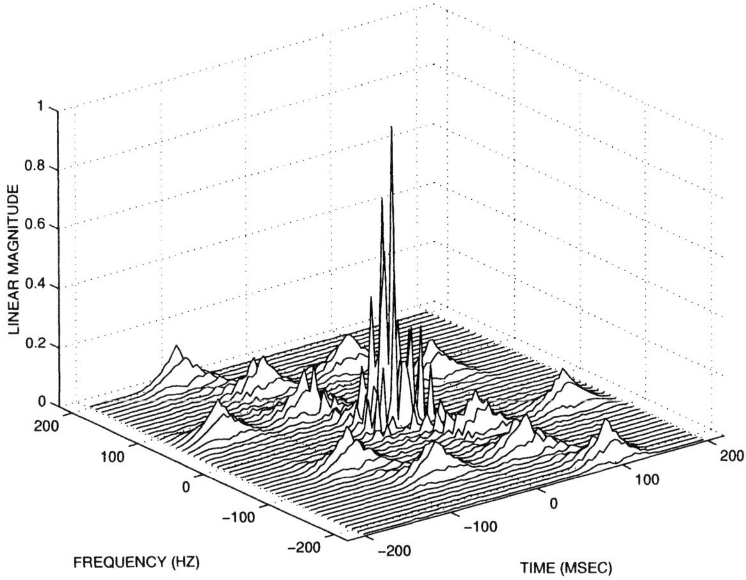

Figure 6.4: Ambiguity function of a sparse FSK code of Welch construction with GF(5)2, T=.2 (sec), W=200 (Hz)

6.4.3 Receiver Outputs and G_p for CW, LFM and Sparse FSK Waveforms

Closed form estimates of MF output are computed from the convolutions (6.12, 6.13) and unit height approximations to CW, LFM, and FSK waveform ambiguity functions examples of which are shown in Figures 5.2 and 6.4. The volume and origin magnitude of a narrowband AF are unity and since the volume of CW and LFM waveform AF's is concentrated in the mainlobe, the unit height approximations over the mainlobe support shown in Figure 6.5 are used. The FSK waveform is assumed to be a single hit Costas or congruence code as discussed in Chapter 5 and is sparse in that $N^2 << TW$. This guarantees that its AF sidelobes are distinct and of nearly uniform amplitude so that $1/N$ of the ambiguity volume is in the mainlobe the remainder being apportioned equally among the $N(N-1)$ sidelobes. The mainlobe support region is approximately that of one CW subpulse with duration $T_p = T/N$ and consequently the mainlobe has a grating or "hairbrush" structure as shown in Figure 6.4 and is modeled as in Figure 6.5. The illustrated

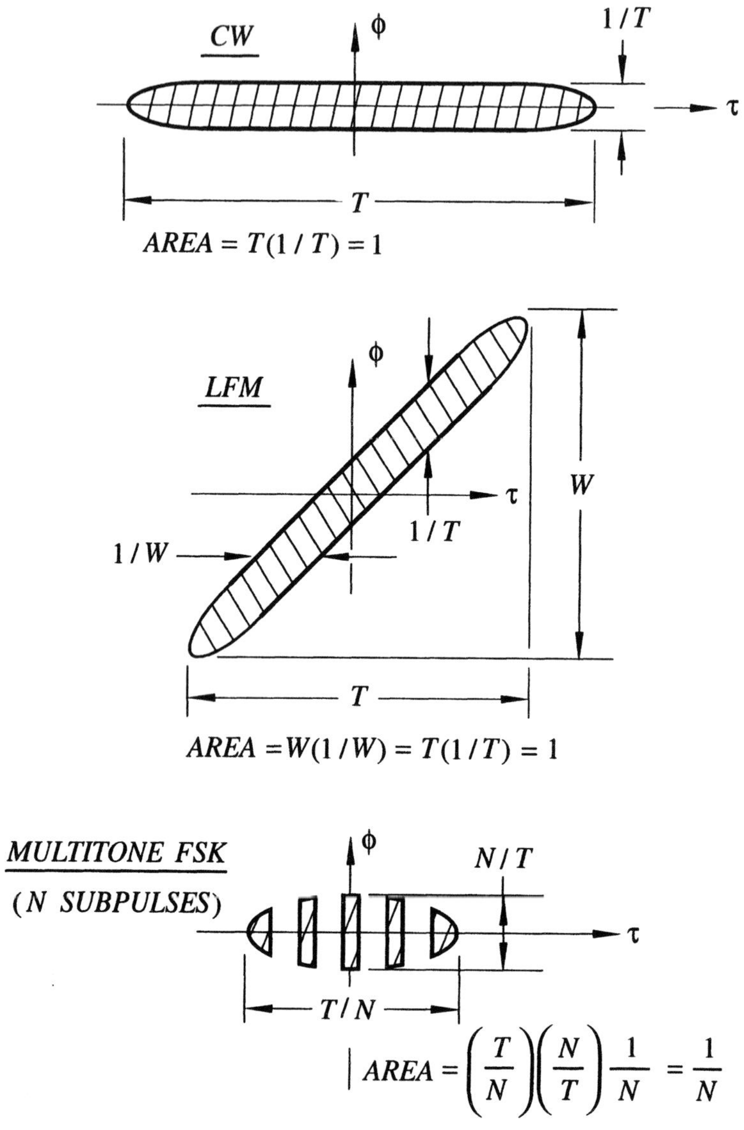

Figure 6.5: Unit volume ambiguity function models with $\Phi_{ff} = 1$ in the cross hatched areas [17] ©ASA 2000

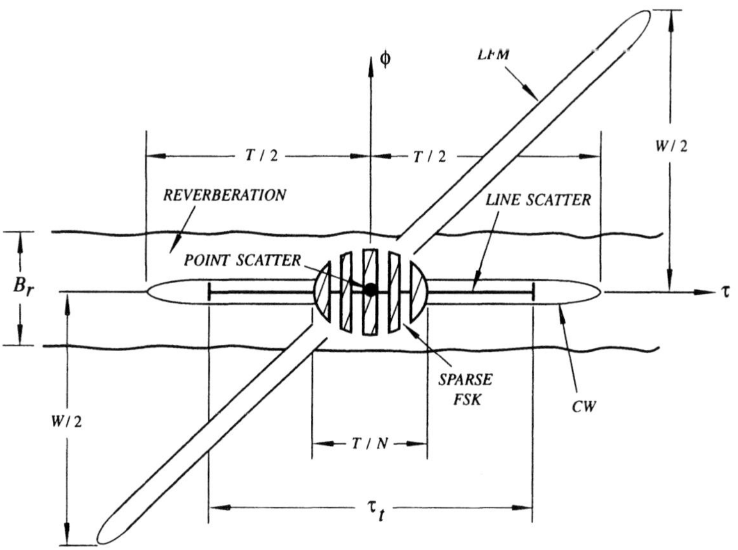

Figure 6.6: Scattering-ambiguity function convolution model for point and line targets in reverberation [17]

FSK waveform is of Welch construction with GF(5)2 with N=4, T=200 (msec), and W=200 (Hz).

The receiver responses are evaluated for zero Doppler conditions wherein the echo and reverberation scattering functions and the waveform ambiguity functions are centered at the origin as in Figure 6.6. The expected MF responses are

$$
\begin{aligned}
E_{fs} &= E_0 \mathcal{R}_{sp} \quad \text{(point scatterer)} \\
&= E_0 \mathcal{R}_{sl} \tau_i \quad \text{(line scatterer)} \\
E_{fr} &= E_0 \mathcal{R}_r A_r \quad \text{(reverberation)}
\end{aligned} \tag{6.28}
$$

and obtained by calculating the SF-AF overlap area or intersection from the constant amplitude models. A_r is the overlap area between the reverberation SF and the AF and τ_i is the intersection of the line scatterer SF and the AF mainlobe as in the figure. It is assumed that the minimum interpulse frequency spacing of the FSK is such that only its ambiguity mainlobe falls within the bandwidth of the reverberation SF as shown. This requires that its bandwidth exceeds $(N-1)B_r$.

The response for a CW waveform of duration T reflected from a line scatterer of length $\tau_t \leq T$ is an example. The intersection is $\tau_i = \tau_t$

and $E_{fs} = E_0\mathcal{R}_{sl}\tau_t$. If $\tau_t > T$, $\tau_i = T$ and $E_{fs} = E_s\mathcal{R}_{sl}T$. The FSK responses are somewhat more complex as the mainlobe area A_r and support τ_i are reduced by the factor $1/N$ and

$$\tau_i = \begin{cases} \tau_t/N, & \text{when } \tau_t \le T_p = T/N \\ T_p/N, & \text{when } \tau_t > T_p = T/N. \end{cases}$$

The FSK line scatter responses are

$$E_{fs} = \begin{cases} E_0 R_{sl}\tau_t/N, & \text{if } \tau_t < T_p \\ E_0 R_{sl} T/N^2, & \text{if } \tau_t > T_p. \end{cases} \tag{6.29}$$

The scatterer and reverberation response expressions for the waveforms under consideration are summarized in Tables 6.1 and 6.2 respectively. The processing gains

$$G_p = \left(\frac{E_{fs}}{E_{fr} + N_0}\right)\left(\frac{N_0 W_s(1+r_n)}{P_s}\right) \tag{6.30}$$

that are summarized in Tables 6.3, 6.4, 6.5 are computed from (6.19, 6.21, 6.22, 6.26) and the responses listed in Tables 6.1, 6.2. They are parameterized by the reverberation to noise input power ratio

$$r_n = \frac{P_r}{P_n} = \frac{E_0\mathcal{R}_r B_r}{N_0 W_s} \tag{6.31}$$

where purely noise and reverberation limited conditions correspond to $r_n = 0, \infty$ respectively.

A low order FSK waveform interrogating a short line scatterer for which $\tau_t < T_p$ and $B_r < 1/T_p$ serves as an example of a processing gain computation. The input SIR is

$$\text{SIR}_{in} = \frac{P_s}{N_0 W_s + P_r} = \frac{P_0\mathcal{R}_{sl}\tau_t}{N_0 W_s(1+r_n)}$$

where the signal term in the numerator is obtained from (6.22) and the interference term is obtained from (6.26). The output SIR is

$$\text{SIR}_{out} = \frac{E_0\mathcal{R}_{sl}\tau_t/N}{N_0 + E_{fr}}$$

where the signal energy term is obtained from (6.29). The reverberation SF bandwidth is less than the subpulse width so that the effective AF-SF overlap area $A_r = B_r T_p/N$. The output reverberation is then $E_{fr} = E_0\mathcal{R}_r B_r T_p/N$ as shown in Table 6.2. The processing gain (6.30) is

$$G_p = \left(\frac{E_0\mathcal{R}_{sl}\tau_t/N}{N_0 + E_0\mathcal{R}_r B_r T_p/N}\right)\left(\frac{N_0 W_s(1+r_n)}{P_0\mathcal{R}_{sl}\tau_t}\right).$$

Scatterer	CW Response	Scatterer	LFM Response	Scatterer	FSK Response
Point	$E_0\mathcal{R}_{sp}$	Point	$E_0\mathcal{R}_{sp}$	Point	$E_0\mathcal{R}_{sp}$
Line $\tau_t \leq T$	$E_0\mathcal{R}_{sl}\tau_t$	Line $\tau_t \leq 1/W$	$E_0\mathcal{R}_{sl}\tau_t$	Line $\tau_t \leq T_p$	$E_0\mathcal{R}_{sl}\tau_t/N$
Line $\tau_t > T$	$E_0\mathcal{R}_{sl}T$	Line $\tau_t > 1/W$	$E_0\mathcal{R}_{sl}/W$	Line $\tau_t > T_p$	$E_0\mathcal{R}_{sl}T_p/N$

Table 6.1: Matched filter scatterer response E_{fs}

Rev. Bwidth	CW Response	Rev. Bwidth	LFM Response	Rev. Bwidth	FSK Response
$B_r \leq 1/T$	$E_0\mathcal{R}_r B_r T$	$B_r < W$	$E_0\mathcal{R}_r B_r/W$	$B_r \leq 1/T_p$	$E_0\mathcal{R}_r B_r T_p/N$
$B_r > 1/T$	$E_0\mathcal{R}_r$	$B_r > W$	$E_0\mathcal{R}_r B_r/W$	$B_r > 1/T_p$	$E_0\mathcal{R}_r/N$

Table 6.2: Matched filter reverberation response E_{fr}

Let $E_0\mathcal{R}_rB_r = r_nN_0W_s$ and $E_0 = P_0T$ by using (6.31). Then,

$$G_p = \left(\frac{P_0T\mathcal{R}_{sl}\tau_t/N}{N_0 + r_nN_0W_sT_p/N}\right)\left(\frac{N_0W_s(1+r_n)}{P_0\mathcal{R}_{sl}\tau_t}\right)$$
$$= \frac{W_sT(1+r_n)}{N + r_nTW_s/N} \quad \text{(FSK with } B_r < 1/T_p,\ \tau_t < T_p)$$

as given in Table 6.5. The table entries may be used to assess the relative

Rev.Bwidth	G_p
$B_r \leq 1/T$	$\frac{W_sT(1+r_n)}{1+W_sTr_n}$
$B_r > 1/T$	$\frac{W_sT(1+r_n)}{1+(W_sr_n/B_r)}$

Table 6.3: CW processing gain for all scatterers

output SIR performance as well as the processing gains. The comparison between CW and LFM waveforms under point scattering conditions is an example. Dividing the output signal response by that of the interference reveals that both waveforms have the same output SIR as well as processing gain under noise limited conditions. However when reverberation is dominant ($r_n \to \infty$), the processing gain of the LFM is TW but that of the CW depends upon the pulse length T in relation to the reverberation bandwidth B_r. If the pulse is relatively short so that $1/T < B_r$, the CW G_p is unity implying that there is no SIR gain obtained by matched filtering with the CW. On the other hand the CW gain is B_rT for a longer pulse for which $B_r < 1/T$ however its output SIR is $\mathcal{R}_{sp}/\mathcal{R}_r$ while that of the LFM is $\mathcal{R}_{sp}W/\mathcal{R}_rB_r$ which is an increase in proportion to the ratio of the signal to reverberation bandwidth. The LFM output SIR remains the same for a shorter pulse such that $B_r < 1/T$ but the CW output is decreased to$\mathcal{R}_{sp}/\mathcal{R}_rB_rT$. The bottom line is that under short pulse reverberation limited conditions, a CW of duration T and an LFM with equivalent bandwidth $W = 1/T$ regardless of its duration will have the same output SIR even though the LFM has a processing gain of TW while that of the CW is only unity. The reason is simply that a short pulse illuminates less of the volume or boundary and hence generates less reverberation in the first place. A longer LFM pulse generates more input reverberation power but its processing gain effectively reduces the interference realized at the output. This indicates

Point scatt.	Line scatt. $0 < \tau_t \leq 1/W$	line scatt $1/W < \tau_t \leq T$	Line scatt. $T < \tau_t$
$\frac{W_s T(1+r_n)}{1+(W_s r_n/W)}$	$\frac{W_s T(1+r_n)}{1+(W_s r_n/W)}$	$\frac{W_s T(1+r_n)}{\tau_t(W+W_s r_n)}$	$\frac{1+r_n}{r_n+(W/W_s)}$

Table 6.4: LFM processing gains for $B_r \leq W$

Rev. Bwidth	Point scatt.	Line scatt. $\tau_t \leq T_p$	Line scatt. $T_p < \tau_t \leq T$	Line scatt. $T < \tau_t$
$B_r \leq 1/T_p$	$\frac{W_s T(1+r_n)}{1+(W_s T r_n/N^2)}$	$\frac{W_s T(1+r_n)}{N+(W_s T r_n/N)}$	$\frac{W_s T^2(1+r_n)}{\tau_t(N^2+W_s T r_n)}$	$\frac{W_s T(1+r_n)}{N^2+W_s T r_n}$
$B_r > 1/T_p$	$\frac{W_s T(1+r_n)}{1+(W_s r_n/N B_r)}$	$\frac{W_s T(1+r_n)}{N+(W_s r_n/B_r)}$	$\frac{W_s T^2(1+r_n)/\tau_t}{N^2+(N W_s r_n/B_r)}$	$\frac{W_s T(1+r_n)}{N^2+(N W_s r_n/B_r)}$

Table 6.5: FSK processing gains for sparse codes with $W_{fsk} > (N-1)B_r$

that when assessing waveform matched filter performance it is important to consider the realized output SIR and not just the processing gain.

The Performance of "Thumbtack" Waveforms

Highly resolvent dense FSK and PRN waveforms have ambiguity functions such as shown in Figures 5.19, 5.23 that approach the ideal thumbtack AF shown in Figure 5.4. They are rarely used as detection waveforms because of the "needle in the haystack" problem involved in searching the parameter domain with a highly resolvent waveform. They can be used for delay-Doppler estimation or imaging purposes once the scatterer is detected and roughly located by a lower resolution processor. The energy normalized thumbtack AF mainlobe is a narrow spike of unity height and its sidelobes approach a continuous pedestal. The pedestal height and mainlobe area are approximately $1/TW$ but unlike sparse FSK waveforms for which the zero Doppler reverberation response involves only the mainlobe, the sidelobes in the form of the pedestal contribute significantly to the reverberation response. Thumbtack G_p is estimated for point and line scattering as for the other waveforms where the input SIR is computed from $\text{SIR}_{in} = P_s/P_i$ using (6.27). The output point scatter response at the ambiguity origin is $E_0\mathcal{R}_{sp}$ (6.28) and the line scatter response assuming that $\tau_t > 1/W$ is

$$
\begin{aligned}
E_{fs} &= E_0\mathcal{R}_{sl}\left(\frac{1}{W} + \frac{(\tau_t - 1/W)}{TW}\right) \\
&= \frac{E_0\mathcal{R}_{sl}}{W}\left(1 + \frac{\tau_t}{T} - \frac{1}{TW}\right) \quad \text{(line scatt. } \tau_t \leq T) \\
&= \frac{E_0\mathcal{R}_{sl}}{W}\left(2 - \frac{1}{TW}\right) \quad \text{(line scatt. } \tau_t > T).
\end{aligned}
$$

The first term within the bracket is the mainlobe response with unity weight and delay resolution $1/W$. The second is the pedestal scatterer response with weight $1/TW$ over the scatter or pedestal duration. The factor of $1/TW$ appearing within the brackets accounts for the mainlobe area when evaluating the pedestal response. It can be neglected if $\tau_t >> 1/W$. The reverberation output assuming that $B_r < W$ is

$$
E_{fr} = E_0\mathcal{R}_r\left(\frac{(B_rT - 1/TW)}{TW} + \frac{1}{TW}\right) = \frac{E_0\mathcal{R}_r}{TW}(B_rT + 1) - \frac{E_0\mathcal{R}_r}{(TW)^2}.
$$

The first bracketed term is the pedestal component of area B_rT and weight $1/TW$, the second is the mainlobe component with unity weight and area $1/TW$, and the term involving $1/(TW)^2$ accounts for the main-lobe area as above. It is neglected if $TW >> 1 + B_rT$.

The resulting thumbtack AF processing gain estimates for point and line scattering assuming $\tau_t >> 1/W >$ and large TW are:

$$\begin{aligned} G_p &= \frac{B_rT^2WW_s(1+r_n)}{B_rTW + r_nW_s(1+B_rT)} \quad \text{(point scattering)} \\ &= \frac{B_rW_sT(1+T/\tau_t)(1+r_n)}{W_sr_n(1+B_rT)+B_rTW} \quad \text{(line scattering)}\tau_t \leq T \\ &= \frac{2B_rW_sT(1+r_n)}{W_sr_n(1+B_rT)+B_rTW} \quad \text{(line scattering)}\tau_t > T. \end{aligned}$$

The pedestal contribution reduces thumbtack reverberation G_p somewhat in relation to that of an LFM with the same TW product. The point scatter thumbtack reverberation processing gain for $r_n \to \infty$ is

$$G_p = \left(\frac{B_rT}{1+B_rT}\right)TW < TW$$

and only approaches the LFM gain given in Table 6.4 for large B_rT or when the reverberation bandwidth becomes large and begins to resemble white noise.

6.5 RECEIVER OPTIMIZATION

Optimization is interpreted as maximizing detection probability for fixed probability of false alarm. Maximization of the detection index (6.16) is a good qualitative measure of correlation receiver performance because it relates directly to the ROC's (6.17). When the scattering is point-like and the interference is white Gaussian noise (WGN), a correctly parameterized replica of the transmitted waveform is all that is necessary to form the optimum receiver. When the echo signal emanates from a spread process in WGN, it is necessary to estimate the echo time series or equivalently its spreading function in order to generate replicas with which to form a coherent filter for each realization of the echo. The receiver that synthesizes an echo replica from an estimated spreading function $\hat{\mathcal{S}}_s(\mathbf{p})$ is called the echo matched filter. An energy normalized replica

$$\hat{f}(t) = \frac{\hat{s}(t)}{\int_{-\infty}^{\infty}|\hat{s}(t)|^2dt}$$

is synthesized from a signal estimate

$$\hat{s}(t) = \sqrt{E_0} \int_{\Omega} \hat{\mathcal{S}}_s(\mathbf{p}) f(\mathbf{p}, t) d\mathbf{p}$$

and used as the MF processing function with perhaps some parameter variation to test delay and Doppler. Correlator performance is proportional to the fidelity of the estimates in relation to the stochastic realizations of the signal $\mathsf{s}(t;\eta)$. If the spreading function is random and changing from ping to ping the receiver is called an ideal or "prescient" receiver because it assumes foreknowledge of a random event. It is of course unrealizable but serves as a useful benchmark for the comparison of other approaches. Were it realizable on a ping to ping basis, the prescient receiver would reduce the spread scattering problem to a point scatter problem since the replica is perfectly matched to each realization of the echo return. In effect it is equivalent to transmitting the echo time series against a point scatterer and processing the return with the echo estimate which is the optimum detector in WGN.

The expected value of the matched filter response using the replica estimate in WGN is

$$\begin{aligned} E_{fs} &= \mathcal{E}\left\{ \left| \int_{-\infty}^{\infty} \mathsf{s}(t) \hat{f}^*(t) dt \right|^2 \right\} + N_0 \\ &\leq \int_{-\infty}^{\infty} \mathcal{E}\{|\mathsf{s}(t)|^2\} dt \int_{-\infty}^{\infty} |\hat{f}(t)|^2 dt + N_0 \\ &= E_s + N_0 \end{aligned}$$

by the Schwarz inequality. E_s is the total echo energy and the only way that the receiver can recover all of it on every ping is if the random spreading function is known for <u>every</u> stochastic realization of the scattering process. In other words, the replica must be perfectly coherent with each realization $\mathsf{s}(t;\eta)$ of the random signal time series such that the integrated coherence on each ping is

$$C_s = \frac{\left| \int_{-\infty}^{\infty} \mathsf{s}(t) \hat{f}^*(t) dt \right|^2}{\int_{-\infty}^{\infty} |\mathsf{s}(t)|^2 dt \int_{-\infty}^{\infty} |\hat{f}(t)|^2 dt} = 1$$

implying that $\hat{f}(t) = \mathsf{s}(t)$ to within a constant.

The time series or equivalently the signal spreading function usually cannot be reliably predicted on a ping to ping basis at high frequency and short ranges when the scatterer is spread and interrogating platform and scatterer are moving rapidly in relation to the interrogation cycle. The spreading associated with long range interrogation at low frequency is often dominated by multipath rather than scatterer spreading. An acoustic propagation ray path model such as the Generic Sonar Model (GSM) [26] can be used to form an estimate $\hat{\mathcal{S}}_s(\mathbf{p})$ of the channel spreading function or impulse response because the propagation channel is deterministic or at most slowly varying in time with little change over several interrogation cycles. The model is used to synthesize a model based matched filter and has been successfully applied to interrogation over long range (30 nm) shallow water channels at low frequency [27].

There are two issues associated with the use of a signal estimate as a replica aside from its coherence with the actual signal on a ping to ping basis. First, matched filtering with the replica is only optimum in WGN but not in correlated interference such as a combination of reverberation and uncorrelated noise with covariance $R_i(t,u) = R_r(t,u) + N_0\delta(t-u)$. Secondly, the transmitted waveform $f(t)$ is not specified and remains to be chosen.

The first issue was discussed in Chapter 3 and is resolved by synthesizing the optimum mismatched receiver (3.74) using the processing function $g(t)$ that is the solution to the integral equation (3.70)

$$g(t) = \frac{k}{N_0}\hat{f}(t) - \frac{1}{N_0}\int_{\Omega_u} g(u)R_r(u,t)du.$$

The reverberation covariance function is related to its scattering function by

$$R_r(t,u) = E_0\int_{\Omega} f(\mathbf{p},t)\mathcal{R}_r(\mathbf{p})f^*(\mathbf{p},u)d\mathbf{p}$$

and the equivalent relationship for wide sense stationary interference is (3.90)

$$G(\omega) = \frac{k\hat{F}(\omega)}{N_r(\omega)+N_0}$$

where $N_r(\omega)$ is the reverberation power spectral density. It is related to the reverberation SF by (6.7)

$$N_r(\omega) = \left(\frac{1}{2\pi}\right)\int_{-\infty}^{\infty}\int_{-\infty}^{\infty} S_f(\omega')\mathcal{R}_s(\tau,\omega-\omega')d\tau d\omega'.$$

The second issue is that of the transmitted waveform $f(t)$ which is the only free choice available to the sonar designer [28,29] because the signal $\hat{s}(t)$ and hence the replica $\hat{f}(t)$ are determined if it is assumed that the spreading function $\hat{S}_s(\mathbf{p})$ is known or can be modeled. The optimum maximum likelihood receiver is then defined by either a matched filter using the replica directly in WGN or by a mismatched filter using $g(t)$ in correlated interference. Optimum processor performance is affected by the choice of waveform and under specific conditions it is often possible to mitigate interference effects without materially decreasing the echo strength. The choice of a CW waveform rather than an LFM for use in a reverberant environment when some scatterer Doppler is expected is a good example [4,15]. Figure 6.7 illustrates a point scatterer that experiences reverberation interference over a Doppler range of approximately $\Delta\phi = B_r + W$ when interrogated by an LFM but only $B_r + 1/T$ when interrogated by a CW waveform. Consequently, the CW echo is clear of the reverberation over a much larger portion of the delay-Doppler search map and a simple matched filter is all that is required. A phase plane plot of simultaneous CW and LFM echo returns derived from actual data is shown in Figure 6.8. The Doppler width of the reverberation bands corresponds to the convolution of the respective ambiguity and reverberation scattering functions as depicted in Figure 6.7. The target is a slowly moving point or short line scatterer with sufficient Doppler to be clear of the CW reverberation between 2.85 to 2.95 sec. delay at the bottom of the figure but is not visible in the LFM reverberation at the top. A simple strategy as suggested by the figure is to combine CW and LFM waveforms in a single interrogation and process the reverberation and noise limited regions with LFM and CW matched filters respectively.

6.5.1 The Optimal Mismatched Receiver in WGN

The echo matched filter represents the optimum coherent receiver given complete spreading function information about the scattering process for each ping. It is the only correlation receiver configuration that can recover all of the received echo energy. If the system is constrained to use a fixed replica given only scattering function information, an optimum receiver can still be constructed. But, because it uses less information about the scattering process, its performance will be less than that of the echo matched filter and in that sense may be considered as a suboptimum but perhaps a more realizable solution. When the scatterer velocity decreases and the CW echo becomes reverberation limited, the LFM provides more processing gain.

Let $g(t)$ be a fixed, complex, deterministic, and unit energy process-

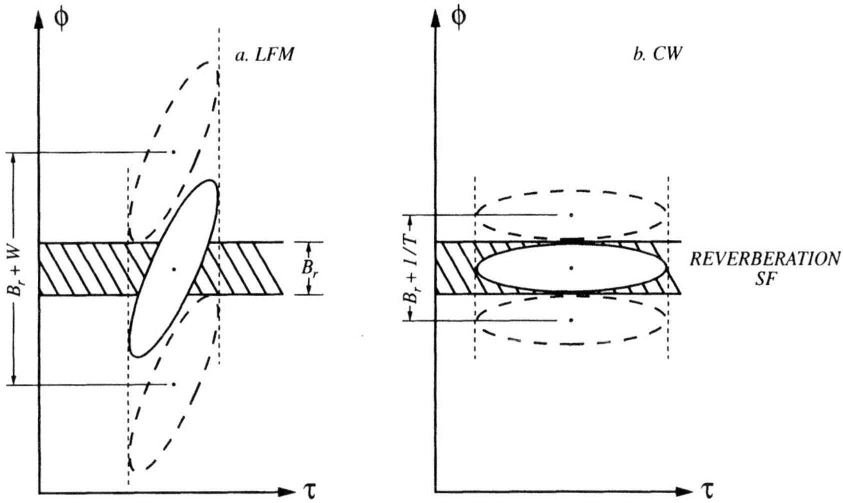

Figure 6.7: CW/LFM reverberation delay-Doppler comparison

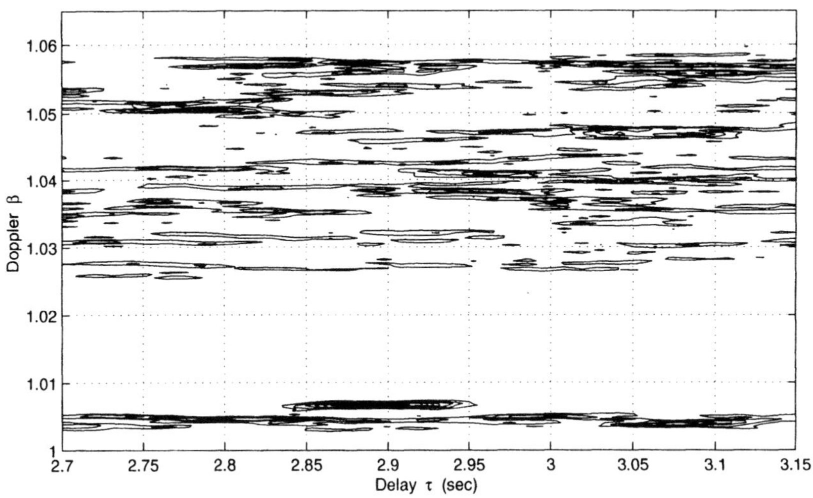

Figure 6.8: In water simultaneous LFM/CW reverberation and moving scatterer phase plane (delay-Doppler) plot

ing function that is correlated with the received signal in place of the transmitted waveform or unrealizable echo replica. The domain of the parameters is denoted by Ω and defined in Chapter 4 by the bounds (4.3). Let the signal SF occupy a bounded support region $\Omega_s \in \Omega$ about the origin. The expected signal output E_{fs} at $\hat{\mathbf{p}} = \mathbf{p}_0$ is (6.12)

$$\begin{aligned} E_{fs} &= \mathcal{E}\left\{\left|\int_{-\infty}^{\infty} s(t)g^*(t)dt\right|^2\right\} \\ &= E_0 \int_{\Omega_s} \mathcal{R}_s(\mathbf{p})\Phi_{fg}(\mathbf{p}, \mathbf{p}_0)d\mathbf{p} \end{aligned} \tag{6.32}$$

for uncorrelated spreading. The parameter origin $\mathbf{p}_0$ is defined by $\tau = 0, \beta = 1, \phi = 0$ as for (4.4, 4.5). $\Phi_{fg}(\mathbf{p}, \mathbf{p}_0)$ is the cross-ambiguity function (4.1) between the the transmitted and processing waveforms $f(\mathbf{p}, t)$ and $g(t) \equiv g(\mathbf{p}_0, t)$. Its volume is unity for normalized functions so that the available echo energy E_s is obtained by integrating over the whole parameter domain (6.14).

The processor output (6.32) is bounded by the Schwarz inequality so that $E_{fs} \leq E_s$ and it is clear that the fixed processor $g(t)$ cannot recover all of the input energy of the spread channel since it cannot be made coherent with each stochastic realization of the spreading process. In order to determine and assess the performance of the optimum processing waveform $g(t)$ that maximizes E_{fs} note that the CAF for the energy normalized functions f and g has unit volume and that $0 \leq \Phi_{fg}(\mathbf{p}, \mathbf{p}_0) < 1$ for $f(t) \neq g(t)$ and all values of $\mathbf{p}$. As a consequence, $\Phi_{fg}^2 < \Phi_{fg} \forall(\mathbf{p})$ and

$$0 \leq \int_{\Omega} \Phi_{fg}^2(\mathbf{p}, \mathbf{p}_0)d\mathbf{p} < \int_{\Omega} \Phi_{fg}(\mathbf{p}, \mathbf{p}_0)d\mathbf{p} = 1. \tag{6.33}$$

Assume that the scattering and ambiguity functions overlap with $\hat{\mathbf{p}} = \mathbf{p}_0$ as illustrated by the "high response" cartoon of Figure 6.2. The recovered signal energy (6.32) is bound by the Schwarz inequality

$$\left(\frac{E_{fs}}{E_0}\right)^2 = \left(\int_{\Omega} \mathcal{R}_s(\mathbf{p})\Phi_{fg}(\mathbf{p}, \mathbf{p}_0)d\mathbf{p}\right)^2 \leq \left(\int_{\Omega} \mathcal{R}_s^2(\mathbf{p})d\mathbf{p}\right) \cdot \left(\int_{\Omega} \Phi_{fg}^2(\mathbf{p}, \mathbf{p}_0)d\mathbf{p}\right).$$

Since both the SF and CAF are non negative functions, the CAF has unit volume, and the right hand side is bounded by (6.33), equality is achieved and the output maximized if

$$\Phi_{fg}(\mathbf{p}, \mathbf{p}_0) = \Phi_s(\mathbf{p}) = kR_s(\mathbf{p})$$
$$k = \left(\int_{\Omega_s} \mathcal{R}_s(\mathbf{p}) d\mathbf{p} \right)^{-1} \tag{6.34}$$

where Φ_s denotes a CAF that is equivalent to the normalized signal SF.

The performance of the optimal incoherent receiver is determined by substituting (6.34) into (6.32) yielding

$$\begin{aligned} E_{fs}(opt) &= \frac{E_0 \int_{\Omega_s} \mathcal{R}_s^2(\mathbf{p}) d\mathbf{p}}{\int_{\Omega_s} \mathcal{R}_s(\mathbf{p}) d\mathbf{p}} \\ &= \left(\frac{E_0^2}{E_s} \right) \int_{\Omega_s} \mathcal{R}_s^2(\mathbf{p}) d\mathbf{p}. \end{aligned} \tag{6.35}$$

The ratio of the CAF output energy to the total available energy is

$$\frac{E_{fs}}{E_s} = \frac{\int_{\Omega_s} \mathcal{R}_s^2(\mathbf{p}) d\mathbf{p}}{\left(\int_{\Omega_s} \mathcal{R}_s(\mathbf{p}) d\mathbf{p} \right)^2} \leq 1$$

because scattering functions are non-negative with bounded integrals. The scattering function and its square have the same support Ω_s in the delay-Doppler plane. Consequently

$$\int_{\Omega_s} \mathcal{R}_s(\mathbf{p}) d\mathbf{p} = A_s \bar{\mathcal{R}}_s$$

by the mean value theorem of calculus where $\bar{\mathcal{R}}_s$ represents the mean value of the scattering function and

$$A_s = \int_{\Omega_s} d\mathbf{p}$$

is the delay-Doppler phase plane "area" associated with the support region Ω_s. The variance of the scattering function over its support is

$$\sigma_{sf}^2 = \frac{1}{A_s} \int_{\Omega_s} (\mathcal{R}_s(\mathbf{p}) - \bar{\mathcal{R}}_s)^2 d\mathbf{p}$$

and

$$\frac{E_{fs}}{E_s} = \frac{1}{A_s} \left(1 + \frac{\sigma_{sf}^2}{\bar{\mathcal{R}}_s^2} \right) \leq 1. \tag{6.36}$$

This represents the fraction of the available signal energy that is recoverable by the optimal fixed receiver formulated from knowledge of the scattering function. Its performance decreases in proportion to the scattering function support which indicates the amount of spreading that the transmitted signal experiences due to propagation and scattering. The minimum performance ratio $1/A_s$ occurs when $\sigma_{sf}^2 = 0$ and the SF is constant over its support. The available energy (6.32) is fixed and at the other extreme, all of the signal energy is recovered when the scattering approaches a point. The scattering function collapses to a two dimensional delta function and

$$\frac{E_{fs}}{E_s} = \frac{\int_{\Omega_s} \delta(\mathbf{p} - \mathbf{p}_0)\mathcal{R}_s^2(\mathbf{p})d\mathbf{p}}{\left(\int_{\Omega_s} \delta(\mathbf{p} - \mathbf{p}_0)\mathcal{R}_s(\mathbf{p})d\mathbf{p}\right)^2} = \frac{\mathcal{R}_{sp}^2(\mathbf{p}_0)}{\mathcal{R}_{sp}^2(\mathbf{p}_0)} = 1$$

corresponding to the performance of the ideal but unrealizable coherent matched filter processor that utilizes the random spreading function to form its replicas.

The CAF of the transmitted and processing waveforms is the mechanism by which the white noise detection index is maximized for correlation receivers utilizing scattering function information. However a cross-ambiguity function cannot be associated uniquely with a particular waveform pair and there may be many pairs that will yield the same CAF. Drumheller [14, 30] has developed a gradient projection algorithm that recursively generates a transmit/processing waveform pair $f(t), g(t)$ maximizing (6.32) for the narrowband model. The recursion seeks to maximize the output energy goal functional

$$J(f,g) = E_{fs} = E_0 \int_{-\infty}^{\infty}\int_{-\infty}^{\infty} \mathcal{R}_s(\phi,\tau)\Phi_{fg}(\phi - \phi_0, \tau - \tau_0)d\phi d\tau$$

for overlapping SF and CAF subject to a unit energy constraint on the waveforms. This is done by fixing either function and maximizing with respect to the other or, by simultaneously varying both functions. The algorithm, using the processing function as an example, updates by

$$g_{k+1}(t) = \frac{g_k(t) + \gamma \triangle J}{||g_k(t) + \gamma \triangle J||}$$

where $\triangle J$ is the gradient of the goal functional with respect to g, γ is a gain constant, and

$$||z|| = \int_{-\infty}^{\infty} |z(t)|^2 dt$$

is the norm of a function z. The best results as measured by the magnitude of the goal function after a fixed number of iterations occurs when both functions are varied simultaneously. This also incurs the greatest computation time.

Iteration of $f(t)$ and $g(t)$ to achieve a match between the CAF and the signal SF (6.34) maximizes correlation receiver performance only under a white noise assumption. It does not address the more general problem of maximizing $D_{\mathbf{p}\hat{\mathbf{p}}}$ (6.16) when the interference includes backscattered reverberation as well as WGN. This is a joint optimization problem that has marginal practical utility in part because the variable Doppler and delay characteristics of echoes with respect to the reverberation would require a computation of the optimum ambiguity distribution for each hypothesis. However when both the echo and reverberation scattering functions occupy fixed overlapping regions Ω_s and Ω_r of the parameter domain it is possible to improve the performance of the optimum mismatched receiver by optimizing its response with respect to a composite signal/reverberation SF.

6.5.2 The Composite Cross Ambiguity Function

The signal and reverberation scattering functions $\mathcal{R}_s$ and $\mathcal{R}_r$ occupy bounded support regions Ω_s, Ω_r in the parameter domain Ω defined by (4.3). The mismatched receiver response is given by (6.32) and the associated detection index (6.16) at the parameter origin $\mathbf{p}_0$ is

$$\begin{aligned} D_{\mathbf{p}_0} &= \frac{E_0 \int_{\Omega_s} \mathcal{R}_s(\mathbf{p}) \Phi_{fg}(\mathbf{p}, \mathbf{p}_0) d\mathbf{p}}{E_0 \int_{\Omega_r} \mathcal{R}_r(\mathbf{p}) \Phi_{fg}(\mathbf{p}, \mathbf{p}_0) d\mathbf{p} + N_0} \\ &= \frac{\int_{\Omega_s} \mathcal{R}_s(\mathbf{p}) \Phi_{fg}(\mathbf{p}, \mathbf{p}_0) d\mathbf{p}}{\int_{\Omega} (\mathcal{R}_r(\mathbf{p}) + N_0/E_0) \Phi_{fg}(\mathbf{p}, \mathbf{p}_0) d\mathbf{p}}. \end{aligned} \qquad (6.37)$$

The term N_0/E_0 is included as part of the reverberation SF because the constituent transmitted and processing functions are assumed to be normalized, $\Omega_r \in \Omega$, and

$$\int_{\Omega} \Phi_{fg}(\mathbf{p}, \mathbf{p}_0) d\mathbf{p} = 1$$

by the ambiguity volume property. The detection index (6.37) is written as

$$D_{\mathbf{p}_0} = \frac{\int_{\Omega_s} \mathcal{R}_s(\mathbf{p}) \Phi_{fg}(\mathbf{p}, \mathbf{p}_0) d\mathbf{p}}{\int_{\Omega} \mathcal{R}_i(\mathbf{p}) \Phi_{fg}(\mathbf{p}, \mathbf{p}_0) d\mathbf{p}}$$

by defining the total interference SF

$$\mathcal{R}_i(\mathbf{p}) = \mathcal{R}_r(\mathbf{p}) + 1/S_0$$

that includes the constant signal to noise energy ratio $S_0 = E_0/N_0$. The normalized signal SF (6.34) is the CAF that maximizes $D_{\mathbf{p}0}$ when $\mathcal{R}_r(\mathbf{p}) = 0$ and the interference is WGN. The associated detection index obtained by substituting $\Phi_s(\mathbf{p})$ for $\Phi_{fg}(\mathbf{p}, \mathbf{p}_0)$ is

$$\begin{aligned} D_{\mathbf{p}0} \equiv D_s &= \frac{\int_{\Omega_s} \mathcal{R}_s^2(\mathbf{p})d\mathbf{p}}{\int_{\Omega_s} \mathcal{R}_s(\mathbf{p})\mathcal{R}_i(\mathbf{p})d\mathbf{p}} \quad (\text{when } \mathcal{R}_r \neq 0) \\ &= \frac{E_0 \int_{\Omega_s} \mathcal{R}_s^2(\mathbf{p})d\mathbf{p}}{N_0 \int_{\Omega_s} \mathcal{R}_s(\mathbf{p})d\mathbf{p}} = \frac{E_0^2 \int_{\Omega_s} \mathcal{R}_s^2(\mathbf{p})d\mathbf{p}}{N_0 E_s} \quad (\text{when } \mathcal{R}_r = 0) \end{aligned} \tag{6.38}$$

where the second equation is (6.35) divided by the noise psd.

D_s represents the performance of the correlation receiver using the CAF (6.34) determined only by the signal scattering distribution. When the interference is likewise spread and $\mathcal{R}_r(\mathbf{p}) \neq 0$ it is possible to account for reverberation spreading by synthesizing a composite receiver utilizing a waveform-processing function pair with the CAF

$$\Phi_{fg}(\mathbf{p}, \mathbf{p}_0) \equiv \Phi_c(\mathbf{p}) = \frac{\mathcal{R}_s(\mathbf{p})/\mathcal{R}_i(\mathbf{p})}{\int_{\Omega_s} (\mathcal{R}_s(\mathbf{p})/\mathcal{R}_i(\mathbf{p}))d\mathbf{p}}. \tag{6.39}$$

The composite CAF Φ_c like Φ_s has some of the attributes of a cross-ambiguity function because it is non-negative, has unit volume, its amplitude is bounded since $\mathcal{R}_i \geq 1/S_0$, and its phase space support is likewise bounded by Ω_s. The associated detection index is

$$D_{\mathbf{p}0} \equiv D_c = \frac{\int_{\Omega_s} (\mathcal{R}_s^2(\mathbf{p})/\mathcal{R}_i(\mathbf{p}))d\mathbf{p}}{\int_{\Omega_s} \mathcal{R}_s(\mathbf{p})d\mathbf{p}} \tag{6.40}$$

and reduces to the second equation of (6.38) when $\mathcal{R}_r = 0$. In that case, D_c and D_s are the same and represent the optimum spread echo detection index for a correlation receiver because the signal term is maximized and the spread interference is removed leaving only the noise psd.

The performance of the composite receiver relative to that of the receiver (6.38) based only upon the signal SF is defined by the ratio

$$r_{cs} = \frac{D_c}{D_s} = \int_{\Omega_s} \mathcal{R}'_{sn}(\mathbf{p})/\mathcal{R}_i(\mathbf{p})d\mathbf{p} \int_{\Omega_s} \mathcal{R}_{sn}(\mathbf{p})\mathcal{R}_i(\mathbf{p})d\mathbf{p}. \tag{6.41}$$

where

$$\mathcal{R}'_{sn}(\mathbf{p}) = \frac{\mathcal{R}_s^2(\mathbf{p})}{\int_{\Omega_s} \mathcal{R}_s^2(\mathbf{p})d\mathbf{p}}$$

$$\mathcal{R}_{sn}(\mathbf{p}) = \frac{\mathcal{R}_s(\mathbf{p})}{\int_{\Omega_s} \mathcal{R}_s(\mathbf{p})d\mathbf{p}}$$

and $r_{cs} > 1$ implies a relative gain. The performance relative to that which would be achieved by eliminating the effect of the reverberation SF $\mathcal{R}_r$ is the ratio

$$r_{co} = \frac{1}{S_0} \int_{\Omega_s} \mathcal{R}'_{sn}(\mathbf{p})/\mathcal{R}_i(\mathbf{p})d\mathbf{p}$$

of (6.40) and the second equation of (6.38). The parameter space area associated with Ω_s is A_s so that $\mathcal{R}'_{sn} = \mathcal{R}_{sn} = 1/A_s$ when $\mathcal{R}_s(\mathbf{p})$ is constant over its support. $\mathcal{R}_i > 0$ and by the Schwarz inequality

$$\begin{aligned} 1 &= \left(\frac{1}{A_s} \int_{\Omega_s} \frac{1}{\sqrt{\mathcal{R}_i(\mathbf{p})}} \sqrt{\mathcal{R}_i(\mathbf{p})} d\mathbf{p} \right)^2 \\ &\leq \frac{1}{A_s} \int_{\Omega_s} \frac{1}{\mathcal{R}_i(\mathbf{p})} d\mathbf{p} \frac{1}{A_s} \int_{\Omega_s} \mathcal{R}_i(\mathbf{p}) d\mathbf{p} = r_{cs} \end{aligned}$$

with equality for constant $\mathcal{R}_i(\mathbf{p})$. Consequently a composite receiver performance gain is realized for constant $\mathcal{R}_s$ only if $\mathcal{R}_i$ varies within Ω_s. Moreover, $r_{cs} = 1$ and no gain is realized if $\mathcal{R}_i$ is constant over the signal support regardless of the variation or lack thereof of $\mathcal{R}_s$.

When the reverberation SF varies within the echo signal parameter domain Ω_s, correlation receiver performance is improved relative to that achieved by using only the signal SF (6.34) by constructing a waveform-processing function pair $f(t), g(t)$ such that $\Phi_{fg} = \Phi_c$ (6.39). The ratio r_{co} of the composite detection index to that achieved by totally eliminating the effects of $\mathcal{R}_r$ when $\mathcal{R}_s$ is constant is

$$r_{co} = \frac{N_0}{A_s E_0} \int_{\Omega_s} \frac{1}{\mathcal{R}_r(\mathbf{p}) + 1/S_0} d\mathbf{p} \leq 1$$

with equality for $\mathcal{R}_r = 0$.

It is not possible to specify the pair $f(t)$, $g(t)$ that will duplicate (6.39) exactly as ambiguity functions constructed from finite duration waveforms have unbounded frequency (Doppler) support. Moreover, the delay support of a CAF is the sum of the constituent pulse durations and it may not be practical to attempt to duplicate scattering functions with large delay extent as that might require excessively long pulse durations. The gain that is achieved depends upon the scattering function distributions and the amount of overlap of Ω_s and Ω_r. No improvement over the WGN case (6.38) is realized if the support regions are disjoint. The following example illustrates the effect of SF overlap for constant amplitude scattering functions.

Example 6.1 Scattering Function Overlap

Figure 6.9 illustrates four conditions of SF overlap. The SF's have constant levels in each defined by

$$\mathcal{R}_s(\mathbf{p}) = \begin{cases} \mathcal{R}_s, & \mathbf{p} \in \Omega_s \\ 0, & \text{elsewhere} \end{cases}$$

and

$$\mathcal{R}_i(\mathbf{p}) = \begin{cases} \mathcal{R}_r + 1/S_0, & \mathbf{p} \in \Omega_r \\ 1/S_0, & \text{elsewhere.} \end{cases}$$

The support regions Ω_s, Ω_r have areas A_s and A_r respectively. Figures 6.9a,b represent complete and partial overlap of the interference SF by the signal SF. Figures 6.9c,d represent complete overlap of the signal SF by the interference SF and disjoint SF's respectively and there is no performance gain over the WGN case for either because in both, the interference is constant within the signal SF support Ω_s.

The composite receiver detection index (6.40) for condition (a) is

$$\begin{aligned} D_c &= \frac{E_0}{E_s}\left(\frac{\mathcal{R}_s^2 A_r}{\mathcal{R}_r + 1/S_0}\right) + \frac{E_0}{E_s}\left(\frac{\mathcal{R}_s^2(A_s - A_r)}{1/S_0}\right) \\ &= \frac{E_s}{A_s N_0}\left(1 - \frac{A_r}{A_s}\left[\frac{S_0 \mathcal{R}_r}{1 + s_0 \mathcal{R}_r}\right]\right) \end{aligned}$$

where use is made of the relationship $E_s = E_0 A_s \mathcal{R}_s$. When the scattered interference vanishes by virtue of A_r or $\mathcal{R}_r \to 0$,

$$D_c = \frac{E_s}{A_s N_0} = \frac{E_{fs}}{N_0}$$

which is equivalent to the optimal prescient receiver detection index reduced by the signal SF support and is the same as (6.36) when the signal SF variance is zero.

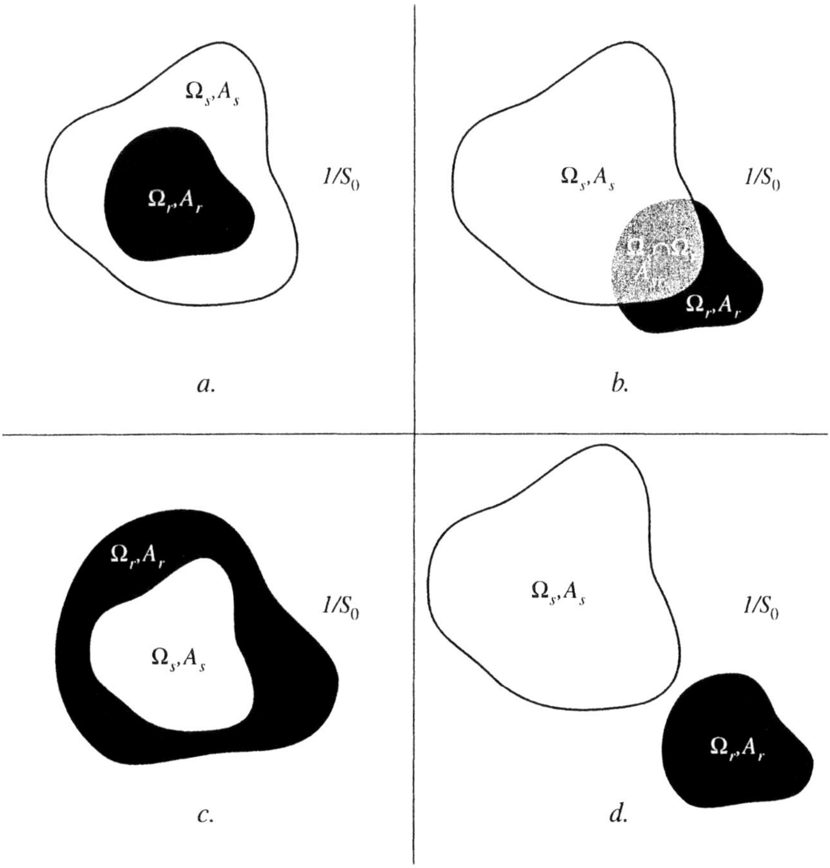

Figure 6.9: Constant signal and interference scattering function overlap regions in the delay-Doppler phase plane

The gain r_{cs} for condition (a) is computed as the product of the two integrals in (6.41) and

$$\int_{\Omega_s} \mathcal{R}'_{sn}(\mathbf{p})/\mathcal{R}_i(\mathbf{p})d\mathbf{p} = \frac{1}{A_s}\left(\frac{A_r}{\mathcal{R}_r + 1/S_0} + S_0(A_s - A_r)\right)$$

$$\int_{\Omega_s} \mathcal{R}_{sn}\mathcal{R}_i d\mathbf{p} = \frac{1}{A_s}\left(A_r(\mathcal{R}_r + 1/S_0) + \frac{(A_s - A_r)}{S_0}\right).$$

Multiplying these together yields

$$r_{cs} = 1 + \left(\left[\frac{1}{1 + S_0\mathcal{R}_r} + 1 + S_0\mathcal{R}_r\right] - 2\right)\frac{A_r(A_s - A_r)}{A_s^2}.$$

S_0 and $\mathcal{R}_r$ are non-negative and the quantity within the square brackets is greater than 2 hence $r_{cs} > 1$ as long as $A_r < A_s$ and the interference is contained within the signal support. The performance ratio of the composite receiver with respect to that achieved by eliminating $\mathcal{R}_r$ for condition (a) is

$$r_{co} = 1 - \frac{A_r}{A_s}\left(\frac{S_0\mathcal{R}_r}{1 + S_0\mathcal{R}_r}\right).$$

This is unity for $\mathcal{R}_r = 0$ and is reduced in proportion to the relative areas occupied by the reverberation and signal SF's when $\mathcal{R}_r$ is large. The interested reader can verify that a similar results occurs for the partial overlap condition (b) as long as the overlap area $A_{sr} < A_s$.

The direct utilization of scattering functions to form the optimum correlation receiver requires that a transmit and processing waveform pair be generated for each arrangement of the scattering functions in the parameter domain. Formulation of the optimum correlation receiver becomes of only academic interest unless only a few specific SF distributions are possible for which the signal pairs can be precomputed. Since most sonar systems are configured to transmit a limited sonar waveform suite, the problem becomes more the problem of generating the optimum processing function given the transmit waveform. This has been found to be a less effective approach than the simultaneous variation of both functions [30]. The process of generating the optimum correlation receiver for each situation becomes prohibitively expensive for practical application.

The utilization of a limited suite of waveforms that are processed by multiple matched filters the outputs of which are incoherently combined is a more practical solution. This approach is called post detection integration or incoherent combining and takes different forms depending upon the available information regarding the signal and reverberation scattering functions. It is called *diversity* processing for communication under fading condition such as high frequency radio links where multiple channels operating at different frequencies transmit the same information that is combined after detection. Diversity processing is also used extensively to combat frequency selective fading in sonar and radar systems [31]. A single matched filter hypothesizes a point scatterer at a position $\hat{\mathbf{p}}$ in the parameter domain. When scattering is spread either in the form of multiple discrete reflectors or as a continuum such as reverberation, the incoherent magnitude square responses of multiple matched filters can be combined to form a composite statistic. The operations of forming the magnitude and magnitude square in matched or mismatched filtering have historically been called envelope and square law detection respectively [32]. Both operations destroy phase information (coherence) hence the combining of such detected outputs is thus often called post detection or incoherent integration.

The filters are realized at a different points $\hat{\mathbf{p}}_i$ in the phase plane corresponding to the underlying scattering functions of the signal and interference spreading processes. Like the formation of the optimum mismatched receiver discussed in the previous section, the incoherent combiner relies upon accurate SF information to properly form its statistic. The difference between the two approaches is that the mismatched receiver is a coherent process. The magnitude square (detection) operation occurs after coherent combining is achieved by processing the received time series generated by the waveform $f(t)$ with the processing function $g(t)$. These functions are chosen to have a CAF that optimally places ambiguity volume in the phase plane according to the signal and interference SF distributions. The incoherent or post-detection combiner forms the weighted sum of multiple real correlation processor outputs after envelope or square law detection has taken place. Waveforms are chosen for their processing gain in reverberation and potential sidelobe interactions with adjacent scatterers and interference as in Figure 6.7. Incoherent combining of LFM matched filter realizations of multipath or otherwise delay spread echoes is a typical example.

The concept is illustrated by the distribution of three discrete scatterers shown in Figure 5.1 that represents a signal SF. The corresponding matched filter responses for three parameter hypotheses that match the SF are shown in Figure 5.3 and summing the incoherent (magnitude square) MF responses at the correct positions generates a statistic with the combined average energy of all three. The performance of such an incoherent combiner depends strongly upon the quality of the SF information because the summation of incorrect MF hypotheses misses some of the signal energy and adds only interference. The optimum combiner that correctly sums the signal energy and weights it according to the local SIR is called the *estimator-correlator* (EC). It is an implementation of the likelihood function for Gaussian data and takes into account both the signal and reverberation distributions as described by their representative scattering functions [33–37].

6.6 POST-DETECTION COMBINING

6.6.1 The Estimator-Correlator (EC)

The EC concept arose from early investigations into the optimum detection of signals propagating in random channels. Kailath [38] was among the first to show that the optimum receiver for randomly perturbed signals computes the maximum *a posteriori* probability of the signal given the received data and its covariance structure. The resulting signal estimate is the least square or conditional mean estimate that is computed from the data itself and correlated with the input hence the name estimator-correlator. The EC can be developed from continuous or discrete representations of the signal and interference data however the discrete representation offers more insight into EC performance and is developed in detail. Discussions regarding continuous EC representations are found in [39, 40].

A discrete representation of continuous random echo data is obtained by the expansion of the continuous random data time series $\mathsf{r}(t) = \mathsf{s}(t) + \mathsf{n}(t)$ with a suitable vector $\boldsymbol{\xi}(t)$ of normalized basis functions $\xi_i(t)$ (3.26). These may be discrete samples, Fourier kernels, wavelets, or replications of the transmitted waveform $\xi_i(t) = f(\mathbf{p}_i, t)$ at discrete delay-Doppler parameter locations $\mathbf{p}_i$ so that the expansion is a sequence of matched filtering operations. The resulting data components (3.31)

$$\mathsf{r}_i = \int_{-\infty}^{\infty} \mathsf{r}(t) f^*(\mathbf{p}_i, t) dt \quad i = 1, N \tag{6.42}$$

are then random realizations on a discrete grid in the parameter domain (delay-Doppler phase plane). Regardless of the kernels used, each realization of (6.42) results in a set of random scalar components comprising the random vector $\mathbf{r}$ (3.30). The discrete form of the EC is then a straightforward implementation of the log likelihood ratio for

$$\begin{aligned} \mathbf{r} &= \mathbf{s} + \mathbf{n} \quad (H_1) \\ &= \mathbf{n} \quad (H_0). \end{aligned} \tag{6.43}$$

$\mathbf{s}$ and $\mathbf{n}$ are random signal and interference vectors with zero mean complex Gaussian components characterized by covariance matrices $\mathcal{E}\{\mathbf{s}\mathbf{s}^{\mathrm{H}}\} = \mathbf{R}_s$ and $\mathcal{E}\{\mathbf{n}\mathbf{n}^{\mathrm{H}}\} = \mathbf{R}_n$. The signal and interference processes are assumed to be uncorrelated with each other and the interference may have white as well as correlated noise components that emanate from backscattered (reverberation) and other ambient sources described by covariance matrices $\mathbf{R}_r$, $\mathbf{R}_c$ respectively. The data covariance under hypotheses H_1 and H_0 is

$$\mathcal{E}\{\mathbf{r}\mathbf{r}^{\mathrm{H}}\} = \begin{cases} \mathbf{R}_1 = \mathbf{R}_s + \mathbf{R}_0, & \text{for } (H_1) \\ \mathbf{R}_0 = \mathbf{R}_n + N_0\mathbf{I}, & \text{for } (H_0). \end{cases} \tag{6.44}$$

The Gaussian pdf's under conditions H_1 and H_0 are proportional to the quantities $\exp -(\mathbf{r}^{\mathrm{H}}\mathbf{R}_1^{-1}\mathbf{r})$ and $\exp -(\mathbf{r}^{\mathrm{H}}\mathbf{R}_0^{-1}\mathbf{r})$ respectively. The log likelihood ratio and the EC detection statistic is

$$\begin{aligned} \ell_r &= \mathbf{r}^{\mathrm{H}}(\mathbf{R}_0^{-1} - \mathbf{R}_1^{-1})\mathbf{r} - (\ln|\mathbf{R}_1| - \ln|\mathbf{R}_0|) \\ \ell_r &= \ell_d - \eta \end{aligned} \tag{6.45}$$

which is the vector equivalent of the scalar llr (3.55). ℓ_d can be expressed as

$$\begin{aligned} \ell_d &= \mathbf{r}^{\mathrm{H}}\mathbf{R}_1^{-1}(\mathbf{R}_1 - \mathbf{R}_0)\mathbf{R}_0^{-1}\mathbf{r} = ((\mathbf{R}_1 - \mathbf{R}_0)\mathbf{R}_1^{-1}\mathbf{r})^{\mathrm{H}}\mathbf{R}_0^{-1}\mathbf{r} \\ &= (\mathbf{R}_s(\mathbf{R}_s + \mathbf{R}_0)^{-1}\mathbf{r})^{\mathrm{H}}\mathbf{R}_0^{-1}\mathbf{r} \\ &= \hat{\mathbf{s}}^{\mathrm{H}}\mathbf{R}_0^{-1}\mathbf{r} \end{aligned} \tag{6.46}$$

where $\hat{\mathbf{s}}$ represents the estimated signal vector derived from the data that is correlated with the whitened data vector $\mathbf{R}_0^{-1}\mathbf{r}$, hence the implementation of ℓ_d is called the estimator correlator (EC). The EC is configured as two parallel branches each operating on the data $\mathbf{r}$ as shown in Figure 6.10 [41]. The lower branch generates the signal estimate $\hat{\mathbf{s}} = \mathbf{r}^{\mathrm{H}}(\mathbf{R}_s + \mathbf{R}_0)^{-1}\mathbf{R}_s$ and the upper branch whitens the data which is then correlated with the estimate. Both branches utilize prior information in the form of the signal and interference covariances. Note

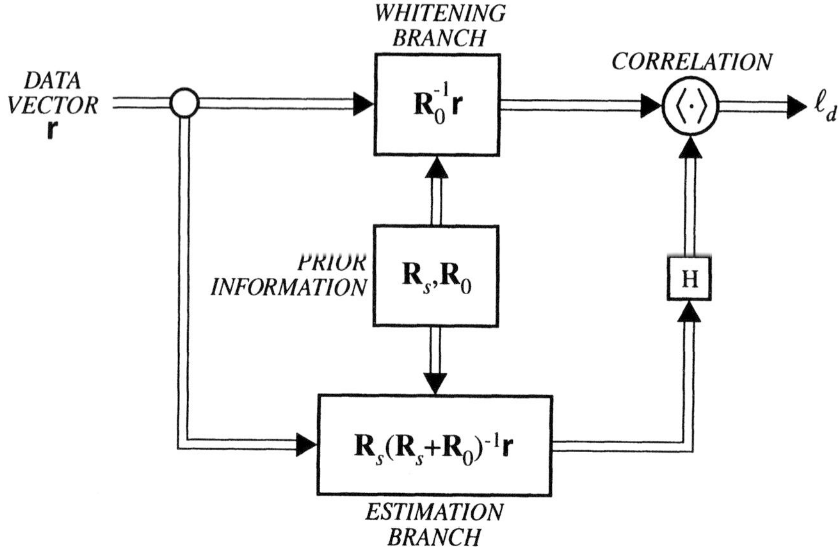

Figure 6.10: The estimator correlator implemented as parallel signal estimation and data whitening branches

that ℓ_r reduces to the scalar form (3.55) when $\mathbf{r}$ is the scalar sufficient statistic r_1. $\hat{\mathbf{s}}$ is the least square estimate (LSE) of the signal vector $\mathbf{s}$ obtained by a linear transformation $\hat{\mathbf{s}} = \mathbf{L}\mathbf{r}$ of the data vector [42]. The orthogonality principle [35,43] states that the expected error $\epsilon = \mathcal{E}\{\mathbf{s}-\hat{\mathbf{s}}\}$ must be orthogonal to the space of the data defined by any linear transformation $\tilde{\mathbf{L}}$ of $\mathbf{r}$ hence

$$\epsilon = \mathcal{E}\{(\mathbf{s} - \hat{\mathbf{s}})^{\mathrm{H}}\tilde{\mathbf{L}}\mathbf{r}\} = \mathcal{E}\{(\mathbf{s} - \mathbf{L}\mathbf{r})^{\mathrm{H}}\tilde{\mathbf{L}}\mathbf{r}\} = 0.$$

The inner product of two vectors $\mathbf{x}$, $\mathbf{y}$ can be expressed as the trace of the outer product

$$\mathbf{x}^{\mathrm{H}}\mathbf{y} = T_r(\mathbf{y}\mathbf{x}^{\mathrm{H}})$$

and the error becomes

$$\begin{aligned}\epsilon &= T_r\left(\tilde{\mathbf{L}}\mathcal{E}\{\mathbf{r}(\mathbf{s} - \mathbf{L}\mathbf{r})^{\mathrm{H}}\}\right)\\ &= T_r\left(\tilde{\mathbf{L}}\left(\mathbf{R}_{rs} - \mathbf{R}_r\mathbf{L}^{\mathrm{H}}\right)\right) = 0.\end{aligned}$$

This is required for any $\tilde{\mathbf{L}}$ therefore $\mathbf{L}^{\mathrm{H}} = \mathbf{R}_r^{-1}\mathbf{R}_{rs}$ and

$$\mathbf{L} = \mathbf{R}_{sr}\mathbf{R}_r^{-1}.$$

The signal and interference are assumed to be uncorrelated so that $\mathbf{R}_s = \mathbf{R}_{sr} = \mathcal{E}\{\mathbf{s}\mathbf{r}^{\mathrm{H}}\}$, $\mathbf{R}_r = \mathcal{E}\{\mathbf{r}\mathbf{r}^{\mathrm{H}}\}_{H_1} = \mathbf{R}_1$, and $\mathbf{R}_s = \mathbf{R}_1 - \mathbf{R}_0$. The signal estimate in (6.46) is

$$\begin{aligned} \hat{\mathbf{s}} &= \mathbf{R}_{sr}\mathbf{R}_r^{-1}\mathbf{r} \\ &= \mathbf{R}_s\mathbf{R}_r^{-1}\mathbf{r} \\ &= \mathbf{R}_s\mathbf{R}_1^{-1}\mathbf{r} \end{aligned} \tag{6.47}$$

and (6.45) is

$$\ell_r = \mathbf{r}^{\mathrm{H}}\mathbf{R}_1^{-1}\mathbf{R}_s\mathbf{R}_0^{-1}\mathbf{r} - \eta \tag{6.48}$$

which is a function of only the data $\mathbf{r}$ and the known signal and interference covariances $\mathbf{R}_s$, $\mathbf{R}_0$. These are used directly if they are available otherwise they are obtained from the signal and reverberation scattering functions $\mathcal{R}_s(\mathbf{p})$ and $\mathcal{R}_r(\mathbf{p})$ by first defining the corresponding continuous time signal and interference covariance functions $R_s(t,v)$, $R_n(t,v)$. The interference covariance function is

$$\begin{aligned} R_n(t,u) = E_0 &\int_\Omega \int_\Omega \mathcal{R}_r(\mathbf{p},\hat{\mathbf{p}}) f(\mathbf{p},t) f^*(\hat{\mathbf{p}},v) d\mathbf{p}\, d\hat{\mathbf{p}} \\ &+ R_c(t,u) + N_0\delta(t-u) \end{aligned} \tag{6.49}$$

and composed of a reverberation term plus ambient correlated and white noise terms (3.24). The signal covariance function is

$$\begin{aligned} R_s(t,v) &= \mathcal{E}\{\mathsf{s}(t)\mathsf{s}^*(u)\} \\ &= E_0 \int_\Omega \int_\Omega \mathcal{E}\{\mathcal{S}_s(\mathbf{p})\mathcal{S}_s^*(\hat{\mathbf{p}})\} f(\mathbf{p},t) f^*(\hat{\mathbf{p}},u) d\mathbf{p}\, d\hat{\mathbf{p}} \\ &= E_0 \int_\Omega \int_\Omega \mathcal{R}_s(\mathbf{p},\hat{\mathbf{p}}) f(\mathbf{p},t) f^*(\hat{\mathbf{p}},u) d\mathbf{p}\, d\hat{\mathbf{p}} \end{aligned} \tag{6.50}$$

and depends upon the SF of the combined reflection and propagation processes. [1] It simplifies to

$$R_s(t,u) = E_0 \int_\Omega \mathcal{R}_s(\mathbf{p}) f(\mathbf{p},t) f^*(\mathbf{p},u) d\mathbf{p}.$$

under uncorrelated scattering conditions. The covariance matrix $\mathbf{R}_s$ is obtained from the signal expansion

$$\mathbf{s} = \int_{-\infty}^{\infty} \mathsf{s}(t)\boldsymbol{\xi}^*(t) dt$$

[1] Only the derivation for $\mathbf{R}_s$ is presented because the derivation of the interference covariance is similar.

and

$$\mathbf{R}_s = \mathcal{E}\{\mathbf{s}\mathbf{s}^{\mathrm{H}}\} = \mathcal{E}\left\{\int_{-\infty}^{\infty}\int_{-\infty}^{\infty} \mathrm{s}(t)\boldsymbol{\xi}^*(t)\boldsymbol{\xi}^t(u)\mathrm{s}^*(u)\,du\,dt\right\}$$
$$= \int_{-\infty}^{\infty}\int_{-\infty}^{\infty} \boldsymbol{\xi}^*(t)R_s(t,u)\boldsymbol{\xi}^t(u)\,du\,dt.$$

Substituting (6.50) yields

$$\mathbf{R}_s = E_0 \int_\Omega\int_\Omega \mathcal{R}_s(\mathbf{p},\hat{\mathbf{p}})$$
$$\cdot \int_{-\infty}^{\infty}\int_{-\infty}^{\infty} [\boldsymbol{\xi}^*(t)f(\mathbf{p},t)]\left[\boldsymbol{\xi}^t(u)f^*(\hat{\mathbf{p}},u)\right]\,du\,dt\,d\mathbf{p}\,d\hat{\mathbf{p}}.$$

The square bracketed terms within the integral are the expansion coefficient vectors for the deterministic waveform functions denoted by $\mathbf{f}(\mathbf{p})$, $\mathbf{f}(\hat{\mathbf{p}})$ and

$$\mathbf{R}_s = E_0 \int_\Omega\int_\Omega \mathcal{R}_s(\mathbf{p},\hat{\mathbf{p}})\mathbf{R}_f(\mathbf{p},\hat{\mathbf{p}})\,d\mathbf{p}\,d\hat{\mathbf{p}}$$
$$\mathbf{R}_f(\mathbf{p},\hat{\mathbf{p}}) = \mathbf{f}(\mathbf{p})\mathbf{f}^{\mathrm{H}}(\hat{\mathbf{p}}).$$

This reduces to

$$\mathbf{R}_s = E_0 \int_\Omega \mathcal{R}_s(\mathbf{p})\mathbf{R}_f(\mathbf{p})\,d\mathbf{p} \tag{6.51}$$

for uncorrelated scattering. If the expansion kernels are replications of the parameterized waveform function $\xi_i(t) = f(\mathbf{p}_i', t)$ defined at N discrete locations $\mathbf{p}_i'$ in the parameter domain, the elements of the waveform vector $\mathbf{f}(\mathbf{p})$ are the uncertainty functions

$$\chi_{ff}(\mathbf{p}_i',\mathbf{p}) = \int_{-\infty}^{\infty} f(\mathbf{p}_i',t)f^*(\mathbf{p},t)dt$$

where

$$\Phi_{ff}(\mathbf{p}_i',\mathbf{p}) = \left|\chi_{ff}(\mathbf{p}_i',\mathbf{p})\right|^2$$

is the associated AF . The $N\times N$ waveform expansion matrix $\mathbf{R}_f(\mathbf{p},\hat{\mathbf{p}})$ then has components

$$R_f(\mathbf{p},\hat{\mathbf{p}})_{ij} = \chi_{ff}(\mathbf{p}_i',\mathbf{p})\chi_{ff}^*(\mathbf{p}_j',\hat{\mathbf{p}})$$

and under the uncorrelated scattering assumption

$$R_f(\mathbf{p})_{ij} = \chi_{ff}(\mathbf{p}_i',\mathbf{p})\chi_{ff}^*(\mathbf{p}_j',\mathbf{p})$$

with diagonal terms equal to the ambiguity functions $\Phi_{ff}(\mathbf{p}'_i, \mathbf{p})$. The elements of the signal matrix (6.51) are

$$R_{sij} = E_0 \int_\Omega \mathcal{R}_s(\mathbf{p})\chi_{ff}(\mathbf{p}'_i, \mathbf{p})\chi^*_{ff}(\mathbf{p}'_j, \mathbf{p})d\mathbf{p}.$$

The role of the transmitted waveform and its associated AF is illustrated by considering a discrete scattering process with M scattering sites described by

$$\mathcal{R}_s(\mathbf{p}) = \sum_{k=1}^{M} \mathcal{R}_{sk}\delta(\mathbf{p} - \mathbf{p}_k)$$

where M is not necessarily equal to the number of data vector components or parameter hypotheses N. The signal matrix elements are

$$R_{sij} = E_0 \sum_{k=1}^{M} \mathcal{R}_{sk}\chi_{ff}(\mathbf{p}'_i, \mathbf{p}_k)\chi^*_{ff}(\mathbf{p}'_j, \mathbf{p}_k), \ i, j = 1, N$$

and on the diagonal

$$R_{sii} = E_0 \sum_{k=1}^{M} \mathcal{R}_{sk}\Phi_{ff}(\mathbf{p}'_i, \mathbf{p}_k). \tag{6.52}$$

This represents the sum of matched filter realizations and the i-th signal element is the contribution of all of the point scatterers weighted by the AF in proportion to the parameter hypothesis mismatch between $\mathbf{p}'_i$ and $\mathbf{p}_k$. When the parameters are drawn from the same set, the weighting will be unity when $k = i$. If $M \neq N$ the situation is regarded as either over or under-determined because there are either too many or too few basis functions or matched filters for the number of scatterers.

The χ_{ff} are generally complex quantities with magnitude less than unity and the discrete SF is real non-negative. Consequently the matrix elements are bounded by

$$|R_{sij}| \leq E_0 \sum_{k=1}^{M} \mathcal{R}_{sk}|\chi_{ff}(\mathbf{p}'_i, \mathbf{p}_k)||\chi^*_{ff}(\mathbf{p}'_j, \mathbf{p}_k)|.$$

The "thumbtack" ambiguity function typical of a PRN sequence or dense Costas code such as shown in Figure 5.23 serves to illustrate the sidelobe contribution to the signal matrix. The ideal thumbtack model has a spike-like mainlobe surrounded by a sidelobe pedestal of approximate area TW with height $1/TW$ as shown in Figure 5.4. Assuming that all parameter mismatch exceeds the mainlobe bounds, the AUF has magnitude

$$|\chi^*_{ff}(\mathbf{p}'_j, \mathbf{p}_k)| : \begin{cases} = 1, & \text{if } \mathbf{p}'_j = \mathbf{p}_k \\ \approx 1/\sqrt{TW}, & \text{if } \mathbf{p}'_j \neq \mathbf{p}_k \end{cases}$$

and if none of the hypothesized parameters match the actual ones,

$$|\hat{R}_{sij}| \leq \frac{E_0}{TW} \sum_{k=1}^{M} \mathcal{R}_{sk} = \frac{E_s}{TW} \forall i, j$$

where E_s is the total backscattered energy. When the parameters and the scattering amplitudes are known, $M = N$, $\mathbf{p}'_k = \mathbf{p}_k \, \forall k$, $\hat{R}_{sij} = R_{sij}$, and $\chi_{ff}(\mathbf{p}_k, \mathbf{p}_k) = 1$. Then, the off-diagonal and diagonal terms are bounded by

$$|R_{sij}| \leq \frac{E_0}{\sqrt{TW}} (\mathcal{R}_{si} + \mathcal{R}_{sj}) + \frac{E_0}{TW} \sum_{k}^{M} \mathcal{R}_{sk}, \quad k \neq i \neq j$$

and

$$|R_{sii}| \leq E_0 \left(\mathcal{R}_{si} + \frac{1}{TW} \sum_{k \neq i}^{M} \mathcal{R}_{sk} \right).$$

Thumbtack AF sidelobes decrease with the TW product and the matrix approaches the diagonal form with elements given by (6.52). The AF is unity at its origin and if the parameters hypothesized by waveform functions $f(\mathbf{p}'_i, t)$ match those of the SF, the diagonal signal matrix elements (eigenvalues) will be the energy weighted point scatter magnitudes

$$R_{sii} = E_0 \mathcal{R}_{si} = E_{si}.$$

The effect of the sidelobes is illustrated by the scattering example of Figures 5.1, 5.2, 5.3. All of the waveforms reproduce the three point scattering images but the VFM and the hop code exhibit considerable sidelobe reinforcement resulting in higher sidelobes than are exhibited by the original ambiguity functions. The CW and LFM images appear as reasonably close reproductions of their respective AFs at the discrete scattering positions. The reason is that the scattering distribution is spread in such a way that the VFM and hop code AF sidelobes interact strongly but those of the CW and LFM do so only minimally.

The EC likelihood ℓ_r (6.48) is easier to implement if the covariances are diagonal suggesting that waveform choice with respect to the scattering functions is a design consideration. Accurate SF information is crucial to EC performance because the signal components are attenuated in proportion to the misregistration between the actual SF and the hypothesized $\mathbf{p}_i'$. Consequently AF delay-Doppler resolution should not exceed the accuracy with which the SF can be estimated. If the SF is confined to a single dimension i.e., delay or Doppler, the problem is less severe because AF resolution can be relaxed in the other dimension. Ideally, the actual number, parameter locations, and the amplitude distribution of the SF coefficients are known a-priori. For example if $N > M$ or vice-versa, the EC is over (under) determined because there are more (fewer) hypothesized scattering locations than necessary and detection performance is degraded. The Receiver Operating Characteristics (ROC's) are the best measure of EC detector performance and the degradation that occurs when the signal and interference scattering functions are not perfectly known. In order to compute them it is first necessary to derive the detection and false alarm probabilities for the ideal estimator correlator with $M = N$ and complete prior information regarding the scattering processes. The likelihood function (6.48) is expressed as the incoherent sum of transformed data components weighted by the eigenvalues of a generalized signal to interference ratio matrix. The pdf's of the sums under H_0, H_1 are then found by straightforward application of characteristic functions. Once the ideal ROC's are obtained it is a relatively simple matter to compute them under less than ideal conditions.

Generalized Signal to Interference Ratio

The likelihood ℓ_d 6.46) is the weighted inner product of the complex data vector (6.43) and is expressed as an incoherent summation of positive real quantities by simultaneously diagonalizing $\mathbf{R}_1$, $\mathbf{R}_0$ as the generalized eigenvalue problem [44, 45]

$$\mathbf{R}_1\mathbf{x}_i = \lambda_{mi}\mathbf{R}_0\mathbf{x}_i \tag{6.53}$$

or equivalently

$$\mathbf{R}_0^{-1}\mathbf{R}_1\mathbf{x}_i = \lambda_{mi}\mathbf{x}_i \quad i = 1,\, M = N. \tag{6.54}$$

The solution provides the eigenvalues λ_{mi} and corresponding eigenvectors $\mathbf{x}_i$ of $\mathbf{R}_0^{-1}\mathbf{R}_1$.

To start, $\mathbf{R}_0$ is diagonalized as

$$\mathbf{R}_0 = \mathbf{U}_0\mathbf{\Lambda}_0\mathbf{U}_0^{\mathrm{H}}$$

where and $\mathbf{\Lambda}_0 = \text{diag}[\lambda_{0i}]$, $i = 1, M$ is the diagonal matrix of eigenvalues and $\mathbf{U}_0 = \text{cols}[\mathbf{u}_{0i}]$, $i = 1, M$ is the corresponding unitary eigenvector matrix. The $\mathbf{R}_0$ is Hermitian, assumed to be full rank and its eigenvalues are positive. Consequently,

$$\begin{aligned}\mathbf{R}_0 &= \mathbf{R}_0^{\mathrm{H}/2}\mathbf{R}_0^{1/2} \\ \mathbf{R}_0^{1/2} &= \mathbf{U}_0\sqrt{\mathbf{\Lambda}_0}\mathbf{U}_0^{\mathrm{H}}\end{aligned} \tag{6.55}$$

and $\sqrt{\mathbf{\Lambda}_0} = \text{diag}[\sqrt{\lambda_{0i}}]$, $i = 1, M$. The generalized eigenvalue problem (6.54) becomes an ordinary eigenvalue problem $\mathbf{Mu} = \lambda_m\mathbf{u}$ with

$$\begin{aligned}\mathbf{M} &= \mathbf{R}_0^{-\mathrm{H}/2}\mathbf{R}_1\mathbf{R}_0^{-1/2} \\ &= \mathbf{R}_0^{-\mathrm{H}/2}(\mathbf{R}_s + \mathbf{R}_0)\mathbf{R}_0^{-1/2} \\ &= \mathbf{S} + \mathbf{I}\end{aligned} \tag{6.56}$$

by substituting (6.55) in (6.53) and linearly transforming the eigenvectors by $\mathbf{x}_i = \mathbf{R}_0^{-1/2}\mathbf{u}_i$. $\mathbf{S} = \mathbf{R}_0^{-\mathrm{H}/2}\mathbf{R}_s\mathbf{R}_0^{-1/2}$ is called the generalized signal to interference matrix [35]. $\mathbf{M}$ is diagonalized by

$$\begin{aligned}\mathbf{\Lambda}_m &= \mathbf{U}^{\mathrm{H}}\mathbf{M}\mathbf{U} \\ &= \text{diag}[\lambda_{mi}] \\ \lambda_{mi} &= 1 + \mu_i,\ i = 1, M\end{aligned} \tag{6.57}$$

and the μ_i are the eigenvalues of $\mathbf{S}$ and represent generalized signal to interference ratios. The sum $\mu_t = \sum \mu_i = Tr\{\mathbf{S}\}$ is a measure of the overall SIR. The eigenvectors $\mathbf{x}_i$ and $\mathbf{u}_i$ are related by

$$\mathbf{X} = \mathbf{R}_0^{-1/2}\mathbf{U} \tag{6.58}$$

where $\mathbf{U} = \text{cols}[\mathbf{u}_i]$ and $\mathbf{X} = \text{cols}[\mathbf{x}_i]$ and $\mathbf{X}$ is related to the interference covariance by

$$\begin{aligned}\mathbf{X}^{\mathrm{H}}\mathbf{X} &= \mathbf{U}^{\mathrm{H}}\mathbf{R}_0^{-1}\mathbf{U} \\ \mathbf{X}\mathbf{X}^{\mathrm{H}} &= \mathbf{R}_0^{-1/2}\mathbf{U}\mathbf{U}^{\mathrm{H}}\mathbf{R}_0^{-\mathrm{H}/2} \\ &= \mathbf{R}_0^{-1/2}\mathbf{R}_0^{-\mathrm{H}/2} = \mathbf{R}_0^{-1}.\end{aligned} \tag{6.59}$$

$\mathbf{X}$ simultaneously diagonalizes $\mathbf{R}_1$ and $\mathbf{R}_0$ since

$$\mathbf{X}^{\mathrm{H}}\mathbf{R}_0\mathbf{X} = (\mathbf{R}_0^{-1/2}\mathbf{U})^{\mathrm{H}}(\mathbf{R}_0^{\mathrm{H}/2}\mathbf{R}_0^{1/2})(\mathbf{R}_0^{-1/2}\mathbf{U}) = \mathbf{I} \tag{6.60}$$

and using (6.58, 6.56, 6.57) to express $\mathbf{R}_1$ in terms of $\mathbf{M}$ and $\mathbf{R}_0^{1/2}$,

$$\mathbf{X}^{\mathrm{H}}\mathbf{R}_1\mathbf{X} = \mathbf{X}^{\mathrm{H}}(\mathbf{R}_0^{\mathrm{H}/2}\mathbf{M}\mathbf{R}_0^{1/2})\mathbf{X} = \mathbf{\Lambda}_m.$$

$\mathbf{R}_1$ and its inverse are then expressed in terms of $\mathbf{X}$ and $\mathbf{\Lambda}_m$ as

$$\mathbf{R}_1 = \mathbf{X}^{-\mathrm{H}}\mathbf{\Lambda}_m\mathbf{X}^{-1} \quad \mathbf{R}_1^{-1} = \mathbf{X}\mathbf{\Lambda}_m^{-1}\mathbf{X}^{\mathrm{H}}. \tag{6.61}$$

Recalling that $\mathbf{R}_1 = \mathbf{R}_s + \mathbf{R}_0$, $N = M$, and using (6.60, 6.57),

$$\begin{aligned}\mathbf{\Lambda}_m &= \mathbf{X}^{\mathrm{H}}(\mathbf{R}_s + \mathbf{R}_0)\mathbf{X} \\ &= \mathbf{X}^{\mathrm{H}}\mathbf{R}_s\mathbf{X} + \mathbf{I} \\ &= \mathrm{diag}[\mu_i + 1]_{i=1,M}\end{aligned}$$

where $\mathbf{X}^{\mathrm{H}}\mathbf{R}_s\mathbf{X}$ is the diagonalized SIR matrix $\mathbf{S}$. The estimated signal vector (6.47) is

$$\hat{\mathbf{s}} = (\mathbf{I} - \mathbf{R}_0\mathbf{R}_1^{-1})\mathbf{r}.$$

By substituting (6.59, 6.61), and making the transformations $\tilde{\mathbf{s}} = \mathbf{X}^{\mathrm{H}}\hat{\mathbf{s}}$ and $\tilde{\mathbf{r}} = \mathbf{X}^{\mathrm{H}}\mathbf{r}$, the transformed covariances become $\mathcal{E}\{\tilde{\mathbf{r}}\tilde{\mathbf{r}}^{\mathrm{H}}\} = \mathbf{\Lambda}_m|H_1$, $\mathbf{I}|H_0$ and

$$\tilde{\mathbf{s}} = (\mathbf{I} - \mathbf{\Lambda}_m^{-1})\tilde{\mathbf{r}}.$$

The transformed signal estimate components for $N = M$ are then

$$\tilde{\mathsf{s}}_i = \left(\frac{\mu_i}{1+\mu_i}\right)\tilde{\mathsf{r}}_i \quad i = 1, \cdots, M.$$

Under the transformation, the estimated EC signal vector is the data vector with components weighted according to their expected signal to interference ratios. The optimal linearly transformed signal estimate is merely the data itself when the component SIR's are much greater than unity and the signal is dominant.

The statistic (6.46) is expressed as the weighted incoherent sum

$$\begin{aligned}\ell_d &= \tilde{\mathbf{s}}^{\mathrm{H}}\tilde{\mathbf{r}} \\ &= \mathbf{r}^{\mathrm{H}}\mathbf{X}\left(\mathbf{I} - \Lambda_m^{-1}\right)\mathbf{X}^{\mathrm{H}}\mathbf{r} \\ &= \sum_{i=1}^{M} |\tilde{\mathsf{r}}_i|^2 \left(\frac{\mu_i}{1+\mu_i}\right)\end{aligned}$$

by using (6.59, 6.61) and the transformed signal and data components. The matrix determinant is equivalent to the product of its eigenvalues hence

$$\begin{aligned}\eta &= \ln|\mathbf{R}_1| - \ln|\mathbf{R}_0| \\ &= \ln\prod_{i=1}^{M}(1+\mu_i) - \ln(1) = \sum_{i=1}^{M}\ln(1+\mu_i)\end{aligned}$$

and

$$\begin{aligned}\ell_r &= \ell_d - \eta \\ &= \sum_{i=1}^{M}\left[|\tilde{\mathsf{r}}_i|^2\left(\frac{\mu_i}{1+\mu_i}\right) - \ln(1+\mu_i)\right].\end{aligned} \tag{6.62}$$

The expectations of the transformed data components are

$$\mathcal{E}\{|\tilde{\mathsf{r}}_i|^2\} = \begin{cases} 1+\mu_i, & \text{for } (H_1) \\ 1, & \text{for } (H_0)\end{cases} \tag{6.63}$$

and consequently

$$\begin{aligned}\mathcal{E}\{\ell_r|H_1\} &= \sum_{i=1}^{M}[\mu_i - \ln(1+\mu_i)] \\ \mathcal{E}\{\ell_r|H_0\} &= \sum_{i=1}^{M}\left[\frac{\mu_i}{1+\mu_i} - \ln(1+\mu_i)\right].\end{aligned} \tag{6.64}$$

The EC for Uncorrelated Data

Uncorrelated data occurs when the scattering processes are themselves uncorrelated in the parameter domain and the expansion kernels comprise a basis set. If they are MF realizations the ambiguity functions must have negligible or non interacting sidelobes as discussed earlier. Assume that the signal is correlated but the interference is WGN with psd N_0, $\mathbf{R}_0^{1/2} = \sqrt{N_0}\mathbf{I}$, $\mathbf{X} = \mathbf{U}/\sqrt{N_0}$, $\tilde{\mathsf{s}} = \mathbf{U}/\sqrt{N_0}\,\hat{\mathsf{s}}$, $\tilde{\mathsf{r}} = \mathbf{U}/\sqrt{N_0}\,\mathsf{r}$, and

$$\begin{aligned}\mathbf{S} = \mathbf{X}^{\mathrm{H}}\mathbf{R}_s\mathbf{X} &= \mathbf{U}^{\mathrm{H}}\left(\mathbf{U}_s\left(\frac{\mathbf{\Lambda}_s}{N_0}\right)\mathbf{U}_s^{\mathrm{H}}\right)\mathbf{U} \\ &= \mathbf{\Lambda}_m - \mathbf{I}\end{aligned}$$

where $\mathbf{R}_s = \mathbf{U}_s\mathbf{\Lambda}_s\mathbf{U}_s^{\mathrm{H}}$. In this case the unitary matrices $\mathbf{U}$ and $\mathbf{U}_s$ are equivalent so that

$$\mathbf{S} = \mathbf{X}^{\mathrm{H}}\mathbf{R}_s\mathbf{X} = \frac{\mathbf{\Lambda}_s}{N_0} = \mathrm{diag}[\mu_i]_{i=1,M}$$

and

$$\mu_i = \mathrm{SNR}_i = [\frac{\lambda_{si}}{N_0}]_{i=1,M}.$$

The component signal to noise ratios are not functions of the data but of the covariances that are assumed known a-priori. They are equal to the actual SIR's under H_1 if the prior information is correct. The presence or absence of the signal under H_1 or H_0 is reflected only in the data vector (6.43) and its linearly transformed counterpart $\tilde{\mathbf{r}}$ that has covariance

$$\begin{aligned}\mathcal{E}\{\tilde{\mathbf{r}}\tilde{\mathbf{r}}^{\mathrm{H}}\} &= \frac{\mathbf{U}_s^{\mathrm{H}}\mathbf{R}_1\mathbf{U}_s}{N_0} \\ &= \begin{cases} \mathbf{\Lambda}_s/N_0 + \mathbf{I}, & \text{under } H_1 \\ \mathbf{I}, & \text{under } H_0. \end{cases}\end{aligned}$$

The elements $\mathcal{E}\{|\tilde{r}_i|^2\} = (1 + \mathrm{SNR}_i)$ under H_1 so that the expectations of the log-likelihood ratio obtained from (6.62, 6.63, 6.64) are

$$\begin{aligned}\mathcal{E}\{\ell_d|H_1\} &= \sum_{i=1}^{M} \frac{(1+\mathrm{SNR}_i)}{(1+\mathrm{SNR}_i)}\mathrm{SNR}_i = \sum_{i=1}^{M} \mathrm{SNR}_i \\ &= \frac{\sum_{i=1}^{M}\lambda_{si}}{N_0} = \frac{E_s}{N_0} = \mu_t \\ \mathcal{E}\{\ell_r|H_1\} &= \frac{E_s}{N_0} - \sum \ln\left(\frac{1+\lambda_{si}}{N_0}\right). \end{aligned} \tag{6.65}$$

This is the maximum possible EC output SNR in WGN because

$$E_s = \sum_{i=1}^{M} E_{si} = \sum_{i=1}^{M} \lambda_{si}$$

is the total signal energy. The receiver (6.62) implementing ℓ_r is fixed and a function of the presumed known covariances with $\mu_i = \lambda_{si}/N_0$. The expected values of the data vector elements are merely the noise power density N_0 when the signal is not present. ℓ_r is given by (6.64) and

$$\mathcal{E}\{\ell_r|H_0\} = \sum_{i=1}^{M} \frac{\mathrm{SNR}_i}{(1+\mathrm{SNR}_i)} - \eta$$

where the SNR_i represent the prior information used to weight the incoherent summation. When the EC implementation is matched to the actual prevailing conditions and the scattering function and covariance information is correct, the μ_i and SNR_i will reflect the actual conditions under H_1.

When both the signal and interference have uncorrelated components, $\mathbf{R}_0 = \text{diag}\,[\lambda_{0i}]$, $\mathbf{R}_s = \text{diag}\,[\lambda_{si}]$, $\lambda_{si} = E_{si}$, $\mathbf{X} = \text{diag}\,[1/\sqrt{\lambda_{0i}}]$, $\tilde{\mathsf{r}}_i = \mathsf{r}_i/\sqrt{\lambda_{0i}}$, $\tilde{\mathsf{s}}_i = \hat{\mathsf{s}}_i/\sqrt{\lambda_{0i}}$, $\mu_i = \lambda_{si}/\lambda_{0i}$ and

$$\ell_d = \sum_{i=1}^{M} \frac{|\mathsf{r}_i|^2}{\lambda_{0i}} \left(\frac{\lambda_{si}/\lambda_{0i}}{1 + (\lambda_{si}/\lambda_{0i})} \right). \tag{6.66}$$

In this case the maximum expected response is

$$\mathcal{E}\{\ell_d|H_1\} = \mu_t = \sum_{i=1}^{M} \frac{\lambda_{si}}{\lambda_{0i}}, \quad \eta = \sum_{i=1}^{M} \ln\left(1 + \frac{\lambda_{si}}{\lambda_{0i}}\right).$$

Uncorrelated interference may consist of reverberant as well as ambient components (6.49) and $\lambda_{0i} = E_0\mathcal{R}_{ri} + \lambda_{ci} + N_0$. Note that μ_t (6.65) is the ratio of the total signal energy to noise energy only when the noise is uniform. It is a function of the signal and interference distributions and is dominated by those terms with the greatest μ_i when the λ_{0i} are not uniform across all parameter hypotheses. It has been suggested that a simpler reduced rank EC detector might be constructed by eliminating from the summation (6.62) those terms with low μ_i that do not contribute significantly to the likelihood [35]. The expected EC signal energy realized under uncorrelated signal and white noise conditions (6.65) matches that of the ideal matched filter or prescient receiver but it does not mean that the EC detection performance (ROC) is the same. The EC energy is realized as the incoherent summation of individual filter responses each matched to a discrete echo not as the coherent integration of a single matched filter. Consequently while the realized signal energy may be the same, the probabilities of detection and false alarm and hence the ROC's are not the same as those of an echo matched filter (6.17) with $D_{\mathbf{pp}} = E_s/N_0$.

EC Receiver Operating Characteristics

ROC curves are plots of detection vs false alarm probability and are computed by evaluating the EC probabilities parametrically with a threshold

γ [36,37]. The detection and false alarm probabilities denoted by $\wp_d$ and $\wp_f$ are computed from the generic relationship for a statistic

$$\mathsf{z} = \sum_{i=1}^{M} b_i |\mathsf{x}_i|^2$$

corresponding to the incoherent summation for ℓ_d given by (6.62). The b_i represent the EC SIR weights and the x_i are complex uncorrelated zero mean Gaussian quantities with real and imaginary components distributed as

$$\mathsf{f}_\mathsf{x}(x) = \frac{1}{\sigma\sqrt{2\pi}} \exp - \frac{x^2}{2\sigma^2}.$$

The pdf is found from the product of the characteristic functions (CF's) of the components where for each,

$$\phi_x(\omega) = \int_{-\infty}^{\infty} \mathsf{f}_\mathsf{x}(x) \exp - j\,\omega g(x) dx$$

with $g(x) = bx^2$. Substituting for the pdf yields

$$\begin{aligned}\phi_x(\omega) &= \sqrt{\frac{2}{\pi\sigma^2}} \int_0^{\infty} \exp - \left(\frac{1}{2\sigma^2} + j\omega b\right) x^2 dx \\ &= \frac{1}{\sqrt{1 + j2\sigma^2\omega b}}.\end{aligned}$$

The CF of $|\mathsf{x}_i|^2$ is the product of the identical real and imaginary component CF's

$$\phi_{x^2} = \frac{1}{1 + j2\sigma^2\omega b_i}$$

with weights

$$b_i = \frac{\mu_i}{1 + \mu_i} \quad i = 1, \cdots, M$$

of the incoherent sum (6.62). The variances of the real and imaginary parts of the transformed data components $\tilde{\mathsf{r}}_i$ under H_1 and H_0 are half of the expectations (6.63)

$$\begin{aligned}\sigma_{i1}^2 &= \frac{1 + \mu_i}{2} \text{ under } H_1 \\ \sigma_{i0}^2 &= \frac{1}{2} \text{ under } H_0\end{aligned}$$

and the product

$$\phi_z(\omega) = \frac{1}{\prod_{i=1,M}(1 + j\alpha_i\omega)} \tag{6.67}$$

is the CF of the generic statistic z with $\alpha_i = 2\sigma_i^2 b_i$ so that

$$\alpha_i = \begin{cases} \mu_i, & (H_1) \\ \frac{\mu_i}{1+\mu_i}, & (H_0). \end{cases} \tag{6.68}$$

This is equivalent to

$$\phi_z(\omega) = \sum_{i=1}^{M} \frac{\kappa_i}{1 + j\alpha_i\omega} \tag{6.69}$$

where the constants κ_i are determined by the partial fraction expansion of (6.67)

$$\kappa_i = (1 + j\omega\alpha_i)\phi_z(\omega)|_{j\omega=-1/\alpha_i}$$

that yields

$$\kappa_i = \frac{1}{\prod_{j=1,M}\left(1 - \frac{\alpha_j}{\alpha_i}\right)_{j\neq i}} = \prod_{j=1}^{M}\left(\frac{\alpha_i}{\alpha_i - \alpha_j}\right).$$

The expressions (6.67, 6.69) must be equivalent at all ω and particularly at $\omega = 0$. Consequently,

$$\sum_{i=1}^{M} \kappa_i = 1.$$

The pdf is obtained by computing the inverse Fourier transform of (6.69) and is the sum of the components

$$\mathsf{f}_z(z) = \sum_{i=1}^{M} \mathsf{f}_i(z)$$

with

$$\begin{aligned} \mathsf{f}_i(z) &= \left(\frac{1}{2\pi}\right)\int_{-\infty}^{\infty} \frac{\kappa_i}{(1 + j\omega\alpha_i)} \exp\ j\omega z\, d\omega \\ &= \left(\frac{\kappa_i}{\alpha_i}\right) \exp -\frac{z}{\alpha_i}. \end{aligned}$$

The pdf for non-uniform weights is then

$$\mathsf{f}_z(z) = \sum_{i=1}^{M} \left(\frac{\exp -z/\alpha_i}{\alpha_i \prod_{j=1,M}(1-\alpha_j/\alpha_i)_{j\neq i}} \right) \quad \text{(pdf for non-uniform } \alpha_i\text{)}$$

with the α_i given by (6.68). The corresponding probability of exceeding a threshold γ is

$$\begin{aligned}
\wp(z \geq \gamma) &= \int_{\gamma}^{\infty} \mathsf{f}_z(z)dz \\
&= \sum_{i=1}^{M} \left(\frac{\exp -\gamma/\alpha_i}{\prod_{j=1,M}(1-\alpha_j/\alpha_i)_{j\neq i}} \right) \\
&= \sum_{i=1}^{M} \prod_{j=1,M} \left(\frac{\alpha_i}{\alpha_i - \alpha_j} \right)_{j\neq i} \exp -\gamma/\alpha_i \\
&= \sum_{i=1}^{M} \kappa_i \exp -\gamma/\alpha_i \quad \text{(non-uniform } \mu_i,\ \alpha_i\text{)}. \qquad (6.70)
\end{aligned}$$

The CF (6.67) is

$$\phi_z(\omega) = \frac{1}{(1+j\alpha\omega)^M}$$

when the weight coefficients are all the same and $\alpha_i = \alpha\ \forall\ i$. The corresponding pdf is evaluated from

$$\mathsf{f}_z(z) = \left(\frac{1}{2\pi}\right) \int_{-\infty}^{\infty} \frac{\exp +j\omega z}{(1+j\omega\alpha)^M} d\omega$$

and Fourier transform tables as the χ^2_M pdf

$$\mathsf{f}_z(z) = \left(\frac{z^{M-1}}{\alpha^M (M-1)!} \right) \exp -z/\alpha.$$

The corresponding threshold crossing probability is

$$\wp(\mathsf{z} \geq \gamma) = \exp -(\gamma/\alpha) \sum_{i=0}^{M-1} \left(\frac{1}{i!}\right) \left(\frac{\gamma}{\alpha}\right)^i \qquad (6.71)$$

for uniform $\mu_i = \mu_t/M$ with $\alpha_i = \alpha\ \forall i$. The probability of a threshold crossing of the statistic ℓ_r associated with a detection or false alarm is $\wp(\ell_r > \gamma') = \wp(\ell_d > \gamma)$ and $\gamma = \gamma' + \eta$ and is interpreted as either $\wp_d$ or $\wp_f$ depending upon the α_i (6.68). Substituting ℓ_d for z in (6.70) and using (6.68) produces

$$\begin{aligned}
\wp_d(\gamma') &= \text{Prob}(\ell_d > \gamma)|_{H_1} \\
&= \sum_{i=1}^{M} \prod_{j=1,M} \left(\frac{\mu_i}{\mu_i - \mu_j}\right)_{j \neq i} \exp -\gamma/\mu_i \\
\wp_f(\gamma') &= \text{Prob}(\ell_d > \gamma)|_{H_0} = \\
\sum_{i=1}^{M} \prod_{j=1,M} & \left(\frac{\mu_i(1+\mu_j)}{\mu_i - \mu_j}\right)_{j \neq i} \exp \left(\frac{\gamma(1+\mu_i)}{\mu_i}\right). \quad (6.72)
\end{aligned}$$

EC $\wp_d$, $\wp_f$ for non-uniform weights μ_i

The uniform case corresponds to a situation in which each of the M bins has the same SIR denoted by $\mu = \mu_t/M$. This implies that the signal energy is equally apportioned among the reflectors and $\mu = \text{SIR} = E_s/(MN_0)$ when the interference is uniform with psd N_0. The EC $\wp_d$ and $\wp_f$ are then computed with $\alpha = \mu_t/M$ and $\mu_t/(M+\mu_t)$ respectively. The uniform scatter pdf's computed from (6.71) are

$$\begin{aligned}
\wp_d(\gamma') &= \text{Prob}(\ell_d > \gamma)|_{H_1} \\
&= \exp -(\gamma M/\mu_t) \sum_{i=0}^{M-1} \left(\frac{1}{i!}\right)\left(\frac{\gamma M}{\mu_t}\right)^i \\
\wp_f(\gamma') &= \text{Prob}(\ell_d > \gamma)|_{H_0} \\
&= \exp -(\gamma(M+\mu_t)/\mu_t) \\
&\quad \cdot \sum_{i=0}^{M-1} \left(\frac{1}{i!}\right)\left(\frac{\gamma(M+\mu_t)}{\mu_t}\right)^i. \quad (6.73)
\end{aligned}$$

EC $\wp_d$, $\wp_f$ for uniform weights $\mu_i = \mu_t/M$

If there is only a single point scatterer, $M = 1$ and

$$\begin{aligned}
\wp_d &= \exp -(\gamma/\mu_t) \\
\wp_f &= \exp -(\gamma(1+\mu_t)/\mu_t).
\end{aligned}$$

Eliminating the threshold yields the familiar slowly fluctuating matched filter relationship for $\mu_t = E_s/N_0$ (6.65).

$$\wp_f = \wp_d^{(1+\mu_t)}. \quad (6.74)$$

6.6.2 EC Mismatch Error

(6.72, 6.73) determine the ROCs of the optimum incoherent combiner (EC) under the assumption that the scattering functions and hence the weights μ_i are known. They reflect EC performance when the prior scattering function information is correct and it is possible to construct the required signal and interference covariances directly from them. EC performance is degraded when insufficient prior information is available to register the matched filter realizations at the proper parameter values and define the covariances and the weighting coefficients μ_i required to form the detection statistic. Mismatch occurs because:

1. The EC is over or under determined such that the number of receiver realizations N is greater or less than the number of discrete scattering points M.

2. Matched filter receiver hypotheses are misregistered in relation to the scattering sites in the delay-Doppler plane. The scatterers do not occur at the ambiguity origin and are attenuated in proportion to the misregistration between the hypothesized parameters $\hat{\mathbf{p}}$ and the actual ones $\mathbf{p}$.

3. The scattering function magnitudes and hence the signal or interference component energies are not accurately known. The generalized SIR eigenvalues μ_i and the resulting EC weighting factors are incorrect.

Clearly these effects can occur in combination.

ROC curves are a useful way to evaluate misinformation error and are easy to derive because they are obtained directly from the weighted sums characterizing the EC statistic. Misinformation changes the weights or the number of terms in the summation but the approach is the same as that used for (6.72, 6.73) because all terms originate from Gaussian data. An over determined situation serves to illustrate the effect of partial information. Matched filters are formed on a grid of N parameter hypotheses $\hat{\mathbf{p}}$ in the delay-Doppler plane but only $M < N$ are actual correctly registered scattering sites. The remaining $N - M$ contain only white noise interference. Assume that the scattering is uncorrelated but no information regarding the scattering strengths of the M sites is available hence the λ_{si} are unknown. The noise psd N_0 is known and is the same at all filter realizations and the full signal energy at each of the M sites is realized because it is assumed that no misregistration (ambiguity) error occurs. While this represents a serious deficit in the prior information required for EC implementation, a more severe situation occurs when in addition, information regarding the delay-Doppler positions of the M scatterers is lacking or in error. Then, the corresponding receiver hypotheses are misregistered so that $\hat{\mathbf{p}} \neq \mathbf{p}$ reducing the filter signal responses. Clearly it would be possible to generate a sufficiently fine grid in relation to the ambiguity resolution so as to include all possible scattering sites but that would increase the amount of over determination and computational load. It would also increase the correlation between the signal components because of ambiguity sidelobe interaction (6.52).

The ROC curves for this mismatch situation are derived by modifying the EC ROC derivations used for (6.70, 6.71). Since neither the M bins containing the signal nor the amplitudes λ_{si} are known it is only possible to form a uniformly weighted incoherent sum of filter realizations with $b_i = 1,\ i = 1, \cdots, N$. The pdf coefficients $2\sigma_i^2 b_i$ (6.68) are then

$$\alpha_i = \begin{cases} 1 + \mu_i, & \text{for } H_1 \text{ and } i = 1, \cdots, M \\ 1, & \text{for } H_1 \text{ and } i = M+1, \cdots, N \\ 1, & \text{for } H_0 \text{ and for all } i. \end{cases}$$

The CF under H_1 is the product of a signal portion corresponding to the M components for which $\mu_i \neq 0$ and a noise only portion with $N - M$ components. It is

$$\phi_z(\omega) = \frac{N_c}{(1 + j\omega)^{N-M} \prod_{i=1,\cdots,M} (c_i + j\omega)} \tag{6.75}$$

where $c_i = 1/(1 + \mu_i)$ and $N_c = \prod_{i=1,L} c_i$. The CF under H_0 is obtained by setting all of the c_i to unity and the probabilities for both cases are determined by integrating from a threshold using

$$\wp(\ell_d > \gamma) = \left(\frac{1}{2\pi}\right) \int_{-\infty}^{\infty} \phi_z(\omega) \int_{\gamma}^{\infty} \exp j\omega\, x dx d\omega. \tag{6.76}$$

Example 6.2 ROC Comparison

A ROC comparison of the ideal (prescient) receiver, the EC, and the mismatched uniformly weighted detector provides an example illustrating relative performance and the mismatch effect. The same delay spread scatter model of Figure 6.11 and fixed μ_t is used for each and a waveform with sufficient resolution to individually resolve the reflectors without sidelobe overlap defines the signal levels. The reflector and noise process are uncorrelated with diagonal covariances so that the actual μ_i represent reflector SIR's E_{si}/N_0 realized at the matched filter outputs and $\mu_t = \sum E_{si}/N_0 = 5$ (7 dB). A total of $N = 20$ sequential constant Doppler delay hypotheses $\hat{\tau}_i$ bins are available of which $M = 12$ contain signal energy and 8 contain only noise. The reflector and signal ambiguity functions are aligned in those bins containing reflectors avoiding attenuation due to ambiguity misregistration. This is a situation of partial information in which the bin positions are known, but those that contain the delayed echoes and their levels are unknown.

The prescient receiver is the matched filter that uses the normalized received signal as the replica. It provides an upper bound for the achievable performance but is unrealizable because it implies foreknowledge of a random event. The EC statistic

$$\ell_d(\text{EC}) = \frac{1}{N_0}\sum_{i=1}^{12} |\mathsf{r}_i|^2 \left(\frac{\lambda_{si}/N_0}{1+\lambda_{si}/N_0}\right)$$

is the properly weighted sum of only those 12 bins containing signal energy. The uniformly weighted statistic apportions the total energy equally among all 20 bins, $\mu_i = \mu_t/20 \; \forall \, i$ and

$$\ell_d(\text{unif}) = \frac{.2}{N_0}\sum_{i}^{20} |\mathsf{r}_i|^2$$

for this example. The resulting ROC curves of Figure 6.12 are derived from

$$\wp_f = \wp_d^{(1+\mu_t)}$$

for the prescient receiver and (6.72, 6.75, 6.76) for the EC, and mismatched uniform weight detectors respectively. The curves clearly illustrate the advantage of correct prior information and the detection performance difference between coherent and incoherent integration. The ideal receiver utilizes a phase coherent replica of the actual signal, forms its statistic from a single integration and square law detection of the received data time series, and has the best performance. The EC has complete second order (scattering function) information regarding signal and interference at the discrete the parameter locations $\mathbf{p}_i$, and forms its statistic from the sum of the $M = 12$ properly weighted time series integrations (6.42)

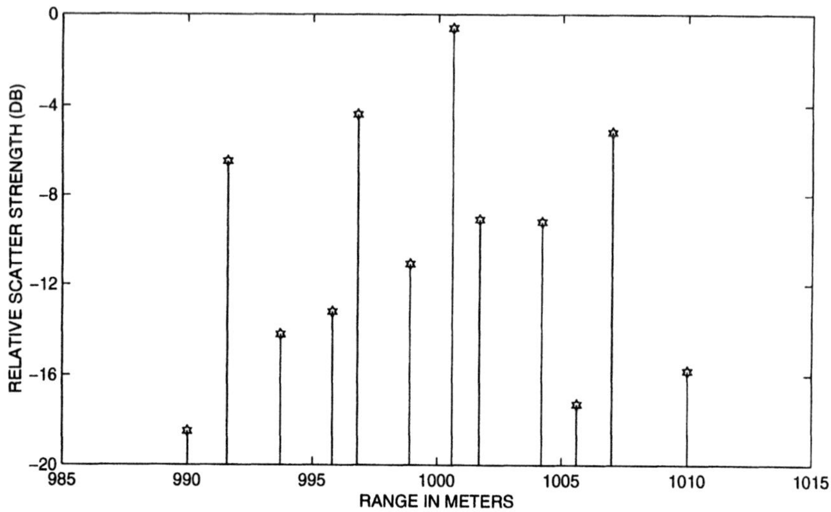

Figure 6.11: Delay spread scatter distribution example [37] (©IEEE 2000)

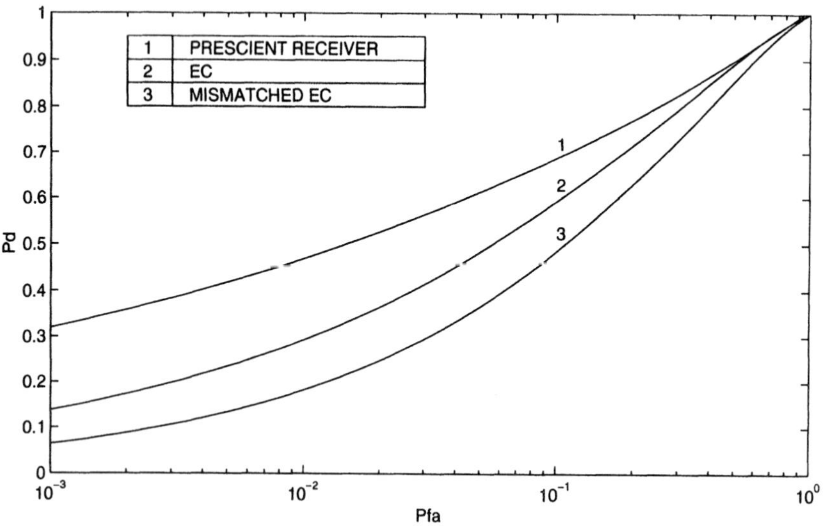

Figure 6.12: ROC comparison with $\mu_t = 5$ [37] (©IEEE 2000)

that contain echo energy. It performs less well even though like the matched filter, it realizes all of the signal energy (6.65). The uniform detector can only perform an incoherent summation of all $N = 20$ available bins because it has only information regarding the total number and positions of the bins. Information about which are occupied and the reflector energies λ_{si} is lacking. It would have better performance if only the correct 12 bins containing the signal were included in the sum excluding the 8 that contribute only interference [46].

EC post detection integration of weighted matched filter responses represents the optimum strategy assuming the random spread scatter model and the availability of the underlying covariances or the signal and interference scattering functions. It is particularly suited to the detection of spread echoes emanating from discrete highlights in non-uniform interference such as reverberation given the required prior SF information.

Sonar echoes that emanate from moving scatterers and are free of reverberation interference are an important subclass. They are most effectively interrogated with a long CW waveform because it minimizes the reverberation limited Doppler region as shown at the bottom of Figure 6.8. Consequently they often have high SIR as they compete only with ambient noise. High Doppler implies a rapidly moving object with the possibility of rapidly changing scattering properties due to changes in orientation and propagation conditions. It may be impossible to track these changes accurately enough to take advantage of EC processing but the echoes are often highly detectable by simple matched filtering. The matched filter ROC is given by (6.74) and is a function of the output SIR that is determined from the input SIR and the processing gain G_p. Processing gains for matched filtered CW, LFM, or FSK waveforms in uniform ambient noise are estimated from the Tables 6.3, 6.4, 6.5 by setting the reverberation to noise power parameter $r_n = 0$. Detector performance is obtained from the ROC curves for a single pulse derived from (6.19, 6.16, 6.17, 6.74).

While the average SIR realized by a single MF filter may be high under these circumstances, the receiver operating characteristics (6.17, 6.74) reflect the fluctuating model. This means that aside from the random interference effects, the signal level on any one interrogation has a non-zero probability of fading below the threshold and evading detection. $\wp_d$ for a single interrogation is therefore reduced in relation to that of a deterministic (non-fluctuating) echo [32]. It can be improved by transmitting a composite waveform and incoherently combining the outputs of multiple receivers "tuned" to the waveform subfunctions.

6.6.3 Diversity Combining

The slowly fluctuating point scatter model is based upon the premise that an individual highlight echo emanates from a multitude of independent randomly fluctuating subreflectors that are clustered in a region that is small in relation to the waveform ambiguity resolution yet may be large in relation to its wavelength. These subreflectors may be comprised of facets of a complex reflecting surface as well as multiple realizations of individual reflectors propagating via different acoustic paths that include forward scatter from the random sea surface and bottom. The echo will be comprised of the contribution of multiple phase delayed replicas of the transmitted signal the sum of which will fluctuate according to the random phase re-enforcement occurring with each interrogation.

A CW echo generated by a multitude of discrete random highlights or multipaths is

$$\mathsf{s}(t) = \sqrt{E_0}\sum_{i=1}^{M} \mathcal{S}_s(\tau_i)\exp j\,\omega(t-\tau_i)$$

and if the waveform duration is greater than the total delay support of the scattering process and multiple approximately equal strength highlights contribute to the echo, the signal magnitude or envelope will exhibit frequency selective *Rayleigh* fluctuation or fading [4, 15, 32, 47]. The fluctuation correlation between interrogating frequencies in a delay spread channel is described by the TFCF (6.5) and signal frequency components separated by more than the inverse of the scattering delay support will on average exhibit uncorrelated fading as discussed in Section 6.2.1. The ROC curves for a matched filter operating in a fluctuating Rayleigh channel are given by (6.74) and $\wp_d$ for constant $\wp_f$ is reduced in comparison to that of a non-fluctuating channel.

The effect of fading is mitigated by incoherently combining the outputs of several independently fluctuating receivers. It is called post detection *diversity combining* and can be used whenever signal components fluctuate as a function of time, frequency, or spatial position. It is only necessary that the overall echo delay and Doppler be common to all channels and is most effective when the fluctuation is uncorrelated between receivers. This occurs when signal components are separated by more than the inverse of the Doppler and delay spread for time and frequency selective fading respectively [4].

Spatial fluctuation occurs between signals received at different locations and commonly occurs in radio communications. High frequency radio signals that skip between the earth and the ionosphere over long distances or bounce between buildings in an urban environment are a form of multipath propagation and often fluctuate over distances measurable in wavelengths. The signal fluctuation of a moving car radio as it navigates city streets is a familiar example . When it is important that critical radio communications not be adversely affected by spatial fading an "antenna farm" of spatially separated antennas may be erected. The detected outputs of the associated receivers are combined to reduce the effects of fluctuation. The idea being that when a signal at one location has faded or has a null, another will be at a peak and the aggregate signal level will be more nearly constant. On average, the signal variance is reduced by the number of contributing receivers if the fading is uncorrelated between channels. This is an example of spatial diversity and is most effective if the antennas are sufficiently separated to ensure uncorrelated fading. Likewise it is possible to combine the outputs of several receivers that are "tuned" to signals at different frequencies or time delays to reduce the effects of frequency or time selective fading.

Diversity combining has a long history in radar and is most often employed to combat the fading that occurs between the sequential subpulses of a radar scan [32]. Radars commonly transmit a scan or burst of hundreds if not thousands of pulses and diversity is used to incoherently combine the detected time separated echo pulses. In radar parlance slow fluctuation occurs when all of the pulses in a scan are correlated so that they all fluctuate together. Fast fluctuation occurs when the fluctuation is uncorrelated between pulses. The four commonly used models used to describe pulse fluctuations within a radar scan are the *Swerling* models defined by Peter Swerling [48] and based in part upon the results obtained by Marcum [49] for non-fluctuating echoes. These are conveniently assembled in the book by DiFranco and Rubin [32]. The models are designated as Swerling I, II, III, and IV. Models I and II represent pulse trains in which the pulse fluctuations are Rayleigh and are correlated and uncorrelated with correlation coefficients $\rho = 1, 0$ respectively. Models III and IV are the same but the fluctuation pdf is that of a "one dominant plus Rayleigh" model corresponding to the contribution of one dominant plus many smaller approximately equal energy reflectors.

These represent a less severe fading problem than the Swerling I and II models [50]. The fully correlated models I and III and the uncorrelated models II and IV represent the extreme cases but partially correlated models for which $0 < \rho < 1$ are discussed in the literature [50–52]. As a rule the $\wp_d/\wp_d$ characteristics of a diversity combiner improve with decreasing ρ because fluctuation variance is reduced when averaging over uncorrelated samples.

Diversity combining is applicable to sonar situations involving delay and/or Doppler spreading with associated frequency and time selective fading respectively. Echos are more likely to exhibit frequency selective fading because delay spreading due to scatter and multipath spreading is the more common problem. They rarely have significant Doppler spread in comparison to the reverberation background. Figure 6.13 is a delay-Doppler map of an in-water three tone FSK fluctuating scatter echo produced by Fourier transforming (matched filtering) the echo time series in concatenated time windows equal to the subpulse duration. The individual CW echo amplitude fluctuations are typical of frequency selective fading and each has slight positive Doppler in relation to its respective reverberation ridge. A frequency diversity sonar is implemented by transmitting a sparse FSK waveform composed of several identical time concatenated CW pulses [31]. The distinct CW subpulses are individually detected by CW matched filters and the time registered outputs incoherently combined with uniform interpulse weighting. This is efficiently implemented by sequentially Fourier transforming the echo time series using the FFT in segments corresponding to the subpulse durations. If the code is designed so that the frequency differences are unique and exceed the inverse of the expected scattering function support Ω_T, the receiver output fluctuations will tend to be uncorrelated. This is called the correlation frequency by Barton [50]. The whole N-subpulse waveform can be simultaneously coherently matched filtered yielding a reverberation processing gain (6.5). This is a *dual processing* concept and permits the simultaneous search for both high and low Doppler echoes with one interrogation. An FSK code designed with contiguous pulses is in reality a time and frequency diversity waveform because the subpulses are transmitted at different times as well as frequencies. If there is little Doppler spread in the signal scattering function, the subpulses will be highly time correlated and may be considered as having been transmitted simultaneously. This is possible if a loss of source level due to amplitude modulation is acceptable.

The expected output SIR of each diversity channel is the same if

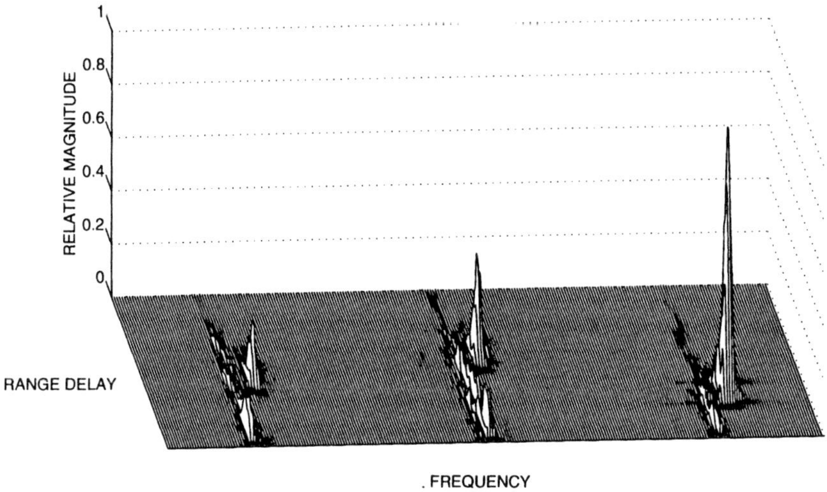

Figure 6.13: An in-water example of the frequency selective fading experienced by a concatenated three tone FSK waveform. The time delay origin is in the foreground and frequency increases to the right. The echoes have positive Doppler in relation to their respective reverberation ridges.

the waveform subpulses are the same and have identical ambiguity functions, are interrogating the same WSSUS scattering function, and the background interference is uniform Gaussian noise. The diversity channel fluctuations are uncorrelated and Swerling II if the interpulse frequencies are sufficiently separated so that the time-frequency correlation function R_H is negligible and the individual subpulses exhibit Rayleigh fading. The Swerling II model is not necessarily accurate if the overall scattering process is dominated by a few strong highlights, there is little multipath, and the subpulses are not sufficiently separated in frequency. A partially correlated model [50–52] may be more appropriate assuming the frequency domain correlation characteristics of the channel are known.

The receiver operating characteristic for the incoherent sum of N FSK subpulses assuming the uncorrelated Swerling II model is derived in Chapter 11 of [32] and is closely related to that of the EC with uniform weighting hypothesizing $N = M$ scatterers given by (6.73). The N individual CW echoes are uncorrelated if the frequencies are sufficiently separated and the output of the incoherent diversity combiner is equivalent to that of the EC with $M = N$ uncorrelated components given by (6.66). The expected signal and interference energy is the same in each receiver so that $\lambda_{0i} = \lambda_0 = N_0$, $\lambda_{si} = \lambda_s = E_s/N \,\forall i$ and

$$\begin{aligned}\ell_d &= \sum_{i=1}^{N} \frac{|\mathsf{r}_i|^2}{\lambda_0}\left(\frac{\lambda_s/\lambda_0}{1+(\lambda_s/\lambda_0)}\right)\\ &= \left(\frac{\mu_t}{\lambda_0(N+\mu_t)}\right)\sum_{i=1}^{n}|\mathsf{r}_i|^2\end{aligned}$$

where $\mu_t = N\lambda_s/\lambda_0 = E_s/N_0$. This simplifies the ROC calculation as it is only necessary to evaluate the $\wp_d$, $\wp_f$ for the statistic

$$\tilde{\ell}_d = \sum_{i=1}^{N}|\mathsf{r}_i|^2 = \left(\frac{\lambda_0(N+\mu_t)}{\mu_t}\right)\ell_d.$$

Its threshold crossing probability is

$$\text{Prob}(\tilde{\ell}_d > \tilde{\gamma}) = \text{Prob}\left(\ell_d > \frac{\tilde{\gamma}\mu_t}{\lambda_0(N+\mu_t)}\right).$$

and evaluated by substituting the modified threshold for γ in (6.73) yielding

$$\begin{aligned}\wp_d(\gamma') &= \text{Prob}(\tilde{\ell}_d > \tilde{\gamma})\Big|_{H_1}\\ &= \exp -(\tilde{\gamma}N/(NN_0+E_s))\sum_{i=0}^{N-1}\left(\frac{1}{i!}\right)\left(\frac{\tilde{\gamma}N}{NN_0+E_s}\right)^i\\ \wp_f(\gamma') &= \text{Prob}(\tilde{\ell}_d > \tilde{\gamma})\Big|_{H_0}\\ &= \exp -(\tilde{\gamma}/N_0)\sum_{i=0}^{N-1}\left(\frac{1}{i!}\right)\left(\frac{\tilde{\gamma}}{N_0}\right)^i. \qquad (6.77)\end{aligned}$$

$\wp_d$, $\wp_f$ for the Swerling II incoherent frequency diversity combiner

These are also determined directly from (6.71) by substituting $\alpha = E_s/N + N_0$, N_0 for $\wp_d$, $\wp_f$ respectively. When $N = 1$

$$\wp_d = \exp -(\tilde{\gamma}/(E_s+N_0))$$

$$\wp_f = \exp -\tilde{\gamma}/N_0$$

and eliminating the threshold produces the same matched filter relationship (6.74) obtained from (6.73) for the uniform EC $\wp_d$, $\wp_f$. The difference between $\tilde{\ell}_d$ and ℓ_d is nothing more than a change of variables and both representations yield the same ROC curves. Only the thresholds in relation to the background are different. The EC relationships are more general as they admit a nonuniform energy distribution and correlation between receiver realizations. The uncorrelated equal energy Swerling II diversity model is a simplification that is justified assuming FSK interrogations with sufficient frequency offset and reasonably dense scattering. It also permits the setting of a false alarm threshold for $\wp_f$ from (6.77) based only upon the noise background level N_0.

The detection performance of the Swerling II diversity combiner with individual pulse SIR's of E_s/NN_0 relative to that of a single matched filter with SIR=E_s/N_0 is assessed by comparing the ROC's (6.77) for $N > 1$ and $N = 1$ respectively. Figure 6.14 is a plot of $\wp_d$ vs total echo SNR (E_s/N_0) as a function of N with $\wp_f = 10^{-5}$. It reveals that $\wp_d$ improves inversely with N at a low SNR but increases with N at high SNR. This occurs because each diversity receiver is an independent matched filter and the signal energy is apportioned equally among the N pulses. Each filter realizes a maximum average energy of E_s/N and at low SNR the energy and $\wp_d$ loss obtained by apportioning the available energy among more receivers outweighs the $\wp_d$ improvement obtained by averaging the fluctuations over N receivers. Each receiver is apportioned more energy at high SNR. The $\wp_d$ improvement via fluctuation averaging outweighs the energy and $\wp_d$ loss in the individual receivers with increasing N. The intersections of the curves with the abscissa at $N = 1$ in Figure 6.15 correspond to the required SNR that a matched filter must have to meet the goal $\wp_d$. Above about $\wp_d = .5$ there is an optimum N that minimizes the required SIR. The ratio of the SNR at the optimum N to that at $N = 1$ expressed in dB is called the *integration loss* [32] and is positive at low $\wp_d$ when incoherent combining results in an increase in required SNR. It is clear from the figure that the optimum N increases with $\wp_d$ but the greatest incremental gain is obtained for just 3 or 4 subpulses. It becomes impractical to increase the number of pulses beyond this without significantly shortening the subpulses if the overall waveform duration is to be constrained. Bandwidth rapidly increases with N because it is desirable that the subpulse frequency differences sufficiently large in relation to the expected inverse

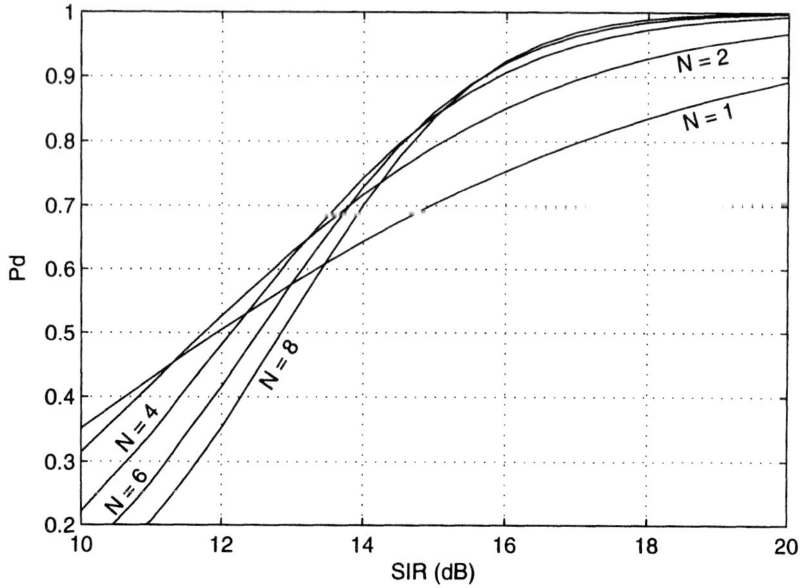

Figure 6.14: Swerling II diversity $\wp_d$ vs SIR (E_s/N_0) as a function of the number of subpulses for $\wp_f = 10^{-5}$

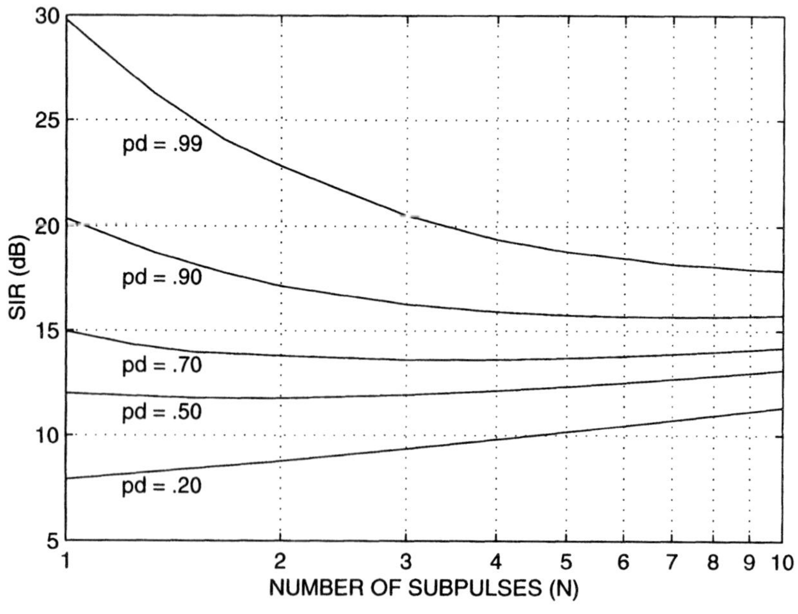

Figure 6.15: Swerling II SIR (E_s/N_0) vs the number of subpulses as a function of $\wp_d$ for $\wp_f = 10^{-5}$

of the scatter spreading Ω_τ to insure minimal interpulse correlation.

Frequency diversity code design

The FSK Costas and congruence codes discussed in Chapter 5 are designed to be efficient in the allocation of bandwidth and for the most part have equispaced subpulses with every frequency slot filled. [2] Frequency diversity is most effective when the CW subpulses are sufficiently spectrally separate so that (ideally) their delay induced fluctuations are uncorrelated. The minimum separation $\triangle\zeta = 1/\Omega_\tau$ is called the correlation frequency [50] and $W_{min} = (N-1)\triangle\zeta$ is the minimum bandwidth required to accommodate N equispaced CW pulses neglecting the subpulse bandwidth N/T. The two frequency correlation function (6.5) depends only upon the frequency differences under the WSSUS assumption. If the minimum frequency difference is sufficiently large so that the expected correlation is negligible, the allocation is immaterial from a diversity point of view. However interpulse correlation may occur if delay spreading is relatively small and bandwidth constraints force the frequency separation to be less than the ideal. In that case an allocation that increases diversity by maximizing the number of frequency differences may have a better chance of reducing fluctuation than a linear code with repeated differences. A code for which $|f_i - f_j|$ is unique for all $i \neq j$ will have $N(N-1)/2$ differences while a linear code covering the same bandwidth realizes only $(N-1)$ differences.

Additionally, a linear allocation is not the best approach when dealing with sonar echoes that must compete with reverberation. When N subpulse frequencies are equispaced at intervals $\triangle f = k\triangle\zeta$ Hz, Doppler shifts of $\phi = i\triangle f,\ 1 = 1, 2, \cdots, N-1$ cause echoes to overlap their neighboring reverberation ridges severely reducing their SIR and limiting the effectiveness of FSK diversity combining. This is evident from the echoes generated by the linearly allocated code shown in Figure 6.13. With sufficient Doppler, the first echo would fall into the reverberation generated by the second and so on. Clearly it would be advantageous if the frequencies were allocated so that there is a minimum of reverberation interaction regardless of the Doppler.

An FSK code such that all of the subpulse frequency differences $f_i - f_j$ are unique guarantees that only one reverberation overlap occurs for any Doppler shift. $N-1$ reverberation free diversity channels are then avail-

[2] The logarithmic allocation 5.27 is an exception but is merely the wideband time dilation version of a linear allocation.

able regardless of the Doppler. An allocation that adjoins a frequency f_{N+1} to a code of order N such that

$$(f_{N+1} - f_N) > \max|f_i - f_j|, \ ,i,j = 1.\cdots, N$$

forms a code of order $N+1$ and guarantees unique differences since the new difference does not belong to the set of existing differences. This can be accomplished by using the narrowband allocation algorithm (5.26) with $j(i) = 2^{i-1} - 1$, $i = 1, \cdots, N$. The frequencies increase exponentially and $N_b = j(N) - j(1)$ is the total number of frequency bins each of width $\Delta\zeta$. $N_b = 31$ for $N = 6$ which represents a five-fold increase in bandwidth over a linear allocation requiring only 6 bins. The required bandwidth is $W = N_b \, \Delta \, \zeta$ and the frequencies are allocated according to (5.26) as

$$f_i = f_0 + W J(i)$$
$$J(i) = \left[\frac{2^{i-1} - 1}{N_b} - .5 \right].$$

For example if Ω_τ=20 msec, $\Delta\zeta$=50 Hz, f_0=10 kHz, and $N = 6$, $W =$ 1550 Hz, the subpulse frequencies are 9225, 9275, 9375, 9575, 9975, 10775 Hz.

The exponential allocation is needlessly wasteful of bandwidth especially for large N. An alternative code allocation that is more bandwidth efficient uses a difference triangle algorithm and is illustrated in Figure 6.16a. The array entries $d_{i,j}$ are the code difference indices indicating the number of discrete frequency bins of width $\Delta\zeta$ separating f_i and f_j and $|f_i - f_j| = d_{i,j} \, \Delta \, \zeta$. At each code increment corresponding to the addition of a new row to the array, the minimum frequency difference not already appearing is chosen assuring that all possible differences are realized for a code of order N. i, $j = 1, \cdots .N$ are the subpulse indices ordered as $i = 1, N-1$ on the abscissa and as $j = N, 1$ in descending order on the ordinant. The entry 0 appears at the position (1,1) and along the diagonal because entries with i=j represent differences of zero. The triangle is started by specifying $d_{1,2} = 1$ meaning that the first and second pulse indices are separated by unity and the corresponding frequencies by $\Delta\zeta$. The process proceeds row wise with each row started from the right by specifying the difference index $d_{i,i+1}, i = 1, \cdots, N-1$ as the minimum integer for which

$$d_{i,i+1} \neq d_{m,n} \; m = 1, \cdots, i-1, \; n = 2, \cdots, i.$$

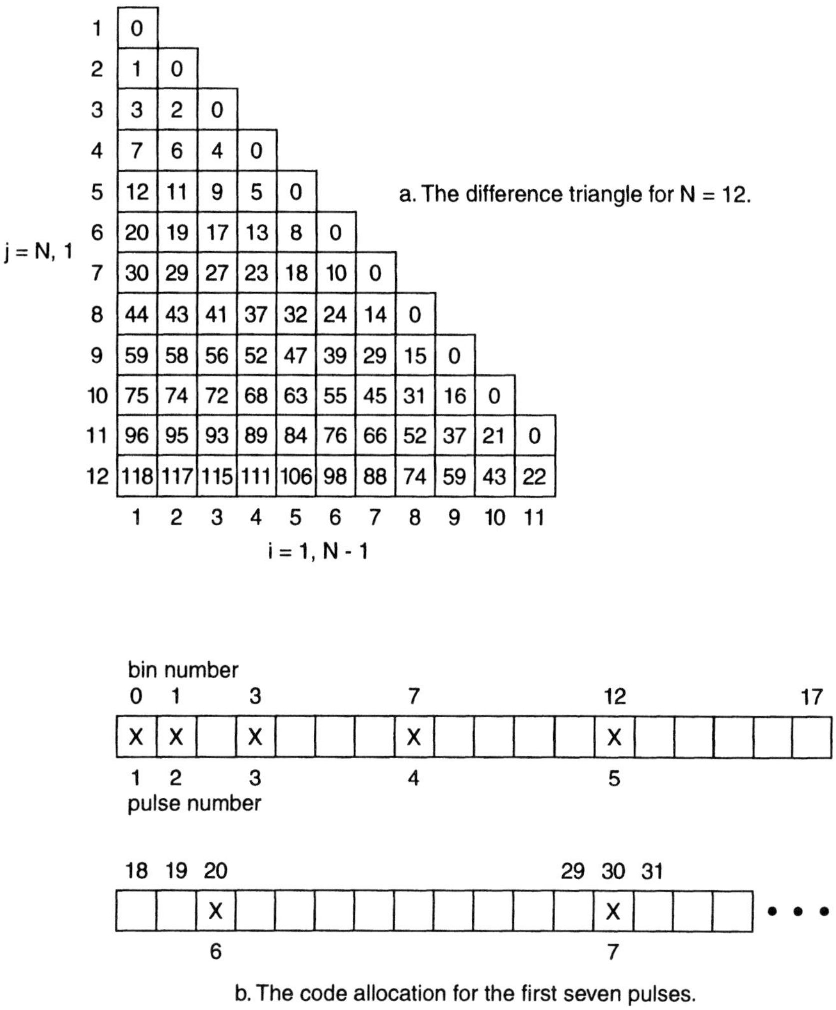

Figure 6.16: (a) Illustration of the difference triangle for the generation of a maximally bandwidth efficient diversity code with unique differences. (b) The frequency bin code allocation for first 7 pulses with $N = 12$.

In other words it is the minimum integer that does not already appear in the array. The integer satisfying this for $d_{2,3}$ is 2. Once this is chosen, the row is filled out working to the left by

$$d_{i,j} = d_{j-1,i} + d_{i,j-1} \; i = 1, \; N-1, \; j = 2, \cdots, N.$$

For example, the row corresponding to j=8 starts at the right with $d_{7,8} =$ 14 because it is the smallest integer that does not appear in rows 1-7. The next entry is $d_{6,8} = d_{7,8} + d_{6,7}$=14+10=24 and so on. The first column corresponding to $d_{1,j}$, $j = 1, N$ is the allocation with respect to the first (lowest frequency) pulse and generates the monotonically increasing sequence shown in Figure 6.16b. A diversity code is not required to be monotonic. The sequence can be time shuffled because the unique frequency differences are preserved regardless of the time order. This is a consideration when the code is used in a dual processing mode as a matched filtered FSK waveform. The ambiguity sidelobes are more uniformly spread in delay-Doppler if the code is rearranged so that the sequence is not monotonic.

The sequence generated by exponential code $d_{1,j} = 2^{j-1} - 1$ is (0, 1, 3, 7, 15, 31, 63, $\cdots$) and is the same as that of the efficient code up to the fourth pulse but expands rapidly thereafter. A six pulse diversity code corresponding to the exponential example requires only $N_b = 20$ bins, $W = 1000$ Hz, and the frequencies may be specified by either the narrow or wideband algorithms (5.26, 5.27). The placement operator for the efficient code of order 6 is j(i)= 0, 1, 3, 7, 12, 20, i=1, 6 and the wideband generating function 5.27) is

$$\log f_i = \log f_c + J(i) \log R_b$$
$$J(i) = \left[\frac{j(i) - j_{min}}{j_{max} - j_{min}} - .5 \right] = \frac{j(i) - 10}{20}$$

with:
$f_0 = 10$ kHz,
$\log R_b = \log(10500/9500) = .043466$,
$\log f_c = \log \sqrt{(10500)(9500)} = 3.9995$.
The resulting pulse frequencies are 9500, 9549, 9645, 9840, 10089, 10501 Hz.

6.7 SUMMARY

The discussion regarding spread scattering relies upon the linear system models that are derived from time varying linear filter theory. They are time-frequency representations interrelated by Fourier transformations, are inherently narrowband, and as pointed out by Ziomek [5], should only be used when the bandwidths involved are sufficiently narrow so that linearity is maintained. For example this would preclude their use to describe scattering and propagation when absorption loss α varies significantly across the bandwidth of the interrogating waveform. This does not prohibit the use of the wideband time dilation model for Doppler as long as the physical propagation and scattering mechanisms are linear. Of the various time frequency representations, the spreading function is the most compatible model for wideband Doppler dilation because it describes the signal as a discrete or continuously weighted representation of the Doppler dilated and delayed waveform

$$\mathsf{s}(t) = \sqrt{E_0}\sum_{i=1}^{M}\sqrt{\beta_i}S(\beta_i,\tau_i)f(\beta_i(t-\tau_i)) \quad \text{or}$$

$$\mathsf{s}(t\ ) = \sqrt{E_0}\int_{\Omega_\beta}\int_{\Omega_\tau}\sqrt{\beta}S_s(\beta,\tau)f(\beta(t-\tau))d\beta d\tau.$$

The scattering function (SF) is the expectation of the spreading function at two different points in phase space. If the scattering is uncorrelated (WSSUS) the SF becomes a function of the local scattering properties and permits the prediction of correlation receiver performance as a 2-d convolution with the waveform ambiguity function. This is one of the most important results of linear system theory for active sonar because it directly relates the expected performance of correlation detectors to signal scattering properties and the waveform and processor functions (ambiguity functions).

The WSSUS assumption that is widely invoked in linear scattering theory is fundamental to high frequency sonar because it establishes a direct linkage between time and range delay. It is most easily envisioned by comparing the impulse responses of a bell or similar mechanical structure and a range spread WSSUS scatterer with no internal communication. A bell which is a localized object will generate a time extended response $h(\tau)$ when impacted that cannot be discriminated by a distant observer from that of an extended WSSUS object with the same response. The difference between the two being the range delay dependence $\tau = 2R/c$ of the extended scatterer. The uncorrelated scattering assumption that established this dependence implies that no energy is transmitted within the body of the object being interrogated. In other words, impacting (interrogating) one side of a bell causes the rim on the other side to vibrate and emit acoustic energy. This cannot occur under the WSSUS assumption because all scattering phenomena are localized. This is why the WSSUS assumption is not generally applicable to low frequency sonar.

Detector robustness is a central issue when interrogating in a spreading environment. It is a reasonably straightforward process to devise an optimal detector given sufficient information about the scattering and propagation conditions. The prescient receiver, EC, and optimally mismatched receivers are examples of detectors designed to be optimal in some sense for which performance is proportional to the quality of the prior information. Receivers designed to be optimal under specific conditions are not robust because they suffer degraded performance when the scattering conditions do not match the assumptions. It is for this reason that the simple matched filter using carefully selected waveforms is widely used even though it is mismatched in a spreading environment. Sonar echoes subject to multipath or reflector elongation are usually delay rather than Doppler spread. and the competing interference is either reverberation at low Doppler or ambient noise at high Doppler. Diversity processed FSK waveforms with subpulse durations that exceed the scatter duration are robust for the detection of range spread echoes at high Doppler. No assumption needs to be made regarding the details of the distribution of scattering function beyond its assumed delay extent.

Low Doppler detection is the more difficult problem because the SIR is lower, and often, boundary reflections generate false echoes that must be sorted out from desired echoes. Detection in this regime benefits from any prior information that can be brought to bear such as a prediction of the multipath from environmental measurements. Doppler tolerant waveforms such as the LFM and HFM that have processing gain against point like scattering are often employed under the assumption that most echoes have some discrete (highlight) components. Both high and low Doppler regimes can be interrogated by transmitting a dual processing waveform that combines FSK and Doppler tolerant components.

Bibliography

[1] W.S. Burdic. *Underwater Acoustic System Analysis* . Prentice-Hall, 1984.

[2] S.C. Clay and H. Medwin. *Acoustical Oceanography* . Wiley Interscience, 1977.

[3] R.J. Urick. *Principles of Underwater Sound* . McGraw-Hill, 1975.

[4] H.L. Van Trees. *Detection, Estimation, and Modulation Theory-Part III* . Wiley, 1971.

[5] L.J. Ziomek. *A Scattering Function Approach to Underwater Acoustic Detection and Signal Design* . PhD Thesis in Acoustics, Penna. State Univ., Nov. 1981. Applied Research Laboratory TR 81-144.

[6] R. Laval. Time-Frequency-Space Generalized Coherence and Scattering Functions . In G. Tacconi, editor, *Aspects of Signal Processing, Part 1*, pages 69–87. D. Reidel, Dordrecht-Holland, 1977.

[7] R. Laval and Y. Labasque. Medium Inhomogenieties: Effects on Spatial and Temporal Processing . In *Underwater Acoustics and Signal Processing.* D. Reidel, Dordrecht-Holland, 1981.

[8] D. Middleton. A Statistical Theory of Reverberation and Similar First Order Scattered Fields, I Waveforms and the General Process. *IRE Trans. Info. Theory*, IT-13(3):372–412, July 1967.

[9] L.J. Ziomek. *Underwater Acoustics, A Linear Systems Approach* . Academic Press, Orlando, FL, 1985.

[10] L.G. Weiss. *Wideband Inverse Scattering and Wideband Deconvolution of Acoustic Signals Using Wavelet Transforms* . PhD Thesis in Acoustics, Penna. State Univ., May 1993.

[11] L.G. Weiss. Wavelets and Wideband Correlation Processing . *IEEE Sig. Proc. Magazine*, Jan. 1994.

[12] K.L. Hillsley and R.K. Young. Linear Time-Varying System Characterization using Wavelet Transforms . In *Proc. Symposium on Progress in Electromagnetic Research*, Seattle WA, July 24-28 1995.

[13] P. Maas. Wideband Radar: the Hyp (hyperbolic) Transform . *Inverse Problems*, 5:849–857, 1989.

[14] D.M. Drumheller and D.W. Ricker. Receiver-Transmitter Optimization for Detection in Doubly Spread Channels . *Journ. of the Acoust. Soc. Amer.*, 89(4):1714–1723, April 1991.

[15] W.C. Knight, G.G. Pridham, and S.M. Kay. Digital Signal Processing for Sonar . *Proceedings of the IEEE*, 69(11):1451–1507, Nov. 1981.

[16] J.J. Kisenwether. *The Characterization of Delay Spread Channels Using Cross Correlation Processing.* PhD Thesis in Acoustics , Penna. State Univ., Aug. 1998.

[17] D.W. Ricker and A.J. Cutezo. Estimation of Coherent Detection Performance for Spread Scattering in Reverberation-Noise Mixtures. *Journ. of the Acoust. Soc. Amer.*, 107(4):1978–1986, April 2000.

[18] L.J. Ziomek and L.H. Sibul. Broadband and Narrowband Signal-to-Interference ratio Expressions for a Doubly Spread Target . *Journ. of the Acoust. Soc. Amer.*, 72(3):804–819, 1982.

[19] S.T. McDaniel and A.D. Gorman. Acoustic and Radar Sea Surface Backscatter. *Journ, Geophys. Res.*, 87:4127–4136, 1982.

[20] S.T. McDaniel. Sea Surface Reverberation: A review. *Journ. of the Acoust. Soc. Amer.*, 94:1905–1922, 1993.

[21] R.P. Goddard. The Sonar Simulation Toolset. In *Proc. Oceans '89, The Global Ocean*, volume 4, pages 1217–1222, 1989. IEEE Pub. no. 89CH 2780-5.

[22] W.S. Hodgkiss. An Oceanic Reverberation Model . *IEEE Journ. of Oceanic Engineering*, OE-9(2), April 1984.

[23] G.R. Valenzuela and M.B. Laing. A Study of Doppler spectra of Radar Sea Echoes. *Journ. Geophys. Res*, 75:551–563, 1970.

[24] R.L. Swarts and C.J. Eggen. Simplified Model of the Spectral Characteristics of High Frequency Surface Scatter . *Journ. of the Acoust. Soc. Amer.*, 59(4):846–851, April 1976.

[25] S.O. McConnel. Remote Sensing of the Air-Sea Interface using Microwave Acoustics. In *Proc. Oceans '83*, pages 85–92. IEEE and Marine Tech. Soc., 1983.

[26] H. Weinberg. Generic Sonar Model. Technical report, Naval Underwater Syst. Center, 1985. TR 5971D.

[27] J.P. Hermand and W.H. Roderick. Acoustic Model-Based Matched Filter Processing for Fading time-Dispersive Ocean channels: Theory and Experiment. *IEEE Journ. of Oceanic Engineering*, 18(4):447–465, October 1993.

[28] L.H. Sibul and E.L. Titlebaum. Signal Design for Detection of Targets in Clutter . *Proceedings of the IEEE*, 69(4):481–482, April 1981.

[29] Altes R.A. Suppression of Radar Clutter and Multipath Effects for Wideband Signals . *IEEE Trans. Info. Theory*, 17:344–345, May 1971.

[30] D.M. Drumheller. Receiver Optimization for Detection in Doubly Spread Communication Channels . Technical report, Applied Research Lab., Penna. State Univ., Dec. 1984. TM-84-185.

[31] D.K. Barton. *Radars: Frequency Agility and Diversity* , volume 6. Artech House, Dedham, Mass., 1977.

[32] J.V. DiFranco and W.L. Rubin. *Radar Detection* . Artech House, Dedham, MA, 1980.

[33] S.C. Schwartz. The Estimator-Correlator for Discrete Time Problems . *IEEE Trans. Info. Theory*, IT-23:93–100, 1977.

[34] J.A. Tague. *Estimation-Correlation, Modeling, and Identification in Adaptive Array Processors* . PhD Thesis in Electrical Engineering, Penna. State Univ., Univ. Park, PA, Dec. 1987.

[35] L.L. Sharf. *Statistical Signal Processing* . Addison-Wesley, 1991.

[36] M.J. Roan and L.H. Sibul. Performance Quantification for the Wavelet Transform Domain Estimator-Correlator . In *Proc. 32nd Intnl. Conf.*, Princeton Univ., March 1998. 32nd Intnl. Conf. on Info. Sciences and Systs.

[37] D.W. Ricker and A.J. Cutezo. A Model Based Estimator Correlator (EC) Structure. *IEEE Trans. Sig. Proc.*, 48(10):2733–2742, Oct. 2000.

[38] T. Kailath. Correlation Detection of Signals Perturbed by a Random Channel . *IRE Trans. Info. Theory*, IT-6:361–366, 1960.

[39] L.H. Sibul, L.G. Weiss, and R.K. Young. Group theoretic Aspects of Signal Processing in the Time-frequency and Time-Scale Domains . In *Proc. 30th Ann. Conf.*, Princeton Univ., March 1996. 30th Ann. Conf. on Info Sciences and Systems.

[40] L.H. Sibul, L.G. Weiss, and T.L. Dixon. Wavelet Transform Domain Implementation of the Estimator Correlator for Detection of Distributed Objects in Stochastic Media . In *Proc. 1997 Conf.*, Johns Hopkins Univ., March 1997. 1997 Conf. on Info. Sciences and Systs.

[41] L.H. Sibul and J.A. Tague. Estimation-Correlation, Modeling and Identification in Adaptive Array Processors. In *Proc. 2nd IFAC Workshop*, Lund Sweden, July 1986. 2nd IFAC Workshop on Adaptive Systems.

[42] A.D. Whalen. *Detection of Signals in Noise* . Academic Press, 1971.

[43] A. Papoulis. *Probability, Random Variables, and Stochastic Processes* . McGraw-Hill, New York NY, 1965.

[44] G. Strang. *Linear Algebra and its Applications* . Academic Press, 1980.

[45] C.F. Van Loan G.H. Golub. *Matrix Computations* , volume 6. Johns Hopkins Press, Baltimore, MD, 1983.

[46] D.W. Ricker and A.J. Cutezo. Detection by Incoherent Recombination with Partial Information. *IEEE Trans. on Aerosp. and Elect. Systs.*, 37(1):2733–2742, Jan. 2001.

[47] C.W. Helstrom. *Elements of Signal Detection and Estimation* . Prentice-Hall, Englewood Cliffs N.J., 1995.

[48] P. Swerling. Probability of Detection for Fluctuating Targets . Technical report, Rand Corp., April 1954.

[49] J.I. Marcum. A Statistical Theory of Target Detection by Pulsed Radar. *IRE Trans. Info. Theory*, IT-6(2):59–267, April 1960.

[50] D.K. Barton. Simple Procedures for Radar Detection Calculations . *IEEE Trans. on Aerosp. and Elect. Systs.*, aes-5(5):837–846, Sept. 1969.

[51] Irving Kanter. Exact Detection Probability for Partially Correlated Rayleigh Targets . *IEEE Trans. on Aerosp. and Elect. Systs.*, aes-22(2):184–196, March 1986.

[52] M.A. Weiner. Detection Probability for Partially Correlated Chi-Square Targets. *IEEE Trans. on Aerosp. and Elect. Systs.*, 24(4):411–416, July 1988.

Chapter 7

THE SPATIAL REPRESENTATION

7.1 ARRAYS AND BEAM FORMATION

In addition to the estimation of Doppler and delay, bearing or angle with respect to the sonar's orientation is crucial for scatterer localization. Acoustic energy at the transmitter is usually introduced into the medium by a distributed source (projector array) and echo energy is received via a distributed array of individual transducer elements that convert the acoustic energy back into an electrical signal. Sonar arrays are spatially distributed and when in a line, plane, on the surface of a body, or when distributed throughout a volume, they are called line, planar, conformal, and volumetric arrays respectively. The process of summing the weighted responses of the individual receive array elements or more generally integrating the response over a continuous array is called *beam formation.* The integrated receive array response to acoustic energy depends upon the direction from which the energy arrives and is described by a *beam or pattern function.* Likewise the transmit pattern function describes the directional distribution of transmitted intensity from a projector array.

The generic volume backscatter geometry is shown in Figure 7.1. It defines a configuration in which a waveform that is a function of time and location within a transmitter aperture defined over a volume V_t is scattered from a point or points in the scatter region V_s and observed within the receiver aperture V_r. The source, scattering process, and receiver may be moving. The aggregate motion of any or all the volumes induces time dilation via the raycone relationships (2.8, 2.9). Relative motion between scattering points within V_s induces Doppler spreading and the combined effects may be incorporated within the definition of a suitable spreading function (6.8).

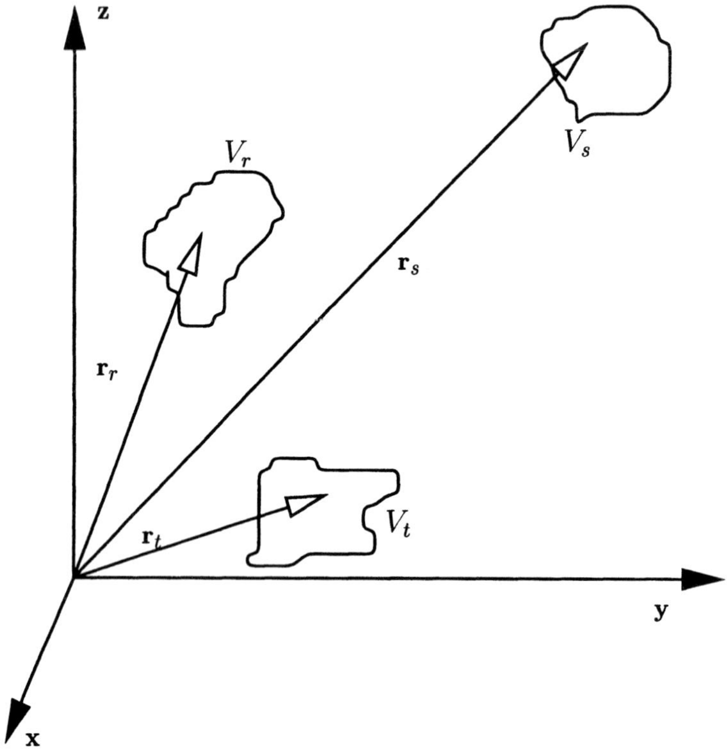

Figure 7.1: Generalized backscatter geometry for signals transmitted, scattered, and received in continuously distributed volumes.

The transmit, scattering, and receive regions are defined by vectors $\mathbf{r}_t$, $\mathbf{r}_s$, $\mathbf{r}_r$ with magnitudes r_t, r_s, r_r respectively. The corresponding unit vectors $\mathbf{u}_t$, $\mathbf{u}_s$, $\mathbf{u}_r$ define the directions from the origin to points in V_t, V_s, V_r and have components that are the direction cosines. For example, a point in the volume V_s is described by

$$\begin{aligned}\mathbf{r}_s &= r_s\mathbf{u}_s = r_s[u_{sx}, u_{sy}, u_{sz}] \\ &= r_s[\cos\theta_{sx}, \cos\theta_{sy}, \cos\theta_{sz}]\end{aligned}$$

in Cartesian coordinants where

$$\cos^2\theta_{sx} + \cos^2\theta_{sy} + \cos^2\theta_{sz} = 1.$$

It is assumed that V_t and V_r are close to the origin and V_s is sufficiently distant so that $r_s >> A_t/\lambda$, A_r/λ. λ is the nominal wavelength of the waveform and A_t, A_r are the areas of V_t and V_r projected in the direction $\mathbf{u}_s$. This is called the far field assumption [1, 2] and permits the signal delays between the transmitter, scattering region, and receive array to be expressed as projections on the unit vectors defining the regions.

The transmitter generates a waveform that is a function of time and position within a volume V_t.[1] The waveform time function $f(t)$ describes the time variation of acoustic pressure at the transmitter and is operated upon by an *aperture function*

$$\mathcal{A}_t(\mathbf{u}) = a_t(\mathbf{r}_t)d_t(\mathbf{u})dV_t$$

that appropriately weights and delays it at each point within V_t such that the aggregate transmitted plane wavefront propagates in the direction $\mathbf{u}$. $a_t(\mathbf{r}_t)$ is a weighting function of position within the aperture, $d_t(\mathbf{u})$ is a delay operator, and the elemental waveform function at each point $\mathbf{r}_t$ is

$$\mathcal{A}_t(\mathbf{u})[f(t)] = a_t(\mathbf{r}_t)f\left(t - \frac{\mathbf{u} \bullet \mathbf{r}_t}{c}\right)dV_t.$$

The notation $\mathbf{x} \bullet \mathbf{y}$ represents the vector dot (inner) product and $\mathbf{u} \bullet \mathbf{r}_t$ is the projection of the position vector in the direction of desired plane wave propagation. Dividing by the propagation velocity assures that the function is appropriately delayed so that all of the elemental waveform contributions in V_t are in phase at a distant point corresponding with the direction $\mathbf{u}$. $a_t(\mathbf{r}_t)$ is called a *shading function* that represents the source density of a continuous volume array or the weighting of a discrete element at the position $\mathbf{r}_t$ and is usually a function of frequency. Frequency dependence arises because the transduction process that transforms electrical into acoustic energy and vice versa involves hardware and circuit elements that have frequency dependent transfer functions. This is often neglected if the transmitted waveforms are confined to a narrow frequency band within which the transfer functions are constant. Delay of the time function at each point in V_t to enhance propagation in a particular direction is called *beam steering*.

[1] Sonars and radar apertures are usually composed of a collection of discrete hydrophone (antenna) elements.

The round trip delay between the origin and the point defined by $\mathbf{r}_s$ is $\tau = 2r_s/c$. The round trip delay between a point $\mathbf{r}_t$ in the transmitter aperture V_t, a scattering element at $\mathbf{r}_s$, and a point $\mathbf{r}_r$ in the receiver aperture V_r is

$$\tau - \frac{\mathbf{u}_s \bullet \mathbf{r}_t}{c} - \frac{\mathbf{u}_s \bullet \mathbf{r}_r}{c}.$$

The delay from the origin to $\mathbf{r}_s$ and back to $\mathbf{r}_r$ is

$$\tau - \frac{(\mathbf{u}_s - \mathbf{u}) \bullet \mathbf{r}_t}{c} - \frac{\mathbf{u}_s \bullet \mathbf{r}_r}{c}$$

when the transmit array is steered in the direction $\mathbf{u}$. The receive aperture is characterized by an aperture function

$$\mathcal{A}_r(\mathbf{u}') = a_r(\mathbf{r}_r)d_r(\mathbf{u}')$$

that weights and steers the receive array response in the direction $\mathbf{u}'$. The total delay translated to the origin is

$$\tau_{tot} = \tau - \frac{(\mathbf{u}_s - \mathbf{u}) \bullet \mathbf{r}_t}{c} - \frac{(\mathbf{u}_s - \mathbf{u}') \bullet \mathbf{r}_r}{c}. \tag{7.1}$$

and neglecting attenuation and assuming a stationary geometry, $f(t - \tau_{tot})$ is the waveform time function that is realized at a point in the receive aperture. Note that $\tau_{tot} = \tau = 2r_s/c$ which is the two-way range delay from the origin when the steering directions $\mathbf{u}$, $\mathbf{u}'$ coincide with $\mathbf{u}_s$.

The linear spreading model discussed in Chapter 6 is extended to the spatial domain by defining the space-time spreading function $\mathcal{S}_s(\beta, V_s) \equiv \mathcal{S}_s(\beta, \tau, \mathbf{u}_s)$ to describe the random spreading process [2]. It is a continuous density function of Doppler dilation, delay, and bearing in a scattering volume V_s at the position defined by the unit spatial vector $\mathbf{u}_s$ and the range delay τ. The signal increment emanating from the transmit volume element at a point $\mathbf{r}_t$ in V_t, scattered in V_s, received at a receive volume element at the point $\mathbf{r}_r$ in V_r, and weighted by the receiver shading function $a_r(\mathbf{r}_r)$ is

$$\begin{aligned}\delta s(t, \mathbf{u}, \mathbf{u}', \mathbf{r}_t, \mathbf{r}_r) &= \sqrt{E_0}\, a_r(\mathbf{r}_r)\, a_t(\mathbf{r}_t)\\ &\cdot \int_{\Omega_\beta} \int_{V_s} \sqrt{\beta} \mathcal{S}_s(\beta, V_s) f(\beta(t - \tau_{tot})) d\beta\, dV_s\, dV_t\, dV_r\end{aligned} \tag{7.2}$$

and is a spatial generalization of the scalar model (6.8, 6.9). It is expressed in the parametric form (6.8) as

$$\delta s(t, \mathbf{u}, \mathbf{u}', \mathbf{r}_t, \mathbf{r}_r) = \sqrt{E_0} a_r(\mathbf{r}_r) a_t(\mathbf{r}_t) \int_{\Omega} \mathcal{S}_s(\tilde{\mathbf{p}}) f(\tilde{\mathbf{p}}, t) d\tilde{\mathbf{p}}\, dV_t\, dV_r$$

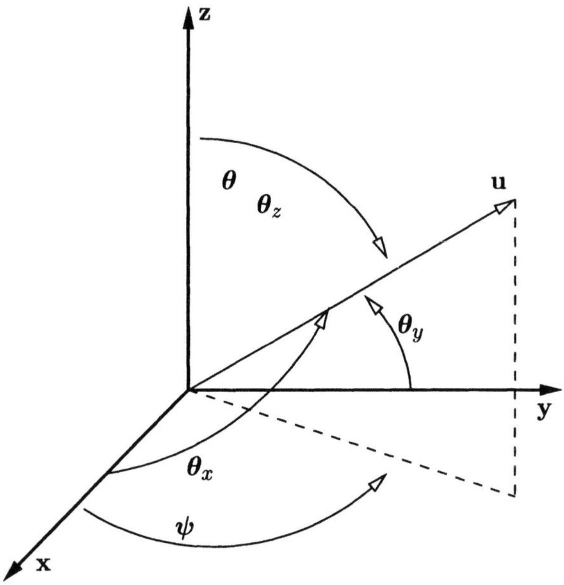

Figure 7.2: Spherical and rectangular coordinant convention

with the augmented parameter set $\tilde{\mathbf{p}} = [\mathbf{p}, \mathbf{u}_s]$ where $\mathbf{p} = [\beta, \tau]$. The aggregate signal transmitted in V_t and received in V_r is

$$\mathrm{s}(t, \mathbf{u}, \mathbf{u}') = \int_{V_t}\int_{V_r} \delta \mathrm{s}(t, \mathbf{u}, \mathbf{u}'\mathbf{r}_t, \mathbf{r}_r) dV_r dV_t \tag{7.3}$$

and obtained by integrating over the transmit and receive volumes. The spreading function has the equivalent representations

$$\mathcal{S}_s(\beta, V_s) \equiv \mathcal{S}_s(\beta, r, \theta, \psi) \equiv \mathcal{S}_s(\beta, \tau, \theta, \psi)$$

in spherical coordinants as defined in Figure 7.2. The integration (7.2) over the volume V_s represents integration over the spatial parameters (τ, θ, ψ) or (r_s, θ, ψ) with $\tau = 2r_s/c$ and

$$\begin{aligned} \int_{V_s} (\,)\, dV_s &\equiv \int_{\Omega_r}\int_{\Omega_\theta}\int_{\Omega_\psi} (\,)\, r_s^2 \sin\theta \, dr_s \, d\theta \, d\psi \\ &\equiv \int_{\Omega_\tau}\int_{\Omega_\theta}\int_{\Omega_\psi} (\,) \frac{\tau^2 c^3}{8} \sin\theta \, d\tau \, d\theta \, d\psi. \end{aligned} \tag{7.4}$$

The variables $r_s, \tau, c, \sin\theta$ defining the differential volume increments are incorporated within the respective spreading function definitions and all of the representations are consistent with the definition of the spreading function as a volume density.

The frequency domain representation of the Doppler dilated and delayed waveform is obtained by applying the Fourier transform operator $\mathcal{F}$ to the dilated waveform with delay τ_{tot} (7.1) and

$$\mathcal{F}\left(\sqrt{\beta}f(\beta(t-\tau_{tot}))\right) = \frac{1}{\sqrt{\beta}}F\left(\frac{\omega}{\beta}\right)\exp -j\omega\tau_{tot}$$
$$= \frac{1}{\sqrt{\beta}}F\left(\frac{\omega}{\beta}\right)[\exp -j\omega\tau][\exp j(\boldsymbol{\eta}_s - \boldsymbol{\eta}')\bullet \mathbf{r}_r][\exp j(\boldsymbol{\eta}_s - \boldsymbol{\eta})\bullet \mathbf{r}_t].$$

The quantity $\omega/c = 2\pi\nu/c = 2\pi/\lambda$ is called a *wave number* and

$$\boldsymbol{\eta} = \frac{\omega \mathbf{u}}{c}, \quad \boldsymbol{\eta}' = \frac{\omega \mathbf{u}'}{c} \quad \boldsymbol{\eta}_s = \frac{\omega \mathbf{u}_s}{c}.$$

are *wave number vectors* where for example

$$\boldsymbol{\eta} = [\eta_t, \eta_y, \eta_z] = \frac{2\pi}{\lambda}[\cos\theta_x, \cos\theta_y, \cos\theta_z].$$

$\nu = \omega/2\pi$ is the frequency in Hertz, $\lambda = \nu/c$ is the wavelength, and $\cos_x/\lambda$ etc. are the *spatial frequency* components corresponding to the Cartesian direction cosines of the unit vector $\mathbf{u}$ with units of cycles per meter. Wave number vectors are the spatial analog of radian frequency and the shading functions $a_t(\mathbf{r}_t)$ and $a_r(\mathbf{r}_r)$ are related to *pattern, beam, or directivity functions* Θ_t and Θ_r by the spatial Fourier transform [2, 3]

$$\Theta(\Delta\boldsymbol{\eta}) = \int_V a(\mathbf{r})\exp j\,\Delta\,\boldsymbol{\eta}\bullet\mathbf{r}dV.$$

This is written in terms of the Cartesian components $\mathbf{r} = [x, y, z]$ as

$$\Theta(\Delta\eta_x, \Delta\eta_y, \Delta\eta_z) =$$
$$\int_{-\infty}^{\infty}\int_{-\infty}^{\infty}\int_{-\infty}^{\infty} a(x,y,z)\exp j[x\,\Delta\,\eta_x + y\,\Delta\,\eta_y + z\,\Delta\,\eta_z]dxdydz$$
$$\Delta\eta_x = \frac{2\pi}{\lambda}(\cos\theta_x - \cos\theta'_x) \text{ etc.} \tag{7.5}$$

The pattern function is also expressed as the function of the spherical angles θ, ψ as shown in Figure 7.2 where

$$\cos\theta_x = \cos\psi\sin\theta$$
$$\cos\theta_y = \sin\psi\sin\theta$$
$$\cos\theta_z = \cos\theta$$
$$\cos^2\theta_x + \cos^2\theta_y + \cos^2\theta_z = 1 \tag{7.6}$$

and

$$\boldsymbol{\eta} = \frac{2\pi}{\lambda}[\cos\psi\sin\theta,\ \sin\psi\sin\theta,\ \cos\theta].$$

θ' and ψ' are the beam steering angles and

$$\Theta(\Delta\eta_t, \Delta\eta_y, \Delta\eta_z) \equiv \Theta(\theta, \theta', \psi, \psi')$$

is the pattern function at θ, ψ for a beam steered to θ', ψ'.

The time average transmitted energy intensity in the direction corresponding to $\boldsymbol{\eta}$ is proportional to $|\Theta(\boldsymbol{\eta})|^2$ hence the pattern function describes the ability of the aperture to concentrate the transmitted energy in the direction corresponding to $\boldsymbol{\eta}$. A receive pattern function describes the ability of a receiver array to discriminate spatially (angularly) distributed energy such as ambient noise [4, 5] .

The frequency domain representation of the point to point scattered signal (7.2) is

$$\begin{aligned}\delta\mathsf{S}(\omega, \mathbf{u}, \mathbf{u}', \mathbf{r}_t, \mathbf{r}_r) = \int_{\Omega_\beta}\int_{V_s} \sqrt{\frac{E_0}{\beta}} a_r(\mathbf{r}_r) a_t(\mathbf{r}_t) \mathcal{S}_s(\beta, V_s) \\ \cdot\ F\left(\frac{\omega}{\beta}\right)[\exp -j\omega\tau_{tot}] d\beta\, dV_s\, dV_t\, dV_r \qquad (7.7)\end{aligned}$$

and

$$\begin{aligned}\mathsf{S}(\omega, \mathbf{u}, \mathbf{u}') = \int_{\Omega_\beta}\int_{V_s} \sqrt{\frac{E_0}{\beta}} \mathcal{S}_s(\beta, V_s) F\left(\frac{\omega}{\beta}\right) \\ \cdot(\exp -j\omega\tau)\left[\Theta_t(\boldsymbol{\eta}_s - \boldsymbol{\eta})\right]\left[\Theta_r(\boldsymbol{\eta}_s - \boldsymbol{\eta}')\right] d\beta\, dV_s\end{aligned}$$

is the aggregate frequency domain signal corresponding to (7.3) that is obtained by substituting (7.1) for τ_{tot} and integrating over the transmit and receive apertures. The corresponding steered transmit and receive pattern functions relative to the vector $\mathbf{u}_s$ are

$$\begin{aligned}\Theta_t(\boldsymbol{\eta}_s - \boldsymbol{\eta}) &= \int_{V_t} a_t(\mathbf{r}_t) \exp j[(\boldsymbol{\eta}_s - \boldsymbol{\eta}) \bullet \mathbf{r}_t] dV_t \\ \Theta_r(\boldsymbol{\eta}_s - \boldsymbol{\eta}') &= \int_{V_r} a_r(\mathbf{r}_r) \exp j[(\boldsymbol{\eta}_s - \boldsymbol{\eta}') \bullet \mathbf{r}_r] dV_r \qquad (7.8)\end{aligned}$$

and represent the frequency dependent plane wave array responses (beam patterns) in the directions defined by the wave number vectors. The spatial Fourier kernel (7.5) for the steered transmitter array is

$$\exp j[(\eta_s - \eta) \bullet \mathbf{r}_t] = \exp j\frac{2\pi}{\lambda}[x(\cos\theta_{sx} - \cos\theta_x) + y(\cos\theta_{sy} - \cos\theta_y) + z(\cos\theta_{sz} - \cos\theta_z)]$$

and θ_{sx}, θ_{sy}, θ_{sz} and θ_x, θ_y, θ_z are the angles of the scattering and steering vectors with respect to the x, y, z axis. It is unity when the angles coincide and the beam steering angle is aligned with the scattering site ($\mathbf{u}_s = \mathbf{u}$). The transmit pattern response and energy intensity is then maximized and

$$\begin{aligned}|\Theta|^2_{max} = |\Theta_t(0)|^2 &= \left|\int_{V_t} a_t(\mathbf{r}_t) dV_t\right|^2 \\ &= \left|\int_{-\infty}^{\infty}\int_{-\infty}^{\infty}\int_{-\infty}^{\infty} a_t(x,y,z)dxdydz\right|^2 . \end{aligned} \quad (7.9)$$

The same holds for the receiver response.

Point scattering is characterized by the collapse of the region V_s to a point at $\tilde{\mathbf{p}} = [\beta, \tau, \mathbf{u}_s]$ and the random spreading function becomes the slowly fluctuating scalar

$$\mathcal{S}_{sp} \equiv \mathcal{S}_s(\tilde{\mathbf{p}}) = \mathcal{S}_s(\tilde{\mathbf{p}}')\delta(\tilde{\mathbf{p}}' - \tilde{\mathbf{p}})$$

as in (3.49). The time domain signal increment (7.2) originating at $\mathbf{r}_t$, reflected at the point $\mathbf{r}_s$, and observed at the point $\mathbf{r}_r$ is

$$\begin{aligned}\delta \mathsf{s}(t, \mathbf{u}, \mathbf{u}', \tilde{\mathbf{p}}, \mathbf{r}_t, \mathbf{r}_r) &= \sqrt{E_0} a_r(\mathbf{r}_r) a_t(\mathbf{r}_t) \mathcal{S}_{sp} f(\tilde{\mathbf{p}}, t)) dV_t\, dV_r \\ f(\tilde{\mathbf{p}}, t) &= \sqrt{\beta} f(\beta(t - \tau_{tot})) \\ &= \sqrt{\beta} f\left(\beta\left[t - \tau + \frac{(\mathbf{u}_s - \mathbf{u}) \bullet \mathbf{r}_t}{c} + \frac{(\mathbf{u}_s - \mathbf{u}') \bullet \mathbf{r}_r}{c}\right]\right) . \end{aligned} \quad (7.10)$$

The total received energy is

$$E_{sc} = \int_{-T/2}^{T/2} \mathcal{E}\{|\delta \mathsf{s}(t)|^2\} dt = E_0 \mathcal{R}_{sp}$$

and $\mathcal{R}_{sp}$ represents the point scattering function (6.21) defined with the parameters $\tilde{\mathbf{p}}$. It accounts for all of the transmitted energy propagation, scattering, and conversion losses. The aggregate point scatter frequency domain signal

$$\mathsf{S}(\omega, \mathbf{u}, \mathbf{u}', \tilde{\mathbf{p}}) = \sqrt{\frac{E_0}{\beta}} \mathcal{S}_{sp} F\left(\frac{\omega}{\beta}\right) (\exp -j\omega\tau) \left[\Theta_t(\eta_s - \eta)\right] \left[\Theta_r(\eta_s - \eta')\right]$$

is obtained by integrating over the transmit and receive apertures and describes the spectrum of the echo signal reflected at the point $\mathbf{r}_s$ in V_s moving with line of sight speed

$$v_{los} = \left(\frac{\beta - 1}{\beta + 1}\right) c.$$

The transmitter pattern function Θ_t determines the degree to which the transmitted intensity is concentrated in a particular direction. The intensity concentration is measured by the directivity index D_I that relates the transmitted energy E_0 (Joules) to the energy flux or intensity E_x (Joules/m^2) in the direction in which the beam is steered referred to the surface of the unit sphere at one meter. Likewise, the receiver pattern function Θ_r defines the degree to which the array discriminates against isotropic ambient noise. The transmitter relationship in decibel notation is

$$E_x = E_0 + D_I - 11 \text{ (dB re 1 Joule/m}^2\text{) @ 1meter)}$$

and defines the source level (2.1) required to perform the sonar equation analysis of input SIR discussed in Chapter 2. D_I is defined as $10 \log D$ where the directivity

$$D = \frac{4\pi E_x}{\int_0^{4\pi} E_x |\Theta_n(\eta - \eta')|^2 d\Omega}$$

is the ratio of the total energy integrated on the unit sphere assuming a uniform energy flux density to that transmitted (or received) by a normalized pattern function and $d\Omega$ is a solid angle element. Integration with respect to the wave number vector η represents integration with respect to a vector $\mathbf{u}$ the locus of which defines the unit sphere. An element of solid angle is $d\Omega = \sin\theta d\theta\psi$ in spherical coordinants so that the directivity index is equivalently written as

$$\begin{aligned} D_I &= 10 \log D \\ D &= \left(\frac{1}{4\pi}\int_0^{2\pi}\int_0^{\pi} |\Theta_n(\omega, \theta, \theta', \psi, \psi')|^2 \sin\theta d\theta d\psi\right)^{-1}. \end{aligned} \quad (7.11)$$

The pattern function may be normalized with respect to its maximum and

$$|\Theta_n(\eta - \eta')|^2 = \frac{|\Theta(\eta - \eta')|^2}{|\Theta|^2_{max}}$$

is the normalized beam so that

$$D = \frac{|\Theta|^2_{max}}{\frac{1}{4\pi}\int_0^{4\pi}|\Theta(\eta-\eta')|^2 d\Omega} = \frac{|\Theta|^2_{max}}{|\Theta|^2_{avg}}.$$

The maximum steered beam pattern response (7.9) occurs on the beam boresight defined as the direction ($\eta = \eta'$). The maximum unsteered response occurs in the normal (z) direction or when $\theta = 0$ for a planar aperture confined to the x-y plane with real aperture weighting. The maximum of (7.5) is then the integrated 2-d shading function and

$$\Theta(max) = \int_{-\infty}^{\infty}\int_{-\infty}^{\infty} a(x,y)dxdy.$$

Note that when the aperture collapses to a point at $\mathbf{r}_{t0}$ or $\mathbf{r}_{r0}$,

$$\begin{aligned} a_t(\mathbf{r}_t) &= a_t(\mathbf{r}_t)\delta(\mathbf{r}_t - \mathbf{r}_{t0}) \\ a_r(\mathbf{r}_r) &= a_r(\mathbf{r}_r)\delta(\mathbf{r}_r - \mathbf{r}_{r0}). \end{aligned}$$

The pattern functions (7.8) then become constant or omni-directional with respect to η, $|\Theta_t|^2 = |a_t(\mathbf{r}_t)|^2$, $|\Theta_r|^2 = |a_r(\mathbf{r}_r)|^2$, and the directivity index $D_I = 0$ dB because the maximum and average responses are equal. Conversely if the pattern function is uniform and equal to unity over a solid angle segment (beamwidth) $\delta\Omega$, the directivity is $D = 4\pi/\delta\Omega$ and is inversely proportional to the beamwidth defined by the segment.

7.1.1 The Rectangular Aperture

Planar apertures are used when physical conditions permit their installation because they present the greatest aperture and hence can synthesize the narrowest beams in the direction normal to the surface. The rectangular piston transducer is a continuous rectangular array in which all points on the surface move in phase and its pattern function is easily determined in closed mathematical form. The normalized uniform shading function of a frequency independent rectangular aperture of dimension $L_x \times L_y$ is $a(x,y) = 1/L_xL_y$ so that $\Theta(max) = 1$. The corresponding normalized pattern function with its origin at x=y=0 is computed from the 2-d version of (7.5) as

$$\Theta_n(\Delta\eta_t, \Delta\eta_y) = \int_{-L_x/2}^{L_x/2}\int_{-L_y/2}^{L_y/2}(1/L_xL_y)\exp j[x\,\Delta\,\eta_x + y\,\Delta\,\eta_y]dxdy \qquad (7.12)$$

which is separable into individual integrations in x and y. The result is the product of sinc(x)=(sin x/x) functions

$$\begin{aligned}\Theta_n(\Delta\eta_t, \Delta\eta_y) &= \text{sinc}\left(\frac{\Delta\eta_x L_x}{2}\right) \text{sinc}\left(\frac{\Delta\eta_y L_y}{2}\right) \\ &= \Theta_{nx}(\Delta\eta_t)\Theta_{ny}(\Delta\eta_y) \qquad (7.13)\end{aligned}$$

where Θ_{nx}, Θ_{ny} are the patterns corresponding to line arrays defined on the x and y axis and

$$\begin{aligned}\Delta\eta_t &= \frac{2\pi}{\lambda}[\cos\psi\sin\theta - \cos\psi'\sin\theta'] \\ \Delta\eta_y &= \frac{2\pi}{\lambda}[\sin\psi\sin\theta - \sin\psi'\sin\theta']\end{aligned}$$

are obtained from (7.5, 7.6) and corresponds to the spherical angle convention of Figure 7.2.

A planar array that is mounted in such a way that it radiates only from its front side (upper half plane) is said to be *baffled* [1] in which case its minimum directivity index is 3 dB. Examples of baffled beam patterns for a square planar aperture with various dimensions are shown in Figure 7.3. The beams are not steered so that $\theta' = \psi' = 0$ and D_I (7.11) is computed by numerical integration using (7.13). D_I is a measure of the effective beamwidth and increases with aperture size as shown. The beam pattern becomes hemispherical (omni-directional) for $L/\lambda << 1$ and the half plane D_I approaches 3 dB as shown by Figure 7.3a. Distances along the axes represent the projections of the 3-d pattern function and $\Theta_{nx} = \Theta\cos\theta_x$ etc. The pattern becomes omni-directional as the aperture is reduced and $\Theta \to 1$ in which case distances along the axes represent the direction cosines. The projections of the unsteered pattern function (7.13) on the x-z and y-z planes are

$$\begin{aligned}\Theta_{nxz} &= \text{sinc}\left(\frac{\pi L_x \sin\theta}{\lambda}\right), \quad \theta' = 0,\ \psi = 0 \\ \Theta_{nyz} &= \text{sinc}\left(\frac{\pi L_y \sin\theta}{\lambda}\right), \quad \theta' = 0,\ \psi = 90^\circ \qquad (7.14)\end{aligned}$$

and are not functions of ψ'. The first null of the sinc function occurs when its argument is π or when $\sin\theta = \lambda/L$. $\sin\theta \approx \theta$ for small angles and the null points define a rough measure of beamwidth $\theta_b = 2\lambda/L$ that is specific to uniform arrays. The half power or 3 dB beamwidth is a more common beamwidth definition that is applicable to nonuniform

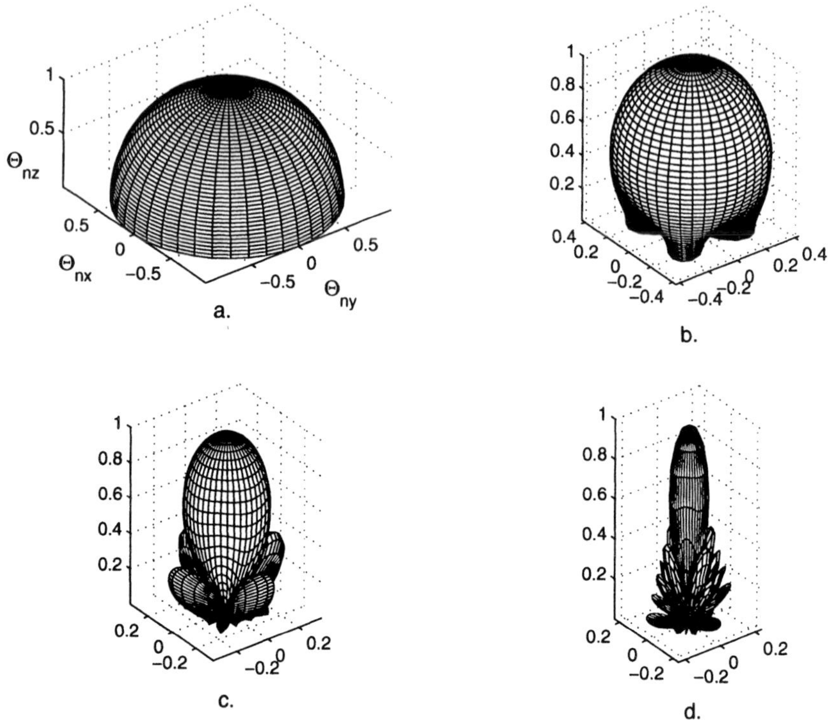

Figure 7.3: Normalized beam patterns Θ_n (7.13) of a uniform square planar unsteered aperture with dimensions ($L_x = L_y$) and directivity indices [D_I]; a. (.1λ),[3.2], b. (1λ),[12.7], c. (2λ), [18.8], d. (5λ),[26.7]

shading functions and is defined as the angular segment between the 3 dB points of the squared pattern function. These points are defined by $\text{sinc}^2(x) = .5$ or $x = \pi L \sin\theta/\lambda = 1.3917$ and the half power projected beam width for small angles ($\theta \leq 10°$) is

$$\theta_b = 2\sin^{-1}\left(\frac{.443\lambda}{L}\right) \approx 2\left(\frac{.443\lambda}{L}\right) \quad \text{for } \frac{\lambda}{L} \leq .39. \qquad (7.15)$$

The corresponding solid angle element is $\delta\Omega = (.886\lambda/L)^2$ for $L_x = L_y = L$ and the directivity and D_I are

$$D = \frac{4\pi}{\delta\Omega} = 16(L/\lambda)^2$$
$$D_I = 12 + 20\log(L/\lambda) \text{ (dB)}$$

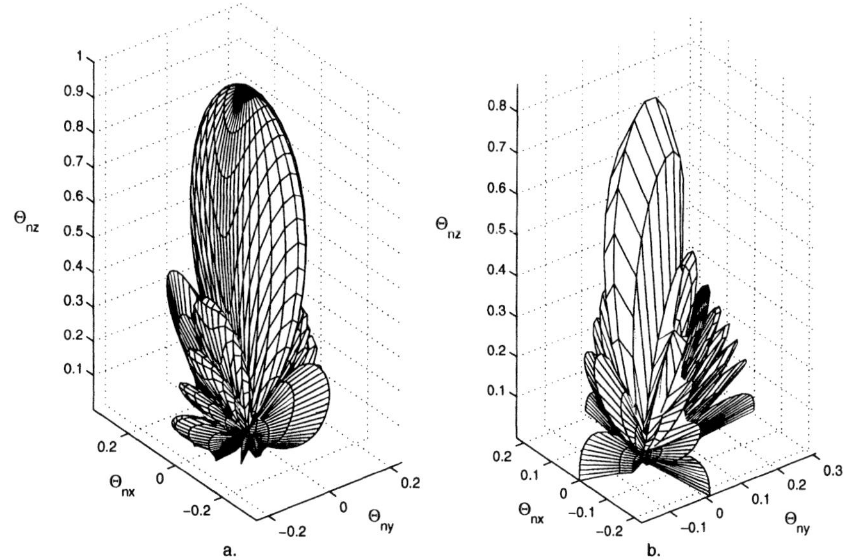

Figure 7.4: a. $2\lambda \times 5\lambda$ beam (DI=22.8 dB.), b. $5\lambda \times 5\lambda$ beam tilted 30 deg. (DI=26.1 dB.)

assuming that $\Theta = 1$ within $\delta\Omega$ and zero elsewhere. This approximation is reasonably accurate as it predicts D_I=12, 18, and 26 dB for the patterns (b,c,d) of Figure 7.3 which are within a dB of the numerically computed values. L^2 is the area of the baffled planar array and more generally

$$D_I \approx 12 + 10 \log(A/\lambda^2) \text{ (dB)} \tag{7.16}$$

approximates the D_I for other uniform planar apertures with minimum dimension larger than the wavelength [5].

The pattern function examples for non-square and tilted rectangular arrays are shown in Figures 7.4a,b respectively. Their width is inversely proportional to the aperture as illustrated by Figure 7.4a for $L_y/L_x = 5/2$. The numerically computed DI is 22.8 dB compared to 22 db predicted by (7.16) which is consistent with the predictions for the square arrays. The beam becomes distorted when it is tilted and the DI decreases in approximate proportion to the aperture area projected in the steering direction. The beam of Figure 7.4b is generated from a square aperture with $L_x/\lambda = L_y/\lambda = 5$ corresponding to that of Figure

7.3d but steered at an angle of 30 deg. from the z axis in the y direction. This a fairly severe steering angle as it is about three times the nominal 10 deg. 3 dB beamwidth of the 5 by 5 array. The beam is distorted, the sidelobes increase in relation to the mainlobe, and the computed DI=26.1 dB is about .6 dB less than that of the untilted beam reflecting an increase in beamwidth. The approximation (7.16) using the projected area $A/\lambda^2 = 25\cos(30)$ predicts DI=25.4 dB which is likewise .6 dB less than that predicted for the untilted beam.

The approximate 3 dB beamwidth increase associated with beam tilt is found from

$$\sin\theta - \sin\theta' \approx \cos\theta'\delta\theta = \frac{.443\lambda}{L}$$

and the tilted beamwidth is

$$\theta_b(\theta') = 2\delta\theta \approx \frac{.886\lambda}{L\cos\theta'}.$$

This amounts to an increase of about 1.5 deg. for the 30° tilted 5λ beam.

The uniform rectangular window provides the narrowest mainlobe for a fixed aperture at the expense of the rather high sidelobes of the sinc function as shown in Figure 7.3. The sidelobes can be significantly reduced at the expense of a somewhat wider mainlobe by using tapered shading functions a_t, a_y that avoid the first order discontinuities of the rectangular window at the aperture edges. The Hanning (cosine squared) window

$$a(x) = 1 - \cos^2(2\pi x/L) \quad -L/2 \leq x \leq L/2$$

shown in Figure 7.5a is an example. The corresponding beam pattern for the rectangular array with $L_x = L_y = 5\lambda$ is shown in Figure 7.5b. The first sidelobe of the sinc function is down 13 dB from the mainlobe peak while that of the Hanning window is down by 32 dB [6]. The reduction is apparent from a comparison of the beam sidelobes with those of the uniform window with the same aperture shown in Figure 7.3d. The characteristics of other shading functions are discussed extensively in [7–9].

7.1.2 The Discrete Aperture

Sonar arrays are usually composed of an arrangement of discrete hydrophones (array elements) that transduce or convert electrical signals to acoustic pressure and vice versa. Modern array elements are usually composed of piezoelectric materials in various forms the most common

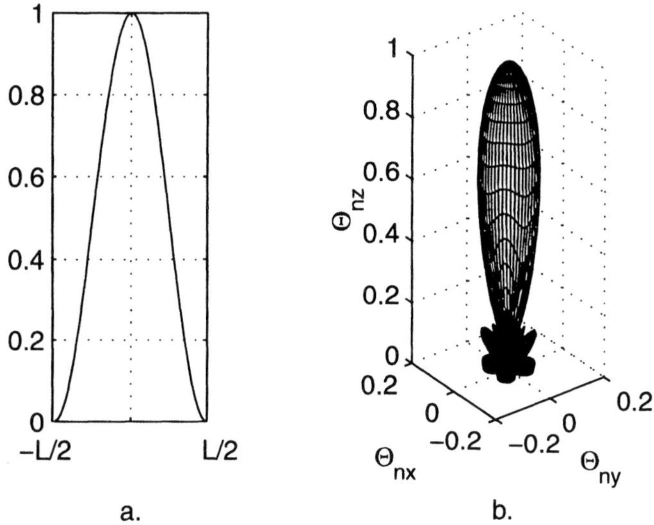

Figure 7.5: a. The Hanning window function, b. The Hanning beam pattern with $L_x = L_y = 5\lambda$.

being the piezoelectric stack that acts as a continuous piston and is small with respect to the wavelength. It is used to build planar or conformal arrays for medium to high frequency sonars and is usually mounted in a rigid baffled structure. The individual element beam patterns resemble that of the uniform rectangular piston illustrated by Figure 7.3a [5]. Cylindrical shaped elements are often used as elements of low frequency arrays that are not necessarily rigidly mounted and may not be baffled.

Beam formation is accomplished by the weighted summation of the individually phased elements

$$\Theta(\triangle\eta_t, \triangle\eta_y, \triangle\eta_z) = \sum_m \sum_n \sum_l a(x_m, y_n, z_l) \cdot \exp j[x_m \triangle \eta_x + y_n \triangle \eta_y + z_l \triangle \eta_z] \quad (7.17)$$

that is the discrete analog of the spatial Fourier transform (7.5). The corresponding normalized beam is

$$\Theta_n = \frac{\Theta}{\Theta_{max}} = \frac{\Theta}{\sum_m \sum_n \sum_l a(x_m, y_n, z_l)}. \quad (7.18)$$

The relationship between beams formed from continuous and discrete arrays is the same as that for scalar time functions and their associated frequency spectra. A discrete array with evenly spaced elements can be considered as a sampling of a continuous array and the resulting beams will repeat (alias) in spatial frequency at intervals corresponding to the inverse of the spatial sampling interval. This is illustrated by considering the discrete version of the continuous uniformly weighted planar array with the pattern function (7.13) that is illustrated in Figure 7.3.

The array dimension is $L_x \times L_y$ and is composed of even numbers $N_x = 1 + L_x/d$ and $N_y = 1 + L_y/d$ of equispaced columns and rows with inter-element spacing d. The array can also have odd numbers of rows and columns but an even number facilitates the division of the array into half apertures [10] which is an advantage when estimating bearing as will be discussed later. The center of the array is defined as the origin so that the rows and columns are at distances $\pm d/2$, $\pm 3d/2$, $\cdots$. $\pm (N-1)d/2$ where N represents either N_x or N_y. The unsteered ($\theta' = 0$) normalized beam function is

$$\Theta_n(\Delta\eta_t, \Delta\eta_y) = \frac{1}{N_x N_y} \sum_{m=1}^{N_x} \sum_{n=1}^{N_y} \exp -j\left[\left(\frac{N_x - 2m + 1}{2}\right) d \bigtriangleup \eta_t\right] \cdot \exp -j\left[\left(\frac{N_y - 2n + 1}{2}\right) d \bigtriangleup \eta_y\right] \tag{7.19}$$

which is the discrete version of (7.12). The discrete array unsteered beam pattern projected upon the x-z plane corresponding to (7.14) for the continuous array is

$$\Theta_{nxz}(\theta) = \frac{1}{N_x} \sum_{m=1}^{N_x} \exp -j\left[\left(\frac{N_x - 2m + 1}{2}\right) \eta d\right]$$
$$\eta = 2\pi \sin\theta/\lambda \tag{7.20}$$

and obtained from (7.19) by setting $\psi = 0$. This is the 2-d pattern response of a discrete line array with N_x elements and spacing d. Combining terms corresponding to the elements at $\pm kd$ produces

$$\Theta_{nxz}(\theta) = \frac{2}{N_x}[\cos(\eta d/2) + \cos(3\eta d/2) + \cdots + \cos((N_x - 1)\eta d/2)]$$
$$= \frac{2}{N_x} \sum_{m=1}^{N_x/2} \cos\left[(N - 2m + 1)\left(\frac{\pi d \sin\theta}{\lambda}\right)\right]. \tag{7.21}$$

The pattern magnitude peaks when the arguments of the cosines are multiples of π which occurs at intervals corresponding to $d\sin\theta/\lambda = k$ an integer when $d \geq \lambda/2$. This is the spatial equivalent of the aliasing effect

$$\mathcal{F}(a(t)) \to A(\nu)$$

$$\mathcal{F}\left(\sum_k a(k)\delta(t - k \triangle t)\right) \to \sum_k A(\nu - k/\triangle t)$$

that occurs for the spectra of an undersampled time series [11]. The spatial frequency is $\sin\theta/d$ and the sampled version of the continuous array beam pattern magnitude (7.14) is

$$\Theta_{nxz}(\theta) = \sum_{k=-\infty}^{\infty} \operatorname{sinc}\left(\pi L\left(\frac{\sin\theta}{\lambda} - \frac{k}{d}\right)\right).$$

It is equivalent to (7.21) and repeats at bearing angles

$$\theta = \sin^{-1}(k\lambda/d)$$

when $d \geq \lambda/2$. The beam response for small angles defines the pattern function mainlobe. The subsidiary aliased peaks are called *grating lobes* and occur at angular increments that are inversely proportional to the element separation d. Good array design seeks to avoid grating lobes by specifying that $d < \lambda/2 < L$.

Note that the uniform array weighting of (7.19) is real, even, and symmetric hence the resulting beam patterns (7.20,7.21) are real and symmetric. This is a general property of Fourier transforms as indicated in Table 1.1 and holds for any set of real coefficients $a_m = a_{N_x+1-m}$ that weight an array symmetrically about its centroid.

7.2 ARRAY PROCESSING

The maximum likelihood (ml) receiver for Gaussian data received at multiple points of a discrete aperture is an *array processor* and is a spatial generalization of the matched filter structures defined for scalar data in Chapter 3 and the estimator-correlator discussed in Chapter 6.

Spatial in this context means that the echo signal and interference have directional (bearing) characteristics as in (7.2) in addition to the usual delay-Doppler parameters. The discrete unsteered N element line

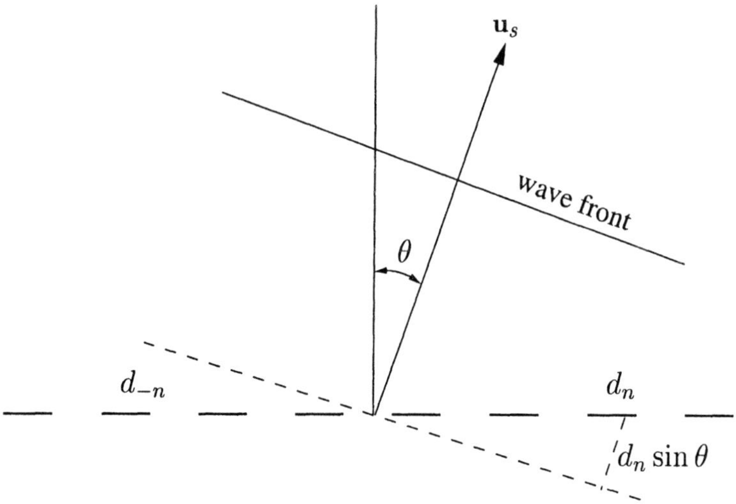

Figure 7.6: The discrete line array

array shown in Figure 7.6 that receives echo data generated by a slowly fluctuating random point scatterer serves to illustrate the form of the array processor. The received random time series at the n-th element consisting of the echo signal and additive interference is

$$\mathsf{r}_n(t, \tilde{\mathbf{p}}) = \sqrt{E_0} S_{sp} a_n \sqrt{\beta} f(\beta(t - \tau_{tot})) + \mathsf{n}_n(t), \quad n = 1, \cdots, N$$

and a_n is the normalized aperture weighting. The signal portion corresponds to (7.1, 7.10) and the effects of the transmit array are incorporated in the definition of the point spreading function $\mathcal{S}_{sp}$. $\tau_{tot} = \tau - d_n \sin\theta / c$ for the line array with $\psi = 0$ and a_n and d_n are the weight and distance from the origin of the n-th element. θ represents the angle of the arriving plane wavefront with respect to the array normal as defined in Figures 7.2 and 7.6. The distance d_n is that indicated in (7.19) if the elements are equally spaced with even N. The interference at each sensor element consists of reverberation and noise and may be correlated in both space (between elements) and in time. It is assumed to be a zero mean Gaussian process with inter-element covariance

$$\mathcal{E}\{\mathsf{n}_m(t)\mathsf{n}_n(u)\} = R_{mn}(t, u)$$

that reduces to $R_{mn}(t-u)$ for wide sense stationarity (WSS) in the time domain. The WSS condition permits the simplification of the likelihood relationships by passing to the frequency domain as done for the scalar detection problem in Chapter 3.

$K = T/t_s$ frequency samples for each element are generated by sampling the element time series for an interval T at intervals t_s, multiplying the time series by $1/\sqrt{\mathrm{T}}$, and Fourier transforming. The normalization assures that the magnitude square of the transformed data has units of energy and represents the sampled power spectral density. The discrete input signal plus interference data components at the K frequencies $\omega_k = 2\pi k/T$ and N elements are

$$\begin{aligned} F_k(\mathbf{p}) &= (F(\omega_k/\beta)/\sqrt{\beta})(\exp -j\omega_k\tau) \\ \mathsf{S}_{kn}(\tilde{\mathbf{p}}) &= \sqrt{E_0}S_{sp}F_k(\mathbf{p})a_n \exp jd_n\eta_k \\ \mathsf{R}_{kn}(\tilde{\mathbf{p}}) &= \mathsf{S}_{kn}(\tilde{\mathbf{p}}) + \mathsf{N}_{kn}, \quad n = 1, \cdots, N, \; k = 1, \cdots, K \end{aligned} \tag{7.22}$$

with $\eta_k = \omega_k \sin\theta/c$, $\mathsf{N}_{kn} = \mathsf{N}_n(\omega_k)$, and $\mathsf{R}_{kn} = \mathsf{R}_n(\omega_k, \tilde{\mathbf{p}})$ denoting the k-th frequency and n-th element. $\tilde{\mathbf{p}} = [\beta, \tau, \theta]$ is the delay-Doppler parameter vector $\mathbf{p} = [\beta, \tau]$ augmented by the bearing parameter that may be the unit direction vector $\mathbf{u}$, wave number vector η, or simply θ. The interference time samples at the n-th element are $\mathsf{n}_{in} \equiv \mathsf{n}_n(it_s)$ and the corresponding spectral samples are

$$\mathsf{N}_{kn} = \sum_{i=1}^{K} \mathsf{n}_{in} \exp -j(it_s\omega_k).$$

It is convenient to formulate the point scatter log-likelihood ratio in the vector-matrix form (6.45) in order to treat the multidimensional spatial-temporal nature of the data. The input spectral components (7.22) are assembled into K individual vectors

$$\begin{aligned} \mathbf{r}_k(\tilde{\mathbf{p}}) &= \mathbf{f}_k(\tilde{\mathbf{p}})\mathsf{s} + \mathbf{n}_k \\ \mathbf{r}_k(\tilde{\mathbf{p}})^{\mathrm{T}} &= [\mathsf{R}_{k1}, \mathsf{R}_{k2}, \cdots, \mathsf{R}_{kN}] \\ \mathbf{n}_k^{\mathrm{T}} &= [\mathsf{N}_{k1}, \mathsf{N}_{k2}, \cdots, \mathsf{N}_{kN}] \quad k = 1, \cdots, K \end{aligned} \tag{7.23}$$

corresponding to the frequencies ω_k each of which has dimension N corresponding to the number of elements. $\mathsf{s} = \sqrt{E_0}S_{sp}$ is the complex, slowly fluctuating, scalar point scatter echo amplitude with the expected echo energy

$$E_s = \mathcal{E}\{|\mathsf{s}|^2\} = E_0\mathcal{E}\{|S_{sp}|^2\} = E_0\mathcal{R}_{sp}.$$

The signal vector ($\mathbf{f}_k(\tilde{\mathbf{p}})s$) with parameters $\tilde{\mathbf{p}} = [\beta, \tau, \theta]$ is comprised of the random point spreading function, the waveform function, and the deterministic steering vector $\mathbf{e}_k$ where

$$\begin{aligned}
\mathbf{f}_k(\tilde{\mathbf{p}}) &= [F(\omega_k/\beta)/\sqrt{\beta}](\exp - j\omega_k\tau)\mathbf{e}_k = F_k(\mathbf{p})\mathbf{e}_k \\
\mathbf{e}_k^{\mathrm{T}} &= [a_1 \exp jd_1\eta_k,\ a_2 \exp jd_2\eta_k,\ \cdots,\ a_N \exp jd_N\eta_k] \\
\eta_k &= \frac{\omega_k \sin\theta}{c}.
\end{aligned} \tag{7.24}$$

The components of the random vector $\mathbf{n}_k$ are the interference spectral realizations at ω_k for each element that may be correlated between elements because of anisotropic ambient noise and reverberation, internal vibration, water flow across the array (flow noise), or inter-element electronic noise coupling [4]. They are asymptotically uncorrelated at different frequencies for sufficiently large T as shown by (1.20) and the interference cross spectrum defined at frequencies k, l and elements m, n is

$$\begin{aligned}
W_{klmn} &= \mathcal{E}\{\mathsf{N}_{km}\mathsf{N}_{ln^*}\} = 0,\ \text{for } k \neq l \\
&= \mathcal{E}\{\mathsf{N}_{km}\mathsf{N}_{kn}^*\} = W_{mn}(\omega_k) \equiv W_{kmn}, \\
&\text{for } k = l,\ (m, n) = 1, \cdots, N.
\end{aligned} \tag{7.25}$$

$\mathcal{E}\{\mathbf{n}_k\mathbf{n}_l^{\mathrm{H}}\} = 0,\ k \neq l$ and the elements $W_{kmn} = W_{knm}^*$ comprise the $N \times N$ Hermitian interference cross spectrum matrix at frequency k [12–14]

$$\mathbf{W}_k = \mathbf{W}_k^{\mathrm{H}} = \mathcal{E}\{\mathbf{n}_k\mathbf{n}_k^{\mathrm{H}}\}.$$

The totality of the input data is expressed by adjoining the K individual frequency dependent vectors (7.23, 7.24) of dimension N into one vector of dimension $K \times N$ that contains all of the element and frequency samples [12]. The adjoined data vector is

$$\begin{aligned}
\mathbf{r} &= [\mathbf{r}_1^{\mathrm{T}},\ \mathbf{r}_2^{\mathrm{T}},\ \cdots,\ \mathbf{r}_K^{\mathrm{T}}]^{\mathrm{T}} = \mathbf{f}\mathsf{s} + \mathbf{n} \\
\mathbf{f} &= [\mathbf{f}_1^{\mathrm{T}},\ \mathbf{f}_2^{\mathrm{T}},\ \cdots,\ \mathbf{f}_K^{\mathrm{T}}]^{\mathrm{T}} \\
\mathbf{n} &= [\mathbf{n}_1^{\mathrm{T}},\ \mathbf{n}_2^{\mathrm{T}},\ \cdots,\ \mathbf{n}_K^{\mathrm{T}}]^{\mathrm{T}}
\end{aligned}$$

where $\mathbf{f}_k \equiv \mathbf{f}_k(\tilde{\mathbf{p}})$, $\hat{\mathbf{f}}_k \equiv \mathbf{f}_k(\hat{\tilde{\mathbf{p}}})$, $\tilde{\mathbf{p}} = [\beta, \tau, \theta]$, and $\hat{\tilde{\mathbf{p}}} = [\hat{\beta}, \hat{\tau}, \hat{\theta}]$ represents the hypothesized parameters. The data vector is of the same form as (6.43) and the log-likelihood ratio (llr) corresponding to (6.45) is

$$\begin{aligned}
\ell_r &= \ell_d - \eta = \mathbf{r}^{\mathrm{H}}(\mathbf{W}^{-1} - (\mathbf{W} + \mathbf{P})^{-1})\mathbf{r} \\
&\quad -(\ln|\mathbf{W} + \mathbf{P}| - \ln|\mathbf{W}|).
\end{aligned} \tag{7.26}$$

$\mathbf{W}$ is the $KN \times KN$ block diagonal covariance of the adjoined interference vector

$$\mathbf{W} = \mathcal{E}\{\mathbf{nn}^{\mathrm{H}}\} = \begin{bmatrix} \mathbf{W}_1 & \cdots & \cdots & 0 \\ \cdots & \mathbf{W}_2 & \cdots & \cdots \\ \cdots & \cdots & \ddots & \cdots \\ 0 & \cdots & \cdots & \mathbf{W}_K \end{bmatrix}$$

$E_s = E_0 \mathcal{E}\{|\mathcal{S}_s|^2\}$ is the scalar energy, and $\mathbf{P} = E_s \mathbf{ff}^{\mathrm{H}}$ is the signal covariance. Unlike the interference covariance $\mathbf{W}$, $\mathbf{P}$ is not necessarily block diagonal because the waveform is deterministic and the sub matrices $\mathbf{f}_i \mathbf{f}_j^{\mathrm{H}}$ formed at different frequencies do not vanish. Note that the data term of the llr can be expressed in terms of an estimated signal vector

$$\begin{aligned} \ell_d &= \hat{\mathbf{s}}^{\mathrm{H}} \mathbf{W}^{-1} \mathbf{r} \\ \hat{\mathbf{s}} &= E_s \mathbf{ff}^{\mathrm{H}} (E_s \mathbf{ff}^{\mathrm{H}} + \mathbf{W})^{-1} \mathbf{r} \end{aligned}$$

as was the case for the estimator-correlator (6.46).

A convenient form for the ℓ_d is obtained by invoking the matrix identity [13, 15] that is called the Sherman-Morrison [16] or Bartlett [17] formula

$$\begin{aligned} (\mathbf{W}^{-1} - (\mathbf{W} + \mathbf{P})^{-1}) &= (\mathbf{W}^{-1} - (\mathbf{W} + \mathbf{f} E_s \mathbf{f}^{\mathrm{H}})^{-1}) \\ &= \mathbf{W}^{-1} \mathbf{f} (E_s^{-1} + \mathbf{f}^{\mathrm{H}} \mathbf{W}^{-1} \mathbf{f})^{-1} \mathbf{f}^{\mathrm{H}} \mathbf{W}^{-1} \end{aligned}$$

and

$$\begin{aligned} \ell_d &= \mathbf{r}^{\mathrm{H}} \mathbf{W}^{-1} \hat{\mathbf{f}} (E_s^{-1} + \hat{\mathbf{f}}^{\mathrm{H}} \mathbf{W}^{-1} \hat{\mathbf{f}})^{-1} \hat{\mathbf{f}}^{\mathrm{H}} \mathbf{W}^{-1} \mathbf{r} \\ &= \frac{|\mathbf{r}^{\mathrm{H}} \mathbf{W}^{-1} \hat{\mathbf{f}}|^2}{E_s^{-1} + \hat{\mathbf{f}}^{\mathrm{H}} \mathbf{W}^{-1} \hat{\mathbf{f}}} = \frac{|\sum_k \mathbf{r}_k^{\mathrm{H}} \mathbf{W}_k^{-1} \hat{\mathbf{f}}_k|^2}{E_s^{-1} + \sum_k \hat{\mathbf{f}}_k^{\mathrm{H}} \mathbf{W}_k^{-1} \hat{\mathbf{f}}_k}. \end{aligned} \tag{7.27}$$

The inverse of a block diagonal matrix is the block diagonal matrix of the individual inverses and

$$\begin{aligned} \hat{\mathbf{f}}^{\mathrm{H}} \mathbf{W}^{-1} \hat{\mathbf{f}} &= \sum_k \hat{\mathbf{f}}_k^{\mathrm{H}} \mathbf{W}_k^{-1} \hat{\mathbf{f}}_k \\ \mathbf{r}^{\mathrm{H}} \mathbf{W}^{-1} \hat{\mathbf{f}} &= \sum_k \mathbf{r}_k^{\mathrm{H}} \mathbf{W}_k^{-1} \hat{\mathbf{f}}_k \end{aligned}$$

are scalars. The waveform vector with hypothesized parameters $\hat{\mathbf{f}} \equiv \mathbf{f}(\hat{\tilde{\mathbf{p}}})$ is used to implement the receiver because the sonar signal parameters are unknown and are to be estimated. The data term is

$$\left|\sum_k \mathbf{r}_k^{\mathrm{H}} \mathbf{W}_k^{-1} \hat{\mathbf{f}}_k\right|^2 = \left|\sum_k (\mathbf{f}_k \mathsf{s} + \mathbf{n}_k)^{\mathrm{H}} \mathbf{W}_k^{-1} \hat{\mathbf{f}}_k\right|^2$$

$$= \left|\mathsf{s}^* \sum_k \mathbf{f}_k^{\mathrm{H}} \mathbf{W}_k^{-1} \hat{\mathbf{f}}_k + \sum_k \mathbf{n}_k^{\mathrm{H}} \mathbf{W}_k^{-1} \hat{\mathbf{f}}_k\right|^2 \qquad (7.28)$$

and the random signal component $\mathcal{S}_{sp}$ and the interference are assumed to be uncorrelated. Consequently, the expectation of (7.27) is

$$\mathcal{E}\{\ell_d|H_1\} = \frac{E_s|\sum_k \mathbf{f}_k^{\mathrm{H}} \mathbf{W}_k^{-1} \hat{\mathbf{f}}_k|^2 + \mathcal{E}\left\{|\sum_k \mathbf{n}_k^{\mathrm{H}} \mathbf{W}_k^{-1} \hat{\mathbf{f}}_k|^2\right\}}{E_s^{-1} + \sum_k \hat{\mathbf{f}}_k^{\mathrm{H}} \mathbf{W}_k^{-1} \hat{\mathbf{f}}_k}.$$

The interference term is

$$\mathcal{E}\left\{\left|\sum_k \mathbf{n}_k^{\mathrm{H}} \mathbf{W}_k^{-1} \hat{\mathbf{f}}_k\right|^2\right\} = \sum_k \sum_l \hat{\mathbf{f}}_l^{\mathrm{H}} \mathbf{W}_l^{-1} \mathcal{E}\{\mathbf{n}_l \mathbf{n}_k^{\mathrm{H}}\} \mathbf{W}_k^{-1} \hat{\mathbf{f}}_l$$

$$= \sum_k \hat{\mathbf{f}}_k^{\mathrm{H}} \mathbf{W}_k^{-1} \hat{\mathbf{f}}_k \qquad (7.29)$$

because the noise spectral components at different frequencies are uncorrelated and

$$\mathcal{E}\{\ell_d|H_1\} = \frac{E_s|\sum_k \mathbf{f}_k^{\mathrm{H}} \mathbf{W}_k^{-1} \hat{\mathbf{f}}_k|^2 + \sum_k \hat{\mathbf{f}}_k^{\mathrm{H}} \mathbf{W}_k^{-1} \hat{\mathbf{f}}_k}{E_s^{-1} + \sum_k \hat{\mathbf{f}}_k^{\mathrm{H}} \mathbf{W}_k^{-1} \hat{\mathbf{f}}_k}$$

$$\mathcal{E}\{\ell_d|H_0\} = \frac{E_s \sum_k \hat{\mathbf{f}}_k^{\mathrm{H}} \mathbf{W}_k^{-1} \hat{\mathbf{f}}_k}{1 + E_s \sum_k \hat{\mathbf{f}}_k^{\mathrm{H}} \mathbf{W}_k^{-1} \hat{\mathbf{f}}_k}.$$

The quantity $(\sum_k \hat{\mathbf{f}}_k^{\mathrm{H}} \mathbf{W}_k^{-1} \hat{\mathbf{f}}_k)^{-1} = \sigma_0^2(\hat{\hat{\mathbf{p}}})$ is the expected interference contribution and is analogous to the interference energy defined for the scalar time and frequency domain processors (3.71, 3.89) in Chapter 3. Its product with E_s represents a generalized signal to interference ratio that becomes more evident when the interference is uncorrelated between the N elements as well as the K frequencies. Then,

$$\mathbf{W}_k = diag\{W_{knn}\} \quad k = 1, K \;\; n = 1, N$$

and (7.29) is

$$\sum_k \hat{\mathbf{f}}_k^{\mathrm{H}} \mathbf{W}_k^{-1} \hat{\mathbf{f}}_k = \sum_{k=1}^{K} |F_k(\hat{\mathbf{p}}|^2 \sum_{n=1}^{N} \frac{|a_n|^2}{W_{knn}}$$

$$= \frac{1}{N_0} \sum_k |F_k(\hat{\mathbf{p}}|^2 \sum_{n=1}^{N} |a_n|^2 = \frac{1}{N_0} \quad \text{for WGN.}$$

The second relationship results when the interference is WGN with $W_{knn} = N_0; \forall n, k$, the array weights are energy normalized such that $\sum |a|^2 = 1$, and the waveform function is normalized by

$$\sum_{k=1}^{K} |F_k(\mathbf{p})|^2 = 1 \quad \forall \mathbf{p}. \tag{7.30}$$

Under these latter conditions

$$\mathcal{E}\{\ell_d|H_1\} = \frac{\text{SNR}^2 \Phi_{ff}(\tilde{\mathbf{p}}, \hat{\tilde{\mathbf{p}}}) + \text{SNR}}{1 + \text{SNR}}$$

where $\text{SNR} = E_s/N_0$ and

$$\Phi_{ff}(\tilde{\mathbf{p}}, \hat{\tilde{\mathbf{p}}}) = \left| \sum_{k}^{K} \mathbf{f}_k^{\text{H}}(\tilde{\mathbf{p}}) \mathbf{f}_k(\hat{\tilde{\mathbf{p}}}) \right|^2 = 1 \text{ for } \hat{\tilde{\mathbf{p}}} = \tilde{\mathbf{p}} \tag{7.31}$$

is the discrete frequency domain auto-ambiguity function of the waveform vectors. When the hypothesized Doppler, delay, and bearing parameters $\hat{\tilde{\mathbf{p}}} = [\hat{\beta}, \hat{\tau}, \hat{\theta}]$ match those of the point echo, $\tilde{\mathbf{p}} = [\beta, \tau, \theta]$, $\hat{\mathbf{f}} = \mathbf{f}$, and

$$\begin{aligned} \mathcal{E}\{\ell_d|H_1\} &= \text{SNR} \\ \mathcal{E}\{\ell_d|H_0\} &= \frac{\text{SNR}}{1 + \text{SNR}}. \end{aligned}$$

The detection index

$$D_{\tilde{\mathbf{p}}\tilde{\mathbf{p}}} = \frac{\mathcal{E}\{\ell_d|H_1\}}{\mathcal{E}\{\ell_d|H_0\}} - 1 = \text{SNR}$$

is then maximized which is consistent with the results for the scalar receiver discussed in Chapter 3.

Continuous Passive Signals

The data term ℓ_d in (7.26) takes the particular form (7.27) because the received signal arises from the slowly fluctuating random scatter model interrogated by a deterministic signal. The signal covariance $\mathbf{P}$ is not block diagonal because its components are outer products of the deterministic frequency domain vectors $\mathbf{f}_k(\mathbf{p})$ that are not necessarily zero between frequencies. The signal is random by virtue of the assumed slowly fluctuating point scatter spreading function $\mathcal{S}_{sp}$ that changes only between interrogations. The correlation between frequency components of a backscattered signal is determined by the time-frequency correlation function (6.5) which is the Fourier transform of the scattering function. Components separated by $\Delta\nu >> 1/\tau$ are uncorrelated for scattering functions with delay support τ hence scattering from point like sources is correlated over a wide frequency interval.

When compensated for inter-element delay, the same coherent signal realization is received at each array element for both active interrogation of a point scatterer and passive reception of a continuous random signal $\mathsf{S}(\omega_k) \equiv \mathsf{S}_k$ emanating from a spatially localized source [18–20]. A random passive signal observed on an interval T is equivalent to the reception of an active echo from a random range spread scattering source with support $\tau = \mathrm{T}$ hence both active and passive signal frequency components separated by more than 1/T are uncorrelated.

Processing with a reasonably long integration time justifies the assumption that the passive signal covariance $\mathbf{P}$ and interference covariance $\mathbf{W}$ are block diagonal. $\mathbf{W}$ has components (7.25) and the individual signal covariance subblocks are N×N matrices

$$\begin{aligned}\mathbf{P}_k &= \mathbf{e}_k S_k \mathbf{e}_k^{\mathrm{H}} \\ S_k &= \mathcal{E}\{|\mathsf{S}_k|^2\}.\end{aligned}$$

S_k is the scalar signal power density spectrum component at frequency k received at each of the N elements and the $\mathbf{e}_k$ are the frequency dependent steering vectors defined in (7.24). When both the signal and interference covariance matrices are block diagonal, (7.26) becomes the sum over the frequency components

$$\ell_r = \sum_{k=1}^{K} \left[\mathbf{r}_k^{\mathrm{H}}(\mathbf{W}_k^{-1} - (\mathbf{W}_k + \mathbf{P}_k)^{-1})\mathbf{r}_k - \ln|\mathbf{W}_k + \mathbf{P}_k| + \ln|\mathbf{W}_k|\right]$$

for the data vectors $\mathbf{r}_k = S_k \mathbf{e}_k + \mathbf{n}_k$. (7.27) is then

$$\ell_d = \sum_k \ell_{dk} = \sum_k \frac{|\mathbf{r}_k^{\mathrm{H}} \mathbf{W}_k^{-1} \mathbf{e}_k|^2}{S_k^{-1} + \mathbf{e}_k^{\mathrm{H}} \mathbf{W}_k^{-1} \mathbf{e}_k}.$$

This is expressed as a combined spatial-temporal (frequency) processor by first defining the average interference psd over the N elements as

$$\bar{W}_k = \frac{\mathrm{tr}(\mathbf{W}_k)}{N} = \frac{1}{N}\sum_n W_{knn}$$

so that the cross spectrum is

$$\mathbf{W}_k = \bar{W}_k\left(\frac{\mathbf{W}_k}{\bar{W}_k}\right) = \bar{W}_k\tilde{\mathbf{W}}_k.$$

$\tilde{\mathbf{W}}_k$ represents the normalized cross spectrum matrix and

$$\begin{aligned} \ell_d &= \sum_k H_k |\mathbf{r}_k^{\mathrm{H}}\tilde{\mathbf{W}}_k^{-1}\mathbf{e}_k|^2 \\ H_k &= \frac{S_k/\bar{W}_k^2}{1+G_k(S_k/\bar{W}_k)}. \end{aligned} \tag{7.32}$$

$G_k = \mathbf{e}_k^{\mathrm{H}}\tilde{\mathbf{W}}_k^{-1}\mathbf{e}_k$ is the *array gain* at frequency ω_k achieved by the likelihood receiver that forms the scalar statistic $\mathbf{e}_k^{\mathrm{H}}\mathbf{W}_k\mathbf{r}_k$. H_k is the frequency domain *Eckart filter* at ω_k [5,18,19] that weights the contribution of the spatial components according to their input $\mathrm{SIR}_{k,in} = S_k/\bar{W}_k$.

Array gain is analogous to the processing gain (6.19) defined for the scalar problem. It is the gain in SIR achieved by processing the data vector $\mathbf{r}_k = S_k\mathbf{e}_k + \mathbf{n}_k$ at each frequency k with a weighting vector $\mathbf{g}_k$ to form an output statistic $r_{1,k} = \mathbf{g}_k^{\mathrm{H}}\mathbf{r}_k$ [21,22]. The resulting output SIR at frequency k is

$$\begin{aligned} \mathrm{SIR}_{k,out} &= \frac{\mathcal{E}\{|r_1|^2|_{H_1}\}}{\mathcal{E}\{|r_1|^2|_{H_0}\}} - 1 = G_k\,\mathrm{SIR}_{k,in} \\ G_k &= \left(\frac{|\mathbf{g}_k^{\mathrm{H}}\mathbf{e}_k|^2}{\mathbf{g}_k^{\mathrm{H}}\tilde{\mathbf{W}}_k\mathbf{g}_k}\right) \end{aligned}$$

which is equivalent to the detection index (3.62) for that particular frequency. The optimum weight vector is obtained as the solution to

$$\max_{\mathbf{g}_k} G_k = \max_{\mathbf{g}_k}\left(\frac{|\mathbf{g}_k^{\mathrm{H}}\mathbf{e}_k|^2}{\mathbf{g}_k^{\mathrm{H}}\tilde{\mathbf{W}}_k\mathbf{g}_k}\right)$$

resulting in

$$\mathbf{g}_{opt} = \alpha\tilde{\mathbf{W}}_k^{-1}\mathbf{e}_k$$

within a constant α. Substituting $\mathbf{g}_{opt}$ into the expression for G_k produces

$$G_{k,max} = \mathbf{e}_k^{\mathrm{H}} \tilde{\mathbf{W}}_k^{-1} \mathbf{e}_k$$

which is the same as that defining the Eckart filter in (7.32) for a particular frequency. This proves that the likelihood receiver maximizes the output SIR. The expected value of (7.32) computed with the optimum array gain is

$$\begin{aligned}
\mathcal{E}\{\ell_d\}|_{H_1} &= \sum_k H_k \mathcal{E}\{|\mathbf{r}_k^H \tilde{\mathbf{W}}_k^{-1} \mathbf{e}_k|^2\} \\
&= \sum_k H_k G_k (S_k G_k + \bar{W}_k) \\
&= \sum_k G_k (S_k / \bar{W}_k) = \sum_k \mathrm{SIR}_{k,out}.
\end{aligned}$$

Likewise,

$$\mathcal{E}\{\ell_d\}|_{H_0} = \sum_k \frac{\mathrm{SIR}_{k,out}}{1 + \mathrm{SIR}_{k,out}}.$$

A spatially uncorrelated interference process is described by cross spectrum elements (7.25) with $\mathbf{W}_{kmn} = 0$, $m \neq n$ and

$$\tilde{\mathbf{W}}_k^{-1} = diag(1/W_{knn}) \sum_n W_{knn}$$

$$G_{max} = \left(\sum_n W_{knn} \right) \left(\sum_n (a_n^2 / W_{knn}) \right).$$

When the interference is both spatially and temporally white with $W_{knn} = N_0 \; \forall \, k, n$

$$G_{max} = (N N_0) \left(\sum_n a_n^2 / N_0 \right) = N \sum_n a_n^2$$

or simply N if the weights are normalized. The output SIR is then the input SIR multiplied by the number of elements.

The direction of arrival θ is the only measurable parameter hence the passive receiver is implemented with only the hypothesized steering vectors $\mathbf{e}_k(\hat{\theta}) \equiv \hat{\mathbf{e}}_k$. Each data vector $\mathbf{r}_k$ is processed by computing its inner product with the spatial filter $\hat{\mathbf{g}}_k^{\mathrm{H}} = \hat{\mathbf{e}}_k^{\mathrm{H}} \tilde{\mathbf{W}}_k^{-1}$ hypothesizing $\hat{\theta}$ to form the square law statistics $|\hat{\mathbf{e}}_k^{\mathrm{H}} \mathbf{W}_k^{-1} \mathbf{r}_k|^2$. ℓ_d (7.32) is then computed by summing these over all $k = 1, \cdots, K$ frequency components weighted by H_k. The maximum likelihood bearing estimate corresponds to the $\hat{\theta}$ providing the maximum ℓ_d.

Both the active and passive receivers form spatial filters from the parameter hypothesis. The active receiver processes the data elements with $\hat{\mathbf{f}}_k = F_k(\hat{\mathbf{p}})\mathbf{e}_k(\hat{\theta})$ reflecting the additional Doppler and delay parameter hypothesis $\hat{\tilde{\mathbf{p}}} = [\hat{\beta}, \hat{\tau}, \hat{\theta}]$ inherent in an echo from a deterministic waveform. The deterministic waveform does not permit the formulation of the likelihood as a block diagonal structure hence the statistic is expressed as the coherent summation across frequency (7.27). The continuous stochastic nature of the passive signal permits the block diagonal detector formulation that results in a receiver that computes the incoherent sum of individual spatial statistics over frequency (7.32). However, only the direction of arrival $\hat{\theta}$ can be ascertained.

7.2.1 Active Receiver Implementation and Parameter Estimation

The multidimensional receiver bank corresponding to the llr (7.27) can be implemented by comparing the data term (7.28) to a threshold. It is desirable that the probability of false alarm be constant for all parameter hypotheses $\hat{\tilde{\mathbf{p}}} = [\hat{\beta}, \hat{\tau}, \hat{\theta}]$ (CFAR processing) and that the receiver with the greatest expected response correspond to the correct parameter hypothesis $\hat{\tilde{\mathbf{p}}} = \tilde{\mathbf{p}}$. This is accomplished by implementing the receiver bank as an *array processor* with the vector version of the scalar detection statistic (3.93)[2]

$$\mathsf{d}_{1m}(\hat{\tilde{\mathbf{p}}}) = \frac{|\mathsf{r}_{1m}(\hat{\tilde{\mathbf{p}}})|^2}{\sigma_0^2(\hat{\tilde{\mathbf{p}}})} - 1 = |\sigma_0^2(\hat{\tilde{\mathbf{p}}}) \sum_k \mathbf{r}_k^{\mathrm{H}} \mathbf{W}_k^{-1} \hat{\mathbf{f}}_k|^2 \sum_k \hat{\mathbf{f}}_k^{\mathrm{H}} \mathbf{W}_k^{-1} \hat{\mathbf{f}}_k - 1$$
$$= \frac{|\sum_k \mathbf{r}_k^{\mathrm{H}} \mathbf{W}_k^{-1} \hat{\mathbf{f}}_k|^2}{\sum_k \hat{\mathbf{f}}_k^{\mathrm{H}} \mathbf{W}_k^{-1} \hat{\mathbf{f}}_k} - 1 \qquad (7.33)$$

corresponding to that of Figure 3.5 for the scalar processor. The random scalar

$$\mathsf{r}_{1m} = \sum_k \mathbf{r}_k^{\mathrm{H}} \hat{\mathbf{g}}_k$$
$$\hat{\mathbf{g}}_k = \mathbf{g}_k(\hat{\tilde{\mathbf{p}}}) = \sigma_0(\hat{\tilde{\mathbf{p}}})^2 \mathbf{W}_k^{-1} \hat{\mathbf{f}}_k$$

[2]The notation $(\mathsf{d}_{1m}, \mathsf{r}_{1m})$ distinguishes the array statistics from their scalar counterparts $(\mathsf{d}_1, \mathsf{r}_1)$.

is comparable to the mismatched receiver (3.91), the processing vector $\hat{\mathbf{g}}_k$ is analogous to the scalar frequency domain processing function (3.90), and $\sigma_0^2(\hat{\tilde{\mathbf{p}}}) = (\sum_k \hat{\mathbf{f}}_k^{\mathrm{H}} \mathbf{W}_k^{-1} \hat{\mathbf{f}}_k)^{-1}$ is the local interference level. The normalized cross ambiguity function (CAF) corresponding to (3.92) is

$$\Phi_{fq}(\tilde{\mathbf{p}}, \hat{\tilde{\mathbf{p}}}) = \left| \frac{\sum_k \mathbf{f}_k^{\mathrm{H}} \mathbf{W}_k^{-1} \hat{\mathbf{f}}_k}{\sum_k \hat{\mathbf{f}}_k^{\mathrm{H}} \mathbf{W}_k^{-1} \hat{\mathbf{f}}_k} \right|^2 = \frac{\sigma_0^2(\hat{\tilde{\mathbf{p}}}) D_{\tilde{\mathbf{p}}\hat{\tilde{\mathbf{p}}}}}{E_s} \tag{7.34}$$

and is unity when $\hat{\tilde{\mathbf{p}}} = \tilde{\mathbf{p}}$. It is often called the *generalized* CAF because it involves spatial as well as temporal quantities [10, 23].

The waveform vectors for the actual and hypothesized parameters $\tilde{\mathbf{p}} = [\beta, \tau, \theta]$, $\hat{\tilde{\mathbf{p}}} = [\hat{\beta}, \hat{\tau}, \hat{\theta}]$ at frequency k are defined by

$$\begin{aligned} \mathbf{f}_k &\equiv \mathbf{f}_k(\tilde{\mathbf{p}}) = \mathbf{f}_k(\beta, \tau, \theta) \\ \hat{\mathbf{f}}_k &\equiv \mathbf{f}_k(\hat{\tilde{\mathbf{p}}}) = \mathbf{f}_k(\hat{\beta}, \hat{\tau}, \hat{\theta}). \end{aligned}$$

Maximization of $\mathsf{d}_{1m}(\hat{\tilde{\mathbf{p}}})$ with respect to $\hat{\tilde{\mathbf{p}}}$ assures that the estimate is unbiased and the receiver with the maximum expected response corresponds to the correct data parameters as shown by Figure 3.8. This requires that the detection index be bounded such that

$$\begin{aligned} D_{\tilde{\mathbf{p}}\hat{\tilde{\mathbf{p}}}} = \mathcal{E}\{\mathsf{d}_{1m}(\hat{\tilde{\mathbf{p}}})\} &= E_s \frac{|\sum_k \mathbf{f}_k^{\mathrm{H}} \mathbf{W}_k^{-1} \hat{\mathbf{f}}_k|^2}{\sum_k \hat{\mathbf{f}}_k^{\mathrm{H}} \mathbf{W}_k^{-1} \hat{\mathbf{f}}_k} \\ &\leq E_s \frac{|\sum_k \mathbf{f}_k^{\mathrm{H}} \mathbf{W}_k^{-1} \mathbf{f}_k|^2}{\sum_k \mathbf{f}_k^{\mathrm{H}} \mathbf{W}_k^{-1} \mathbf{f}_k} = E_s \sum_k \mathbf{f}_k^{\mathrm{H}} \mathbf{W}_k^{-1} \mathbf{f}_k \end{aligned}$$

or equivalently

$$\begin{aligned} E_s \left| \sum_k \mathbf{f}_k^{\mathrm{H}} \mathbf{W}_k^{-1} \hat{\mathbf{f}}_k \right|^2 &\leq \left(E_s \sum_k \hat{\mathbf{f}}_k^{\mathrm{H}} \mathbf{W}_k^{-1} \hat{\mathbf{f}}_k \right) \left(\sum_k \mathbf{f}_k^{\mathrm{H}} \mathbf{W}_k^{-1} \mathbf{f}_k \right) \\ &\Rightarrow D_{\tilde{\mathbf{p}}\hat{\tilde{\mathbf{p}}}} \leq D_{\tilde{\mathbf{p}}\tilde{\mathbf{p}}} \end{aligned}$$

which is a form of the Schwarz inequality for weighted inner products (see appendix F of [24]). It is proven by making use of the fact that the covariance matrices $\mathbf{W}_k$ are Hermitian and positive definite. Note that when the interference is WGN and uncorrelated between elements with psd N_0, the cross ambiguity function (CAF) (7.34) becomes the auto ambiguity function (7.31) and $D_{\tilde{\mathbf{p}}\hat{\tilde{\mathbf{p}}}} = \mathrm{SNR}\Phi_{ff}(\tilde{\mathbf{p}}, \hat{\tilde{\mathbf{p}}}) \leq \mathrm{SNR}$.

The CR Bound for Spatial Processing

The Cramér-Rao bound (CRB) is a measure of the accuracy with which parameter estimates can be made. The scalar receiver discussed in Chapters 3 and 4 hypothesized delay and Doppler parameters and for these the CRB depends upon the attributes of the waveform. The extension to the multidimensional receiver and the estimation of bearing requires that the array configuration as well as the waveform attributes be considered.

The C-R lower bounds for the parameter estimate variances and the corresponding degree of coupling between the estimates are proportional to the elements of the inverse Fisher matrix **J** (3.112) and given by (3.115) for two parameters in Example 3.6. **J** is given by (3.116) for the slowly fluctuating point scatterer when the interference is uncorrelated. It is obtained from the second partial derivatives of the auto-ambiguity function (7.31) with respect to the hypothesized parameters $\hat{\mathbf{p}}$ evaluated at the origin $\hat{\mathbf{p}} = \mathbf{p}$. The estimation accuracy for the Doppler parameters β or ϕ depends primarily upon the waveform as discussed in Chapter 4. The spatial parameters τ, θ depend upon both the waveform and array characteristics and in order to simplify the discussion, the CRB for τ and θ are derived in terms of the ambiguity function (7.31) and signal vector (7.24) assuming that the Doppler dilation parameter $\beta = 1$.

$\tau_e = \tau - \hat{\tau}$ and $\theta_e = \theta - \hat{\theta}$ are the delay and bearing errors. Derivatives with respect to $\hat{\tau}$ and τ_e are equivalent for fixed τ so that

$$
\begin{aligned}
J_{\tau\tau} &= -S_n \frac{\partial^2 \Phi_{ff}}{\partial {\tau_e}^2} \\
J_{\theta\theta} &= -S_n \frac{\partial^2 \Phi_{ff}}{\partial \hat{\theta}^2} \\
J_{\theta\tau} &= -S_n \frac{\partial^2 \Phi_{ff}}{\partial \hat{\theta} \partial \tau_e} \\
S_n &= \frac{\mathrm{SNR}^2}{1 + \mathrm{SNR}}
\end{aligned}
\qquad (7.35)
$$

are the elements of the 2×2 Fisher matrix **J** (3.116) for the spatial parameters evaluated at ambiguity origin $\tau_e = \theta_e = 0$. The ambiguity function $\Phi_{ff} = |\chi_{ff}|^2$ the derivatives of which are given by (4.19), and

$$
\begin{aligned}
\chi_{ff}(\tau, \hat{\tau}, \theta, \hat{\theta}) &= \sum_k^K \mathbf{f}_k^{\mathrm{H}}(\tilde{\mathbf{p}}) \mathbf{f}_k(\hat{\tilde{\mathbf{p}}}) \\
&= \sum_k \sum_n |F_k|^2 a_n^2 (\exp -j\omega_k \tau_e)(\exp j((\sin\theta - \sin\hat{\theta}) d_n \omega_k / c))
\end{aligned}
$$

$$= \sum_k |F_k|^2 = \sum_n a_n^2 = 1 \quad (\tau_e = \hat{\theta} = 0) \tag{7.36}$$

is obtained from (7.31,7.24) with $\beta = 1$. The derivatives of (7.36) with respect to τ_e and evaluated at the origin are

$$\frac{\partial \chi_{ff}}{\partial \tau_e} = -\sum_n a_n^2 \sum_k (j\omega_k |F_k|^2)$$
$$\frac{\partial^2 \chi_{ff}}{\partial {\tau_e}^2} = -\sum_n a_n^2 \sum_k (\omega_k^2 |F_k|^2).$$

Substituting these into (4.19, 7.35) and invoking the array and spectral normalizations (7.36) yields

$$J_{\tau\tau} = 2S_n \left[\sum_k |F_k|^2 \omega_k^2 - \left(\sum_k |F_k|^2 \omega_k \right)^2 \right]$$
$$= 2S_n [\overline{\omega^2} - \omega_0^2] = 2S_n W_\omega^2. \tag{7.37}$$

W_ω^2 is the mean square radian bandwidth about the waveform spectral centroid ω_0 and is equivalent to the scalar ambiguity expansion coefficient for delay λ defined by (4.34).

$J_{\theta\theta}$ is similarly derived by taking the $\hat{\theta}$ derivatives of (7.36) at the origin

$$\frac{\partial \chi_{ff}}{\partial \hat{\theta}} = \sum_k \sum_n |F_k|^2 a_n^2 (jd_n \omega_k / c) \cos \theta$$
$$\frac{\partial^2 \chi_{ff}}{\partial \hat{\theta}^2} = -\sum_k \sum_n |F_k|^2 a_n^2 (d_n \omega_k / c)^2 \cos^2 \theta$$

and substituting into (4.19) as for (7.37). This yields

$$J_{\theta\theta} = 2S_n \left[\sum_k |F_k|^2 \omega_k^2 \sum_n (a_n d_n / c)^2 - \left(\sum_k |F_k|^2 \omega_k \sum_n a_n^2 d_n / c \right)^2 \right]$$
$$= 2S_n [\overline{\omega^2}\, \overline{t^2} - \omega_0^2 (\bar{t})^2] \tag{7.38}$$

where

$$\bar{t} = \sum_n a_n^2 d_n \cos \theta / c, \quad \overline{t^2} = \sum_n (a_n d_n \cos \theta / c)^2$$

are the mean and mean square array time delay parameters. $t_n = d_n/c$ is the incremental time delay corresponding to the element at position n with respect to the line array origin at d_0. If $d_0 = 0$ and the array is symmetric such that $a_{-n} = a_n$, $d_{-n} = -d_n$, the mean delay $\bar{t} = 0$ and $J_{\theta\theta} = 2S_n\overline{\omega^2}\,\overline{t^2}$. Bearing accuracy depends upon both the waveform time resolution that is proportional to mean square bandwidth and the array spatial resolution that is proportional to its weighted mean square extent $\overline{d^2} = c^2\overline{t^2}$ projected at the angle θ. (7.38) also represents the Fisher matrix element for a deterministic signal in white spatially uncorrelated Gaussian noise (3.117) by defining $S_n = \text{SNR}$ [10, 25].

The joint estimates of delay and bearing are coupled when $J_{\theta\tau} \neq 0$ indicating that a systematic error in one tends on average to induce a systematic error in the other. It was shown in Chapter 4 that uncoupled delay-Doppler estimation using a scalar receiver requires symmetry of the ambiguity function that in turn requires even symmetry of the waveform time-frequency function. The analogous conditions for θ, τ are obtained from

$$\begin{aligned}
J_{\theta\tau} &= -S_n \frac{\partial}{\partial\hat{\theta}}\left\{\frac{\partial|\chi_{ff}|^2}{\partial\tau_e}\right\} \\
&= -2S_n\, \Re\frac{\partial}{\partial\hat{\theta}}\left\{\chi_{ff}^*\frac{\partial\chi_{ff}}{\partial\tau_e}\right\}\Bigg|_{\tau_e=\hat{\theta}=0} \\
&= 2S_n\left[\sum_k |F_k|^2\omega_k^2 - \left(\sum_k |F_k|^2\omega_k\right)^2\right]\sum_n a_n^2 d_n \cos\theta/c \\
&= 2S_n W_\omega^2 \bar{t} \\
&= 0 \quad \text{for } d_{-n} = -d_n,\ a_{-n} = a_n.
\end{aligned}$$

This indicates that the array must be symmetrically arranged and weighted about its midpoint in order that the joint estimates of delay and bearing be uncoupled. The lower bounds for estimate variance (3.115) are

$$\boxed{\begin{aligned}
\sigma_\tau^2 &\geq \frac{J_{\theta\theta}}{J_{\theta\theta}J_{\tau\tau} - J_{\theta\tau}^2} = \frac{1}{2S_n W_\omega^2} \quad \text{for } J_{\theta\tau} = 0 \\
\sigma_\theta^2 &\geq \frac{J_{\tau\tau}}{J_{\theta\theta}J_{\tau\tau} - J_{\theta\tau}^2} = \frac{1}{2S_n\overline{\omega^2}\,\overline{t^2}} \quad \text{for } J_{\theta\tau} = 0 \quad (7.39)
\end{aligned}}$$

Line array Cramér Rao lower bounds for the joint estimation of τ and θ

and are minimized for a symmetric array. Furthermore, the bearing bound is minimum at the boresight angle $\theta = 0$. The angle CRB results are derived for two-dimensional arrays in [26].

Split Processing

The waveform vector corresponding to the parameter hypothesis $\hat{\tilde{\mathbf{p}}}$ is $\hat{\mathbf{f}}_k = F_k(\hat{\beta}, \hat{\tau})\mathbf{e}(\hat{\theta})$ and a 3-dimensional receiver bank is required to simultaneously test Doppler, delay, and bearing. If point scattering is assumed it is computationally more efficient to split the problem into two parts by separately thresholding and testing for the Doppler and delay parameters $\mathbf{p} = [\beta, \tau]$ with a bank of scalar receivers hypothesizing $\hat{\mathbf{p}}$. The input is obtained from a fixed beam (or beams) that may be steered and the best (maximum likelihood) parameter estimates of the true parameters $\mathbf{p}$ are the $\hat{\beta} = \beta_{ml}$ and $\hat{\tau} = \tau_{ml}$ that maximize the receiver response. The optimal bearing estimate θ_{ml} is subsequently obtained from a vector receiver bank or array processor that maximizes (7.33) utilizing the fixed $\mathbf{p}_{ml} = [\beta_{ml}, \tau_{ml}]$ obtained from the scalar receiver bank. Bearing only needs to be determined for those echos that exceed the scalar receiver threshold. This approach is appropriate when the receive directivity is low and the beam is not highly resolvent.

The scalar line array receiver steered to θ' generates the shaded sum of the frequency domain element data (7.22) and the Gaussian scalar

$$
\begin{aligned}
\mathsf{R}_k &= = \sum_n \mathsf{R}_{nk} = \sum_n [\mathsf{S}_{nk}(\tilde{\mathbf{p}}) + \mathsf{N}_{nk}] = \mathsf{S}_k(\tilde{\mathbf{p}}) + \mathsf{N}_{\Sigma k} \\
\mathsf{N}_{\Sigma k} &= \sum_n \mathsf{N}_{kn} \\
\mathsf{S}_k(\tilde{\mathbf{p}}) &= \sqrt{E_0 S_{sp}} F_k(\mathbf{p}) \Theta_n(\Delta\eta_k) \\
\Delta\eta_k &= \frac{2\pi}{\lambda_k}(\sin\theta - \sin\theta')
\end{aligned}
$$

is the resulting beamformer output (7.17) at each frequency ω_k. The normalized and steered beam (7.18) is $\Theta_n(\Delta\eta_k) = \Theta(\Delta\eta_k)/\Theta(max)$ and $\Delta\eta_k \approx (2\pi/\lambda_k)(\theta - \theta')$ for small angles. Internal interference is modeled as additive to the individual element outputs as shown hence its aggregate contribution $\mathsf{N}_{\Sigma k}$ is the direct rather than the weighted sum. The contribution of directional external interference such as reverberation from boundaries or anisotropic ambient noise is, like the echo signal, dependent upon its spatial distribution in relation to $\Theta_n(\Delta\eta_k)$. The maximum likelihood WSS frequency domain detector is the mismatched filter that forms the WSS frequency domain detection statistic r_1 (3.91). Rather than a waveform replica, the scalar mismatched filter utilizes the processing function (3.90)

$$G_k(\hat{\mathbf{p}}) \equiv G(\hat{\mathbf{p}}, \omega_k) = \left(\frac{F_k(\hat{\mathbf{p}})}{W_{\Sigma k}}\right) \sigma_0^2(\hat{\mathbf{p}})$$

that is a function of only the hypothesized delay-Doppler parameters $\hat{\mathbf{p}} = [\hat{\beta}, \hat{\tau}]$ with

$$\sigma_0^2(\hat{\mathbf{p}}) = \left(\sum_k \frac{|F_k(\hat{\mathbf{p}})|^2}{W_{\Sigma k}}\right)^{-1} \quad \text{(Joules)} \tag{7.40}$$

which is the normalizing constant and interference energy contribution (3.89) for discrete array data.

The psd (1.18) of the WSS time domain interference sum process is the sum of the cross spectral energy components (7.25)

$$W_{\Sigma k} = \mathcal{E}\{|\mathsf{N}_{\Sigma k}|^2\} = \sum_m \sum_n W_{kmn}, \quad m, n = 1, \cdots, N \ \text{(Joules)}.$$

If the interference is uncorrelated between elements, the cross spectrum components (7.25) are $W_{kmn} = 0,\ m \neq n,\ \forall k$ and using (7.40)

$$\begin{aligned} W_{\Sigma k} &= \sum_m W_{kmm} \\ &= N\, N_0 = \sigma_0^2 \quad \text{(for WGN)}. \end{aligned}$$

The factor N occurs because the unshaded psd at each of the N elements is N_0 (Watts/Hz or Joules).

The scalar square law processor is implemented as a receiver bank that synthesizes the discrete versions of the random statistics r_1 (3.91) and d_1 (3.93) and performs the hypothesis testing of the delay-Doppler parameters $\hat{\mathbf{p}} = [\hat{\beta}, \hat{\tau}]$. The detection statistic d_1 is

$$\mathsf{d}_1(\hat{\mathbf{p}}) = \frac{|\mathsf{r}_1|^2}{\sigma_0^2(\hat{\mathbf{p}})} - 1 = \frac{\left|\sum_k \frac{\mathsf{R}_k F_k^*(\hat{\mathbf{p}})}{W_{\Sigma k}}\right|^2}{\sum_k \frac{|F_k(\hat{\mathbf{p}})|^2}{W_{\Sigma k}}} - 1 \tag{7.41}$$

and obtained from

$$\begin{aligned} |\mathsf{r}_1(\hat{\mathbf{p}})|^2 &= \left|\sum_k \mathsf{R}_k G_k^*(\hat{\mathbf{p}})\right|^2 \\ &= \left|\sigma_0^2(\hat{\mathbf{p}}) \sum_k \left[(\sqrt{E_0} S_{sp} F_k(\mathbf{p}) \Theta_n(\Delta\eta_k) + \mathsf{N}_{\Sigma k}) \left(\frac{F_k^*(\hat{\mathbf{p}})}{W_{\Sigma k}}\right)\right]\right|^2 \end{aligned} \tag{7.42}$$

using (7.40). The echo spreading function $\mathcal{S}_{sp}$ is uncorrelated with the interference process, the interference components are uncorrelated in frequency, and $\mathcal{E}\{|\mathsf{N}_{\Sigma k}|^2\} = W_{\Sigma k}$ so that the expectation of the processor output is

$$\mathcal{E}\{|\mathsf{r}_1|^2|H_0\} = \mathcal{E}\left\{\left|\sigma_0^2 \sum_k \frac{F_k^*(\hat{\mathbf{p}})\mathsf{N}_{\Sigma k}}{W_{\Sigma k}}\right|^2\right\} = \sigma_0^2$$

$$\mathcal{E}\{|\mathsf{r}_1|^2|H_1\} = E_s\Phi_{fg}(\mathbf{p},\hat{\mathbf{p}}) + \sigma_0^2, \quad E_s = E_0\mathcal{R}_{sp} \quad \text{and}$$

$$\Phi_{fg}(\mathbf{p},\hat{\mathbf{p}}) = \left|\sigma_0^2 \sum_k \frac{\Theta_n(\Delta\eta_k)F_k(\mathbf{p})F_k^*(\hat{\mathbf{p}})}{W_{\Sigma k}}\right|^2. \tag{7.43}$$

Φ_{fg} is the discrete frequency domain CAF corresponding to (3.92) with σ_0^2 given by (7.40). The scalar receiver statistic (7.41) is equivalent to (7.33) for the multidimensional receiver and

$$\begin{aligned}
\mathcal{E}\{\mathsf{d}_1(\hat{\mathbf{p}})\} &= D_{\mathbf{p}\hat{\mathbf{p}}} \\
&= \frac{E_s\Phi_{fg}(\mathbf{p},\hat{\mathbf{p}}) + \sigma_0^2}{\sigma_0^2} - 1 = E_s\frac{\Phi_{fg}(\mathbf{p},\hat{\mathbf{p}})}{\sigma_0^2} \\
&= \frac{E_s\left|\sum_k \frac{\Theta_n(\Delta\eta_k)F_k(\mathbf{p})F_k^*(\hat{\mathbf{p}})}{W_{\Sigma k}}\right|^2}{\sum_k \frac{|F_k(\hat{\mathbf{p}})|^2}{W_{\Sigma k}}} \le E_s\sum_k \frac{|F_k(\mathbf{p})|^2}{W_{\Sigma k}} = D_{\mathbf{pp}}
\end{aligned}$$

when $\theta' = \theta$ as for (3.102). When the interference is white and uncorrelated between sensors $W_{\Sigma k} = NN_0$,

$$D_{\mathbf{p}\hat{\mathbf{p}}} = \text{SNR}\left|\sum_k F_k(\mathbf{p})F_k^*(\hat{\mathbf{p}})\right|^2 = (\text{SNR})\,\Phi_{ff}(\mathbf{p},\hat{\mathbf{p}})$$

$$\text{SNR} = \frac{E_s}{NN_0},$$

and $\Phi_{ff}(\mathbf{p},\mathbf{p}) = 1$ because of the waveform normalization (7.30,7.31).

The combined receiver structure is shown in Figure 7.7. The interference is indicated as being introduced on the dry side of the array elements implying that the interference is internal to the processing hardware. This is done only for convenience as interference emanates both internally from electronic noise and vibration, and externally from ambient ocean noise, reverberation, and water flow when the array is moving. The array weighting constants affect the external interference and can be incorporated in the description of the interference samples in the data (7.22).

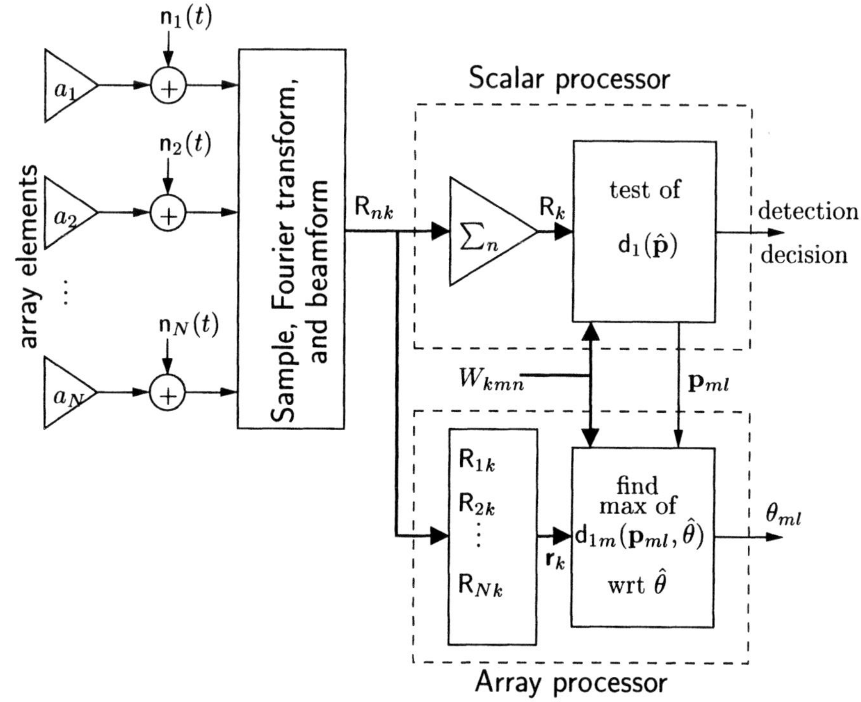

Figure 7.7: Combined scalar and multidimensional receiver structure for detection and the estimation of delay, Doppler, and bearing angle.

7.2.2 Spatially Spread Scattering

The time-frequency correlation between scalar echo components is a function of the spreading characteristics of the scattering process as indicated by the Fourier relationship (6.5) between the scattering and time-frequency correlation functions. Processes with considerable Doppler and delay spreading have commensurately short time and frequency correlation spans. Likewise, spreading that is a function of the angular scattering vector $\mathbf{u}_s$ as described by the spreading function $\mathcal{S}_s(\beta, \tau, \mathbf{u}_s)$ defines the spatial correlation properties of echos received at separated array elements. Reverberation is an example of such a spread process as its spreading function $\mathcal{S}_r(\beta, \tau, \mathbf{u}_s)$ has large delay and angular support. When it is the dominant interference source its delay and angular spreading characteristics determine the correlation and cross spectrum between array elements (7.25).

The point to point scattering relationship (7.7) describes the echo increment transmitted from a single point defined by $\mathbf{r}_t$, scattered from a scattering volume increment dV_s, and received at another position. Receiver positions defined by $\mathbf{r}_r$ can be points in a continuous volume array or more commonly elements of a discrete array. Assuming that the transmitter array has small L/λ so that its beamwidth (7.15) is large and Doppler spreading is negligible with $\beta = 1$, the terms $\mathbf{u}$, $\mathbf{r}_t$ in (7.7) relating to the transmit array are neglected and the scattered interference (reverberation)

$$\mathsf{N}_m(\omega, \mathbf{u}') = \sqrt{E_0} a_{rm} \int_{\Omega_r} \int_{\Omega_\theta} \int_{\Omega_\psi} \mathcal{S}_r(r, \theta, \psi) r^2 \sin\theta F(\omega) \cdot \exp -j\frac{\omega}{c}[2r - (\mathbf{u}_s - \mathbf{u}') \bullet \mathbf{r}_{rm}] dr d\theta d\psi$$

received at element "m" at position $\mathbf{r}_{rm}$ is the integral of (7.7) over the receive aperture with $a_r(\mathbf{r}_r) = a_{rm}\delta(\mathbf{r}_r - \mathbf{r}_{rm})$. The integration over the scattering volume V_s is replaced by the integrations over the individual spherical coordinant components as indicated by (7.4) and illustrated by Figure 7.2. This relationship holds for spread signals as well as scattered interference.

It is reasonable to assume that the scattering process is spatially uncorrelated within the support volume V_s especially if it arises from high frequency reverberation. This being the case, the scattered interference correlation between two specific frequencies (k, l) and array elements (m, n) with equal weighting a is

$$\begin{aligned} W_{klmn} &= \mathcal{E}\{\mathsf{N}_m(\omega_k)\mathsf{N}_n^*(\omega_l)\} \\ &= E_0 a^2 \int_{\Omega_r} \int_{\Omega_\theta} \int_{\Omega_\psi} \tilde{\mathcal{R}}_r(r, \theta, \psi) F(\omega_k) F^*(\omega_l) \left(\exp -j\frac{2r(\omega_k - \omega_l)}{c}\right) \\ &\quad \cdot \left(\exp j\left[\frac{\omega_k}{c}(\mathbf{u}_s - \mathbf{u}') \bullet \mathbf{r}_{rm} - \frac{\omega_l}{c}(\mathbf{u}_s - \mathbf{u}') \bullet \mathbf{r}_{rn}\right]\right) dr d\theta d\psi \\ \tilde{\mathcal{R}}_r(r, \theta, \psi) &= \mathcal{R}_r(r, \theta, \psi) r^2 \sin\theta. \end{aligned} \tag{7.44}$$

Example 7.1 Spatially Spread Signal

The correlation between two array elements symmetrically arranged with respect to a continuous scattering volume as shown in Figure 7.8

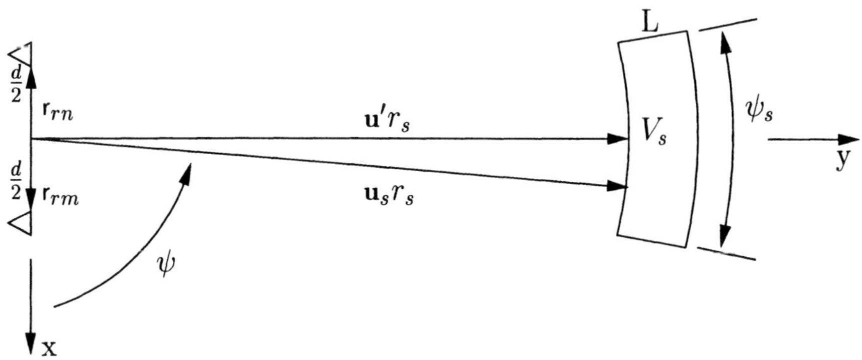

Figure 7.8: Geometry for a 2-element array receiving backscatter from a uniformly distributed scattering volume characterized by a constant scattering function $\tilde{\mathcal{R}}_r$.

illustrates the correlation effects of delay and bearing spread. Two elements with equal weighting (a) lie on the x axis of the coordinant system (7.2) at equal distances $d/2$ from the origin. The continuous scattering volume is an angular segment at a nominal distance r_s from the origin on the y axis with range extent L, bearing extent ψ_s, and depth (in the z direction) $d_z = r_s\delta\theta$ symmetric about the y axis $(\theta = \pi/2)$. The total scattering volume $V_s = L\psi_s d_z r_s \sin(\pi/2)$ and the scattering function is the constant $\tilde{\mathcal{R}}_r$ within it. The steering vector $\mathbf{u}'$ is aligned with the y axis so that it is orthogonal to the element vectors and $\mathbf{u}' \bullet \mathbf{r}_{rm} = \mathbf{u}' \bullet \mathbf{r}_{rn} = 0$ in (7.44). The projections of the scattering vector $\mathbf{u}_s$ on the array vectors are $\pm d\sin\psi/2$ and the correlation (7.44) becomes

$$W_{klmn} = a^2 E_0 \tilde{\mathcal{R}}_r \delta\theta F(\omega_k) F^*(\omega_l) \left(\int_{r_s - L/2}^{r_s + L/2} \exp j \frac{2r(\omega_k - \omega_l)}{c} dr \right) \cdot \left(\int_{-\psi_s/2}^{\psi_s/2} \exp j \frac{\psi d(\omega_k + \omega_l)}{2c} d\psi \right).$$

The integrals reduce to sinc $(\sin x/x)$ functions and

$$W_{klmn} = C_{kl}\text{sinc}\left(\frac{(\omega_l - \omega_k)L}{2c}\right) \text{sinc}\left(\frac{(\omega_l + \omega_k)\psi_s d}{4c}\right)$$

where $C_{kl} \propto F(\omega_k)F^*(\omega_l)$ and $\sin\psi \approx \psi$. The first sinc function represents the relationship between the two frequency correlation function and scatterer range extent which is the same as occurs for the scalar situation as discussed in Chapter 6. The second represents the effect of scatter angular spread. If the $x \leq \pi$ is regarded as the condition for which the sinc function is significant, signals with frequencies $|\triangle \nu| = |\omega_l - \omega_k|/2\pi > c/L$ and array separations $d > 2c/(\psi_s(\nu_k + \nu_l))$ are regarded as uncorrelated. The frequency correlation condition is consistent with the assertion that the cross spectrum components (7.25) are asymptotically uncorrelated for large integration time intervals. L/c is a time delay interval commensurate with the interrogation pulse length T. It is the illumination period for continuous range spread scattering such as reverberation as indicated by (6.15, 6.25) hence spectral components $|\triangle \nu| >> 1/T$ are regarded as uncorrelated. The spectral components of non-backscattered interference such as ambient noise are also asymptotically uncorrelated as the integration interval increases as shown by (1.20).

7.3 MONOPULSE BEARING ESTIMATION

In the early days of radar, echo bearing was estimated by transmitting sequential pulses while physically or electronically scanning a narrow (pencil) transmit/receive beam pattern in azimuth/elevation or in a circular motion. The measured target bearing corresponds to that scan angle yielding the maximum echo output. These multipulse techniques are called sequential lobing and conical scanning in the radar literature [27,28] and are practical for radar because the interrogation interval for each pulse is short. The relatively low propagation speeds associated with acoustic echo-location and the physical size and construction of projectors and transducers precludes the use of scanning techniques for most sonars. However, in some applications where it is possible to construct large L/λ sonars, multiple narrow pencil receive beams can be formed in conjunction with a wide transmit beam projector to estimate bearing on a single ping basis. The major limitation to multipulse scanning methods for either radar or sonar is their reliance upon relative amplitude measurements that are subject to random propagation and scattering induced fluctuation and fading. Moreover, scanning is impractical for sonar because of the low propagation speed and bearing measurement resolution is often limited by the available aperture.

The fluctuation problem and requirements for rapid bearing estimation led to the development of radar and sonar *monopulse* techniques that extract complete bearing information from a single interrogation (pulse) received simultaneously in several beams [27–30]. The simplest monopulse process requires four channels (beams), one pair for each coordinant (azimuth and elevation). Each channel is independently processed with a scalar receiver and angle information is extracted from the relative output magnitude or phase difference between the coordinant pairs. Monopulse is based upon a narrowband assumption whereby the waveform and resulting signal are spectrally concentrated about a nominal carrier frequency that defines the beam pattern Θ and the relative phase between channels unambiguously relates to delay. Monopulse bearing measurement does not depend upon the absolute echo amplitude and is unbiased by echo fading that occurs simultaneously in both channels. On the other hand, the accuracy of bearing measurements is affected by channel unbalance, anisotropic interference, and fading that causes the SIR fluctuation.

The maximum likelihood approach discussed in the previous section is by definition a monopulse technique because the parameter estimates including bearing are obtained by processing the data from a single interrogation. It is an indirect method and resembles radar beam scanning because the estimates are obtained by performing multiple parameter hypothesis tests (scans) with either a 3-d receiver bank or the combined receiver of Figure 7.7. The difference between this and radar scanning is that this is done by repeatedly testing the time series resulting from a single interrogation whereas each radar scan requires a new interrogation with a shifted beam pair

Monopulse bearing measurement is implemented as a combined receiver such as that of Figure 7.7 by replacing the array processor with a monopulse processor. It is invoked only upon detection and prior estimation of the delay-Doppler parameters in a separate channel. These operations are carried out by the scalar receiver that forms the test statistic $\mathsf{d}_1(\hat{\mathbf{p}})$ with an input obtained from a single wide detection beam. The monopulse bearing of detected echos is obtained by comparing the outputs of scalar sub-processors whose inputs are derived from the monopulse beam pairs and are implemented with the maximum likelihood delay-Doppler parameters $\mathbf{p}_{ml}$ obtained from the scalar processor.

The three most common monopulse approaches to direct bearing estimation are:

1. Amplitude comparison which measures bearing from the relative amplitude response of two steered (squinted) beams generated from

the same set of array elements.

2. Phase comparison which measures the phase difference ϕ_m between channels emanating from two offset phase centered (OPC) beams that are formed from sub-apertures of a common array separated by a distance d. This is an interferometer technique and bearing is related to the phase by $\theta = \sin^{-1}(\lambda\phi_m/2\pi d)$.

3. Combination or sum-difference Δ, Σ techniques that measure the relative amplitude and /or phase between beams formed by summing and differencing the outputs of two offset and squinted beams.

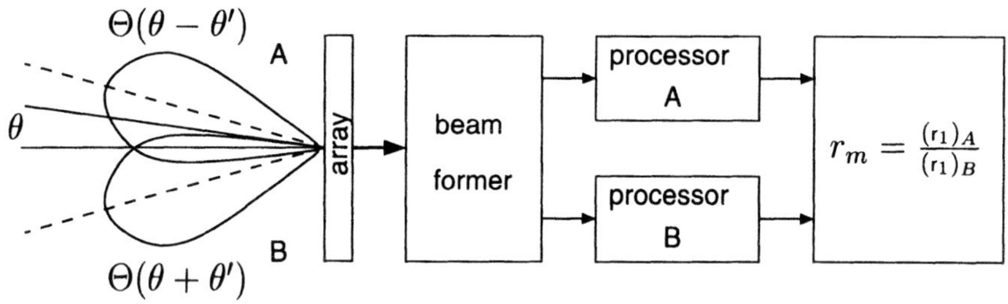

Figure 7.9: Amplitude monopulse multiplicative ratio generated by parallel processors with squinted beams formed from a common array and steered to $\pm\theta'$

7.3.1 Amplitude Comparison

Amplitude comparison monopulse is a bearing estimation process that computes the ratio r_m of the outputs $r_1(\mathbf{p})$ (7.42) of two scalar processing receivers as shown in Figure 7.9[3]. The processor pair replaces the array processor in Figure 7.7 and both channels labeled as A and B for left-right or up-down are provided with the Doppler and delay parameter estmates obtained from the scalar processor shown in the figure. The input to this processor is obtained from a detection beam with sufficient width to span the coverage of both the A and B channels. It is assumed that the signal is narrowband about a nominal wavelength λ and that $\eta_k \equiv \eta = (2\pi/\lambda)\sin\theta \, \forall \, k$. The beam patterns Θ_{nk} defined at each frequency

[3] r_m often is called the multiplicative ratio in the radar literature [27]

ω_k over the signal bandwidth are replaced by the single pattern function and

$$\Theta_n(\eta) = \Theta_{nk}(\theta) \equiv \Theta_n(\eta_k) \,\forall\, k$$

defined at the frequency centroid. The processor inputs are obtained from the beamformer (7.17, 7.18, 7.20) that generates squinted normalized beams

$$\Theta_n(\eta \mp \eta') = \sum_{n=1}^{N} a_n \exp -j\left[\left(\frac{N-2n+1}{2}\right)\left(\frac{\omega d}{c}(\sin\theta \mp \sin\theta')\right)\right]$$

$$a_n = a_{N+1-n}, \quad \sum_{n}^{N} a_n = 1$$

where the the a_n are real and the approximation $(\sin\theta \mp \sin\theta') \approx (\theta \mp \theta')$ may be invoked for small angles when appropriate.

The echo signal emanating from a spatially compact (point) scatterer will, except for beam attenuation, be the same in both channels if the beamformers and their associated electronics are balanced and there is no directional multipath to introduce spatial spreading. Anisotropic interference such as ambient noise, flow noise caused by water passing across the transducer, and reverberation often have highly directional characteristics. Consequently, the interference is not necessarily the same in both channels because the beams are squinted and observing different portions of the water column.

The random A and B channel statistics (7.42) for the constant beam are

$$\begin{aligned}\mathsf{r}_1(\mathbf{p})_{A,B} &= \Theta_n(\eta \mp \eta')\sqrt{E_0}S_{sp}\left(\sigma_0^2(\mathbf{p})\sum_k \frac{|F_k(\mathbf{p})|^2}{W_{\Sigma k}}\right) \\ &\quad + \left(\sigma_0^2 \sum_k \frac{\mathsf{N}_{\Sigma k}F^*(\hat{\mathbf{p}})}{W_{\Sigma k}}\right)_{A,B} \\ &= \mathsf{s}_{A,B} + \mathsf{n}_{A,B}\end{aligned} \tag{7.45}$$

when $\hat{\mathbf{p}} = \mathbf{p}$. Substituting (7.40) for σ_0^2 simplifies the signal term and

$$\mathsf{s}_{A,B} = \Theta_n(\eta \mp \eta')\sqrt{E_0}S_{sp}.$$

The notation $\mathsf{r}_{1A,B} \equiv \mathsf{r}_1(\mathbf{p})_{A,B}$ is adopted for convenience where the dependence upon the delay-Doppler parameters $\mathbf{p}$ is understood. The channel outputs are

$$\begin{aligned} \mathsf{r}_{1A,B} &= \sqrt{E_0}S_{sp}\Theta_n(\eta \mp \eta') + \mathsf{n}_{A,B} \\ \mathcal{E}\{|\mathsf{r}_{1A,B}|^2\} &= E_s\Theta_n^2(\eta \mp \eta') + \sigma_{A,B}^2 \end{aligned}$$

where the same random signal appears in both channels but with different average amplitudes caused by beam squint. The random interference while different, may be correlated depending upon the relative contributions of internal noise and external interference and its spatial characteristics. The interference energies $\sigma_{A,B}^2$ evaluated for the A and B channels are equivalent to σ_0^2 in (7.43, 7.40). The narrowband A and B channel statistics depend on a single beam pattern function squinted at angles $\pm\theta'$ from the boresight direction and the corresponding random multiplicative ratio of Figure 7.9 is

$$\mathsf{r}_m(\theta) = \frac{\mathsf{r}_{1A}}{\mathsf{r}_{1B}} = \frac{\sqrt{E_0}S_{sp}\Theta_n(\eta + \eta') + \mathsf{n}_A}{\sqrt{E_0}S_{sp}\Theta_n(\eta - \eta') + \mathsf{n}_B}. \tag{7.46}$$

In most cases bearing is not estimated unless the signal dominates the interference and the echo has sufficient SIR to exceed the detection threshold. The multiplicative ratio at high SIR is simply the beam pattern ratio

$$\begin{aligned} r_m &\approx \frac{\Theta_n(\eta - \eta')}{\Theta_n(\eta + \eta')} \quad \text{(at high SIR)} \\ \eta \pm \eta' &= \frac{2\pi}{\lambda}(\sin\theta \pm \sin\theta'). \end{aligned}$$

r_m by itself is a poor indicator of bearing because it is unbounded at the nulls of the squinted pattern functions and

$$\begin{aligned} \Theta_n(\eta + \eta') : \theta < 0 &= \Theta_n(\eta - \eta') : \theta > 0 \\ 1/(r_m(-\theta)) : \theta < 0 &= r_m(\theta) : \theta > 0 \end{aligned}$$

indicating that it is highly asymmetric. The usual approach to counter this is to form sum and difference channels from (7.45)

$$\begin{aligned} \mathsf{r}_\Sigma &= \mathsf{s}_\Sigma + \mathsf{n}_\Sigma = \mathsf{r}_{1A} + \mathsf{r}_{1B} \\ \mathsf{r}_\Delta &= \mathsf{s}_\Delta + \mathsf{n}_\Delta = \mathsf{r}_{1A} - \mathsf{r}_{1B} \\ \mathsf{s}_{\Sigma,\Delta} &= \mathsf{s}_A \pm \mathsf{s}_B, \; \mathsf{n}_{\Sigma,\Delta} = \mathsf{n}_A \pm \mathsf{n}_B \end{aligned} \tag{7.47}$$

and form the additive ratio from (7.46) which is

$$\begin{aligned}\mathrm{r}_a &= \frac{\mathrm{r}_\Delta}{\mathrm{r}_\Sigma} = \frac{\mathrm{r}_m - 1}{\mathrm{r}_m + 1} \\ &= \frac{\frac{\Theta_n(\eta-\eta')}{\Theta_n(\eta+\eta)} - 1}{\frac{\Theta_n(\eta-\eta')}{\Theta_n(\eta+\eta')} + 1} = \frac{\mathrm{s}_\Delta}{\mathrm{s}_\Sigma} = \frac{\Theta_\Delta}{\Theta_\Sigma} = r_a(\theta) \qquad (7.48)\end{aligned}$$

in the absence of noise. Θ_Δ is the difference beam and an odd function of the bearing angle. Θ_Σ is the even sum beam that usually serves as the detection beam because it has broad coverage and is very nearly constant for small bearing deviations about boresight.

$\Theta_\Delta/\Theta_\Sigma$ is likewise an odd function of the bearing angle and its sign indicates directional sense [27, 30]. Clearly it would be possible to use Θ_Δ alone as a bearing measure but normalizing it to form r_a removes the effects of signal fluctuation. The squinted beam patterns and corresponding additive ratio for a uniform line array with

$$\Theta_n(\eta \pm \eta') = \mathrm{sinc}\left(\frac{\pi L}{\lambda}(\sin\theta \pm \sin\theta')\right)$$

and $L/\lambda = 4$ are illustrated in Figs. (7.10,7.11).

The amplitude monopulse bearing estimate is obtained by measuring the additive ratio and directly using the $r_a(\theta)$ curve for the appropriate squint angle or by linearizing the curve about the origin and computing the bearing estimate

$$\begin{aligned}\hat{\theta} &= \mathrm{r}_a/k_m \\ k_m &= \frac{d}{d\theta}(\Theta_\Delta/\Theta_\Sigma)\bigg|_{\theta=0}.\end{aligned}$$

k_m is called the monopulse slope and is alternatively expressed as [30]

$$k_m = \left(\frac{1}{\Theta_\Sigma}\right)\frac{d\Theta_\Delta}{d\theta}.$$

If the beam pattern functions represent voltages, k_m has units of difference beam volts per unit angle normalized by the sum pattern voltage at θ. The two forms are virtually identical since the sum beam is nearly constant near boresight.

The sensitivity of the additive ratio to bearing angle variations is proportional to k_m which is a function of the squint angle and L/λ. The

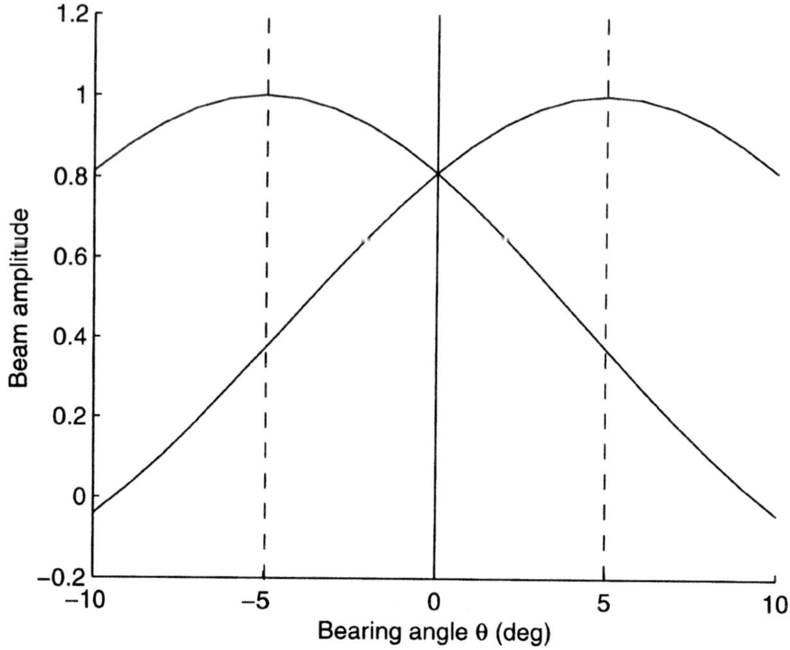

Figure 7.10: Steered $\sin(x)/x$ beams of a uniform line array with L/λ=4 and squint angle θ'=5 (deg)

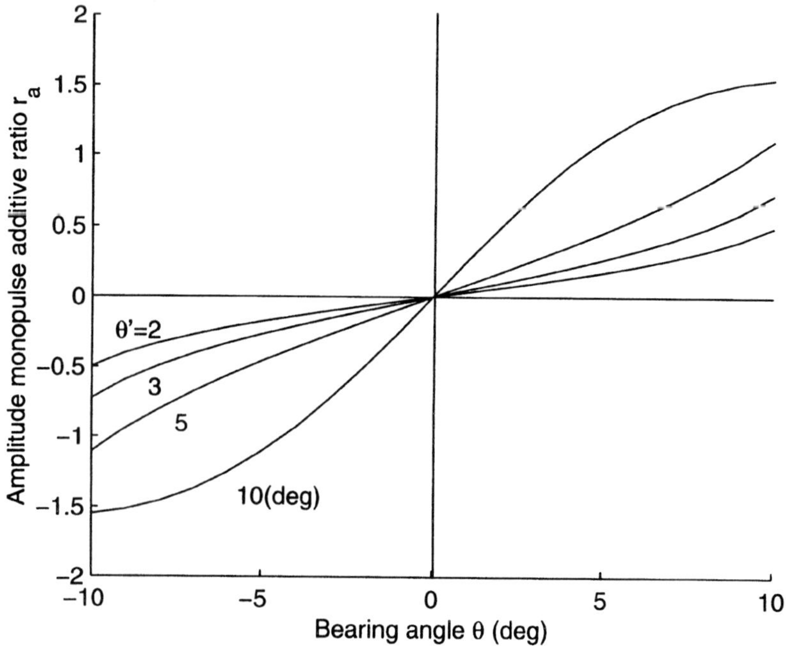

Figure 7.11: Amplitude monopulse additive ratio r_a (7.48) as a function of squint angle θ' for the uniform line array with L/λ=4

slope increases with θ' up to a point as shown but the range of bearing angles that can be measured over the approximately linear portion of r_a curve decreases for fixed L/λ. It is possible to optimize the squint angle for a given aperture by minimizing a measure of bearing estimate variance that depends upon k_m as well as the SIR. This results in $\theta'_{opt} \approx .5\,\theta_b$ [27]. Sherman [30] obtains an optimum squint angle of $\theta'_{opt} = .47\,\theta_b$ for sinc(x) beams by this approach. The half power beamwidth (7.15) of the squinted beams shown in Figure 7.10 for $L/\lambda = 4$ is about 12.7 deg. and the optimum squint angle is about 6 deg.

Interference Error

Non-negligible interference causes error in the measured additive ratio that translates to a bearing error. The error has two components; a systematic error with non-zero expectation called the *monopulse bias error* and a non-systematic (zero mean) error that results in measurement variance about the bias point. The variance is the result of randomness in the A and B channels induced by n_A and n_B that causes r_m (7.46) to fluctuate. This will always be present when the interference level is non-negligible in comparison with the echo signal. Bias error is the mean error that is caused by interference that is correlated between the channels and may be caused by non-isotropic boundary reverberation and directional ambient noise such as that generated by shipping.

The interference in the sum and difference channels is composed of both uncorrelated and correlated components [30] that are denoted by subscripts u, c respectively. The components are

$$\begin{aligned}
\mathsf{n}_\Sigma &= \mathsf{n}_{\Sigma u} + \mathsf{n}_c,\ \mathsf{n}_\Delta = \mathsf{n}_{\Delta u} + k_c \mathsf{n}_c \\
\sigma^2_{\Sigma u} &= \mathcal{E}\{|\mathsf{n}_{\Sigma u}|^2\},\ \sigma^2_{\Delta u} = \mathcal{E}\{|\mathsf{n}_\Delta|^2\},\ \sigma^2_c = \mathcal{E}\{|\mathsf{n}_c|^2\} \\
\mathcal{E}\{\mathsf{n}_{\Sigma u}\mathsf{n}^*_{\Delta u}\} &= \mathcal{E}\{\mathsf{n}_{\Sigma u}\mathsf{n}^*_c\} = \mathcal{E}\{\mathsf{n}_{\Delta u}\mathsf{n}^*_c\} = 0 \\
\sigma^2_\Sigma &= \sigma^2_{\Sigma u} + \sigma^2_c,\ \sigma^2_\Delta = \sigma^2_{\Delta u} + k^2_c \sigma^2_c
\end{aligned} \tag{7.49}$$

where k_c is a constant specifying the amount of the correlated interference n_c shared by the two channels. The interference is uncorrelated if $k_c = 0$ and n_c contributes equally to both channels if $k_c = 1$. It is possible for the proportion of correlated interference to change with range delay because the reverberation can change from uncorrelated isotropic volume at close range backscatter to highly correlated directional boundary backscatter further away.

The error between the measured r_a and the noise free additive ratio r_a

$$\epsilon_a(\theta) = \mathsf{r}_a - r_a(\theta) = \frac{\mathsf{n}_\Delta - r_a(\theta)\mathsf{n}_{\Sigma u} + (k_c - r_a(\theta))\,\mathsf{n}_c}{\mathsf{s}_\Sigma + \mathsf{n}_\Sigma}$$

$$r_a(\theta) = \frac{\mathsf{s}_\Delta}{\mathsf{s}_\Sigma} = \frac{\Theta_\Delta(\theta)}{\Theta_\Sigma(\theta)}. \tag{7.50}$$

is obtained from (7.47, 7.48, 7.49). The expectation of ϵ_a is the systematic bias error in r_a that translates via the inversion of (7.48) to a bias in the measured bearing angle. The bias is denoted by $\bar{\epsilon}_a$ and derived by Sherman [30] as

$$\bar{\epsilon}_a(\theta) = \left[\rho\left(\frac{\sigma_\Delta}{\sigma_\Sigma}\right) - r_a\right] \exp -(E_s/\sigma_\Sigma^2) \tag{7.51}$$

for a fixed realization of the signal with energy E_s. The signal to interference ratio SIR $= E_s/\sigma_\Sigma^2$ in the detection channel and ρ is the correlation coefficient between the sum and difference channel interference processes (7.49)

$$\rho = \frac{\mathcal{E}\{\mathsf{n}_\Delta \mathsf{n}_\Sigma^*\}}{\sigma_\Sigma \sigma_\Delta} = \frac{k_c \sigma_c^2}{\sigma_\Sigma \sigma_\Delta}.$$

The first and second terms of (7.51) represent the effects of the correlated and uncorrelated interference components respectively. The cumulative effect of both upon the additive ratio and the resulting bearing estimate is illustrated by the expected value of r_a

$$\begin{aligned}\bar{\mathsf{r}}_a(\theta) &= \bar{\epsilon}_a(\theta) + r_a = \rho\left(\frac{\sigma_\Delta}{\sigma_\Sigma}\right) \exp -(E_s/\sigma_\Sigma^2) \\ &+ r_a(\theta)(1 - \exp -(E_s/\sigma_\Sigma^2)).\end{aligned} \tag{7.52}$$

Clearly the interference and bias effects are reduced exponentially in proportion to the SIR. A bearing estimate is obtained from a measurement r_a and the appropriate r_a vs θ plot of Figure 7.11. At high SIR the exponential term vanishes and $\bar{\mathsf{r}}_a \to r_a$. In the absence of the signal or at low SIR, the first term of (7.52) which is the correlated component becomes dominant and has a "pulling" effect that tends to bias or pull the bearing estimate in the direction of the correlated interference [30]. The direction is determined by the constant k_c and the ratio of correlated to uncorrelated interference power because

$$\rho\left(\frac{\sigma_\Delta}{\sigma_\Sigma}\right) = k_c\left(\frac{\sigma_c^2}{\sigma_\Sigma^2}\right).$$

The uncorrelated component (second term) tends to pull the estimate back toward the boresight direction effectively reducing the estimation sensitivity by the factor $(1 - \exp(-\mathrm{SIR}))$. The variance of the measured additive ratio is

$$\begin{aligned}\sigma_a^2(\theta) &= \mathcal{E}\{(\mathsf{r}_a - \bar{\mathsf{r}}_a)^2\} \\ &= \frac{\sigma_\Delta^2}{E_s}\left(1 + r_a^2(\theta)\left(\frac{\sigma_\Sigma^2}{\sigma_\Delta^2}\right) - 2\rho r_a\left(\frac{\sigma_\Sigma}{\sigma_\Delta}\right)\right) \qquad (7.53)\end{aligned}$$

and is valid for high SIR when the interference term in the denominator of the error (7.50) is small in relation to the signal term [30]. The additive ratio bias and variance relationships (7.51, 7.53) are derived with the assumption that the signal portions of the statistics $\mathsf{r}_1(\mathbf{p})_{A,B}$ (7.45) are non-random fixed quantities. When the signal fluctuates as implied by its dependence upon the random point scatter spreading function $\mathcal{S}_{sp}$, the bias and variance may or may not increase depending upon the fluctuation statistics, the average signal energy, and the detection threshold. Bearing measurement is not invoked unless the scalar or sum channel processor output exceeds a specified threshold indicating a detection. Consequently, random echos with energy below the threshold do not contribute and the bearing bias and variance can be less than it would be for those with constant energy. This can be illustrated in closed form for the bias (7.51) occurring when the signal is random and the exponentially distributed scalar processor output statistic is d_1 (7.41) or equivalently $|\mathsf{r}_1|^2$ (7.42). The random signal energy component of r_1 is denoted as $\mathsf{z} = E_0|\mathcal{S}_{sp}|^2$ with an exponentially probability density (3.16)

$$\mathsf{f}(z) = 1/E_s \exp -z/E_s$$

assuming Gaussian statistics.

Let [ ] denote the square bracketed term in (7.51). The fluctuating signal bias is

$$\bar{\epsilon}_{af} = [\,]\exp -\mathsf{z}/\sigma_\Sigma^2$$

where the constant energy E_s appearing in (7.51) is now treated as the random variable z with $\mathcal{E}\{\mathsf{z}\} = E_s$. The expectation of the bias with respect to the random signal normalized to those realizations of z exceeding the threshold γ is [30]

$$\begin{aligned}\mathcal{E}\{\bar{\epsilon}_{af}\} &= \frac{\int_\gamma^\infty ([\,]\exp -(z/\sigma_\Sigma^2))(\exp -(z/E_s))dz}{\int_\gamma^\infty \exp -(z/E_s)dz} \\ &= \frac{[\,]\exp -(\gamma/\sigma_\Sigma^2)}{1 + E_s/\sigma_\Sigma^2}.\end{aligned}$$

The fluctuating signal bias decreases exponentially with increasing threshold because only those signal realizations above γ are included in the average. The ratio of the fluctuating signal bias to that of the steady signal (7.51) is

$$\frac{\mathcal{E}\{\bar{\epsilon}_{af}\}}{\bar{\epsilon}_a} = \frac{\exp((E_s - \gamma)/\sigma_\Sigma^2)}{1 + E_s/\sigma_\Sigma^2}$$

and can be made arbitrarily small with increasing γ. It is greater than unity when $\gamma = 0$ and all signal realizations are included.

7.3.2 Phase Comparison.

The bearing angle of a narrowband plane wave emanating from a point scatterer and intercepted by a two element array with spacing d is related to the relative echo phase between the elements by $\theta = \sin^{-1}(\lambda\phi_m/2\pi d)$. This is generalized to the measurement of the relative phase between processors with inputs derived from physically separated sub-apertures as shown in Figure 7.12. The corresponding beams are called *offset phase centered* (OPC) beams and are the basis for phase comparison monopulse processing.

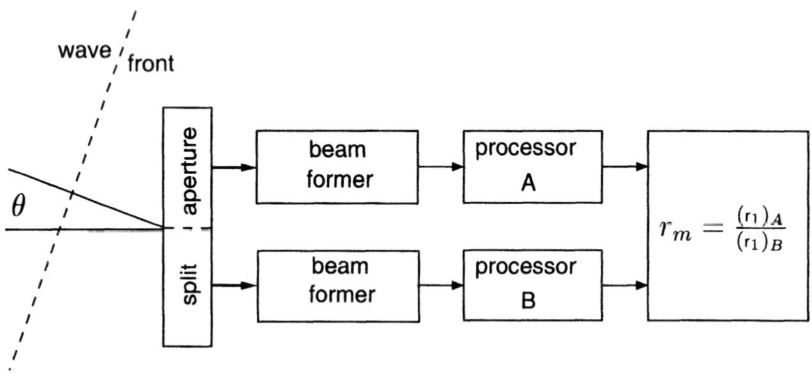

Figure 7.12: Phase monopulse multiplicative ratio generated by parallel processors with offset phase centered beams formed from a split array

OPC beams are formed from identical half apertures (7.21) each with symmetric weighting, an even number of elements, and centroids that are separated by a distance d. The A and B channel processor outputs with $\eta = 2\pi \sin\theta/\lambda$ corresponding to those for amplitude monopulse (7.45)

are

$$\begin{aligned}
\mathsf{r}_{1A,B} &= \mathsf{s}_{A,B} + \mathsf{n}_{A,B} \\
\mathsf{s}_{A,B} &= \sqrt{E_0}\mathcal{S}_{sp}\Theta_n(\eta)\exp \pm jd\eta/2 = \mathsf{s}\exp \pm jd\eta/2 \\
\mathcal{E}\{|\mathsf{r}_{1A,B}|^2\} &= E_s\Theta_n^2(\eta) + \sigma_0^2 \approx E_s + \sigma_0^2 \text{ (near boresight).} \quad (7.54)
\end{aligned}$$

OPC beams may overlap but are usually not squinted and except for their relative phase, the beam signal responses are identical for objects in the far field. The OPC multiplicative ratio corresponding to (7.46) is

$$\mathsf{r}_m(\theta) = |\mathsf{r}_m(\theta)| \exp j\phi_m = \frac{\mathsf{r}_{1A}}{\mathsf{r}_{1B}} = \frac{\sqrt{E_0}\mathcal{S}_{sp}\Theta_n(\eta)\exp jd\eta/2 + \mathsf{n}_A}{\sqrt{E_0}\mathcal{S}_{sp}\Theta_n(\eta)\exp -jd\eta/2 + \mathsf{n}_B}.$$

The interference outputs $\mathsf{n}_{A,B}$ from balanced OPC channels have both correlated and uncorrelated components with the same energy

$$\mathcal{E}\{|\mathsf{n}_{A,B}|^2\} = \sigma_{A,B}^2 = \sigma_0^2$$

which is not necessarily the case for the amplitude monopulse responses (7.45) derived from squinted beams. The signal phase ϕ_m and bearing estimate $\hat{\theta}$ are computed from the ratio of imaginary to real parts of r_m

$$\begin{aligned}
\phi_m &= \tan^{-1}\frac{\Im \mathsf{r}_m}{\Re \mathsf{r}_m} \\
\hat{\theta} &= \sin^{-1}\frac{\phi_m \lambda}{2\pi d}.
\end{aligned}$$

At high SIR

$$\frac{\Im \mathsf{r}_m}{\Re \mathsf{r}_m} = \tan\left(\frac{2\pi d \sin\theta}{\lambda}\right) = \tan(\phi_m)$$

and $\hat{\theta} = \theta$. Note that the relative phase can also be obtained by a phase detector that computes $\arg(\mathsf{r}_{1A}\mathsf{r}_{1B}^*)$ since

$$\arg(\mathsf{r}_{1A}\mathsf{r}_{1B}^*) = \arg(\frac{\mathsf{r}_{1A}}{\mathsf{r}_{1B}}|\mathsf{r}_{1B}|^2) = \arg(\mathsf{r}_m) = \phi_m.$$

ϕ_m is the random phase difference between a correlated pair of stationary narrowband Gaussian processes regardless of how it is computed. Its density function is [31, 32]

$$\begin{aligned}
\mathsf{f}(\phi_m) &= \frac{1-k_0^2}{2\pi}\left[(1-\zeta^2)^{-1} + \zeta(1-\zeta^2)^{-3/2}(\pi/2 + \sin^{-1}\zeta)\right] \\
\zeta &= k_0\cos(\phi_m - \phi_0), \quad -\frac{\pi}{2} \le \sin^{-1}\zeta \le \frac{\pi}{2} \\
\phi_0 &= \tan^{-1}(\mu_{14}/\mu_{13}) \\
k_0^2 &= \frac{\mu_{13}^2 + \mu_{14}^2}{\mu_{11}\mu_{44}} \le 1 \quad (7.55)
\end{aligned}$$

where the μ_{ij} are the elements of the 4×4 covariance matrix $\mathcal{E}\{\mathbf{m}\mathbf{m}^{\mathrm{H}}\}$ and

$$\mathbf{m}^{\mathrm{T}} = [\Re \mathsf{r}_{1A},\ \Im \mathsf{r}_{1A},\ \Re \mathsf{r}_{1B},\ \Im \mathsf{r}_{1B}]$$

is the vector composed of the scalar Gaussian real and imaginary components of the A and B channel outputs. The mean and variance of ϕ_m are [4]

$$\begin{aligned}
\mathcal{E}\{\phi_m\} &= \frac{(k_0 \sin\phi_0)\cos^{-1}(k_0\cos\phi_0)}{\sqrt{1-k_0^2\cos^2\phi_0}} \\
\sigma^2_{\phi_m} &= \frac{\pi^2}{12} - e(k_0) + \frac{(1-k_0^2)(\cos^{-1}(k_0\cos\phi_0))^2}{1-k_0^2\cos^2\phi_0} \\
e(k_0) &= 2\sum_{n=1}^{\infty}\frac{k_0^{2n}}{(2n)^2} = \frac{\pi^2}{12} \quad \text{for } k_0 = 1.
\end{aligned} \tag{7.56}$$

The bearing angle is assumed to be small with respect to the beamwidth of Θ_n so that $\Theta_n(\eta) \approx 1$ and the random signal component of (7.54)

$$\begin{aligned}
\mathsf{s} &= \sqrt{E_0}\mathcal{S}_{sp}\Theta_n(\eta) \approx \sqrt{E_0}\mathcal{S}_{sp} = \Re\mathsf{s} + j\Im\mathsf{s} \\
&\mathcal{E}\{(\Re\mathsf{s})^2\} = \mathcal{E}\{(\Im\mathsf{s})^2\} = E_s/2
\end{aligned}$$

is common to both channels. The complex interference terms $\mathsf{n}_{A,B}$ have components $\Re\mathsf{n}_{A,B}$, $\Im\mathsf{n}_{A,B}$ with energies $\mathcal{E}\{(\Re\mathsf{n}_{A,B})^2\} = \mathcal{E}\{(\Im\mathsf{n}_{A,B})^2\} = \sigma_0^2/2$ (7.40) in both channels.

The real and imaginary parts of the A and B channel outputs (7.54)

$$\begin{aligned}
\mathsf{r}_{1A,B} &= [\Re\mathsf{s}\cos(d\eta/2) \mp \Im\mathsf{s}\sin(d\eta/2) + \Re\mathsf{n}_{A,B}] \\
&\quad + j[\Im\mathsf{s}\cos(d\eta/2) \pm \Re\mathsf{s}\sin(d\eta/2) + \Im\mathsf{n}_{A,B}] \\
&= \Re\mathsf{r}_{1A,B} + j\Im\mathsf{r}_{1A,B}
\end{aligned}$$

are obtained by combining the signal and interference terms. The real and imaginary parts are uncorrelated and the covariance elements of $\mathcal{E}\{\mathbf{m}\mathbf{m}^{\mathrm{H}}\}$ required to generate the parameters of the density (7.55) are

$$\begin{aligned}
\mu_{11} &= \mu_{44} = \mathcal{E}\{(\Re\mathsf{r}_{1A})^2\} = \mathcal{E}\{(\Im\mathsf{r}_{1B})^2\} = (E_s + \sigma_0^2)/2 \\
\mu_{13} &= \mathcal{E}\{\Re\mathsf{r}_{1A}\Re\mathsf{r}_{1B}\} = E_s\cos(d\eta) + \rho_{13}\sigma_0^2 \\
\mu_{14} &= \mathcal{E}\{\Re\mathsf{r}_{1A}\Im\mathsf{r}_{1B}\} = E_s\sin(d\eta) + \rho_{14}\sigma_0^2
\end{aligned} \tag{7.57}$$

[4]The closed forms of these quantities were derived by F.H. Fenlon and appear in Merchant's thesis [32]. Alternate series forms were derived by Rainal [31].

where $|\rho_{13}|, |\rho_{14}| \leq 1$ and

$$\mathcal{E}\{\Re n_A \Re n_B\} = \rho_{13}\sigma_0^2, \quad \mathcal{E}\{\Re n_A \Im n_B\} = \rho_{14}\sigma_0^2.$$

The constituent parameters of the pdf (7.55) are

$$\begin{aligned} k_0^2 &= \frac{(\text{SIR})^2 + \rho_{13}^2 + \rho_{14}^2 + 2(\text{SIR})(\rho_{13}\cos(d\eta) + \rho_{14}\sin(d\eta))}{(1+\text{SIR})^2} \leq 1 \\ \phi_0 &= \tan^{-1}\left[\frac{(\text{SIR})\sin(d\eta) + \rho_{14}}{(\text{SIR})\cos(d\eta) + \rho_{13}}\right] \\ \text{SIR} &= E_s/\sigma_0^2. \end{aligned} \tag{7.58}$$

k_0 is proportional to the signal to interference ratio and ϕ_0 is the most probable phase angle (peak of the pdf). $\phi_0 \to d\eta = 2\pi d \sin\theta/\lambda$ and $k_0^2 \to 1$ as SIR $\to \infty$. The correlation coefficients ρ_{13}, ρ_{14} represent the bias due to correlated inter-channel interference and $\phi_0 = \tan^{-1}(\rho_{14}/\rho_{13})$ in the absence of a signal. The pdf (7.55) evaluated for $\phi_0 = 0$ and $0 \leq k_0 < 1$ is plotted in Figure 7.13. It becomes sharply peaked about the most probable phase ϕ_0 at high SIR ($k_0 \to 1$) and uniformly distributed in uncorrelated noise when $\rho_{13} = \rho_{14} = \text{SIR} = 0$. The condition $\phi_0 = 0$ can occur in two ways assuming the presence of a detected echo. It can correspond either to a phase measurement of a boresight echo with $\eta = 0$ made in the presence of uncorrelated interference such that $\rho_{13} = \rho_{14} = 0$, or a combination of signal and correlated interference such that $\rho_{14} = -(\text{SIR})\sin(d\eta)$. The peak of the pdf (7.55) is shifted away from the origin to correspond with the most probable phase ϕ_0 as shown by Figure 7.14 when the echo bearing is not on boresight or the interference is correlated. The expected value of the phase $\mathcal{E}\{\phi_m\}$ (7.56) shown in Figure 7.15 approaches the most probable phase ϕ_0 for high SIR when $k_0 \to 1$ but is biased toward zero by uncorrelated interference as $k_0 \to 0$. In the presence of correlated interference, $k_0 \neq 0$ (7.58) because ρ_{13} or ρ_{14} are non-zero even if the signal is absent. The expected phase is not necessarily zero when the SIR=0 because of the correlated interference bias $\phi_0 = \tan^{-1}(\rho_{14}/\rho_{13}$. $\mathcal{E}\{\phi_m\} \to 0$ as $\phi_0 \to \pi$ in Figure 7.15 regardless of the value of k_0 because the pdf becomes bimodal and symmetric on the interval $-\pi$, π as shown by Figure 7.14. The bimodality of the pdf as $\phi_0 \to \pi$ is also reflected in the phase standard deviation σ_{ϕ_m} shown in Figure 7.16. $|\phi_0| < \pi$ if the aperture L and aperture offset d are specified such that $d \leq \lambda/2 \leq L$ so as to avoid grating lobes. If $d = \lambda/2$ and $L = 2\lambda$, the 3 dB beamwidth (7.15) of the OPC beams is about 25.6

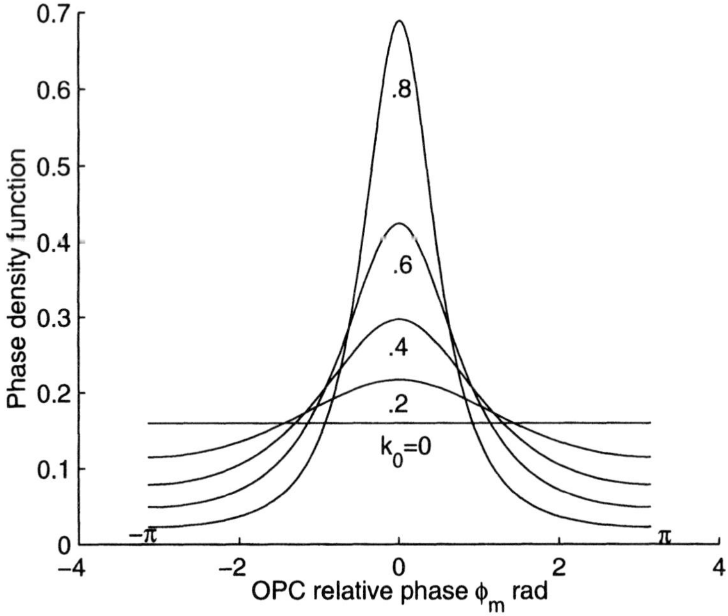

Figure 7.13: Probability density of the OPC phase (7.55) as a function of the signal to interference parameter k_0 with $\phi_0 = 0$.

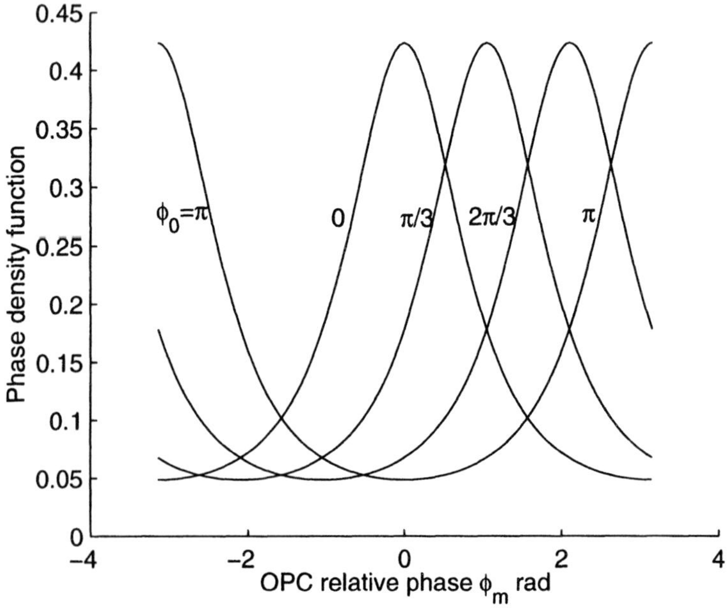

Figure 7.14: Probability density of the OPC phase (7.55) as a function of the ϕ_0 with $k_0 = .6$.

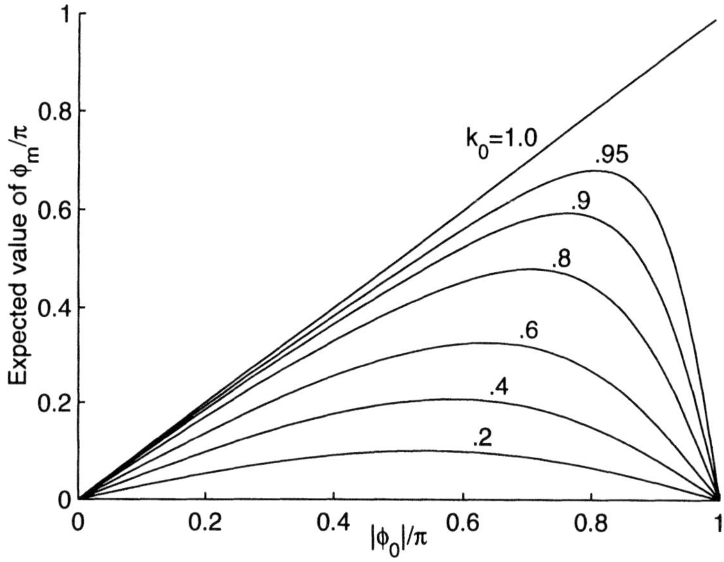

Figure 7.15: Expected value of the relative OPC phase (7.56) as a function of the signal to interference parameter k_0.

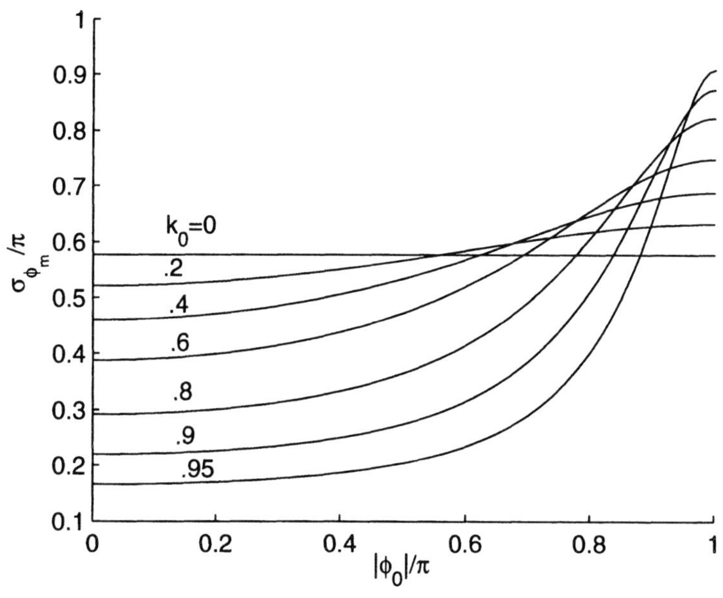

Figure 7.16: Standard deviation of the relative OPC phase (7.56) as a function of the signal to interference parameter k_0.

deg and the bearing $-12.8 \leq \theta \leq +12.8$ deg. The SIR is sufficiently high so that $\phi_0 \approx \phi_m = (2\pi d/\lambda)\sin\theta$ and $-.222\pi \leq \phi_m \leq +.222\pi$. The high SIR assumption is realistic because the signal must exceed a reasonable threshold for detection. If for example the detection and false alarm probabilities of the scalar receiver are specified as $\wp_d = .9$, $\wp_f = 10^{-5}$, the required SIR = 20.3 dB (6.74). This is equivalent to $k_0 = .991$ in the absence of interference correlation (7.58).

Spread Scattering Induced Bearing Bias

The interference correlation between amplitude and phase monopulse channels that biases the estimated bearing can be caused by non-uniform flow around the transducer, internal unbalance and cross talk between channels, and a non-uniform distribution of the backscattered reverberation that lends it a directional characteristic. The existence of more than one backscattered echo in the same range cell also has a biasing effect. Directionality of the echo signal and reverberation for phase monopulse is described by the spreading relationships

$$\begin{aligned} \mathsf{s}_{A,B} &= \sqrt{E_0}\int_{\Omega_\theta} \mathcal{S}_s(\theta)\Theta_n(\eta)\exp \pm j(d\eta/2)d\theta \\ &\approx \sqrt{E_0}\int_{\Omega_\theta} \mathcal{S}_s(\theta)\exp \pm j(d\eta/2)d\theta \\ \mathsf{n}_{A,B} &= \sqrt{E_0}\int_{\Omega_\theta} \mathcal{S}_r(\theta)\Theta_n(\eta)\exp \pm j(d\eta/2)d\theta + \mathsf{n}_{a,b} \\ &\approx \sqrt{E_0}\int_{\Omega_\theta} \mathcal{S}_r(\theta)\exp \pm j(d\eta/2)d\theta + \mathsf{n}_{a,b} \end{aligned}$$

that are generalizations of the point scatter relationships (7.54) for sources that are spread in bearing. $\mathcal{S}_s$ and $\mathcal{S}_r$ represent the signal and reverberation bearing spreading functions with associated scattering functions $\mathcal{R}_s$ and $\mathcal{R}_r$. n_a, n_b are the A and B channel uncorrelated noise components with energy $\mathcal{E}\{|\mathsf{n}_a|^2\} = \mathcal{E}\{|\mathsf{n}_b|^2\} = \sigma_n^2$ and $\mathcal{E}\{\mathsf{n}_a\mathsf{n}_b^*\} = 0$.[5] The output

[5]The most general spreading functions includes Doppler and delay as well as bearing to describe the backscattered signals (7.2,7.3). Delay and Doppler are assumed to be predetermined when monopulse processing is invoked and only the bearing spreading component is discussed in order to simplify the presentation.

signal and interference energies are the same in both OPC channels and

$$\mathcal{E}\{|\mathsf{s}_{A,B}|^2\} = E_0 \int_{\Omega_\theta} \mathcal{R}_s(\theta)d\theta = E_s$$
$$\mathcal{E}\{|\mathsf{n}_{A,B}|^2\} = E_0 \int_{\Omega_\theta} \mathcal{R}_r(\theta)d\theta + \sigma_n^2 = E_r + \sigma_n^2 = \sigma_0^2$$

assuming uncorrelated scattering. The signal and reverberation have energies E_s, E_r and directionality described by their respective scattering functions. The non-backscattered noise with energy σ_n^2 is assumed to be uncorrelated between channels and is a constituent of the total interference output energy σ_0^2 (7.40) in each channel.

The OPC A and B channel outputs written in terms of the real and imaginary parts of the spreading functions are

$$\begin{aligned}
\mathsf{r}_{1A,B} &= \Re[\mathsf{s}_{A,B} + \mathsf{n}_{A,B}] + j\Im[\mathsf{s}_{A,B} + \mathsf{n}_{A,B}] \\
&= \Bigg[\sqrt{E_0}\int_{\Omega_\theta} (\Re\mathcal{S}_s(\theta)\cos(d\eta/2) \mp \Im\mathcal{S}_s(\theta)\sin(d\eta/2))d\theta \\
&\quad +\sqrt{E_0}\int_{\Omega_\theta} (\Re\mathcal{S}_r(\theta)\cos(d\eta/2) \mp \Im\mathcal{S}_r(\theta)\sin(d\eta/2))d\theta + \Re\mathsf{n}_{a,b}\Bigg] \\
&\quad +j\Bigg[\sqrt{E_0}\int_{\Omega_\theta} (\Im\mathcal{S}_s(\theta)\cos(d\eta/2) \pm \Re\mathcal{S}_s(\theta)\sin(d\eta/2))d\theta \\
&\quad + \sqrt{E_0}\int_{\Omega_\theta} (\Re\mathcal{S}_r(\theta)\cos(d\eta/2) \pm \Im\mathcal{S}_r(\theta)\sin(d\eta/2))d\theta + \Im\mathsf{n}_{a,b}\Bigg] .
\end{aligned}$$

The real and imaginary components of the signal and interference are uncorrelated and have equal average energy. The signal, reverberation, and noise are uncorrelated with each other, and the scattering processes are uncorrelated in θ. Under these conditions the covariance elements (7.57) are

$$\begin{aligned}
\mu_{11} = \mu_{44} &= \frac{E_s + E_r + \sigma_n^2}{2} \\
\mu_{13} &= \frac{E_0}{2}\int_{\Omega_\theta} (\mathcal{R}_s(\theta) + \mathcal{R}_r(\theta))\cos(d\eta)d\theta \\
\mu_{14} &= \frac{E_0}{2}\int_{\Omega_\theta} (\mathcal{R}_s(\theta) + \mathcal{R}_r(\theta))\sin(d\eta)d\theta
\end{aligned} \tag{7.59}$$

and obtained by using the trig identities $\cos 2x = \cos^2 x - \sin^2 x$ and $2\sin x \cos x = \sin 2x$. The spread scattering parameters of the pdf (7.55) that are comparable to (7.58) are

$$k_0^2 = \left(\frac{E_0}{E_s + E_r + \sigma_n^2}\right)^2 \left[\left(\int_{\Omega_\theta} (\mathcal{R}_s(\theta) + \mathcal{R}_r(\theta)) \cos(d\eta) d\theta\right)^2 + \left(\int_{\Omega_\theta} (\mathcal{R}_s(\theta) + \mathcal{R}_r(\theta)) \sin(d\eta) d\theta\right)^2\right]$$

$$\phi_0 = \tan^{-1}\left[\frac{\int_{\Omega_\theta} (\mathcal{R}_s(\theta) + \mathcal{R}_r(\theta)) \sin(d\eta) d\theta}{\int_{\Omega_\theta} (\mathcal{R}_s(\theta) + \mathcal{R}_r(\theta)) \cos(d\eta) d\theta}\right]. \tag{7.60}$$

Example 7.2 Monopulse Bias for Two Point Scatterers

The scattering functions are

$$\mathcal{R}_s(\theta) = \mathcal{R}_s \delta(\theta - \theta_s)$$
$$\mathcal{R}_r(\theta) = \mathcal{R}_r \delta(\theta - \theta_r)$$

when the backscattered data consists of a point scatter signal source at bearing θ_s and a specular (point like) interference source at bearing θ_r. $E_s = E_0\mathcal{R}_s$ and $E_r = E_0\mathcal{R}_r$ are the signal and interference energies and the covariance elements (7.59) are

$$\mu_{13} = \frac{E_s \cos(d\eta_s) + E_r \cos(d\eta_r)}{2}$$
$$\mu_{14} - \frac{E_s \sin(d\eta_s) + E_r \sin(d\eta_r)}{2}$$
$$\eta_s = (2\pi/\lambda) \sin\theta_s, \ \eta_r = (2\pi/\lambda) \sin\theta_r.$$

The most probable phase (7.60) becomes

$$\phi_0 = \tan^{-1}\left[\frac{E_s \sin(d\eta_s) + E_r \sin(d\eta_r)}{E_s \cos(d\eta_s) + E_r \cos(d\eta_r)}\right]$$

and illustrates the bias effect occurring when two scatterers occupy the same range cell and the most probable phase becomes the energy weighted phase contributions of both. The parameter k_0^2 (7.60) is

$$k_0^2 = \frac{E_s^2 + E_r^2 + 2E_sE_r \cos(\eta_s - \eta_r)}{(E_s + E_r + \sigma_n^2)^2} \leq 1$$

and σ_n^2 represents the uncorrelated noise energy that reduces the overall signal to interference ratio but does not bias the phase. The total interference energy is $\sigma_0^2 = E_r + \sigma_n^2$ and k_0^2 reverts to (7.58) with $\rho_{13} = \cos(d\eta_r)$ and $\rho_{14} = \sin(d\eta_r)$ when $\sigma_n^2 = 0$.

7.4 SUMMARY

The bearing of echo returns most often involves only one scattering source since it is assumed that echos can be isolated in range and Doppler by the resolution of the finite duration waveform. A second resolved echo source can be considered as a form of interference that is spatially localized and would bias a monopulse processor in proportion to its relative strength. Amplitude and phase comparison monopulse processing with two channels cannot unambiguously resolve two azimuthally separated sources that fall within the same range-Doppler resolution cell. The maximum likelihood line array processor that seeks the parameter hypothesis $\hat{\hat{\mathbf{p}}}$ maximizing the statistic (7.33) can resolve multiple angular sources if the number of sources is less than the number of elements in the array. The multiple source problem is mitigated for active echo-location by using suitably resolvent waveforms and non-overlapping OPC beam pairs.

The problem is more severe for the processing of continuous passive signals that cannot be temporally separated. Simultaneously received narrowband signals that are spectrally separable can be spatially resolved by monopulse processing of the individual spectral bins. Continuous spectrally and temporally overlapping signals cannot easily be spatially resolved by monopulse methods. For these the signal covariance structure between array elements may be exploited to determine the directions of arrival by what are called *subspace* methods. Perhaps the best know is the Multiple Signal Characterization (MUSIC) algorithm developed originally by Schmidt [33–35] that decomposes the data covariance into signal and noise subspaces and exploits their orthogonality. MUSIC works best when the dimension of the signal subspace (number of sources) is known and sufficient time series data is available to permit a good estimate of the data covariance.

Bibliography

[1] S.C. Clay and H. Medwin. *Acoustical Oceanography* . Wiley Interscience, 1977.

[2] L.J. Ziomek. *Underwater Acoustics, A Linear Systems Approach* . Academic Press, Orlando, FL, 1985.

[3] D.C. Swanson. *Signal Processing for Intelligent Sensor Systems* . Marcel Dekker, New York, NY, 2000.

[4] R.J. Urick. *Principles of Underwater Sound* . McGraw-Hill, 1975.

[5] W.S. Burdic. *Underwater Acoustic System Analysis* . Prentice-Hall, 1984.

[6] L.S. Bendat and A.G. Piersol. *Random Data Analysis and Measurement Procedures* . John Wiley & Sons, 1986.

[7] F.J. Harris. On the Use of Windows for Harmonic Analysis with the Discrete Fourier Transform . *Proceedings of the IEEE*, 66(1):51–83, Jan. 1978.

[8] D.C. Rife and G.A. Vincent. Use of Discrete Fourier Transform in the Measurement of Frequencies and Levels of Tones . *Bell Sytem Tech. Journ.*, 49(2):197–228, Feb. 1970.

[9] H. Rohling and J. Schuerman. Discrete Time Window Functions With Arbitrarily Low Sidelobe Level . *Signal Processing*, 5(2):127–138, Mar. 1983.

[10] D.W. Ricker. Small Aperture Angle Measurement for Active Echo Location Systems. *IEEE Trans. on Aerosp. and Elect. Systs.*, 22(4):380–388, 1986.

[11] S.D. Stearns. *Digital signal Analysis* . Hayden, Rochelle Park, NJ, 1975.

[12] H. Cox. Interrelated Problems in Estimation and Detection II. In *Proc. NATO Advanced Study Inst. on Sig. Proc.*, Enschede, Netherlands, Aug. 1968. NATO.

[13] A.D. Whalen. *Detection of Signals in Noise* . Academic Press, 1971.

[14] W.B. Davenport and W.L. Root. *An Introduction to the Theory of Random Signals and Noise* . McGraw-Hill, 1958.

[15] H. Cox. Interrelated Problems in Estimation and Detection I . In *Proc. NATO Advanced Study Inst. on Sig. Proc.*, Enschede, Netherlands, Aug. 1968. NATO.

[16] C.F. Van Loan G.H. Golub. *Matrix Computations* , volume 6. Johns Hopkins Press, Baltimore, MD, 1983.

[17] M.S. Bartlett. An Inverse Matrix Adjustment Arising in Discriminant Analysis . *Annals of Mathematical Statistics*, 22:107–111, 1951.

[18] W. J. Bangs II. *Array Processing with Generalized Beamformers* . PhD Thesis in Statistics, Yale Univ., New Haven, CT, Sept. 1971.

[19] W. J. Bangs and P. M. Schultheiss. *Space-Time Processing for Optimal Parameter Estimation* , pages 577–589. Academic Press, 1973. in NATO edited book "Signal Processing".

[20] Altes R.A. Cross Correlation and Energy Detection in Multiarray Processing . *IEEE Trans. Acoust.,Speech, and Signal Proc.*, 33(3):493–504, 1985.

[21] R.C. Burkhardt. *Robust Adaptive Processing for Application to Underwater Acoustic Arrays* . PhD Thesis in Acoustics , Penna. State Univ., Dec. 1992.

[22] H. Cox, R.M. Zeskind and M.W. Owen. Robust Adaptive Beamforming . *IEEE Trans. Acoust.,Speech, and Signal Proc.*, 35(10):1365–1376, Oct. 1987.

[23] L.H. Sibul and L.J. Ziomek. Generalized Wideband Cross-Ambiguity Function . In *Proc. IEEE Intl. Conf. on Acoust. Spch. and Sig. Proc.*, Atlanta GA, 1981.

[24] R.A. Monzingo and T.W. Miller. *Introduction to Adaptive Arrays.* Wiley Interscience, New York, 1980.

[25] Altes R.A. Target Position Estimation in Radar and Sonar, and Generalized Ambiguity Analysis for Maximum Likelihood Parameter Estimation . *Proceedings of the IEEE*, 67(6):920–930, June 1979.

[26] A.N. Mirkin and L.H. Sibul. Cramér-Rao Bounds on Angle Estimation with a Two-Dimensional Array. *IEEE Trans. Sig. Proc.*, 39(2):515–517, Feb. 1991.

[27] A.I. Leonov and K.I. Fomichev. *Monopulse Radar.* Artech House, Norwood MA, 1986. translated by W.F. Barton.

[28] B.R. Mahafza. *Introduction to Radar Analysis* . CRC Press LLC, Boca Raton Fla, 1998.

[29] D.R. Rhodes. *Introduction to Monopulse* . McGraw-Hill, 1959.

[30] S.M. Sherman. *Monopulse Principles and Techniques* . Artech House, 1984.

[31] A.J. Rainal. Monopulse Radars Excited by Gaussian Signals . *IEEE Trans. on Aerosp. and Elect. Systs.*, 2(3):337–345, May 1966.

[32] C.C. Merchant. Detection of a Dual Channel Differential Phase Modulated Signal in Correlated Noise. MS Thesis in Electrical Engineering, Penna. State Univ., Univ. Park, PA, Aug. 1982.

[33] R.A. Schmidt. *A Signal Subspace Approach to Multiple Emitter Location and Spectral Estimation* . PhD Thesis in Electrical Engineering, Stanford Univ., Nov. 1981.

[34] B. Friedlander. The root-MUSIC Algorithm for Direction finding with Interpolated Arrays. *Signal Processing*, 30(1):15–29, 1993.

[35] L.L. Sharf. *Statistical Signal Processing* . Addison-Wesley, 1991.

Symbols

Font examples and their primary usage

$a, x, A, Z, \alpha, \Omega, \Phi$: deterministic scalar.
$\mathbf{a}, \mathbf{x}, \boldsymbol{\alpha}, \boldsymbol{\zeta}, \boldsymbol{\lambda}$: deterministic vector.
$\mathsf{a}, \mathsf{A}, \mathsf{B}, \mathsf{\Gamma}, \mathsf{\Phi}, \mathsf{\Omega}$: random scalar.
$\mathsf{\textbf{a}}, \mathsf{\textbf{b}}, \mathsf{\textbf{x}}, \mathsf{\textbf{z}}$: random vector.
$\mathbf{A}, \mathbf{B}, \mathbf{R}, \mathbf{Q}, \boldsymbol{\Sigma}, \boldsymbol{\Theta}, \boldsymbol{\Omega}$: deterministic matrix.
$\mathsf{\textbf{A}}, \mathsf{\textbf{B}}, \mathsf{\textbf{R}}, \mathsf{\textbf{X}}, \mathsf{\textbf{Z}}, \boldsymbol{\mathsf{\Theta}}, \boldsymbol{\mathsf{\Omega}}$: random matrix.

Notation

$\mathrm{diag}[\mu_i]$: a diagonal matrix with components μ_i.
$\mathbf{a}^{\mathrm{T}}, \mathbf{A}^{\mathrm{T}}$: transpose of a vector or matrix.
$\mathbf{a}^{\mathrm{H}}, \mathbf{A}^{\mathrm{H}}$: Hermitian (conjugate) transpose.
$T_r\,\mathbf{A}$: diagonal sum or trace of the matrix $\mathbf{A}$.
$\mathbf{A}^{-1}$: matrix inverse.
$\forall$: for all.
$\in$: contained in or belongs to.
$\Rightarrow$: implies that.
$x * y$, $x * *y$: one and two-dimensional convolution.
$< x, y >$, $\mathbf{x}^H\mathbf{y}$: inner products.
$\hat{x}$: hypothesized quantity x.
$\wp$ probability.
$\mathsf{f}(x)$: probability density function.
$\mathsf{F}(x)$: cumulative distribution function.
$\mathcal{E}\{x\}$: expectation operator
$\bar{x}$: average or expected value of x.
$(\,)^*$ complex conjugate of the argument.
$\Re(), \Im()$: real and imaginary parts of a complex argument.
$\mathrm{cols}[\mathbf{u}_i]$: a matrix composed of column vectors $\mathbf{u}_i$.

Abbreviations and Acronyms

AF, UF: ambiguity, uncertainty function.
AAF, AUF: auto ambiguity, uncertainty function.
AM, PM, FM: amplitude, phase, frequency modulation.
BPSK: bi-phase shift keyed.
CAF, CUF: cross ambiguity, uncertainty function.

CF: characteristic function.
CFAR: constant false alarm rate.
CGN: colored (correlated) Gaussian noise.
CON: complete orthonormal; refering to a complete set of orthonormal functions.
C-R, CRB: Cramér-Rao, bound.
CW: continuous wave or sinusoidal waveform.
CWD: Choi-Williams distribution.
DI: directivity index (not to be confused with the directivity index D_I).
EC: Estimator Correlator.
FFT: fast Fourier transform.
FT: Fourier transform.
FSK: frequency shift keyed.
HFM,LPM: hyperbolic-FM, linear period modulation.
HOP: frequency hop or FSK waveforms.
iid: independent identically distributed.
IO: input-output.
K-L: Karhunen Loéve.
llr: log-likelihood ratio
lrt: likelihood ratio test
llrt: log-likelihood ratio test
LFM, VFM: linear-FM, vee-FM.
ML: maximum likelihood.
MLSR: maximal length shift register.
NB, WB: narrowband, wideband.
NBUF, WBUF: narrowband, wideband uncertainty function.
NBAF, WBAF: narrowband, wideband ambiguity function.
OPC: offset phase center, in reference to beams used for phase monopulse processing.
PRN: pseudo random noise.
QD: quadratic distortion, distortion due to an uncompensated quadratic term in the time base.
ROC: receiver operating characteristic.
ROI: resolution of identity.
SF: scattering function.
sinc(x): $\sin(x)/x$
SNR, SIR:signal to noise, interference ratio.
TFCF: time-frequency correlation function.
TW: time-bandwidth product.
WGN,: white Gaussian noise.

WSS: wide sense stationary.
WSSUS: wide sense stationary, uncorrelated scattering.
WT: wavelet transform.
WVD: Wigner-Ville distribution.

Glossary

Chapter 1

$f(t)$, $F(\omega)$: waveform and its spectrum.
$\tilde{f}(t)$: waveform complex envelope.
$\mathcal{F}$, $\mathcal{F}^{-1}$: Fourier transform operator and its inverse.
$H(\omega)$: transfer function.
$N(\omega)$: interference power spectral density (Joules).
N_0: interference power spectral density for a white (uncorrelated) noise process.
ω, ν: radian frequencies.
ω_0: radian carrier frequency.
$\mathsf{f}(x)$, $\mathsf{F}(x)$: probability density and distribution functions.
P_s, E_s: signal power and energy.
$\Psi(t)$, $\psi(t)$: phase functions.
$\mathsf{r}(t) = s(t) + \mathsf{n}(t)$: received time series.
$R_{xx}(t,u)$ (Watts), $S_{xx}(\omega)$ (Joules): autocorrelation and power density spectrum.
s_a, s_r, s_h: analytic signal, real signal, Hilbert transform.
$s(t)$, $S(\omega)$: signal time and frequency functions.
$\mathsf{n}(t)$: random interference time series.
T: signal duration.
T_p: signal period.
$\tilde{x}$: quantity or property associated with the complex envelope unless otherwise noted.
$\mathcal{U}(t)$: unit step function.
$\xi_i(t)$; vector of orthonormal functions or generalized Fourier kernels $\xi_i(t)$.

Chapter 2

Unless otherwise noted, upper case quantities in Section 2 associated with the echo energy cycle and sonar equation represent dB for consistency with extant sonar equation literature. This is not necessarily the case in other chapters.
a_ℓ: relative acceleration $(m/(\text{sec})^2)$.
A_t: backscatter cross section area $(m)^2$.
β, $\tilde{\beta}$: time dilation parameters.
β_e: relativistic dilation parameter.
d_{ts}, d_{sr}: scalar distances.

D_t, D_r: transmit and receive beam attenuation (dB).
E_0: electrical transmitted signal energy (Joules or dB re 1 Joule as noted in text).
E_i, E_r: incident and reflected energy flux (dB).
E_t: energy flux at the transducer (dB).
f_0: carrier frequency (Hz).
$f_r(t)$: received (time mapped) waveform function.
G_t: transducer conversion factor (dB re 1 v/μ Pa).
k_l: Lorentz constant.
L_{xr}: spherical spreading loss (dB).
$P_b = 10 \log p_b$, $P_v = 10 \log p_v$: receiver input boundary and volume reverberation powers.
ϕ: Doppler frequency (Hz).
P_x, E_x: transmitted power and energy flux (dB re 1 Watt, Joule/m^2).
$\mathbf{p}$, $\hat{\mathbf{p}}$, $\hat{\mathbf{p}}'$: Doppler and delay echo parameters.
$\mathbf{p}_0$: Doppler and delay origin parameter set $\beta = 1$, $\phi = \tau = 0$.
r_0, r_ℓ: initial and line of sight ranges.
$\mathbf{r}_t, \mathbf{r}_s, \mathbf{r}_r$: vectors defining the transmitter, scatterer, and receiver positions (m).
S_l: transmitter source level (dB re 1 μ Pa).
$S_b = 10 \log s_b$, $S_v = 10 \log s_v$: boundary and volume scattering coefficients.
T: pulse duration (sec).
t_1, t_2, t_3: times of transmission, reflection, and reception (sec).
$\mathbf{t}_t, \mathbf{t}_s, \mathbf{t}_r$: 4-d space-time position vectors (sec).
$T_r = T/\beta$: dilated received pulse duration (sec).
$T_x = 10 \log T$: pulse length.
T_s: target strength (dB).
$\boldsymbol{\tau}_t = \mathbf{r}_t/c$, $\boldsymbol{\tau}_s = \mathbf{r}_s/c$, $\boldsymbol{\tau}_r = \mathbf{r}_r/c$: delay vectors (sec).
$\boldsymbol{\tau}_w$, $\mathbf{t}_w$: wavefront delay and space-time vectors. (sec)
v_p, v_s, $v_\ell = v_p + v_s$: platform, reflector, and relative line-of-sight (los) speeds (m/sec).
v_t: relativistic speed between objects (m/sec).
v_x: cross range velocity component (m/sec).
W: bandwidth (Hz), unless other wise noted this refers to the design bandwidth or design frequency range of a waveform.
W_s: system bandwidth (Hz) or pass band of the receiver.
W_ω^2: mean square radian bandwidth.
z: acoustic impedance of water.

Chapter 3

C_{ij}: Bayes cost.

$\chi_{fg}(\mathbf{p},\hat{\mathbf{p}})$, $\Phi_{fg}(\mathbf{p},\hat{\mathbf{p}})$: cross-uncertainty and corresponding ambiguity function for waveforms $f(t)$, $g(t)$.

$\chi_{ff}(\mathbf{p},\hat{\mathbf{p}})$, $\Phi_{ff}(\mathbf{p},\hat{\mathbf{p}})$: auto-uncertainty and corresponding ambiguity function for waveform $f(t)$.

d_1: CFAR detection statistic.

$D_{\mathbf{p}\hat{\mathbf{p}}}$: point scatter detection index as a function of the parameters.

$D_0(\mathbf{r})$, $D_1(\mathbf{r}) \in [0,1]$ set membership functions.

$\bar{e}_b$: estimator bias.

$f_{\mathbf{p}}$: shorthand for $f(\mathbf{p},t)$.

$f(\mathbf{p},t), \mathbf{f}(\mathbf{p})$: waveform time function and corresponding expansion vector with parameters $\mathbf{p}$.

G_p: processing gain.

$g(\mathbf{p},u)$, $\mathbf{g}(\mathbf{p})$: correlated interference processing function and corresponding vector with parameters $\mathbf{p}$.

γ, γ_l, γ_r: detection thresholds.

H_0, H_1, H_k: decision hypotheses; denoting presence or absence of the signal or the k-th set of parameters.

$\mathbf{I}$: Identity matrix.

$\mathbf{J}$: Fisher information matrix.

$\boldsymbol{\lambda}$: diagonal matrix of eigenvalues λ_i.

$l(\mathbf{r},\mathbf{p})$: log-likelihood function of the data and parameter vectors.

L_r, ℓ_r: likelihood ratio, log-likelihood ratio (llr).

ℓ_d: data dependent part of the log likelihood ratio.

Ω_x: the domain of x.

p, $\hat{p}$: scalar parameter and its estimate.

$\mathbf{p}_{ml}$: maximum likelihood parameter estimate.

$\wp_d$, $\wp_f$, $\wp_{fr}$: probabilities of detection, false alarm, and false rest.

$\nabla_{\mathbf{p}} l$: gradient of l with respect to parameter vector $\mathbf{p}$.

$\nabla^2_{\mathbf{p}} l$: outer product $\nabla_{\mathbf{p}}(\nabla_{\mathbf{p}} l)^t$.

$\mathsf{n}_c(t)$, $\mathsf{n}_0(t)$: correlated and white noise time functions.

$Q_n(u,t)$: inverse kernel corresponding to the covariance function $R_n(z,u)$.

$\mathcal{Q}$: inverse operator.

r_1: sufficient statistic for the slowly fluctuating point scatterer.

$\mathbf{R}_{\hat{\mathbf{p}}}$: parameter estimate covariance.

$\mathsf{R}(\mathbf{p},\omega)$: spectrum of the random data with parameters $\mathbf{p}$.

R^n Euclidean space of dimension n.

$R_c(t,u)$: covariance function of the correlated interference component .

$\mathbf{R}_n$, $\mathbf{R}_s$: interference and signal covariance matrices.
$R_n t, u$, $R_s(t,u)$: interference and signal covariance functions.
$\mathbf{s}$, $\mathbf{n}$, $\mathbf{r}$: random signal, interference, and data vectors.
σ_e^2: mean square estimation error.
$\sigma_{\hat{p}}^2$: estimate variance, scalar C-R bound.
σ_0^2, σ_1^2: receiver output energy with signal absent and present.
SIR($\mathbf{p}$): signal to interference ratio conditioned on parameters $\mathbf{p}$ (dB).
S_{sp}: random, slowly fluctuating signal coefficient equivalent to the point scatter spreading function.
W_s: system or noise bandwidth. It is equivalent to the sampling frequency for sampled data.

Chapter 4

$a(t)$ real amplitude modulation or window function.
c_x: admissibility constant.
$\mathcal{C}(x), \mathcal{S}(x)$: Fresnel cosine and sine integrals.
$\triangle\phi_{opt}$: Doppler frequency maximizing $\Phi_h(\beta, \triangle\phi)$.
D_s: dilation sensitivity factor.
$\int_{\Omega_p}$: integration over the domain of p, e.g. $\int_{\Omega_t}$ denotes integration over the time support of the integand.
κ_1, κ_2: constituents of the dilation expansion coefficient η^2.
$\lambda^2, \gamma, \eta^2$: ambiguity expansion coefficients for delay, cross coupling, and Doppler.
$\Omega(t) = \omega(t) + \omega_0$: bandpass frequency function, time derivative of $\Psi(t)$.
Ω_p: denotes the support or domain of the parameter p.
$P(t, \xi)$: general class of time-frequency distributions.
$\Phi_h(\beta, \triangle\phi)$: mismatched hybrid ambiguity function using narrowband replica when signal is dilated.
$\Psi(t) = \psi(t) + \omega_0 t$: bandpass phase function.
$S_n = \text{SNR}/(1 + \text{SNR})$: signal to noise parameter.
T_w: amplitude weighted time-frequency product of a waveform.
v_{los}: line of sight velocity.
V_N, V_W, V_{W-N}: ambiguity volume measures.
$V_w(t, \xi)$, $C_w(t, \xi)$: Wigner-Ville, Choi- Williams distributions in time (t) and frequency (ξ).
$W_g f(a, b)$: wavelet transform of f with respect to the analyzing function or mother wavelet g and dilation a and delay b.

Chapter 5

A_a: area of a slice through the ambiguity function.
A_{wp}: area of a slice through the WVD.
C_{jk}: correlation between two CW chips.
χ_{ss}, Φ_{ss}: uncertainty and ambiguity functions of a composite waveform.
d^k: delay operator of order k.
e_p: primitive element.
$GF(q)$: Galois field of order q.
$g(d)$: shift register generating function.
$j(i)$, $I(i)$: frequency placement operators for hop codes.
$J(m)$, $I(m)$: switching operators for PRN sequences.
LC, QC, CC, HC: linear quadratic, cubic, and hyperbolic congruence codes.
$p(d)$: shift register characteristic polynomial.
$R_p(\tau)$: autocorrelation of a PRN sequence.
$s_i(\mathbf{p}_i, t)$: component of a composite waveform.
T_r: Woodward's time resolution constant.
$\mathcal{U}(t)$: unit step function.
V_{ml}, V_{sl}: ambiguity main and sidelobe volume.
ζ: shape parameter for the dilation tolerant waveform.

Chapter 6

A_s: phase plane area associated with the signal support Ω_s.
B_r(Hz): reverberation SF Doppler spread.
$\mathsf{B}(\zeta, \phi)$: random bifrequency function.
D_c: detection index of the correlation receiver constructed using the composite CAF $\Phi_c(\mathbf{p})$.
$D_{\hat{\mathbf{p}}}$: detection index as a function of the parameter hypothesis for spread scattering.
D_s: detection index of the correlation receiver constructed using the signal only CAF $\Phi_s(\mathbf{p})$.
E_{fs}, E_{fr}: expected signal, reverberation or scattered interference energy at the output of a correlation receiver.
$g(t)$: processing function of a mismatched receiver.
$G_p = \mathrm{SIR}_{out}/\mathrm{SIR}_{in}$: processing gain .
$\mathsf{h}(\tau, t)$: random time varying impulse response.
$\mathsf{H}(\zeta, t)$: random time varying transfer function.
$\int_\Omega$: integration over the whole delay-Doppler (phase) plane.
$\int_{\Omega_r}$: integration over the support of the interference SF.
$\int_{\Omega_s}$: integration over the support of the signal SF .

ℓ_r, ℓ_d: log likelihood ratio (llr) and its data dependent part.
M: The actual number of discrete reflectors comprising the signal scattering function.
N: the dimension of the discrete data expansion. It also represents the number of parameter hypotheses made when the expansion is a vector of matched filtering operations. Ideally $N = M$ for the EC.
$N_r(\omega)$: reverberation power spectral density (PSD).
Ω: the whole of the delay-Doppler domain.
Ω_s, $\Omega_r \in \Omega$: delay-Doppler support regions of the signal and interference scattering functions
ϕ_ξ, ϕ_x: characteristic functions.
$\Phi_c(\mathbf{p})$: composite cross ambiguity function (CAF).
$\Phi_s(\mathbf{p})$: CAF constructed from the normalized signal SF.
ϕ: frequencies associated with time variation of the scattering impulse response, Doppler frequencies.
r_{cs}: ratio of D_c to D_s.
R_H: time-frequency correlation function (TFCF).
$\mathbf{R}_f$: waveform expansion matrix.
$\mathbf{R}_1$, $\mathbf{R}_0$: $N \times N$ data covariances under H_1, H_0 respectively.
$\mathbf{R}_s$, $\mathbf{R}_n$: signal and interference covariances.
$\mathbf{S}$: generalized SIR matrix with eigenvalues μ_i.
$\sigma^2_{\phi e}$, $\sigma^2_{\phi v}$: mean square environmental and velocity induced reverberation Doppler spread.
S_s, S_r: random signal, reverberation spreading functions.
$\mathcal{R}_i$: total interference SF that includes the constant noise term N_0. .
$\mathcal{R}_s$, $\mathcal{R}_r$: signal, reverberation scattering functions.
$\mathcal{R}_{sl}$, $\mathcal{R}_{sp}$: line and point scatter scattering functions.
T_p: FSK subpulse duration.
τ_t: line scatterer delay extent.
θ_b: beamwidth.
W, W_s: waveform bandwidth, system or noise bandwidth that is equivalent to the sampling frequency for sampled systems.
ζ: frequency associated with delay (τ) dependent scattering variation.

Chapter 7

a_x, a_r: aperture weighting functions of the transmit and receive arrays.
a_n: discrete aperture weight at the n-th element.
$\mathcal{A}_x(\mathbf{u})$, $\mathcal{A}_r(\mathbf{u})$: transmitter and receiver aperture functions defining the weighting and delay applied at a point defined by the vector $\mathbf{u}$ in a transmit or receive aperture.

d_1: CFAR detection statistic.
D, D_I: directivity and directivity index (not to be confused with the abbreviation for the detection index DI).
$D_{\tilde{\mathbf{p}},\hat{\tilde{\mathbf{p}}}}$: detection index.
$\mathbf{e}_k$: array steering vector at frequency ω_k.
E_0, E_x, E_s: transmitted energy, energy flux density, and received energy.
$\boldsymbol{\eta} = \lfloor \eta_x, \eta_y, \eta_z \rfloor$: wavenumber vector.
$\mathbf{f}_k$: array waveform vector at frequency ω_k. $\hat{\mathbf{f}}_k$ indicates the vector with hypothesized parameters.
$F_k(\mathbf{p})$: spectral component at ω_k of waveform with parameters $[\beta,\ \tau]$.
$\mathcal{F}, \mathcal{F}_k$: Fourier transform operator, subscript indicates evaluation at frequency (k).
$G_k(\hat{\tilde{\mathbf{p}}})$: mismatched receiver processing function at ω_k for WSS interference.
k_c: constant indicating the relative amount of correlated interference in the sum/difference monopulse channels.
k_m: slope of the additive ratio vs bearing angle curve at the origin.
L: array length.
ν: frequency variable.
n_{ni}: random interference time sample at elment n and time increment i.
N_{nk}: random interference spectral component at element n and ω_k.
$\mathsf{N}_{\Sigma k}$: array output interference spectral component at ω_k.
Ω_τ, Ω_β: domains of the delay and Doppler parameters.
$\tilde{\mathbf{p}}, \hat{\tilde{\mathbf{p}}}$: bearing augmented parameter vectors $[\mathbf{p},\ \theta] = [\beta, \tau, \theta]$, hat $\Rightarrow$ hypothesis.
Φ_{ff}, Φ_{fg}: auto and cross ambiguity functions .
r_m, r_a: monopulse multiplicative and additive ratios.
ρ: correlation coefficient between the sum and difference interference processes.
$R_{mn}(t, u), R_{mn}(\tau)$: interference time series covariance between elements m and n.
R_{nk}: random data spectral component at element n and frequency k.
R_k: array signal plus interference output at frequency k.
$\mathcal{S}_s(\tilde{\mathbf{p}})$: random spreading function for parameters $[\beta, \tau, \theta]$.
$\mathcal{S}_{sp}$: random spreading function at a point defining the strength of a slowly fluctuating point scatterer.
$\mathsf{s}(t, \mathbf{u}, \mathbf{u}')$: aggregate time domain signal integrated over the transmit and receive volume apertures.

$\mathsf{s}_c(t, \mathbf{u}, \mathbf{u}', \mathbf{r}_x, \mathbf{r}_r)$: point to point time domain echo signal.
$\mathsf{S}(\omega, \mathbf{u}, \mathbf{u}')$, $\mathsf{S}_c(\omega, \mathbf{u}, \mathbf{u}'\mathbf{r}_x, \mathbf{r}_r)$: frequency domain equivalents of s, s_c.
$\mathsf{S}_{nk}(\tilde{\mathbf{p}})$: random signal spectral component at element n and frequency k.
S_k: array signal output at frequency k.
$\sigma_0^2(\hat{\tilde{\mathbf{p}}})$: interference normalization for a CFAR processor.
τ: range delay.
τ_t: range delay plus the delay to points in the transmit and receive apertures.
θ_b: beam width .
$\Theta(\boldsymbol{\eta})$, $\Theta_n(\boldsymbol{\eta})$: directivity, pattern or beam function, subscript (n) indicates normalized beam.
Θ_Σ, Θ_Δ: sum and difference monopulse beams.
$\mathbf{u}$, $\mathbf{u}'$: unit vector defining the direction of plane wave propagation, prime denotes beam steering.
V_x, V_r, V_s: volumes defining the transmit and receive apertures and the echo scattering region.
W_{klmn}: interference frequency domain covariance between elements m, n and frequencies ω_k, ω_l.
$W_{mn}(\omega_k) \equiv W_{kmn}$: cross spectrum between elements m,n. at frequency ω_k.
$W_{\Sigma k}$: power spectral density (psd) of the scalar receiver sum interference process.
$\mathbf{W}$: interference cross spectrum matrix.

Index